FOOD POWDERS

Physical Properties, Processing, and Functionality

FOOD ENGINEERING SERIES

Food Engineering Series

Jose M. Aguilera and David W. Stanley, *Microstructural Principles of Food Processing and Engineering*, Second Edition (1999)
Stella M. Alzamora, María S. Tapia, and Aurelio López-Malo, *Minimally Processed Fruits and Vegetables: Fundamental Aspects and Applications* (2000)
Gustavo Barbosa-Cánovas and Humberto Vega-Mercado, *Dehydration of Foods* (1996)
Gustavo Barbosa-Cánovas, Enrique Ortega-Rivas, Pablo Juliano, and Hong Yan, *Food Powders: Physical Properties, Processing, and Functionality* (2005)
P.J. Fryer, D.L. Pyle, and C.D. Rielly, *Chemical Engineering for the Food Industry* (1997)
Richard W. Hartel, *Crystallization in Foods* (2001)
Marc E.G. Hendrickx and Dietrich Knorr, *Ultra High Pressure Treatments of Food* (2002)
Lothar Leistner and Grahame Gould, *Hurdle Technologies: Combination Treatments for Food Stability, Safety, and Quality* (2002)
Michael J. Lewis and Neil J. Heppell, *Continuous Thermal Processing of Foods: Pasteurization and UHT Sterilization* (2000)
Rosana G. Moreira, M. Elena Castell-Perez, and Maria A. Barrufet, *Deep-Fat Frying: Fundamentals and Applications* (1999)
Rosana G. Moreira, *Automatic Control for Food Processing Systems* (2001)
M. Anandha Rao, *Rheology of Fluid and Semisolid Foods: Principles and Applications* (1999)
George D. Saravacos and Athanasios E. Kostaropoulos, *Handbook of Food Processing Equipment* (2002)

FOOD POWDERS

Physical Properties, Processing, and Functionality

Gustavo V. Barbosa-Cánovas
Washington State University
Pullman, Washington

Enrique Ortega-Rivas
Autonomous University of Chihuahua
Chihuahua, Mexico

Pablo Juliano
Washington State University
Pullman, Washington

Hong Yan
Washington State University
Pullman, Washington

Kluwer Academic/Plenum Publishers
New York, Boston, Dordrecht, London, Moscow

Library of Congress Cataloging-in-Publication Data

ISBN 0-306-47806-4

233 Spring Street, New York, New York 10013

http://www.kluweronline.com

10 9 8 7 6 5 4 3 2 1

A C.I.P. record for this book is available from the Library of Congress.

Printed in the United States of America

To our families

PREFACE

Food powders represent a large fraction of the many food products available in the food industry, ranging from raw materials and ingredients, such as flours and spices, to processed products like instant coffee or powdered milk. Food powders can be distinguished not only by their composition and microstructure, but also by particle size, size distribution, chemical and physical properties, and functionality. Historically, a number of unit operations have been developed and adopted for the production and handling of different food powders. Information on the physical properties, production, and functionality of food powders has been published, mainly through research and review articles, reports in trade magazines, and symposia presentations. This is likely the first book ever authored that addresses key aspects of food powder technology.

This book was designed and developed as a useful reference for individuals in both the food industry and academia interested in an organized and updated review, from an engineering perspective. The book consists of twelve chapters including several tables, figures, diagrams, and extensive literature citation, and covers as thoroughly as possible a fascinating field of study and practical applications. The first section of the book (Chapters 1–3) deals with food powder characterization. Chapter 1 presents statistical concepts related to powder sampling as well as techniques, equipment, and procedures for optimal sampling. Single particle-related properties and their evaluation are covered in Chapter 2, which includes particle size and shape, density, size distribution, surface area, and moisture. Chapter 3 describes in detail the bulk powder properties, giving special attention to flow, handling, packing, strength, and instant properties.

The second part of the book describes, analyzes, and provides tools needed for the design of a typical unit operation, as related to production, handling, and processing of food powders. Chapter 4 includes useful information about storage alternatives for food powders, as well as flow patterns, together with the analysis of natural and assisted discharge from bins. Chapter 5 covers typical food powder transportation systems utilized during processing, which includes belts, chain, screw and pneumatic conveyors, among other conveying systems. Size reduction, and conversely, size enlargement processes are covered in Chapters 6–8. Reduction of larger food pieces or particles, including energy requirements and equipment used, is described in Chapter 6. Particle enlargement methods, fundamentals, and other design aspects are described in Chapter 7. A specific case on particle size enlargement, i.e., particle encapsulation, can be found in Chapter 8 with focus on the methods used for the production of different food capsules.

Chapter 9 analyzes in depth fundamental aspects and the design of food particle mixing systems, while Chapter 10 deals with dry powder separation and classification technology. The most widely encountered process in food particle production is drying, a subject covered in Chapter 11 that includes relevant drying systems commonly used in the food industry. Last but not least, in Chapter 12 four key undesirable phenomena occurring during food particle handling, processing and testing—namely particle attrition, segregation, bulk caking, and dust explosion—are addressed. A thorough description of each phenomenon is given, including evaluation tests, methods for minimization, and mechanisms of action.

We truly hope this book will be a valuable addition to the food powder technology literature and will promote additional interest in advancing food powders research, development, and implementation.

Gustavo V. Barbosa-Cánovas
Enrique Ortega-Rivas
Pablo Juliano
Hong Yan

CONTENTS

PART I

FOOD POWDERS

CHAPTER 1

SAMPLING

1.1. THEORY AND STATISTICAL ASPECTS

1.1.1. Introduction: Importance of Sampling

The definition of the term "sample" is expressed as "a portion of the whole, selected in such a way as to be truly representative of the whole." Some additional explanations for this definition include: (a) no sample truly represents all the respects of the whole consignment or population; (b) the sample is always different from the whole consignment, even for the parameters of interests; (c) the sample will only be adequate for the determination of certain elements; and (d) the sample will only be adequate for some analytical techniques (Smith and James, 1981). The sole objective of sampling is to reduce the mass of a target material without significantly changing its other properties, either by taking increments from flowing streams of a material or by splitting when the whole lot of the material can be handled (Gy, 1998).

Unlike fluids, the properties of powders are likely to change under an applied load. For example, they may consolidate with time and present phenomena like attrition or segregation, due to handling and transport. In particular, because powders have a size distribution that affects many of their properties, and segregation and stratification by size is so common, representative sampling is absolutely critical for the success and relevance of any subsequent testing. As a general rule, only a very small part of a particulate material is subjected to a given analytical technique. Therefore, it is essential for this part to be representative for the total universe of the material, since it is customary to generalize from test results about physical properties of the whole material (Herdan, 1960). Without a well-prepared representative sample, the result, no matter how good it is, will be meaningless and irrelevant, and may be misleading, no matter how good the utilized characterization method.

Sampling is an important element of powder handling that demands careful scientific design and operation of the sampling systems. The general purpose of sampling is to collect a manageable mass of material which must be representative of the total mass of the sampled powder. This action is achieved by taking many small samples from all parts of the total which, when combined, will represent this total with an acceptable degree of accuracy. All particles in the total must have the same probability of being included in the final sample, so all of them must be equally accessible. To satisfy these requirements, the following basic "golden rules" of sampling should be applied whenever possible:

- Sampling should be done preferably from a moving stream (for both powders and suspensions), but powder on a stopped belt can be sampled.
- A sample of the whole of the stream should be taken for many (equally spaced) periods of time, rather than part of the stream for the whole of the time.

The first rule recommends that the sample should be taken from a flowing powder stream, such as a discharging flow from a belt conveyor or a feeding flow from one storage container to another.

Furthermore, as the second rule mentions, the sampling process should continue for a long series of short time intervals (Masuda, 1997). It is very likely that the recombined, primary sample taken from the whole will be too large for most powder tests. It will therefore be necessary to subdivide the original sample into secondary or even tertiary sub-samples. This subdivision may be built into the primary sampler or it may be achieved with a separate sampling divider. Many methods of sampling and sample splitting have been reviewed and tested by Allen (1981). Some of them will be discussed in the following section.

1.1.2. Sampling Variation Sources

If a sampled material were perfectly homogeneous with respect to its properties, any fraction of the bulk would be exactly the same as far as those properties were concerned. For instance, if a certain powder were heterogeneous in size but homogeneous in density, it would be considered homogeneous if it were sampled to determine density. For this reason, it is often easier to obtain a representative sample from liquid materials, where all their properties are generally homogeneous. For example, when sampling ionic solutions for component concentration, any variation in results would normally be attributed to experimental error. However, when the material is heterogeneous, as usually happens with industrial food powders, difference is expected to be found in some measured properties. The source of these variations, in addition to variations caused by the assaying process, could be attributed to the fact that the smaller fragments of particulate material are themselves heterogeneous, and to segregation of certain types of fragments due to handling of the bulk material, i.e., the separation of fine material from coarse material during the motion of a powder bed (Gy, 1982). The probability of obtaining a perfect unbiased sample from the parent material is remote. If several samples are taken and they are representative, the expected variation may be estimated from statistical analysis (Allen, 1981). With very few exceptions the characterization of particulate material must be made by the examination of a small fraction of the material. Commonly, errors in particle size analysis may be due to incorrect sampling, among other errors such as instruments limitations, or operator errors. The total error in sampling is made up of errors due to primary sampling and subsequent sample dividing errors in the analysis itself. Sampling is said to be accurate when it is free from bias; that is the error of sampling is a random variable close to the true mean. Sampling is precise when the error variation is small irrespective of whether the mean is the true mean or not. Two types of sampling errors are possible:

- errors due to segregation of the bulk in non-cohesive materials
- statistical error.

Segregation errors depend upon the previous history of the powder and can be minimized by suitable mixing and building up of a sample from a large number of increments. Statistical error, however, cannot be prevented. Even for an ideal random mixture the quantitative distribution in samples of a given magnitude is not constant, but is subject to random fluctuations.

The statistical theory of sampling considers a sample of size N selected randomly and independently. A random selection process must be such that every member of the population being sampled would have the same chance to be selected. The independence of specimens implies that the selection of one specimen does not influence the selection of another in the whole mass of the sampled powder. These requirements are not usually met in practice, either due to the two sources of variation previously mentioned or to the nature of the sampling technique, or both. For example, a perfectly mixed batch of a food powder may become segregated during the sampling process due to the nature of the mechanical handling system.

Sampling is a process with statistical properties determined by inherent random variables of the sampled population and the sampling process. Naturally, the apparent variability of the values determined from a certain sample can be greatly influenced by the sampling and measurement techniques. In statistics, a random variable is defined as a function that assigns real numbers to the outcomes of a random experiment. For particulate materials, the random experiment includes three major steps: (a) the selection of a fraction of a given volume of bulk by giving equal chance to all other fractions in the lot; (b) further reduction of volume and selection of a fraction of the specimen; and (c) testing of a final fraction for different properties.

The random variable being observed is defined by the sampling process. A change of the volume of the specimen in the sampling process will correspond to a different experiment, so it will result in the realization of a different random variable. When sampling from a well-mixed bulk of particulate material, it can be postulated that the random variables observed by choosing different specimen volumes would possess the same distribution function and the same mean, but different variances. The theoretical and experimental studies indicate that, keeping other things constant, reduction of specimen volume increases the variance. A practical relationship proposed by Gy (1982) correlates the minimum acceptable specimen weight as a function of the diameter of the coarsest fraction in the total mass of sampled powder. Such relationship can be represented as:

$$\mathrm{M_s} \geq \frac{Cd^3}{\sigma^2} \tag{1.1}$$

where σ^2 is the variance of the tolerated sampling error, C is a constant characterizing the material to be sampled, d is the diameter of the coarsest fragment, and M_s is the weight of specimen.

1.1.3. Minimum Sample Size

Samples are withdrawn from a population in order to estimate certain characteristics of that population and to establish confidence limits for those characteristics. The characteristic may be particle size, composition or quality; a measure of the spread of the distribution may also be required. For example, sampling is desired to set up specification limits between which the quality of a final product is acceptable or to decide whether the characteristics of a given lot meet preset criteria, or it may be to estimate the variability within a lot or between lots.

The arithmetic average of random independent observations of a normally distributed random variable is known to be the best estimator of the unknown population mean. The Central Limit Theorem (Kennedy and Neville, 1976) asserts that the arithmetic average of random independent observations of a random variable will be distributed normally when the sample size is infinitely large. Even for moderate sample sizes, the statistical behavior of the sample average is acceptably close to that of the normal distribution. Also, by increasing the number of specimens (sample size), sample averages closer to the unknown value of the population mean are obtained. The Law of Large Numbers (Larson, 1978) asserts that when the sample size is infinitely large, the sample mean becomes equal to the population mean. In practice, however, small sample amounts are handled for obvious practical and economic reasons. So it is almost impossible for the sample average to become equal to the unknown population mean. For this reason, it is common to establish an interval around the sample average, called a confidence interval, which will contain the unknown mean with a certain predetermined probability. Given the probability $(1 - \alpha)$, the length of the confidence interval becomes a function of the population variance and the sample size. This relationship is derived from the following probability statement:

$$\text{Prob}\left[\bar{X} - \frac{Z_{\left(\frac{1-\alpha}{2}\right)} \cdot \sigma}{\sqrt{N}} \le \mu \le \bar{X} + \frac{Z_{\left(\frac{1-\alpha}{2}\right)} \cdot \sigma}{\sqrt{N}}\right] = 1 - \alpha \tag{1.2}$$

where $\bar{X}$ is the sample arithmetic average, σ is the population standard deviation, μ is the population mean, N is the sample size, and $Z_{(1-\alpha)/2}$ is the $(1 - \alpha)/2$ percentile of standard normal variable.

The length L of the confidence interval is:

$$L = \frac{2Z_{(1-\alpha)/2}\sigma}{\sqrt{N}} \tag{1.3}$$

It is clear from Eq. (1.3) that the length of the confidence interval is linearly proportional to the population standard deviation, and inversely related to the square root of the sample size. If σ were known, Eq. (1.3) could be used to determine the minimum sample size required to obtain a confidence interval, which will contain the unknown mean μ, with a $(1 - \alpha)$ probability. Therefore, an expression for the minimum sample size will be:

$$N \ge \frac{4Z^2_{(1-\alpha)/2}\sigma^2}{L^2} \tag{1.4}$$

Since σ is usually unknown and the sampling process normally has the aim of estimating the standard deviation as well as the mean, Eq. (1.4) has limited applicability but indicates, however, an important relationship between the sample size, the inherent variability of the sampled population, and the precision at which the mean μ is to be estimated. Such precision will be increased by reducing L and/or increasing $(1 - \alpha)$, both resulting in larger sample sizes. Also, as indicated by the power of σ in Eq. (1.4), its variability will increase the sample size quadratically. Reduction of σ is only possible by employing a different sampling method, further crushing or grinding the bulk to be sampled or increasing the specimen volume. Considering that all these measures are not very practical for real purposes, an alternative would be the use of an estimator for σ, which is a well-known practice in statistics. A relationship for estimating the population standard deviation can be represented as follows:

$$S^2 = \frac{1}{N-1}\sum_{i=1}^{N}(X_i - \bar{X})^2 \tag{1.5}$$

where S is the estimator for σ and X_i is the assay value.

Since the variance is not known, but estimated by S, the following probability statement can be established:

$$\text{Prob}\left[\bar{X} - \frac{t_{\left(\frac{1-\alpha}{2}, N-1\right)} \cdot S}{\sqrt{N}} \le \mu \le \bar{X} + \frac{t_{\left[\frac{1-\alpha}{2}, N-1\right]} \cdot S}{\sqrt{N}}\right] = 1 - \alpha \tag{1.6}$$

where S is the sample standard deviation and $t_{(1-\alpha)/(2,N-1)}$ is the $(1 - \alpha)/2$ percentile of the Student's t distribution with $N - 1$ degrees of freedom.

The length of the confidence interval can be expressed by:

$$L = \frac{2t_{(1-\alpha)/(2,N-1)}S}{\sqrt{N}} \tag{1.7}$$

An expression for the minimum sample size can be derived from Eq. (1.7), similar to the one derived from Eq. (1.3):

$$N \ge \frac{4t^2_{(1-\alpha)/(2,N-1)}S^2}{L^2} \tag{1.8}$$

By inspecting Eqs. (1.5) and (1.8), it can be noticed that the values of S and t are not independent of N. Also, a sample of a certain size is needed, i.e., N', to obtain S initially. This value will change as new specimens are sampled, assayed, and their values used in recalculating S. When the population variance is unknown, it is not possible to determine exactly the minimum sample size. An approach can be achieved by taking several samples and calculating successive approximations of N. To accomplish this, using a small sample size, the estimator S is determined by means of Eq. (1.5) and a new value for N is calculated with appropriate values of $t_{(1-\alpha)/(2,N-1)}$, S, and L using Eq. (1.8). If this value is larger than the actual sample size, additional sampling is done, and the procedure is repeated. As the sample size increases, the value of S will approach the unknown population standard deviation with decreasing error. Furthermore, the successive values of S, as sample size increases, may fluctuate around σ with decreasing error and, thus, the calculated values of N will also fluctuate. For this reason, it is advisable to increase the sample size by the smallest possible increments, and to continually calculate N until its value and the actual sample size are sufficiently close.

A practical approximation can be done considering the ratio L^2/S^2 in Eq. (1.8) and estimating the population mean for a given confidence interval, such that its length is equal to a certain multiple of S. In such a manner, an initial determination of the sample size is possible. By rearranging the terms in Eq. (1.8), the following relationship can be obtained:

$$U \geq \frac{4t^2_{(1-\alpha)/(2,N-1)}}{N} \tag{1.9}$$

where $U = L^2/S^2$.

Equation (1.9) can be used to calculate different values of U determined by certain values of α and N. The value of the percentiles of Student's t distribution is available in most standard textbooks on statistics. The results of such calculations for different confidence levels can be presented in graphical or tabular form. Table 1.1 lists U values for sample sizes from 2 to 500 for confidence levels of 0.90, 0.95, and 0.99. It can be used to read the required minimum sample size for a predetermined value of U. Table 1.1 shows that, in agreement with probability principles, as the confidence interval length L shortens with respect to the estimator of the standard deviation S, the unknown mean and standard deviation will be predicted with high certainty so that a large sample size N will be the most appropriate. On the other hand, for a large value of L and a small value of S, the predictions of unknown mean and standard deviation will be uncertain, and the sample size N will be relatively low. A good approximation is to use a confidence interval length L equal to the standard deviation estimator S, so the value of U would equal unity. Taking such value, according to Table 1.1, minimum sample sizes for confidence levels of 0.90, 0.95, and 0.99 would be 13, 18, and 30, respectively.

Considering all theoretical aspects of sampling discussed above as well as field experience for practical purposes, it is recommended to use a sample size between 30 and 50 for most powdered and particulate food materials.

1.1.4. Standard Sampling Deviation

The probability of obtaining a perfect unbiased sample from parent material is remote. If several samples are taken and they are representative, the expected variation may be estimated from statistical analysis theories (Allen, 1981). If a powder is made up of components A and B, the bulk can be represented in terms of component A by the corresponding particle fraction (p_A) (p_A being the ratio between the number of particles of component A and the total number of particles in the bulk). The value of p_A corresponds to the probability of finding a certain number of particles of component A in a perfect sample. The variance of the probability can be computed from the total number of

Table 1.1. Sample size as a function of U.

Sample size, N	$U = L^2/S^2$ $1-\alpha = 0.90$	$1-\alpha = 0.95$	$1-\alpha = 0.99$
2	79.63	323.09	8105.19
3	11.37	24.65	131.21
4	5.52	10.11	34.11
5	3.63	6.18	16.93
6	2.69	4.40	10.83
7	2.15	3.43	7.87
8	1.81	2.78	6.13
9	1.54	2.37	5.02
10	1.34	2.04	4.23
11	1.19	1.81	3.65
12	1.08	1.61	3.22
13	0.97	1.46	2.88
14	0.90	1.33	2.59
15	0.83	1.22	2.37
16	0.77	1.13	2.18
17	0.72	1.06	1.89
18	0.67	0.99	1.77
19	0.63	0.93	1.66
20	0.60	0.87	1.64
21	0.56	0.83	1.54
22	0.54	0.79	1.46
23	0.51	0.75	1.38
24	0.49	0.71	1.32
25	0.47	0.68	1.25
26	0.45	0.65	1.20
27	0.43	0.63	1.14
28	0.41	0.60	1.10
29	0.40	0.58	1.05
30	0.39	0.55	1.02
40	0.28	0.41	0.73
50	0.22	0.32	0.57
60	0.19	0.27	0.47
80	0.14	0.20	0.35
100	0.11	0.16	0.28
200	0.05	0.08	0.14
500	0.02	0.03	0.05

particles A and B in the sample (n particles) and the bulk (N particles):

$$\mathrm{Var}(p_\mathrm{A}) = \frac{p_\mathrm{A}(1-p_\mathrm{A})}{n}\left(1-\frac{n}{N}\right) \tag{1.10}$$

The theoretical standard deviation is equal to the square root of the variance. Furthermore, the bulk and sample compositions in terms of weight fractions p'_A and p'_B can be used to calculate the variance of the probability p'_A (the weight fraction of a perfect sample A) by the following equation:

$$\mathrm{Var}(p'_\mathrm{A}) = \frac{p'_\mathrm{A}(1-p'_\mathrm{A})}{w}(p'_\mathrm{A}w_\mathrm{B} + p'_\mathrm{B}w_\mathrm{A})\left(1-\frac{w}{W}\right) \tag{1.11}$$

where p'_A is the weight fraction of component A in the bulk material, p'_B is the weight fraction of component B in the bulk material, W and w are the bulk and sample weights, respectively, and w_A and w_B are the weights of individual grains of components A and B, respectively.

Example 1.1. Particle A weighs 0.5 g and particle B 0.7 g. Two kg of A are mixed with 7 kg of B. A sample of 0.5 kg is taken from the mixture. Find the number of particles A ($n_{A,S}$) and the weight of particles A ($m_{A,S}$) in the sample and characterize them with the sampling error.

Data: $w_A = 0.5$ g; $w_B = 0.7$ g; $w = 0.5$ kg; $m_A = 2000$ g; $m_B = 7000$ g

Unknowns: $n_{A,S}$; $m_{A,S}$

(1) On a particle number basis:

$$N = \frac{m_A}{w_A} + \frac{m_B}{w_B}$$

then $N = \frac{2000\,\text{g}}{0.5\,\text{g}} + \frac{7000\,\text{g}}{0.7\,\text{g}} = 14000$ particles
$p_A = \frac{m_A/w_A}{N}$ (ratio of total particles of component A and total particles in the bulk)
then $p_A = \frac{2000\,\text{g}/0.5\,\text{g}}{14000} = 0.286$

$$n = \frac{N \times w}{m_A + m_B} = \frac{14000 \times 0.5\,\text{kg}}{(2+7)\,\text{kg}} = 778$$

$n_{A,S} = n \times p_A = 778 \times 0.286$; $\boxed{n_{A,S} = 223 \text{ particles of component A}}$

From Eq. (1.10):

$$\text{Var}(p'_A) = \frac{0.286(1-0.286)}{778}\left(1 - \frac{778}{14000}\right) = 2.4789 \times 10^{-4}$$

$$\sigma_i = \sqrt{\text{Var}(p'_A)} = 0.0157$$

(2) On a particle weight basis:
$w_A = 0.5$ g/particle A, $w_B = 0.7$ g/particle B,

$$p'_A = \frac{m_A}{m_A + m_B} = \frac{2\,\text{kg}}{(2+7)\,\text{kg}} = 0.222,$$

$w = 0.5$ kg, $W = 9$ kg
$m_{A,S} = w \times p'_A = 0.5\text{ kg} \times 0.222$; $\boxed{m_{A,S} = 0.111 \text{ kg of component A}}$

From Eq. (1.11):

$$\text{Var}(p'_A) = \frac{0.222(1-0.222)}{500}[0.222 \times 0.7 + (1-0.222) \times 0.5]\left(1 - \frac{0.5}{9}\right) = 0.0001776$$

$$\sigma_A = \sqrt{\text{Var}(p'_A)} = 0.013$$

Assuming the variance is normally distributed, the probability (p) of representative samples lying within one standard deviation of the mean is 68.3%. The probability (p) is 95.6 or 99.7% for samples lying within two or three standard deviations of the mean, respectively. Thus, the particle amounts of component A in the 0.5 kg sample with a total n of 778 particles at the three probability levels are: 223 ± 12, 223 ± 24 and 223 ± 36 particles A in the sample; the weights of component A at the three probability levels are: 0.111 ± 0.007 kg A, 0.111 ± 0.013 kg A and 0.111 ± 0.020 kg A.

In addition to Eqs. (1.10) and (1.11), a number of similar relationships have been developed to calculate the standard deviation of sampling error under complex situations existing in practical sampling processes, for example, when there are more than two components in the mixture or particles present as a size distribution. Gy's equation is of particular interest in this section, mainly because of its capability of covering the complex situations in practical sampling and its simplicity of use under some empirical approximations (Smith and James, 1981). Assuming that: (a) the material is thoroughly mixed; (b) there is no error inherent in the sampling tool or crushing machine; and (c) individual particles can be selected with equal probability, Gy's equation can be expressed as:

$$S^2_{\text{sampling}} = \rho \cdot D^3 \cdot x \cdot (100 - x) \cdot \left(\frac{1}{w} - \frac{1}{W}\right) \cdot f_{\text{L}} \cdot f_{\text{S}} \cdot f_{\text{D}} \tag{1.12}$$

where S^2_{sampling} is the standard deviation of sampling error, ρ is the weighted mean density of all particles, D is the side length of a cubic particle, x is the weight percentage of element x in the whole material, w is the sample weight, W is the weight of the whole consignment or a lot, and f_{L}, f_{S}, and f_{D} are the factors of mineral liberation, particle shape, and particle size distribution, respectively. It is believed that Gy's main contribution to sampling theory lies in developing useful approximations of those factors and proving them by experiment (Smith and James, 1981).

1.2. SAMPLING TECHNIQUES AND PROCEDURES

Sampling methods are normally designed or selected for collection of primary samples from large-scale processes. In many applications, samplers have to be designed specifically for a particular material and duty. There are usually two main stages in any sampling scheme. A gross sample, which represents the parent lot, is initially collected; a representative laboratory sample is then prepared from this gross sample. These two operations are generally performed by different techniques. Gross samples can range up to 50 kg or more, depending on the size of the bulk quantity, while laboratory samples rarely exceed 1 or 2 kg, or even less considering the wide array of state-of-the-art instrumental methods currently available for powder characterization. This small amount of sample may require further subdivision, determined by the minimum sample size previously discussed. The process of reduction from bulk to measurement sample should be implemented in stages because it often entails a reduction from tons to kilograms (or liters) of material to grams or milligrams (or milliliters) in the second sampling stage, as shown in Fig. 1.1.

Many factors affect the representativeness of the bulk material in each sample preparation stage. For example, when a non-cohesive powder is poured into a heap, size separation or

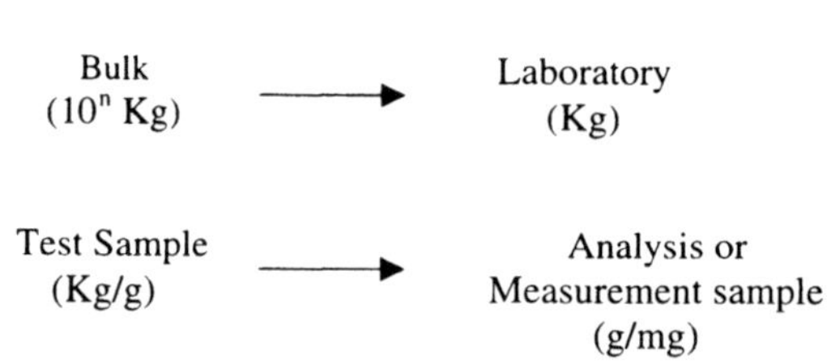

Figure 1.1. Sample amount.

partitioning occurs, with the fine particles being located at the center of the heap. When a container of powder is subjected to vibration, the fine particles percolate through the coarse particles, and when a powder flows out of a hopper it may have a tendency to unmix. Segregation is one of the main problems encountered in the sampling of solid particles. In conical heaps of material, for instance, a large proportion of the coarser particles will generally be found in the lower levels of the pile. On the other hand, coarse particles tend to migrate towards the top of the contents of a container, especially after transportation. Any selected sampling method must take into account these irregularities.

In sample reduction of non-flowing powders, such as fine cohesive solids, sticky or moist materials, or fibrous solids, since these materials do not have a tendency to segregate but may not be uniform, it is necessary to pass these materials through a mixer or to shake the sample in a container using different modes of shaking. Collecting the sample with scoops or shovels is considered a simple alternative for obtaining a good representation of the bulk samples. A large number of random samples may be taken from a heap of the material, or regular samples can be scooped into a sample container when the bulk material is transported manually from one location to another. Regular sampling is most often favored for materials packed in containers. A common sampling implement for this purpose is a piece of metal tubing with a sharp bottom edge and fixed to a long handle, known as sampling "thief" or sampling spear. This sampler is pushed gently, firmly, and vertically into the full depth of the material and then withdrawn for sample removal. When sampling from stored powders, the samples should be withdrawn from small spaces of equal volume within the total, to try to give every member of the population a chance to be sampled. Since this is very difficult to achieve, as not all of the material is equally accessible, sampling from stationary places is generally avoided.

For material flowing down chutes or from hoppers, a sample collector may be placed in the path of the outlet stream. Open-ended scoops or shovels are unsuitable in this case because they allow coarser particles to roll out of the sample and thus bias the sample towards the finer fractions. A ladle is the only recommended device for such sampling (Plowman, 1985), since it does not allow this rolling out to happen, unless overfilled. The falling stream is traversed with the ladle to collect the sample, and if the whole of the stream cannot be sampled in one pass, it is divided into several equal areas with the sample being collected from those incrementally. In agreement with the "golden rules" of sampling mentioned before, the collector should sample the whole of the outflow stream at regular time intervals.

Stopped belt sampling is considered the best manual sampling method. It is often taken as a reference method with others being compared against it. It follows the two rules of sampling in that it samples from a moving stream (with the movement momentarily stopped for taking the sample), and it takes the whole of the stream many times. It is carried out with a suitable profiled sampling frame, which is inserted through the material on a stopped belt conveyor until it comes in contact with the conveyor belt over its full width. The material within the frame is then collected from the belt and represents one increment in the manual sampling series.

At the laboratory level, gross samples can be subdivided into one or more smaller samples manually or mechanically. The most common manual method is known as "coning and quartering," which consists of mixing the sample and piling it into a conical heap, flattening it to about one quarter of its original height, dividing the flattened heap into four equal quarters, and rejecting two opposite quarters while re-mixing the remaining pair. This remaining amount is piled again into a new heap, and the procedure repeated until the required sample quantity is obtained. Figure 1.2 illustrates the various stages of the coning and quartering method of sample division. If symmetry does not exist in the flattening and cutting of the heap, size error will occur. This method is completely dependent on the skill of the operator.

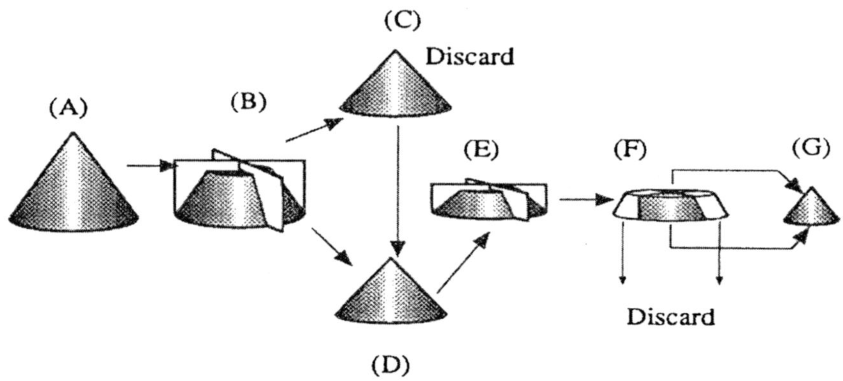

Figure 1.2. Coning and quartering methods (from Allen, 1997).

As mentioned before, sampling methods sometimes have to be designed for specific purposes. Also, the many stages of a sampling scheme that are often required make sampling a technique that has to be carried out very carefully, and multiple sample taking is normally involved in the whole process. For these reasons, many sampling devices and dividers are available to perform sampling in a mechanic or automatic manner. A short review of the many samplers currently available is given below.

1.3. SAMPLERS AND SAMPLE DIVIDERS

Primary samples are usually collected by a primary splitter, which cuts through the full cross-section of the stream. The design of such cutters is subject to guidelines given in several publications (Gy, 1982; Merks, 1985, 1986; Plowman, 1985). Cutters used in mechanical sampling comprise diverted types and bucket types. Both of them are used to cut a stream falling due to gravity off the end of a conveyor belt, or from the discharge end of a pneumatic conveying pipe or a chute. *Diverted cutters* deflect the stream increment clear of the main stream and, when properly designed, do not allow accumulation or sticking of the powder inside. However, they do require considerable headroom and can only deposit the increment below the point of sampling and not very far from it laterally. *Bucket type cutters*, on the other hand, have the advantage of collecting and transporting the sample laterally, without the loss of headroom. However, since they collect and hold the sample, they may allow material build-up within the bucket when sampling sticky powders.

Some designs of these types of cutters are presented in the literature (Merks, 1986; Plowman, 1985). For instance, one special design that is very applicable in food industrial processes is the *cross-belt type cutter* (Fig. 1.3), also known as a rotating hammer sampler. This device swings in a plane perpendicular to the movement of a conveyor belt, scooping a well-cut sample from it. Unlike some other linear cutters, the cross-belt type cutter is not limited in its speed, but it should traverse the bed of powder on the belt in the shortest possible time.

Primary samplers can be operated either in intervals of constant time or constant mass. The constant mass option makes the design and operation of the secondary sampling system simpler. It requires a continuous weighing system, like a belt scale, installed near the primary cutter, preferably before it. Such a system monitors the mass flux of the solids conveyed and adjusts the speed of the primary cutter before each cut, generating a primary increment of constant mass. In such a manner, collection of an excessive sample amount, which would overload the secondary system, is prevented.

During food powder manufacturing, powders are frequently transported as flowing streams, as in the case of hoppers, which are emptied by screw or belt conveyors. For example, one type of flowing-stream sampler is the traversing cutter. A traversing cutter is used as a primary sample, and

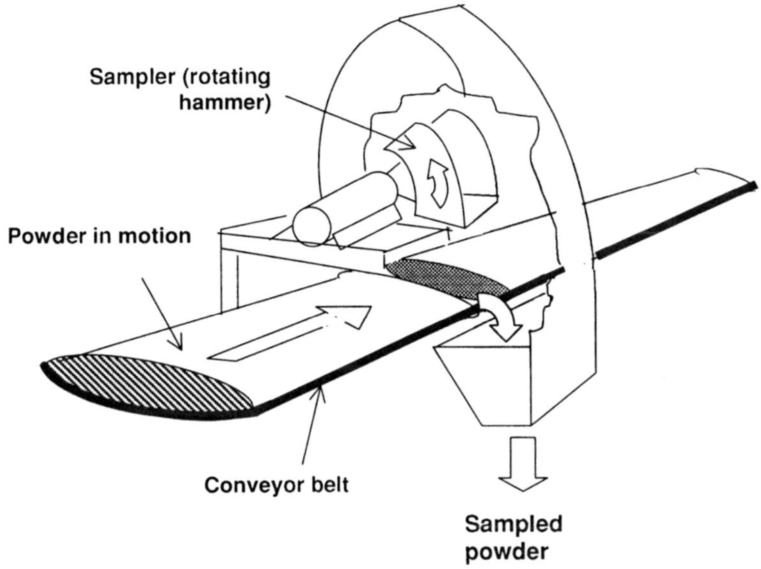

Figure 1.3. Cross-belt type cutter.

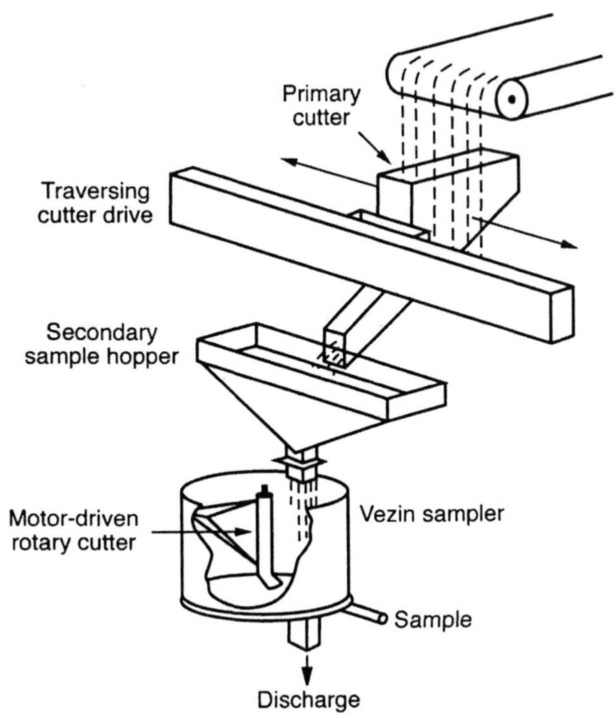

Figure 1.4. Full-stream sampler with secondary Vezin-type sampler (from Mazuda, 1997).

the extracted sample is further cut into a convenient quantity by a secondary sampling device. A particular example of a traversing cutter is the *radial cutter* or *Vezin cutter* (Fig. 1.4). These samplers vary in size from a 15-cm laboratory unit to a 152-cm commercial unit. The cutter is rotated through the stream by a gear motor drive so that representative samples of the flow are collected.

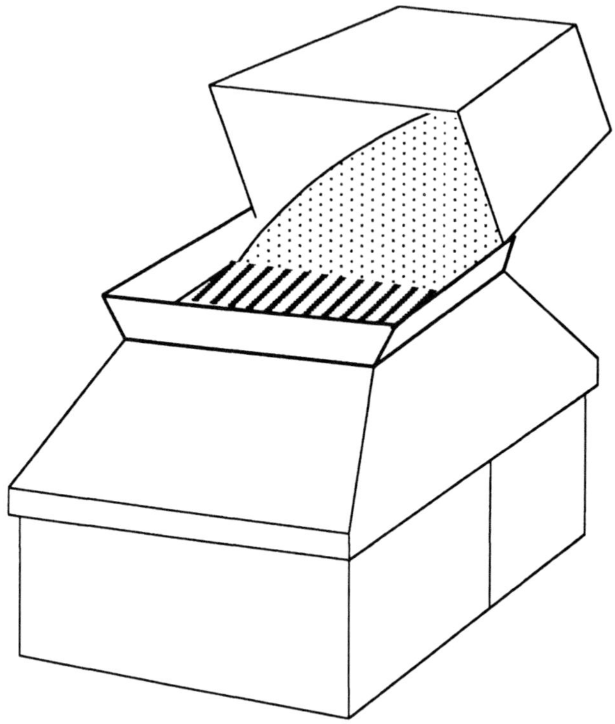

Figure 1.5. Secondary sample divider—example of a common type.

Secondary sample dividers are used to reduce the size of the primary sample and can be classified as intermittent and reciprocating cutters, as well as continuous and rotational dividers. The rotational dividers are usually considered more suitable than linear cutters. A common type of divider, shown in Fig. 1.5, is a box with an open top divided into a number of compartments with their bottoms sloping towards opposite sides of the box. When a quantity of material is poured evenly into the hopper, it is split into equal portions. Thus, by successive repetitions of the procedure, the sample can be reduced to the minimum amount required to perform an assay.

An alternative to reduce a sample to a given proportion in a single step can be performed by doing *table sampling* with a *sample reducer*, such as the one illustrated in Fig. 1.6. By pouring the sample through the hopper at the top of the inclined plane, a fixed reduction of the whole amount of the sample can be obtained. The sample stream flows downward and is broken into fractions as it encounters prisms and holes. The drawback of this device is that it depends upon the initial feed being uniformly distributed and completely mixed after each separation. An efficient sample divider is the *spinning riffler*, which follows both rules of sampling. As can be seen in Fig. 1.7, one such sampling splitter is a rotary sample divider, which comprises a hopper mounted above a vibratory feeder, and a rotating carousel divided into many compartments. The sample fills a mass flow hopper so that no heaps are present and then falls from the hopper into the collection boxes that are set in circular motion. In operation, the sample is slowly conveyed from the hopper by the feeder onto sampling containers fixed to the compartments. A vibratory feeder supplies a constant flow rate. The sub-samples are collected into the containers, and the feed rate is controlled by varying the gap under the hopper and the electromagnetic vibration of the feeder. This method yields the most accurate results among the many sample reduction devices. Many commercial instruments based

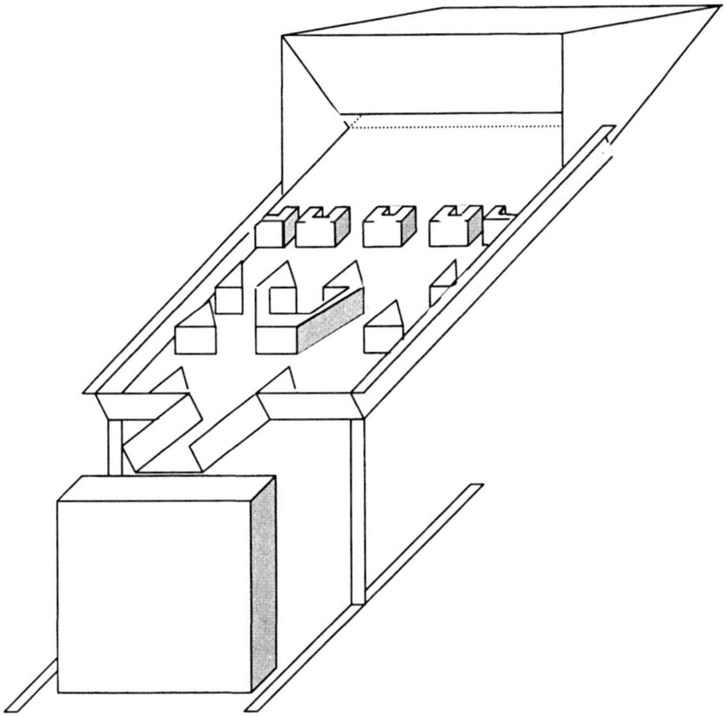

Figure 1.6. Sample reducer.

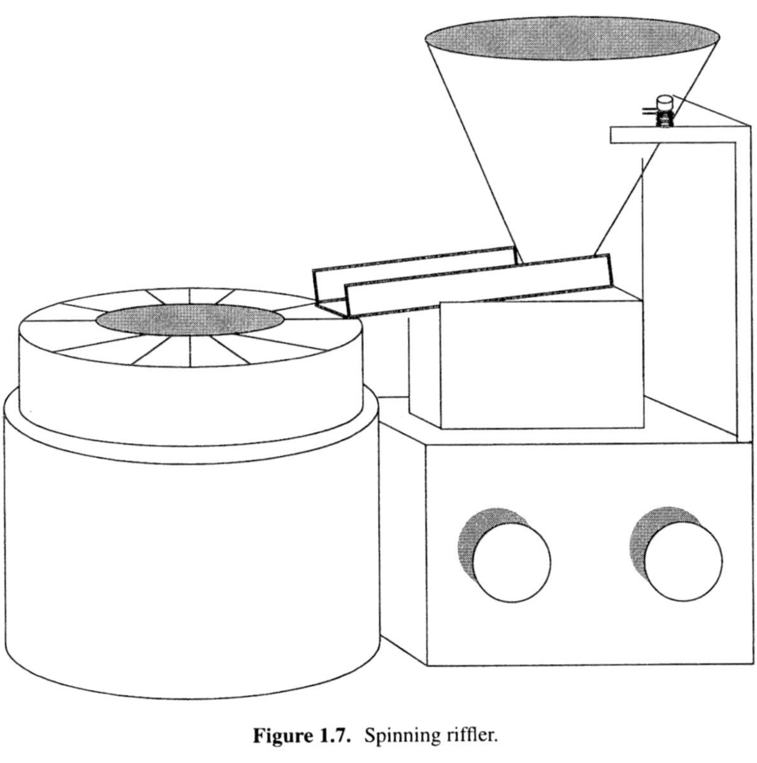

Figure 1.7. Spinning riffler.

on this principle are available, covering sampling amounts as little as 25 g or as large as 40 kg of powder, or even more.

In a chute splitter, the sample is placed in a V-shaped trough and migrates down a series of chutes that alternately feed into two trays placed on each side of the trough. The sample is poured into chutes and repeatedly halved until a desired amount of the sample is obtained. Satisfactory sample division can be obtained only if the operator loads the sample carefully into the trough without segregation occurring.

1.4. SAMPLE DISPERSION

Dispersion is a process in which particles are homogeneously suspended by the medium, after which the particles can be examined individually. Dispersion can be part of the analysis; therefore, sufficient attention must be paid to sample dispersion before introducing the sample into the instrument (e.g., to measure particle size).

Liquid sample dispersion occurs when a dry powder is suspended in a liquid medium before analysis. The medium can be either aqueous or non-aqueous. Characteristics of the dry powder dictate what medium is to be utilized as a diluent. Factors such as solubility, reactivity, suspendibility and intended end-use govern the choice of media. For example, powders such as niacin are water soluble and require the use of non-aqueous media. If the particles dissolve in a diluent, there should be no chemical reaction between the sample and the fluid medium. Physical changes to the particles such as agglomeration, aggregation, clumping, shrinking, or swelling should not exist if the medium is chosen correctly. Another issue in choosing a proper diluent is its suitability for the chosen characterization technology and instrument. For example, the degradation effect on parts of the instrument may need to be checked; the color and refractive index of the diluent must be chosen so as not to affect measurement using light scattering technologies.

Slurry, a watery mixture of insoluble solids (e.g., as found in starch manufacturing), varies its flow rate, solids concentration, and particle size distribution during processing. Any sampling available must be able to cope with these variations without affecting the representativeness of the extracted sample. For batch sampling, automatic devices are available where sampling traverses intermittently across free falling slurry (Xu, 2000). Samplers can be either narrow slots continuously rotated on an axis parallel to the slurry flow or a slotted pipe mounted vertically in the overflow compartment next to the vortex finder of a hydroclone (Fig. 1.8). Solid sample dispersion is the attempt to disperse a dry powder with the intent of suspending it in air. In many instances, the

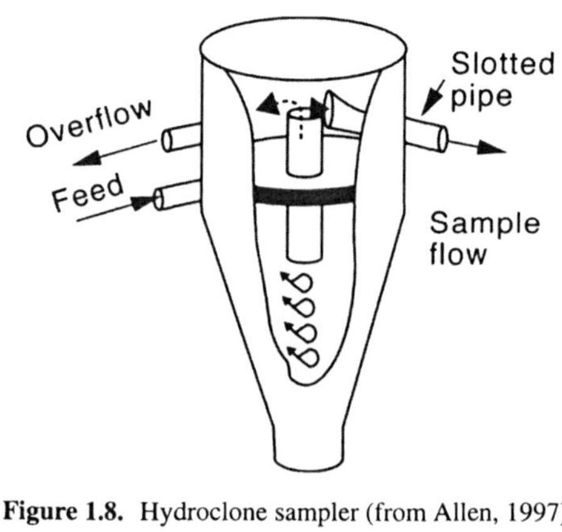

Figure 1.8. Hydroclone sampler (from Allen, 1997).

selection of solid sample dispersion is totally dependent on the end use of the product. For example, particle size measurement of flour must be carried out using a dry system, hence requiring a solid sample dispersion. Dry powders that are free-flowing, anti-caking and anti-clogging require no external dispersion aids. However, if these conditions do not hold true, external dispersion aids (additives having a much smaller or distinguishable size from the sample so that its overall effect on particle size measurement is negligible), such as silica or tricalcium phosphate, are available. Tricalcium phosphate, when added as an anticaking agent in a ratio of 0.5–1 wt.%, can help the flow of dry powder because of its "ball bearing" effect.

REFERENCES

Allen, T. (1981). *Particle Size Measurement*, 3rd ed. Chapman & Hall, London.

Allen, T. (1997). *Particle Size Measurement*, 5th ed. Chapman & Hall, London.

Gy, P.M. (1982). *Sampling of Particulate Materials—Theory and Practice*. Elsevier, Amsterdam.

Gy, P. (1998). *Sampling for Analytical Purpose*. John Wiley & Sons, New York.

Herdan, G. (1960). *Small Particle Statistics*. Butterworths, London.

Kennedy, J. and Neville, A.M. (1976). *Basic Statistical Methods for Engineers and Scientists*. Harper & Row, New York.

Larson, H.J. (1978). *Introduction to Probability Theory and Statistical Interference*. John Wiley & Sons, New York.

Masuda, H. (1997). Powder sampling, Chapter VI.1. In *Powder Technology Handbook*, 2nd ed., Gotoh, K., Masuda, H. and Higashitani, K. (eds.). Marcel Dekker, Inc., New York, pp. 817–824.

Merks, J.W. (1985). *Sampling and Weighing of Bulk Solids*. Trans Tech Publications, Clausthal-Zellerfeld, Germany.

Merks, J.W. (1986). Mechanical sampling for high capacities, Part II. *Bulk Solids Handl.* **6**: 115–119.

Plowman, P.M. (1985). Practical aspects of sampling. *Bulk Solids Handl.* **5**: 1259–1265.

Smith, R. and James, G.V. (1981). *The Sampling of Bulk Materials*. The Royal Society of Chemistry, London.

Xu, R. (2000). *Particle Characterization: Light Scattering Methods*. Kluwer Academic Publishers, The Netherlands.

CHAPTER 2

PARTICLE PROPERTIES

Particle characterization, i.e., description of primary properties of food powders in a particulate system, underlies all work in particle technology. Primary particle properties such as particle shape and particle density, together with the primary properties of a fluid (viscosity and density), and also the concentration and state of dispersion, govern secondary properties such as settling velocity of particles, rehydration rate of powders, resistance of filter cakes, etc. It could be argued that it is simpler, and more reliable, to measure the secondary properties directly without reference to primary properties. Direct measurement of secondary properties can be carried out in practice, but the ultimate aim is to predict secondary properties based on primary properties, as when determining pipe resistance to flow from known relationships, feeding in data from primary properties of a given liquid (viscosity and density), as well as properties of a pipeline (roughness). Since many relationships in powder technology are complex and often are not yet available in many areas, particle properties are mainly used for qualitative assessment of the behavior of suspensions and powders, for example, as an equipment selection guide. Since a powder is considered to be a dispersed two-phase system consisting of a dispersed phase of solid particles of different sizes and a gas as the continuous phase, complete characterization of powdered materials is dependent on the properties of a particle as an individual entity, the properties of the assembly of particles, and the interactions between those assemblies and a fluid.

2.1. PARTICLE SIZE AND SHAPE

2.1.1. Introductory Definitions

Several single particle characteristics are important to product properties (Davies, 1984). These include particle size, shape, surface, density, hardness, adsorption properties, etc. Of these features, particle size is the most essential and important. The term "size" of a powder or particulate material is relative. Since "size" is often used to classify, categorize, or characterize a powder, the term is not clearly defined. Common convention considers that for a particulate material to be considered powder, its approximate median size (50% of the material is smaller than the median size and 50% is larger) should be less than 1 mm. It is also a common practice to talk about "fine" and "coarse" powders; several attempts have been made to standardize particle nomenclature in certain fields. For example, Table 2.1 shows the terms recommended by the *British Pharmacopoeia* referring to standard sieves apertures. Also, by convention, particle size may be expressed in different units, depending on the size range involved. Coarse particles may be measured in centimeters or millimeters, fine particles in terms of screen size, and very fine particles in micrometers or nanometers. Due to recommendations of the International Organization for Standardization (ISO), SI units have been adopted in many countries. Thus, particle size may be expressed in meters when conducting engineering calculations, or in micrometers by virtue of the small range normally covered or when creating graphs. A significant number of food powders may be considered to be in the fine range. Some median sizes of common food commodities are presented in Table 2.2.

Table 2.1. Terms recommended by the British Pharmacopoeia for use with powders.

	B.S. meshes	
Powder type	All passes	Not more than 40% passes
Coarse	10	44
Moderately coarse	22	60
Moderately fine	44	85
Fine	85	—
Very fine	120	—

Table 2.2. Median particle sizes in common food commodities.

Commodity	B.S. mesh	Microns
Rice and barley grains	6–8	2,800–2,000
Granulated sugar	30–34	500–355
Table salt	52–72	300–210
Cocoa	200–300	75–53
Icing sugar	350	45

2.1.2. Selection of Relevant Characteristic Particle Size

The selection of a relevant characteristic particle size to start any sort of analysis or measurement often poses a problem. In practice, the particles forming a powder will rarely have a spherical shape. Many industrial powders are of mineral (metallic or non-metallic) origin and have been derived from hard materials by any sort of size reduction process. In such a case, the comminuted (or size-reduced) particles resemble polyhedrons with nearly plane faces, in a number of 4 to 7, and sharp edges and corners. The particles may be compact, with length, breadth, and thickness nearly equal, but sometimes they may be plate-like or needle-like. As particles get smaller, and by influence of attrition due to handling, their edges may become smoother; thus, they can be considered to be spherical. The term "diameter" is therefore often used to refer to the characteristic linear dimension. All these geometrical features of an important number of industrial powders, such as cement, clay, and chalk, are related to the intimate structure of their forming elements, whose arrangements are normally symmetrical, with definite shapes like cubes, octahedrons, etc. On the other hand, particulate food materials are mostly organic in origin, and their individual grain shapes can have a great diversity of structures, since their chemical compositions are more complex than those of inorganic industrial powders. Shape variations in food powders are enormous, ranging from extreme degrees of irregularity (ground materials like spices and sugar), to an approximate sphericity (starch and dry yeast) or well-defined crystalline shapes (granulated sugar and salt).

Considering the aspects mentioned above, expressing a single particle size is not simple when its shape is irregular. This case is frequent in many applications, mostly when dealing with food powders of truly organic origin. Irregular particles can be described by a number of sizes. There are three groups of definitions, as listed in Tables 2.3–2.5: equivalent sphere diameters, equivalent circle diameters, and statistical diameters. In the first group, the diameters of a sphere that have the same property of the particle itself are found (e.g., same volume, same settling velocity, etc.). In the second group, the diameters of a circle that have the same property of the projected outline of the particle are considered (e.g., projected area or perimeter).

Table 2.3. A list of definitions of "equivalent sphere diameters" (adapted from Allen, 1981).

Symbol	Name	Equivalent property of a sphere
x_v	Volume diameter	Volume
x_s	Surface diameter	Surface
x_{sv}	Surface volume diameter	Surface to volume ratio
x_d	Drag diameter	Resistance to motion in the same fluid at the same velocity
x_f	Free-falling diameter	Free-falling speed in the same liquid, same particle density
x_{st}	Stokes' diameter	Free-falling speed if Stokes' law is used ($Re_p < 0.2$)
x_A	Sieve diameter	Passing through the same square aperture

Table 2.4. A list of definitions of "equivalent circle diameters".

Symbol	Name	Equivalent property of a circle
x_a	Projected area diameter	Projected area if particle is resting in a stable position
x_p	Projected area diameter	Projected area if particle is randomly orientated
x_c	Perimeter diameter	Perimeter of the outline

Table 2.5. A list of definitions of "statistical diameters".

Symbol	Name	Dimension measured
x_F	Feret's diameter	Distance between two tangents on opposite sides of particle
x_M	Martin's diameter	Length of the line which bisects the image of particle
x_{SH}	Shear diameter	Particle width obtained with an image shearing eyepiece
x_{CH}	Maximum cord diameter	Maximum length of a line limited by the contour of the particle

The third group of sizes are obtained when a linear dimension is measured (usually by microscopy) parallel to a fixed direction. The most relative measurements of the diameters mentioned above would probably be the statistical diameters because they are practically determined by direct microscopy observation. Thus, for any given particle, Martin's and Feret's diameters could be radically different and, also, both different from a circle of equal perimeter or equal area (see Fig. 2.1). In practice, most of the equivalent diameters will be measured indirectly to a given number of particles taken from a representative sample and, therefore, it would be more practical to use a quick, less accurate measure on a large number of particles than a very accurate measure on few particles. Also, it would be difficult to perceive the above-mentioned equivalence of the actual particles with an ideal sphericity. Furthermore, such equivalence would depend on the method employed to determine the size. For example, Fig. 2.2 shows an approximate equivalence of an irregular particle depending on different equivalent properties of spheres.

Taking into account the concepts presented above, it is obvious that the measurement of particle size is dependent upon the conventions involved in particle size definition and also upon the physical principles employed in the determination process (Herdan, 1960). When different physical principles are used in particle size determination, it cannot be assumed that they should give identical results. For this reason, it is recommended that a characteristic particle size be selected to be measured according to the property or the process under study. For example, in pneumatic conveying or gas cleaning, it is more relevant to choose to determine the Stokes' diameter, as it represents the diameter of a sphere of the same density as the particle itself, which would fall in the gas at the same velocity as the real particle. In particles flowing through packed or fluidized beds, on the other hand, the surface–volume

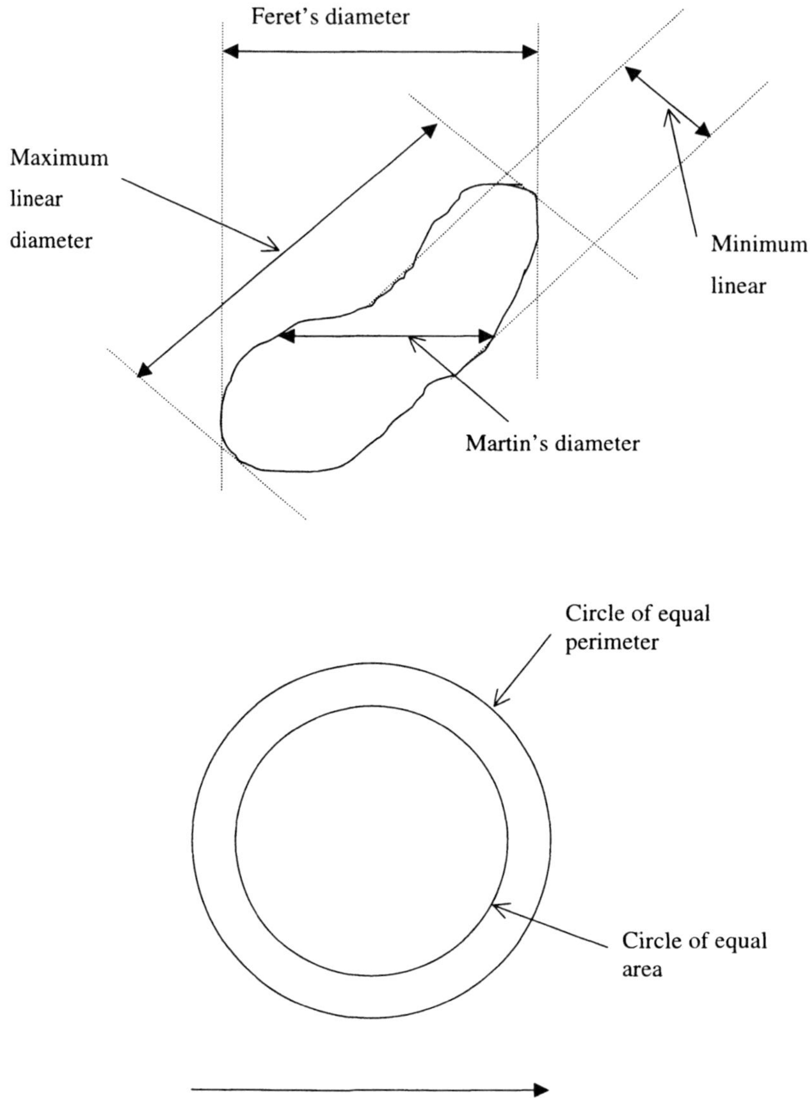

Figure 2.1. Methods used to measure diameter of non-spherical particles.

diameter is more useful, i.e., the diameter of a sphere having the same surface-to-volume ratio as the particle is more relevant to the aerodynamic process.

2.1.3. Shape of Particle Related to Sphericity

General definitions of particle shapes are listed in Table 2.6. Such simple definitions are not accurate enough to evaluate different particle size measurement methods or to incorporate them as parameters into equations where particle shapes are not the same (Herdan, 1960; Allen, 1981). Shape, in its broadest meaning, is important in particle behavior, and just looking at the particle shapes, with

Table 2.6. General definitions of particle shape (adapted from Allen, 1981).

Shape name	Shape description
Acicular	Needle shape
Angular	Roughly polyhedral shape
Crystalline	Freely developed geometric shape in a fluid medium
Dentritic	Branched crystalline shape
Fibrous	Regularly or irregular thread-like
Flaky	Plate-like
Granular	Approximately equidimensional irregular shape
Irregular	Lacking any symmetry
Modular	Rounded irregular shape
Spherical	Global shape

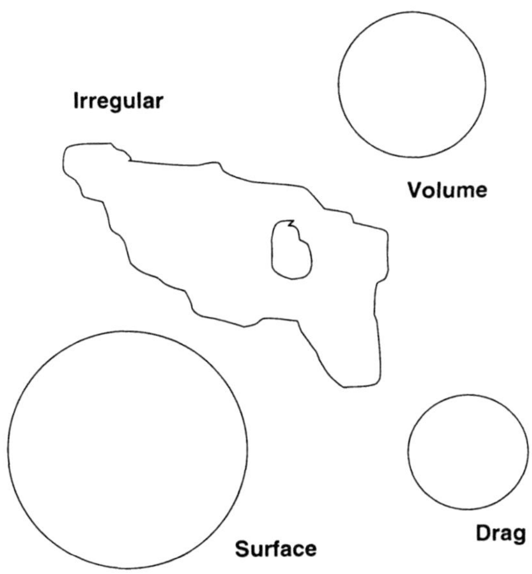

Figure 2.2. Equivalent spheres.

no attempt at quantification, can be beneficial. Shape can be used as a filter before size classification is performed. For example, as shown in Fig. 2.3, all rough outlines could be eliminated by using the ratio (perimeter):(convex perimeter), or all particles with an extreme elongation ratio. The earliest methods of describing the shape of particle outlines used length L, breadth or width B, and thickness T in expressions such as the elongation ratio (L/B) and the flakiness ratio (B/T) (Fig. 2.4).

The drawback of simple, one number shape measurements is the possibility of ambiguity; the same single number could be obtained from more than one shape. Nevertheless, a measurement of this type which has been successfully employed for many years, is the so-called sphericity, Φ_s, defined by the relation:

$$\Phi_s = \frac{6V_p}{x_p s_p} \tag{2.1}$$

where x_p is the equivalent diameter of one particle, s_p is the surface area of one particle, and V_p is the volume of one particle. For spherical particles, Φ_s equals unity, while for many crushed materials

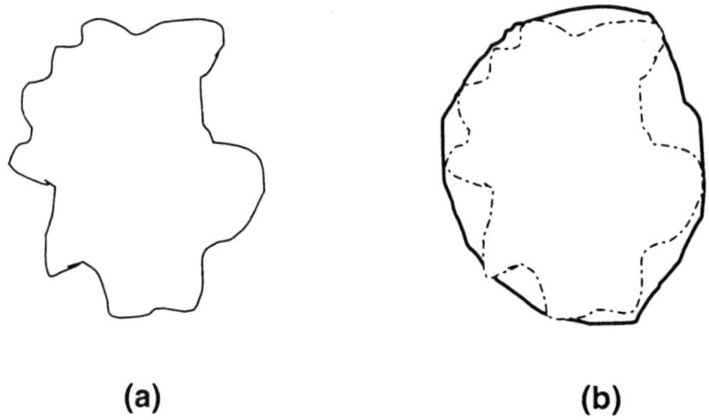

Figure 2.3. Relation between (a) perimeter and (b) convex perimeter of a particle.

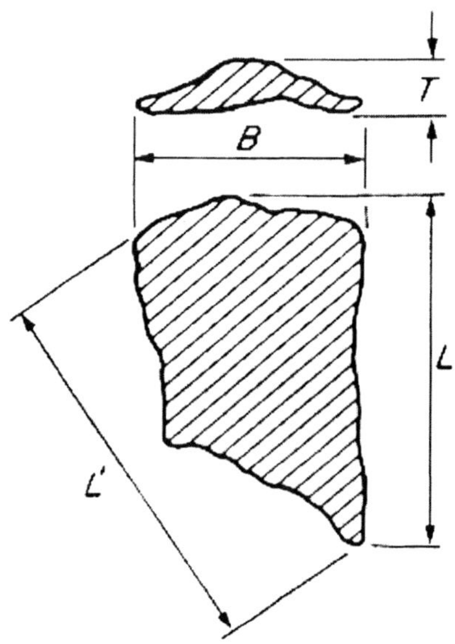

Figure 2.4. Heywood's dimensions (adapted from Allen, 1997).

its value lies between 0.6 and 0.7. Since direct measurement of particle volume and surface is not possible, to evaluate such variables, a specific equivalent diameter should be used to perform the task indirectly. For example, when using the mean projected diameter x_a, as defined in Table 2.4, volume and surface of particles may be calculated using:

$$V_p = \alpha_v x_p^3 \tag{2.2}$$

and

$$s_p = \alpha_s x_p^2 \tag{2.3}$$

where α_v and α_p are the volume and surface factors, respectively, and their numerical values are all dependent on the particle shape and the precise definition of the diameter (Parfitt and Sing, 1976). The projected diameter x_p is usually transferred into the volume diameter x_v of a sphere particle, as defined in Table 2.3, which is used as a comparison standard for the irregular particle size description.

Thus, the sphere with the equivalent diameter has the same volume as the particle. The relationship between the projected and the equivalent diameters in terms of volume is expressed as follows:

$$x_v = x_p \left[\frac{6\alpha_v}{\pi} \right]^{1/3} \tag{2.4}$$

where x_v is the equivalent diameter of the sphere of the same volume as the particle. When the mean particle surface area is known, the relationship between those two diameters is

$$x_v = x_p \left[\frac{\alpha_s}{\pi} \right]^{1/2} \tag{2.5}$$

where all the variables have been previously defined.

2.1.4. Evaluation of Shape Outline

Unambiguous shape representation involves collection and manipulation of a great deal of data. In view of this fact, consideration should be given to mechanical shape sorting before shape analysis. If shape is believed to be the cause of a particular problem or of powder behavior, then the use of size- and shape-sorted material may provide confirmation (Riley, 1968/1969; Shinohava, 1979). In some cases, however, this alternative is not possible and, particularly in investigative work, detailed measure is necessary. Sebestyen (1959) suggested characterization of silhouettes by polar coordinates of their peripheries with the center of gravity of the figure as origin, as shown in Fig. 2.5. When the R, θ readings are plotted, it is possible to represent the trace by a truncated harmonic series (Hatton, 1978). The value of the radius vector R, as it is rotated about the origin, is expressed as a function of the angle of rotation θ, in the truncated harmonic series of the form:

$$R(\theta) = A_0 + \sum_{n=1}^{M} A_n \cos(n\theta - \varphi_n) \tag{2.6}$$

where φ_n is the phase angle of the nth harmonic and $A_n = [(B_n)^2 + (C_n)^2]^{1/2}$. Each term of the harmonic series represents a particular shape. The silhouette is represented by different amplitudes and phases of these individual shapes (Fig. 2.5). Clearly, the system is not ideal because fine detail and protuberances produce difficulties of representation unless an inordinately large number of terms are used. Problems can also arise with silhouettes for re-entrant features since R must remain single valued for all values of θ. The shapes can be reconstructed only by the use of both the amplitudes

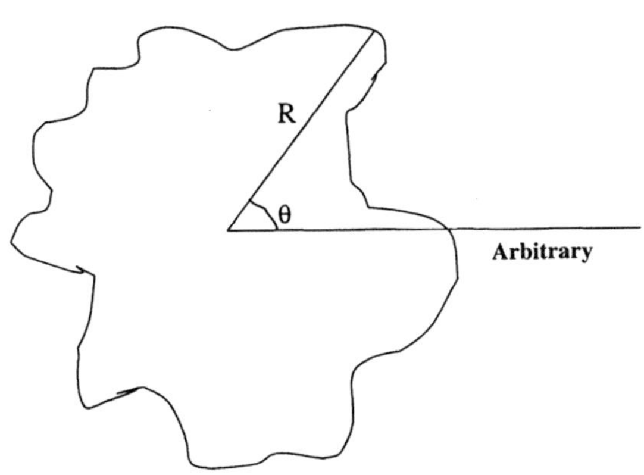

Figure 2.5. Representation of polar coordinates in an irregular silhouette (adapted from Hatton, 1978).

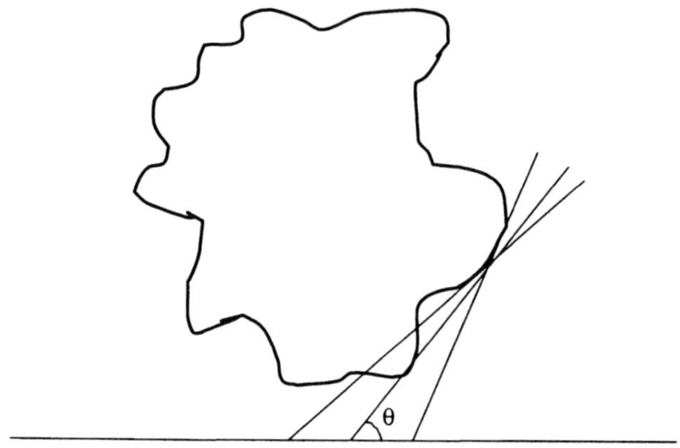

Figure 2.6. Description of a silhouette as a closed curve by its accumulated change in direction.

and the phases, but there are several ways of using the information in the representational truncated harmonic series. Attempts have been made to develop a series of combinations of harmonics to condense the information in the series (Luerkens et al., 1982), as well as using amplitudes of particular harmonics (Ehrlich and Weinberg, 1970; Holt, 1981).

An alternative approach to overcoming shape evaluation difficulties with re-entrant features is to represent any closed curve as a function of arc length by the accumulated change in direction of the curve (Gotoh and Finney, 1975). As represented in Fig. 2.6, the outline is essentially described by taking a tangent around the shape and noting the change in the contour as a function of the angle θ. Re-entrant features are automatically dealt with and the outcome is again a harmonic series, but one which converges very slowly. The regenerate produced with a usefully small number of harmonic terms is not a closed figure, but this drawback has been overcome by a computational technique (Jones, 1983).

The surface of a particle in terms of roughness can be observed in detail by stereo-microscopy, scanning electron microscopy and confocal scanning optical microscopy. However, a particle surface can be understood in further detail with a mathematical description of this surface, which depends upon obtaining a satisfactory silhouette. The most promising approach for surface modeling of a certain image would be the one based on boundary-line analysis by fractal mathematics (Mandelbrot, 1977, 1982), which is concerned directly with the properties of the outline.

In powder technology, the use of fractals was pioneered by Kaye (1981, 1984), among others, but the use of fractal mathematics is now appearing in many other fields as well. When measuring lengths of irregular contours by stepping along them with a pair of dividers, as the step length s decreases, the estimated length L_s keeps increasing without limit. In such a case, L_s is related to s by

$$L_s = K(s)^{1-D} \tag{2.7}$$

or

$$\log L_s = \log K + (1 - D) \log(s) \tag{2.8}$$

where D is known as the fractal dimension, being different for jagged and for smoothly rounded contours. The principle is, in general, applicable to any system where there is a series of estimates that tend towards infinity as the resolution of the estimate improves. The emphasis is changed from the magnitude to the rate at which it is increasing towards infinity. Chapter 12 (Section 12.1.3.5) covers the application of fractals in detail, giving a closer approach to the description of attrition phenomenon in agglomerated powders.

2.2. PARTICLE DENSITY

2.2.1. Density Expressions

The density of a particle is defined as its total mass divided by its total volume. Density is considered quite relevant for determining other particle properties such as bulk powder structure and particle size; so it requires careful definition (Okuyama and Kousaka, 1991). Depending on how the total volume is measured, different definitions of particle density can be given: the *true* particle density, the *apparent* particle density, or the *effective* (or aerodynamic) particle density. Since particles usually contain cracks, flaws, hollows, and closed pores, it follows that all of these definitions may be different.

True particle density represents the mass of the particle divided by its volume excluding open and closed pores, and is the density of the solid material of which the particle is made. For pure chemical substances, organic or inorganic, this is the density quoted in reference books with physical/chemical data. Since most inorganic materials consist of rigid particles, while most organic substances are normally soft, porous particles, true density of many food powders is considerably lower than that of mineral and metallic powders. Typical non-metallic minerals, such as some previously mentioned, have true particle densities well over 2,000 kg/m^3, while some metallic powders can present true densities of the order of 700 kg/m^3. By contrast, most food particles have densities considerably lower of about 1,000–1,500 kg/m^3.

Apparent particle density is defined as the mass of a particle divided by its volume, excluding only the open pores, and is measured by gas or liquid displacement methods such as liquid or air pycnometry.

Effective particle density refers to the mass of a particle divided by its volume, including both open and closed pores. In this case, the volume is within an aerodynamic envelope as "seen" by a gas flowing past the particle. This density is of primary importance in applications involving bulk flow of air around particles like in fluidization, of liquid as sedimentation, or flow through packed beds.

The different values of particle density can also be expressed in a dimensionless form, as relative density, or specific gravity, which is simply the ratio of the density of the particle to the density of water. It is easy to determine the mass of particles accurately, but difficult to evaluate their volume because they have irregular shapes and voids between them. The three particle densities defined above should not be confused with bulk density of materials, which includes the voids between the particles in the volume measured. Section 3.3.1 will discuss the concept of bulk density as a packing property, providing different classifications related to the intended application as well as different measurement methods. Table 2.7 lists typical densities for some food powders. As can be observed, salt (which is of inorganic origin) presents a notably higher density than the other substances listed.

Table 2.7. Densities of common food powders.

Powder	Density (kg/m^3)
Glucose	1,560
Sucrose	1,590
Starch	1,500
Cellulose	1,270–1,610
Protein (globular)	~1,400
Fat	900–950
Salt	2,160
Citric acid	1,540

The apparent particle density, or, if the particles have no closed pores, the true density, can be measured by fluid displacement methods, i.e., pycnometry, which are in common use in industry. The displacement can be carried out using either a liquid or a gas, with the gas employed normally being air. Thus, the two known techniques to determine true or apparent density, when applicable, are liquid pycnometry and air pycnometry.

2.2.2. Liquid Pycnometry

Liquid pycnometry can be used to determine particle density of fine and coarse materials depending on the volume of pycnometer bottle used. For fine powders, a pycnometer bottle of 50 ml volume is normally employed, while coarse materials may require larger calibrated containers. Figure 2.7 shows a schematic diagram of the sequence of events involved in measuring particle density using a liquid pycnometer. The liquid should be a special solvent that does not dissolve, react or penetrate the particulate food solid. The particle density ρ_s is clearly the net weight of dry powder divided by the net volume of the powder, calculated from the volume of the bottle, subtracting the volume of the added liquid, i.e.

$$\rho_s = \frac{(m_s - m_0)\rho}{(m_l - m_0) - (m_{sl} - m_s)} \tag{2.9}$$

where m_s is the weight of the bottle filled with the powder, m_0 is the weight of the empty bottle, ρ is the density of the liquid, m_l is the weight of the bottle filled with the liquid, and m_{sl} is the weight of the bottle filled with both the solid and the liquid. Air bubbles adhering to particles and/or liquid absorbed by the particles can cause errors in density measurement. Therefore, a liquid with low surface tension should be selected in order to avoid gas retention. In order to eliminate the air, the particles can be previously heated in a liquid with a high boiling point where the solid remains insoluble (Okuyama and Kousaka, 1991).

When the density of larger, irregular-shaped solid objects, such as compressed or aggregated bulk powders is needed, a method to evaluate fruit or vegetable volumes might be used. A schematic diagram of a top-loading platform scale for volume and density measurement is shown in Fig. 2.8. A beaker large enough to host the solid is partially filled with a liquid that will not dissolve the solid. The weight of the beaker filled with liquid is recorded. The solid object is completely immersed and suspended at the same time, using a string so that it does not touch the sides or bottom of the beaker. The total weight of this arrangement is recorded, and the volume of the solid V_s can be calculated (Ma et al., 1997) by

$$V_s = \frac{m_{LCS} - m_{LC}}{\rho_L} \tag{2.10}$$

where m_{LCS} is the weight of the container with liquid and submerged solid, m_{LC} is the weight of the container partially filled with liquid, and ρ_L is the density of the liquid.

2.2.3. Air Pycnometry

This density method is based on measuring the displacement of air or other gas (such as helium) using a special instrument. The instrument consists of two cylinders, A and B, both containing a piston, as shown in Fig. 2.9. Cylinder A always remains empty and works as the reference cylinder. Cylinder B has a facility for inserting a cup with the powder sample. When no sample is in the cup, the gas volume in both cylinders is the same. The gas pressure must have the same value in each cylinder at the beginning and at the end of the experiment. Pressure can be checked with a differential

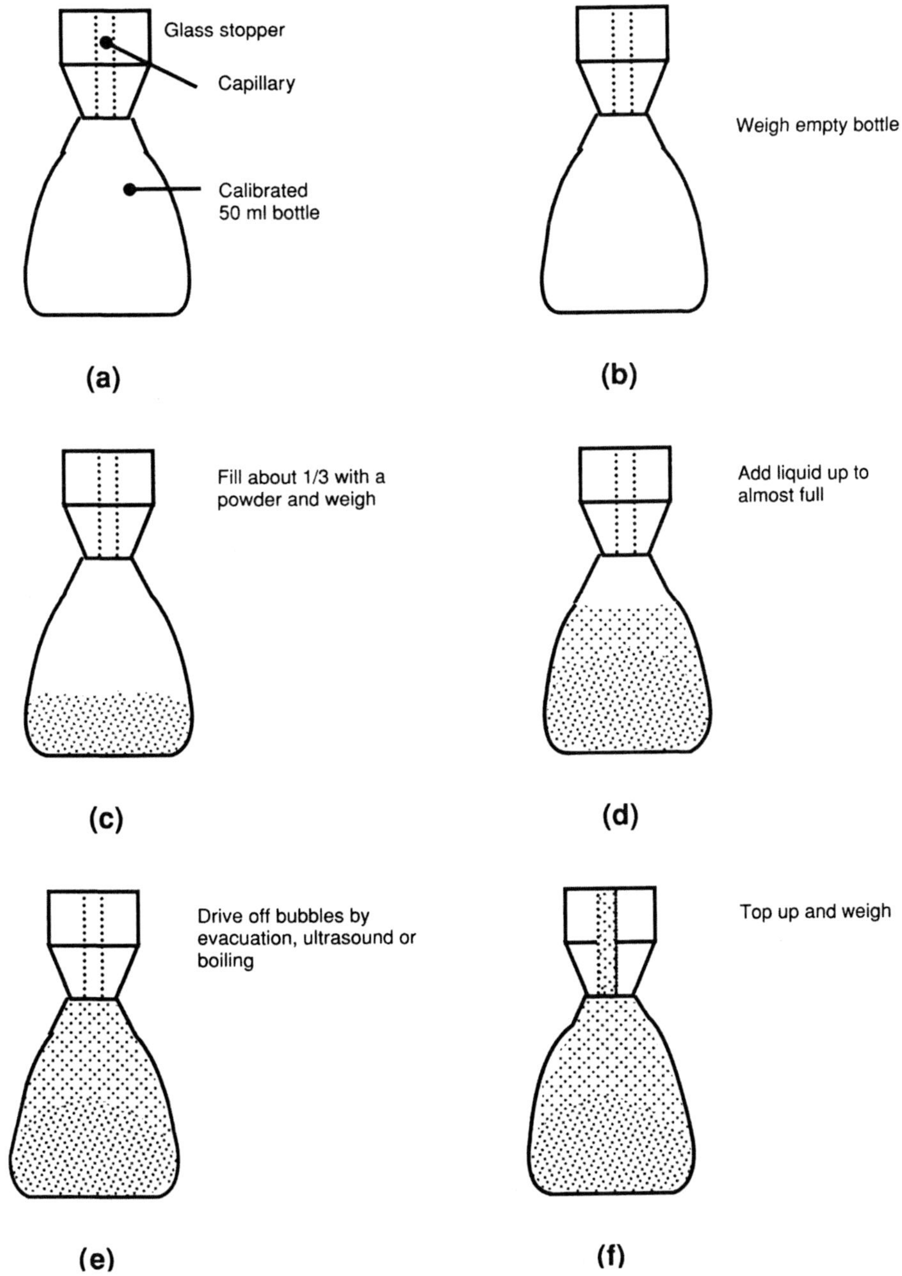

Figure 2.7. Descriptive diagram of density determination by liquid pycnometry: (a) description of pycnometer, (b) weighing, (c) filling to about 1/2 with powder, (d) adding liquid to almost full, (e) eliminating bubbles, (f) topping and final weighing (adapted from Svarovsky, 1987).

pressure indicator. As shown in Fig. 2.9, the powder sample is introduced into cylinder B. The piston in cylinder A is advanced until it reaches the stop at a new pressure value. Then the measuring piston in cylinder B is moved forward in order to reach the same pressure. As the sample is occupying extra volume, the measuring piston B cannot be moved the same distance as the piston in the cylinder A

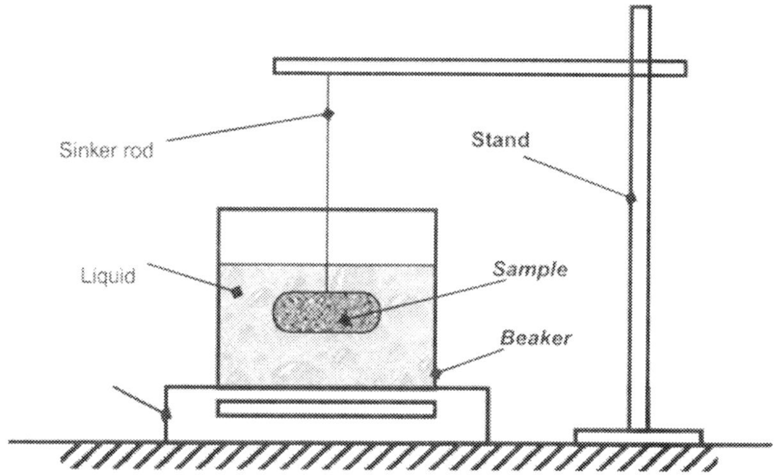

Figure 2.8. Top-loading platform scale for density determination of irregularly shaped objects (adapted from Mohsenin, 1996).

in order to reach the same pressure. This difference in the distance covered by the two pistons is proportional to the sample volume.

The equipment can be calibrated in order to be read directly in cubic centimeters, usually with a digital counter. This method will measure the true particle density if the particles have no closed pores, or the apparent particle density if there are any closed pores, because the volume measured normally excludes any open pores. If, however, the open pores are filled either by wax impregnation or by adding water, the method will also measure the envelope volume. The volume difference between the measurement of impregnated and non-impregnated particles will be the measure of the open pore volume. Neglecting closed pores within each particle in the sample, this volume difference can give a measure of particle porosity.

Considering all the aspects involved in the operation of air pycnometers, densities of most inorganic materials can be reliably determined using the equipment standard procedure with normal air because their particles are normally hard and rigid. However, extreme care must be taken when dealing with food powders, which are mostly organic in origin and their particles may be soft and porous. Materials that are not surface-active and incompressible can be measured using room air, following the equipment standard procedure that exposes the sample to air pressures from 1 to 2 bar. Food compressible materials, however, would need to be tested at lower pressures (0.5–1 bar), and surface-reactive materials may be tested using helium or another inert gas, because they tend to absorb some or all of the air constituents.

There are also other automatic pycnometers available in the market that calculate apparent density of particulate materials, either by having fixed-volume sample chambers of different sizes, or by means of volume-filling inserts placed into the chamber. The accuracy of the measurement is related to the percent of total capacity the sample material occupies in the sample cell. Sample volume is calculated from the observed pressure change that a gas undergoes when it expands from one chamber containing the sample into another chamber without sample. Figure 2.10 is a schematic diagram of an automatic pycnometer having fixed sample chamber size. The system consists of two chambers, a pressure measuring transducer and three valves.

The first requirement of sample preparation is the removal of atmospheric gases and vapors and their replacement with helium gas. This usually can be accomplished with a series of purges. Purging is performed by closing valves 2 and 3 and leaving valve 1 open, filling the sample compartment

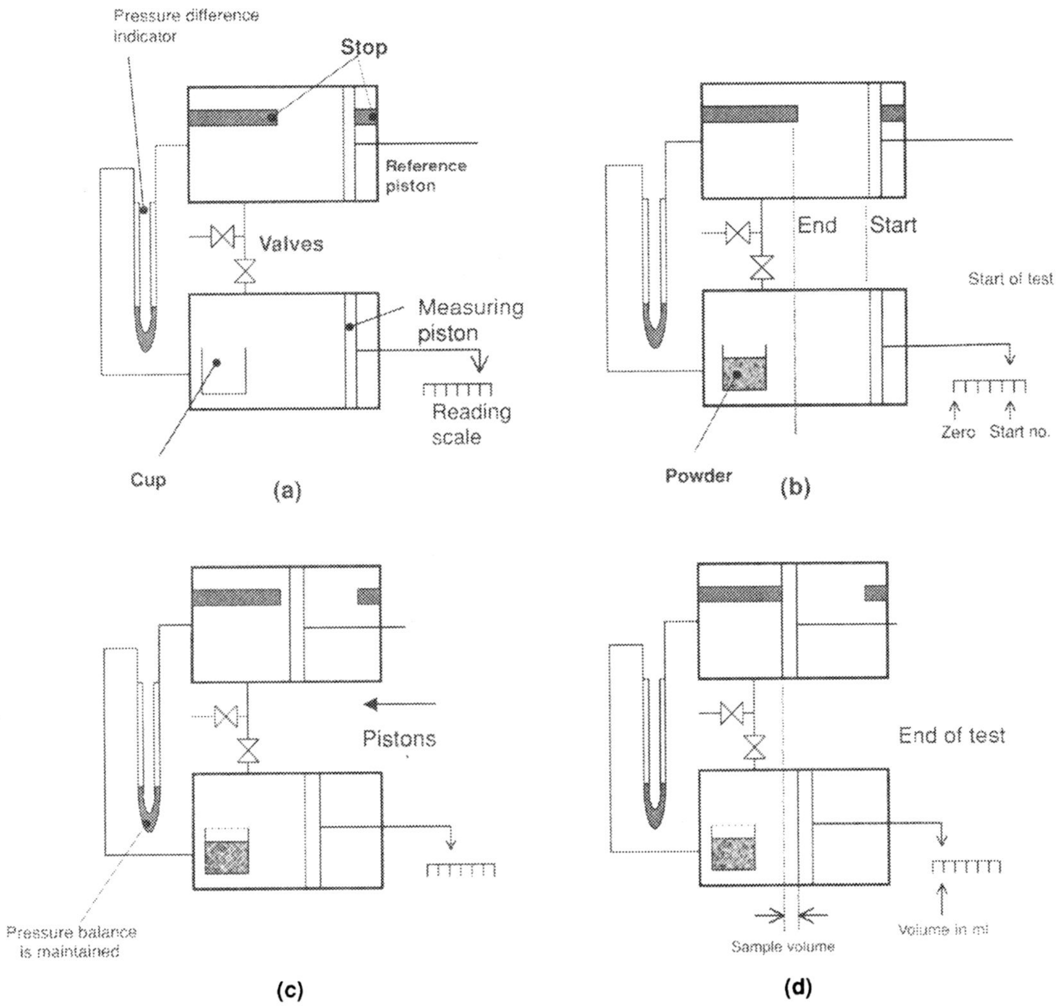

Figure 2.9. Descriptive diagram of density determination by air pycnometry: (a) description of instrument, (b) filling of cup, (c) pistons displacement, (d) reading (adapted from Mohsenin, 1996; Svarovsky, 1987).

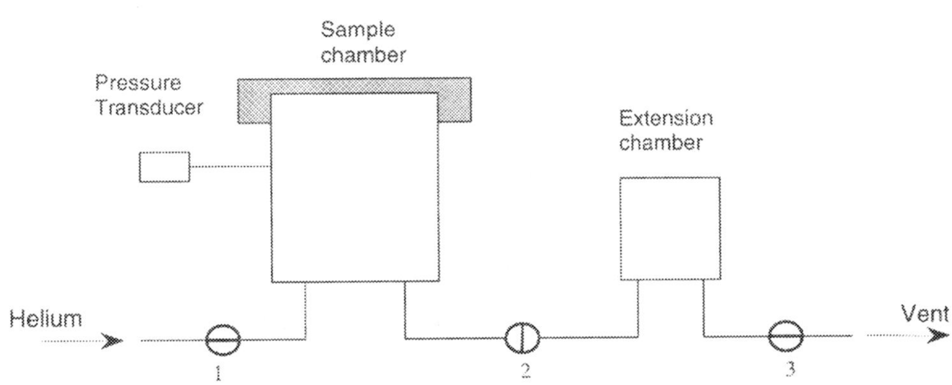

Figure 2.10. Schematic diagram for an automated pycnometer of fixed sample size (adapted from Webb and Orr, 1997).

with helium to an absolute pressure of about 20 atm; then valve 1 closes and valves 2 and 3 open, allowing the gas to escape. This process is repeated several times in order to let unwanted gases escape. Then, the pressure surrounding the sample is measured accurately. Valve 3 now opens, the gas is vented to the atmosphere, and the cycle is repeated until consecutive volume determinations are reproduced within user-prescribed limits. Finally, the sample volume is divided into sample weight to give the apparent density. These results are displayed, and if the instrument is connected to a printer, a complete report of the test is produced. Reproducibility of apparent density results is typically of $\pm 0.01\%$ when the sample volume fills the sample holder. This equipment can be applicable to several food powders, from coffee creamer to black pepper (Webb and Orr, 1997).

2.2.4. Aerodynamic Particle Density

Effective particle density is based on the average density within an aerodynamic envelope around it, including any open or closed pores in the considered volume. One obvious way to measure the volume of the open pores is with a mercury porosimeter, but this is only suitable for coarse solids, and the necessary equipment is very expensive. Alternative choices include the bed voidage method, the bed pressure drop method, and the sand displacement method.

The bed voidage method (Abrahamson and Geldart, 1980) is based on the fact that the minimum packed voidage is virtually the same for particles of similar size and particle shape. The procedure consists of pouring a small amount (0.2–0.25 kg) of a control powder of known particle effective density into a measuring cylinder and tapping it (i.e., packing particles by light vertical vibratory hits on the container) to its minimum volume, corresponding to the maximum bulk density. This procedure is repeated with the unknown powder (ideally, several control powders should be used in order to obtain more accuracy). If the porosity after tapping is assumed to be the same for the control and unknown powders, then their effective densities are in the same ratio as their bulk densities. The drawback of this method is to find control powders of the same shape that can yield the same voidage as the test powder.

The bed pressure method is based on making measurements of bed pressure drop as a function of gas velocity at two different voidages, when gas is passed through the bed of powder in the laminar flow regime. The bed is first fluidized to get it well mixed and then gently settled to its maximum voidage. Pressure drop is measured for four different gas velocities; the bed is then tapped in order to reach the lowest voidage as possible, and more measurements are made. Calling the set of measurements conditions 1 and 2, respectively, s the gradient of pressure drop, ρ_b the bulk density, and ρ_p the particle density, it can be shown (using the Carman–Kozeny equation) that a basic relationship can be derived as follows:

$$\frac{s_1}{s_2} = \left(\frac{\rho_{b1}}{\rho_{b2}}\right)\left(\frac{\rho_p - \rho_{b2}}{\rho_p - \rho_{b1}}\right)^3 \tag{2.11}$$

where ρ_p is the only unknown variable, which can be readily found by trial and error.

The sand displacement method uses fine sand into which a known amount of coarse particles of the sample are mixed. The density of the sample is determined from the difference of the bulk density of the sand alone and that of the mixture. This method is sometimes used for density determinations of coarse bone particles, for example, and it gives lower density than that of the solid bone as measured by pycnometry. The use of aerodynamic (or effective) particle density is largely restricted to fluidization and pneumatic conveying applications and, as such, is potentially a useful tool for food powder characterization due to the importance of these operations in food processes. Many food powders are brittle and fragile, so they are normally transported by pneumatic means in order to avoid attrition.

Also, fluidization processes, such as fluidized bed drying, are important for a number of food products to obtain dehydrated products with required quality attributes and good reconstitution properties.

2.3. PARTICLE SIZE DISTRIBUTION

2.3.1. Relevance of Particle Size Distribution

Particle size distribution measurement is a common method in any physical, mechanical, or chemical process because it is directly related to material behavior and/or physical properties of products. Foods are frequently in the form of fine particles during processing and marketing (Schubert, 1987). The bulk density, compressibility, and flowability of a food powder are highly dependent on particle size and its distribution (Barbosa-Cánovas et al., 1987). Segregation will occur in a free-flowing powder mixture because of the difference in particle size (Barbosa-Cánovas et al., 1985). Size distribution is also one of the factors affecting the flowability of food powders (Peleg, 1977). For quality control or system property description, the need to represent the particle size distribution of food powders becomes paramount. Proper descriptors in the analysis of the handling, processing, and functionality of each food powder will be defined in this section.

There are many different types of instruments available for measuring particle size distribution, but most of them fall into five general methods: sieving, microscope counting techniques, sedimentation, and stream scanning. In particle size measurement, two important decisions must be made before a technique is selected; these are concerned with the two variables measured: the type of particle size and the occurrence of such size. Although particle size has been previously discussed, it is important to bear in mind that great care must be taken when choosing a certain particle size as an equivalent diameter for a powder bulk; especially when a relevant property is to be studied or a certain process is to be controlled. The occurrence of amount of particle matter that belongs to specified size classes may be classified or arranged by diverse criteria so as to obtain tables or graphs. In powder technology, the use of graphs is convenient and customary for a number of reasons that will be discussed as this topic is developed.

2.3.2. Types of Particle Size Distribution

There are four different particle size distributions for a given particulate material, depending on the quantity measured: by number $f_N(x)$, by length $f_L(x)$, by surface $f_S(x)$, or by mass (or volume) $f_M(x)$. From these, the second function is not used in practice, as the length of a particle by itself is not a complete definition of its dimensions. These distributions are related, but conversion from one to another is possible only in cases when the shape factor is constant, i.e., when the particle shape is independent of the particle size. The following relationships show the basis of such conversions:

$$f_L(x) = k_1 \cdot x \cdot f_N(x) \tag{2.12}$$

$$f_S(x) = k_2 \cdot x^2 \cdot f_N(x) \tag{2.13}$$

$$f_M(x) = k_3 \cdot x^3 \cdot f_N(x) \tag{2.14}$$

where constants k_1, k_2, and k_3 contain a shape factor which may often be particle size-dependent, making an accurate conversion impossible without the full quantitative knowledge of its dependence on particle size. If the shape of the particles does not vary with size, the constants can easily be found

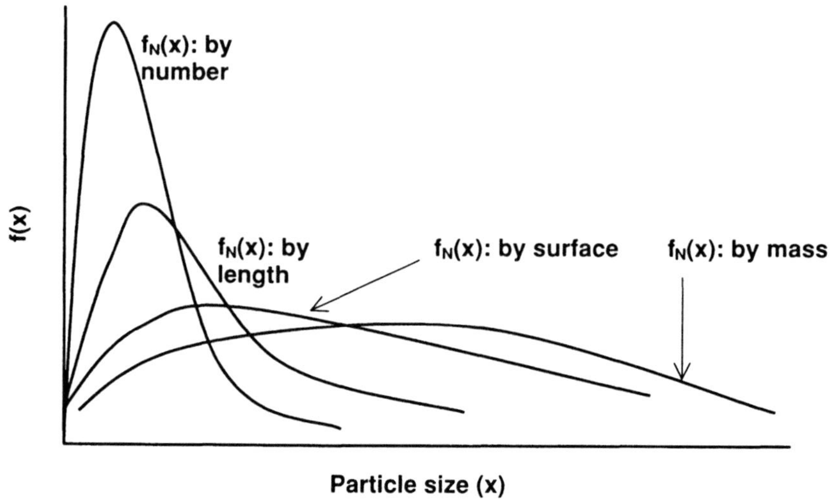

Figure 2.11. Four particle size distributions of a given particle population (modified from Svarovsky, 1981).

because, by definition of distribution frequency:

$$\int_0^\infty f(x)\,\mathrm{d}x = 1 \tag{2.15}$$

so the areas under the curve should be equal to 1.

The conversions mentioned above are to be avoided whenever possible, because of inherent errors in such procedures, by using a method which will give the desired type of distribution directly. Different methods give different types of distributions, and the selection of a method should be done on the basis of both the particle size and the type of distribution required. In food processes, many types of specific distributions might be relevant. For example, when clarifying fruit juices, for primary removal of suspended solids the size distribution by mass should be the one of interest because this particular stage is defined by gravimetric efficiency. Final clarification, however, would be better described by surface, or even number, distribution because of the low concentration of solids that cause turbidity. Ortega-Rivas et al. (1997) successfully described suspended solids removal in apple juice using particle size distributions by mass. Figure 2.11 shows the four types of distribution.

2.3.3. Particle Size Distribution Tendencies

A great number of different average or mean sizes, i.e., measures of central tendency, can be defined for a given particle size distribution. The purpose of such measures of central tendency is to represent a population of particles by a single figure. Although such representation will give no indication of the width of the distribution, it may sometimes provide a useful guide for process control. There are three important measures of central tendency for a given size distribution. These are the *mode*, the *median*, and the *mean*, as shown in Fig. 2.12. The *mode* is the most commonly occurring size, i.e., the size corresponding to the peak on the size distribution frequency curve. Some distributions may have more than one peak, and those are commonly referred as multi-modal. The *median*, or the 50% size, is the size at which half the particles are larger and half are smaller, i.e., the size that splits the area under the distribution frequency curve into two halves. The median is most easily determined from the cumulative percentage curves, where it corresponds to 50%. Many *mean*

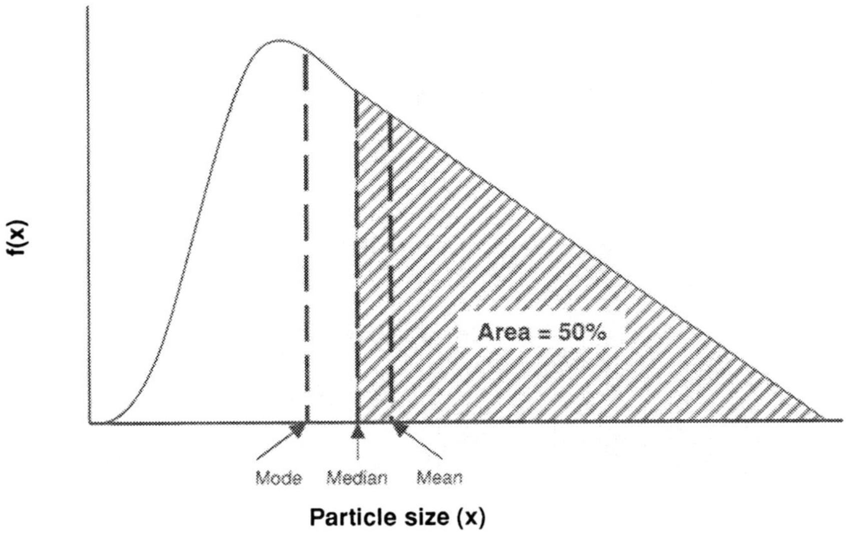

Figure 2.12. The mode, median, and mean of a particle size distribution (modified from Svarovsky, 1981).

diameters can be defined for a given particle size distribution with a definition of the general form:

$$g(\bar{x}) = \int_0^\infty g(x) f(x)\, \mathrm{d}x \tag{2.16}$$

where $f(x)$ is the particle size distribution, either by number, length, area, or mass, whichever may be of interest, and $g(x)$ is a certain function of particle size x; depending on the form of this function there are different mean diameters $\bar{x}$, as listed in Table 2.8.

Evaluation of the various means required for a given particle size distribution is based on Eq. (2.16), which may also be written as

$$g(\bar{x}) = \int_0^1 g(x)\, \mathrm{d}F \tag{2.17}$$

because

$$f(x) = \frac{\mathrm{d}F}{\mathrm{d}x} \tag{2.18}$$

If either $f(x)$ or $F(x)$ are available as analytical functions, the desired mean diameters are evaluated by integration following Eq. (2.16). If, however, no analytical function is fitted and the particle size distribution is in the form of a graph or a table, evaluation of mean diameters can best be shown graphically. Most particle size measurement techniques result in a cumulative percentage $F(x)$. Thus, according to Eq. (2.17), if $F(x)$ is plotted against $g(x)$ for a number of corresponding sizes, $g(x)$ is then represented by the area under the curve with respect to the $F(x)$ axis, as illustrated in Fig. 2.13. The mean is evaluated from this area using the corresponding relation for $g(x)$, as given in Table 2.8, while $F(x)$ is plotted as percentage oversize or undersize. It is always best to represent a population of particles by the actual size distribution curve; only in cases when this is not possible or feasible, resort to using a single number, a measure of central tendency, for characterizing

Table 2.8. Mean diameters as a function of $g(x)$ (modified from Svarovsky, 1981).

Form of $g(x)$	Name of mean diameter, $\bar{x}$
$g(x) = x$	Arithmetic mean, $\bar{x}_a$
$g(x) = x^2$	Quadratic mean, $\bar{x}_q$
$g(x) = x^3$	Cubic mean, $\bar{x}_c$
$g(x) = \log x$	Geometric mean, $\bar{x}_g$
$g(x) = 1/x$	Harmonic mean, $\bar{x}_h$

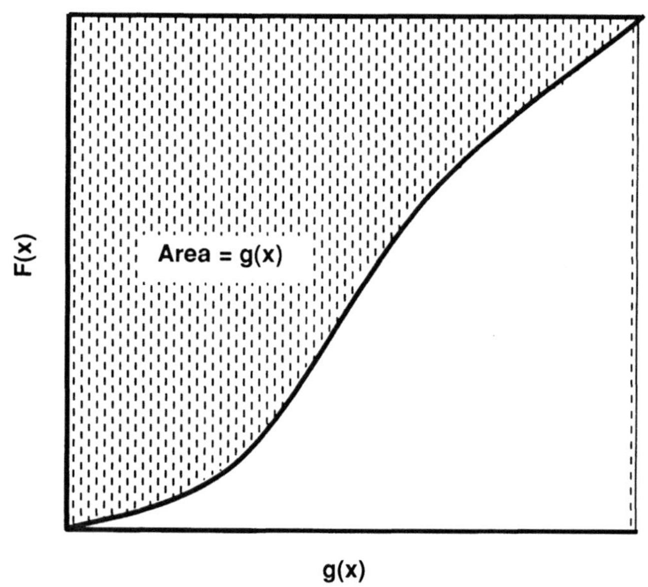

Figure 2.13. Evaluation of a mean $\bar{x}$ from the cumulative percentage $F(x)$ (modified from Svarovsky, 1981).

a particulate system. In such cases, care must be taken to select the type of mean size most relevant to the given application.

2.3.4. Presentation of Data

The common way of presenting particle size data of different powders, including food powders, is in the form of a plot representing particle amount or size frequency vs. particle size. Normally, the particle size axis is the "x" or horizontal, and the particle amount axis is the "y" or vertical. As mentioned above, what is plotted on the particle size axis is a matter of which size property, selected to represent the size of an individual particle, is better suited to a specific problem. The number of particles and mass of particles found in a specific size range is commonly used to represent the amount or size frequency of particle matter, but surface area and volume can be used as well. In general, particle size distributions can be presented as frequency $f(x)$ or cumulative frequencies $F(x)$, which are related to each other by the following equation:

$$f(x) = \frac{dF(x)}{dx} \tag{2.19}$$

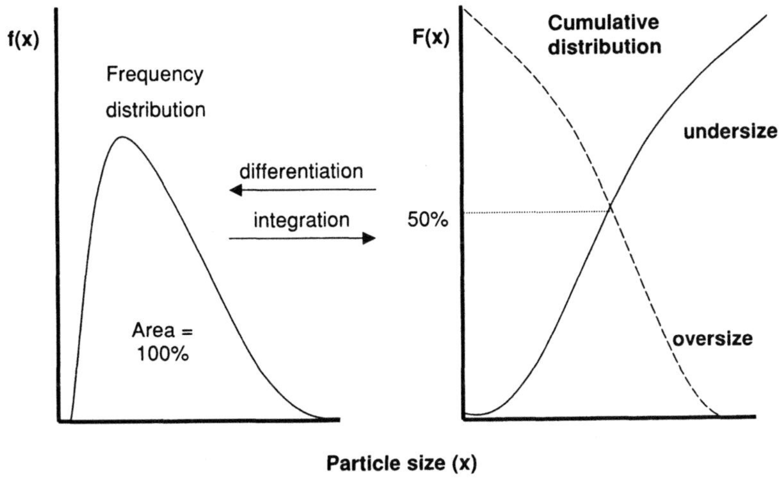

Figure 2.14. Relationship between frequency and cumulative distributions (modified from Svarovsky, 1981).

The graphical representation of a particle size distribution is usually plotted in a cumulative form. In a typical cumulative plot, points are entered showing the amount of particulate material contributed by particles below or above a specified size. Hence, the curve presents a continuously rising or decreasing character. These oversize and undersize distributions, as illustrated in Fig. 2.14, are simply related by

$$F(x)_{\text{oversize}} = 1 - F(x)_{\text{undersize}} \tag{2.20}$$

A cumulative plot will, therefore, include a broad range of particle sizes. It is often convenient, however, to refer to a single characteristic size for the system. Many characteristic sizes have been proposed, most of them involving a mathematical formula. One important formula, which can be read off any cumulative plot of the particle size data, is the median particle size. It is defined as that particle size for which the particle amount equals 50% of the total. If the particle amount is represented by a number, such point is called the number median size. If mass is used as the measure of particle amount, this variable is known as the mass median size. The distinction between number and mass median is important, since they differ by a considerable amount. Such difference means that number and mass cumulative plots do not agree for the same system of particles. The weight of a particle, which varies as the cube of its diameter, accounts for this disagreement.

For practical purposes, it is reasonable to fit an analytical function to experimental particle size distribution data, and then handle this function mathematically in further treatment. It is, for example, very much easier to evaluate mean sizes from analytical functions than from experimental data. Several different distribution functions can be found in the literature. All of them should be treated as empirical equations, as they rarely have any theoretical relation to the process in which the particles were produced. Some of these mentioned functions will be reviewed below.

2.3.5. Size Distribution Functions

There are many different types of size distribution functions, either in cumulative or relative percentage frequency form. These include the normal distribution, log-normal distribution, and Rosin–Rammler functions (Allen, 1981); the Gates–Gaudin–Schumann, Bennett's form, Gaudin–Meloy, and modified Gaudin–Meloy functions (Bergstrom, 1966; Harris, 1968); the Roller and

Svenson functions (Beke, 1964); the error function (Herdan, 1960); the modified beta function (Peleg et al., 1986; Peleg and Normand, 1986); and the Griffith and Johnson's S_B function (Yu and Standish, 1990). Some of these distributions, of common use in characterizing food powders, will be reviewed.

The *Gates–Gaudin–Schumann* function is expressed as (Harris, 1968):

$$Y = \left[\frac{x}{k}\right]^m \tag{2.21}$$

where Y is the cumulative weight fraction under size x, k is the characteristic size of the distribution, and m is the measurement of the distribution spread, also called the Schuhmann slope.

The *Rosin–Rammler* function is widely used in particle size distribution characterization. It was introduced in size reduction studies in 1933 (Harris, 1971, 1972), but was also used to describe the particle size distribution of moon dust (Allen, 1981). Usually, it is a two-parameter function given as a cumulative percentage undersize (Svarovsky, 1981):

$$Y = 1 - \exp\left[-\left(\frac{x}{x_R}\right)^n\right] \tag{2.22}$$

where Y is the weight fraction of material finer than size x, x_R is a constant giving a measure of the present particle size range, and n is a constant characteristic of the analyzed material that gives a measure of the steepness of the cumulative curve. The x_R can easily be found from the plot in the Rosin–Rammler graph because it is the size corresponding to $100/e = 36.8\%$, and n is the slope of the line (Svarovsky, 1981). Lower values of n are associated with a more scattered distribution, while higher values of n will imply an increasingly uniform particle structure.

The modified *Gaudin–Meloy* function is expressed as (Bergstrom, 1966; Harris, 1968):

$$Y = \left[1 - \left(1 - \frac{x}{x_0}\right)^r\right]^m \tag{2.23}$$

where Y is the cumulative weight fraction under size x, x_0 the parameter related to the maximum particle size, m is called the Schuhmann slope, and r is the ratio of x_0 to the size modulus.

Materials with a *normal distribution* of particle size are relatively rare and are found chiefly among the particulate produced by chemical processes like condensation or precipitation. The importance of this function, however, is that it provides an idealized error distribution built upon the assumption that elementary errors or small causes combine at random to produce the observed effect (Herdan, 1960). The normal distribution function is given by

$$y = \frac{1}{\sigma\sqrt{2\pi}} \exp\left[-\frac{(x-a)^2}{2\sigma^2}\right] \tag{2.24}$$

where y is the probability density, x is the diameter of the particle, a is the arithmetic mean, and σ is the standard deviation. From the normal function, if the arithmetic mean a is zero, the probability of occurrence within the interval form the mean (zero) to the value x, i.e., $F(x)$, is given by the *error function* (Herdan, 1960):

$$F(x) = \frac{1}{\sigma\sqrt{2\pi}} \int_0^x \exp\left[-\frac{x^2}{2\sigma^2}\right] dx \tag{2.25}$$

where σ and x have the same meaning as described in the normal function.

As described by Herdan (1960), if a powder is obtained by comminution processes, such as milling, grinding, or crushing, its distribution appears to be governed by the *log-normal* law. Furthermore, the log-normal function is the most useful one among the different types of functions

(Beddow and Meloy, 1980). It can be given in this form:

$$f(x) = \frac{1}{x \ln \sigma_g \sqrt{2\pi}} \exp\left[-\frac{(\ln x - \ln x_g)^2}{2 \ln^2 \sigma_g}\right] \tag{2.26}$$

where $f(x)$ is the size distribution function for particle size x, x_g is the geometric mean of the distribution, and σ_g is the geometric standard deviation of ln x (Peleg et al., 1986).

For many processes in which the population mode and spread vary independently and the size distributions have a finite range, the *modified beta* distribution function is more appropriate than the log-normal function because it has a finite range and the ability to describe symmetric as well as asymmetric distributions skewed to the right or left (Peleg and Normand, 1986). The modified beta distribution $f_{am}(x)$ is defined as

$$f_{am}(x) = \frac{x^{am}(1-x)^m}{\int_0^1 x^{am}(1-x)^m \, dx} \tag{2.27}$$

where a and m are constants, and x^{am} is the normalized length given by

$$x^{am} = \frac{x - x_{\min}}{x_{\max} - x_{\min}} \tag{2.28}$$

where $x_{\min}$ and $x_{\max}$ are the smallest and largest particle sizes, respectively, and therefore $x_{\min} < x < x_{\max}$, $0 < x < 1$.

Most of the functions discussed above can be used to describe particle size distribution of food powders. For example, Yan and Barbosa-Cánovas (1997) successfully fitted the Gates–Gaudin–Schumann, Rosin–Rammler, modified Gaudin–Meloy, log-normal, and modified beta size distribution functions to granular sugar, corn meal, and non-fat instant milk distribution data. They concluded that application of these models could provide useful tools for powder quality and size characterization of different food powders.

2.3.6. Analytical Techniques for Size Measurement

There is an abundance of methods available for measurement of particle size distribution and several textbooks, such as referred in this chapter (Allen, 1997; Kaye, 1981), are available and review the field in great depth. Table 2.9 gives a schematic review of the methods available, size ranges covered and types of particle size and size distribution measured. This information is given in order to assist an engineer faced with the problem of selection of the best method for a given application. Only a preliminary selection can be attempted using Table 2.9, because it is impossible to list all the important factors influencing the choice, such as type of equivalent diameter required, quantity to be measured, size range, quantity of sample available, degree of automation required, etc. It should be noted that many of these factors are interrelated and their relative importance varies in different applications.

As mentioned at the beginning of Section 2.3.1, most of the analytical techniques currently available for particle size measurement can be categorized into five main methods, namely: sieving, microscopy techniques, sedimentation, stream scanning, and on-line measurement techniques. The most representative techniques following this classification will be reviewed in this section.

2.3.6.1. Sieving

Sieving is known as one of the most useful, simple, reproducible, and inexpensive methods of particle size analysis, and belongs to the techniques using the principle of geometry similarity. It is

Table 2.9. Analytical techniques of particle size measurement.

Technique	Approximate size range (μm)	Type of particle size	Type of size distribution
Sieving			
Woven wire	37–4,000	x_A	By mass
Electro formed	5–120	x_A	By mass
Microscopy			
Optical microscopy	3–1,000	x_a, x_F, x_M	By number
Transmission electron microscopy	0.002–1	x_{SH}, x_{CH}	
Scanning electron microscopy	0.02–1,000	x_{SH}, x_{CH}	
Gravity sedimentation			
Incremental	2–100	x_{st}, x_f	By mass
Cumulative	2–100	x_{st}, x_f	By mass
Centrifugal sedimentation			
Two layer-incremental	0.01–10	x_{st}, x_f	By mass
Cumulative			
Homogeneous-incremental			
Flow classification		x_{st}, x_f	
Gravity elutriation (dry)	5–100	x_{st}, x_f	By mass
Centrifugal elutriation (dry)	2–50	x_{st}, x_f	By mass
Impact separation (dry)	0.3–50	x_{st}, x_f	By mass or number
Cyclonic separation (wet or dry)	5–50	x_{st}, x_f	By mass
Particle counters			
Coulter principle (wet)	0.8–200	x_v	By number
Laser refraction			
Low angle laser light scattering	0.1–3,000		By number

considered the only method for giving a particle size distribution based on the mass of particles in each size range. Particle size is defined by the sieve aperture by which a particle may, or may not, pass through. As presented in Table 2.9, all types of sieving cover a range from 5 μm to 4 mm. This lower limit can be achieved using micro-mesh sieves, while the upper limit can be extended to the centimeter range by punched-plate sieves. The minimum applicable particle size range is limited for two main reasons: first, it is not possible to produce sieve cloth fine enough for it and, second, very small powders do not have a strong enough gravity force to resist its tendency to adhere to one another and to the sieve cloth (Allen, 1981; Herdan, 1960).

A standard sieve series usually consists of a set of sieves with apertures covering a wide range from microns to centimeters. The sieve size is defined as the minimum square aperture through which the particles can pass. Sieves are often referred to by their mesh size, i.e., the number of wires per linear inch. Mesh size and the wire diameter determine the aperture size. The ratio of aperture of a given sieve to the aperture of the next one in a sieve series is a constant. Standardized sieve apertures were first proposed by Rittinger in 1867. Modern standards are based on either a $\sqrt{2}$ or $\sqrt[4]{2}$ progression. In the United States the series of sieves with standard opening sizes are called "Tyler" sieves and the openings of successive sieves are based on a $\sqrt{2}$ progression starting at 45 μm. The most common shape of openings is square, but some electroformed and punched-plate sieves have circular openings. Sieves with openings of other shapes (diamond, rectangle, hexagon, slotted) are also in use. Table 2.10 lists the ISO (International Standardization for Organization) and ASTM (American Society for Testing and Materials) standard sieve series.

In Table 2.10 the left column is the part of the sieve series as defined in ISO 565 and ISO 3310 with nominal openings given in millimeters, coinciding with the sieve number. The ASTM series,

Table 2.10. Standard sieve series.

ISO (mm)	ASTM (mesh)	ISO (mm)	ASTM (mesh)
2.80	No. 7	0.250	No. 60
2.50	—	0.224	—
2.36	No. 8	0.212	No. 70
2.24	—	0.200	—
2.00	No. 10	0.180	No. 80
1.80	—	0.160	—
1.70	No. 12	0.150	No. 100
1.60	—	0.140	—
1.40	No. 14	0.135	No. 120
1.25	—	0.112	—
1.18	No. 16	0.106	No. 140
1.12	—	0.100	—
1.00	No. 18	0.090	No. 170
0.900	—	0.080	—
0.850	No. 20	0.075	No. 200
0.800	—	0.071	—
0.710	No. 25	0.063	No. 230
0.630	—	0.056	—
0.600	No. 30	0.053	No. 270
0.560	—	0.050	—
0.500	No. 35	0.045	No. 325
0.450	—	0.040	—
0.425	No. 40	0.038	No. 400
0.400	—	0.036	—
0.355	No. 45	0.032	No. 450
0.315	—	0.025	No. 500
0.300	No. 50	0.020	No. 635
0.280	—	—	—

which is defined in the ASTM Standard E11, is listed in the right column; the nominal openings correspond to openings in the ISO series. Many countries also have their own standard test sieve series corresponding to part of the ISO series. A partial list of other country's standards includes Australia (AS 1152), Britain (BS 410), Canada (CGS-8.2-M88), France (NFX 11-501), Germany (DIN 4188), India (IS 460), Ireland (I.S. 24), Italy (UNI 2331), Japan (JIS Z 8801), and South Africa (SABS 197).

Sieving analysis consists of stacking the sieves in ascending order of aperture size, placing the material concerned on the top sieve, vibrating the sieves by machine or hand for a fixed time, and determining the weight fraction retained on each sieve. Additional forces may also be used to help the sieving process, such as liquid flow, air jet, and vibrating air column. Allen (1997) brought about an update of sieving equipment such as air-jet sieving, the Sonic Sifter, and automatic sieving systems among other types. Figure 2.15 shows the mode of action of a Sonic Sifter, one of the most used sieving methods in both industry and laboratory research. The wet sieving method is useful for very fine powders or when the material is originally suspended in a liquid. It represents an excellent alternative for powders forming aggregates when dry sieving is used. Typically, results from sieve analysis varies with the method of moving the sieve or particles, the geometry of the sieve surface (sieve type, frictional open area, etc.), the time length of operation, the number of particles of the sieve, and the physical properties of the particles (e.g., their shape, stickiness, and brittleness).

There are two main forms in which the results of a sieve test can be presented: tabular and graphical. As previously discussed, graphical methods are preferred in particle size analysis, as they

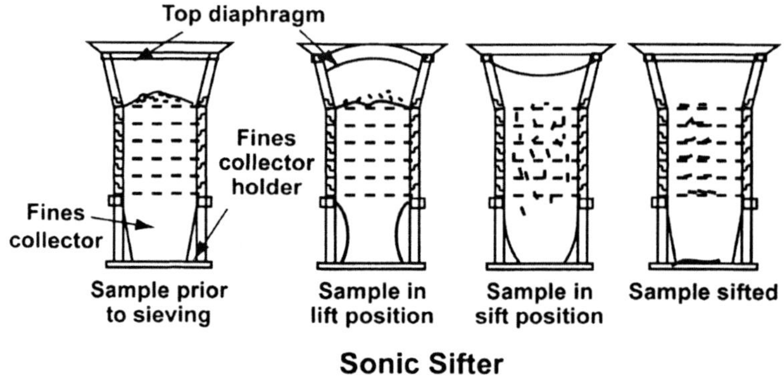

Figure 2.15. Mode of action of a Sonic Sifter (adapted from Allen, 1997).

provide a simple way of identifying a representative size of the powder being analyzed. In sieving, as in all size measurement techniques, cumulative percentages of oversize or undersize material against particle size is plotted to obtain graphs of useful information for powder characterization. Convention commands the use of the sieve diameter x_A, as defined in Table 2.3, to be the particle size plotted on cumulative percentage graphs, but it is still customary to use the mesh number instead. Sometimes the mesh number is plotted progressively, so the normally obtained oversize graph would present the typical rising character of an undersize graph, since mesh number gets larger as the particle size actually gets smaller.

2.3.6.2. Microscopy Techniques

Microscopy counting is the most direct method of particle size evaluation, as the particles are actually counted using an optical microscope. Optical microscopy is used to perform direct counting, while electron microscopy is more often used to characterize shape and morphology of particles due to its range and resolution. Counting particles by the use of a microscope is quite tedious, so the practical limit only reaches the value of 50 μm, although it may be used for limit sizes of the order of 150 μm (see Table 2.9).

For microscope counting, a sample is prepared by suspending particles in a liquid medium (normally with added dispersant) and putting them on slide for examination. The recommended suspension concentration is, as in some other techniques, of about 0.1% by volume in order to avoid overlapping of particles. Once the slides have been prepared, it is necessary to decide whether or not the particles should rest in their most stable position under gravity or be held in the orientation in which they fell by coating the slides with some sticky substance. The counting is done bearing in mind two important aspects: first, a statistical diameter, such as the Martin's or Feret's diameter (see Table 2.5), should be measured keeping the same direction for all counts in every slide; second, at least six particles in every one of a hundred slides ought to be counted. Because of the nature of the technique, the data is normally collected in tabular form, but final graphical representation is recommended.

In transmission electron microscopy (TEM), particles are deposited on a very thin film, supported on metal grids or frames, in which the electron beam can pass through. For scanning electron microscopy (SEM) backscatter measurements, the powder is thoroughly mixed with metal substrate. The powder–substrate dispersion is made conductive by coating with a thin layer of carbon from a vacuum evaporator. After this, sample is ready for particle size measurement. The limit of resolution is the distance at which two particles in close proximity appear as a single particle. Resolution will be proportional to the wavelength of the light source and inversely proportional to both the refractive index of immersion medium and the sine of the angular aperture of the objective.

2.3.6.3. Sedimentation

These methods are probably the most numerous and varied of all particle sizing techniques. They measure the Stokes' diameter of the particle and, therefore, measure the variable which is often most required to describe the behavior of the particles in suspension. There are some limitations with respect to particle shape in sedimentation analysis, as particles that diverge grossly from spheres would not fall in a vertical pattern and, hence, the results when compared with other methods will tend to show oversize material. Two factors can in general lead to incorrect dispersion analyses. The first is incorrect sampling, since analyses are carried out on from a tenth of a gram up to a few grams and these samples must be representative of the bulk for the analyses to be meaningful. The second is dispersion, being fundamental in order to obtain accurate results. The two most common methods of particle size measurement by sedimentation are *gravimetric* and *centrifugal* sedimentation.

Gravimetric sedimentation techniques

Gravity sedimentation is normally carried out by using the "Andreasen" pipette technique. In its simplest form, a suspension of approximately 0.1% by volume of particles is placed in a measuring cylinder, and a pipette is used to withdraw 10 ml from a fixed depth below the surface. Such an apparatus is shown in Fig. 2.16. The concentration of samples are measured and compared with the initial one. Since all the particles will settle at their terminal velocities, the ratio of the two concentrations is the fractional weight of particles to have reached a depth of H cm. Excluded will be the particles with a diameter smaller than x_{st} (Table 2.3) as given in the relationship:

$$x_{st} = \sqrt{\frac{18\mu H}{(\rho_s - \rho)gt}} \tag{2.29}$$

where μ is the liquid viscosity, ρ_s is the particle density, ρ is the liquid density, g is the acceleration due to gravity, and t is the time.

From the fractional weight evaluated by Eq. (2.29) the fraction undersize of a given powder can be determined. Allen (1981) has shown that variations in concentration within a settling suspension can be related to cumulative percentage undersize $F(x)$ by the formula:

$$F(x) = \frac{C(H, t)}{C(H, 0)} \tag{2.30}$$

where $C(H, t)$ is the concentration in weight by volume at a depth H and time t, and $C(H, 0)$ is the concentration in weight by volume at depth H and time 0. The time scale recommended for sample withdrawal (Allen, 1981) is the first sample at 1 min and then continuing in a geometric progression at a ratio equal to 2, i.e., 2, 4, 8 min, and so on, to obtain a smooth curve when plotting cumulative percentage undersize against Stokes' diameter because of the inherent skewed distribution of most powders.

Modern methods such as the photosedimentation technique combine gravitational settling with photoelectric measurement. The principle of the technique is that a narrow horizontal beam of parallel light is projected through the suspension at a known depth on to a photocell. Assuming an initially homogenous suspension, the attenuation at any time will be related to the undersize concentration.

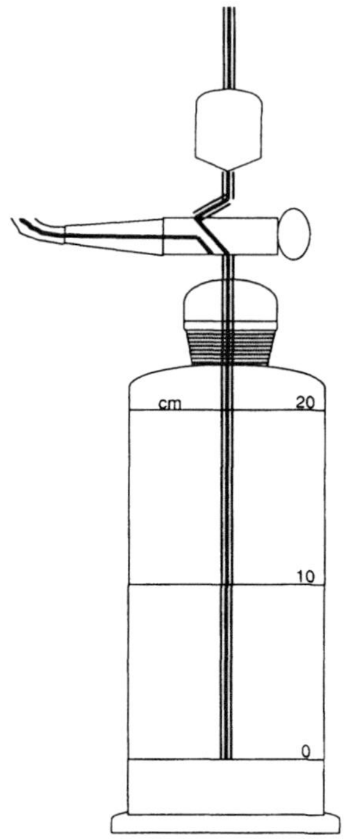

Figure 2.16. Andreasen's sedimentation pipette.

As the particle size increases, the diffracted light is contained in a decreasing solid angle and the effective obscuration area becomes the same as the projected area. Commercial equipment operates with three light sources and an algorithm included in software that obtains a k factor that controls the unequal obstruction effect of the light, corresponding to the cross-sectional area of each particle. Another known method is X-ray sedimentation, where X-rays are attenuated directly proportional to the atomic mass of the suspended particles in the beam, i.e., the mass undersize.

Centrifugal sedimentation techniques

Centrifugal sedimentation can be used to extend the range of applicability of sedimentation to sub-micron size. In addition, most sedimentation devices suffer from the effect of convection, diffusion, and Brownian motion. Speeding up the settling process by centrifuging the suspension may reduce these difficulties. Particle velocity is now not only dependent on particle size, as in gravitational sedimentation, but also depends upon the radial position of the particles. Analytical equations have been derived from the relationship between the concentration at the starting radius (initial concentration of the suspension) and the concentration at the measurement zone.

Centrifugal sedimentation can be carried out using a pipette centrifuge that consists of a bowl centrifuge equipped with a withdrawal device connected to a pipette. The principle of particle size evaluation is similar to that used in the Andreasen pipette, except that the particle is subjected to centrifugal acceleration, instead of gravity force, thus giving the possibility of detecting particles well below the sub-micron range at reasonably short periods of time. A diagram of the apparatus is given in Fig. 2.17. As can be seen, it comprises a shallow bowl disc adapted to a hollow shaft equipped with six radially arranged narrow bore needles (a "spider"). The shaft is connected to a pipette by which samples are extracted following a sampling procedure similar to that used for the

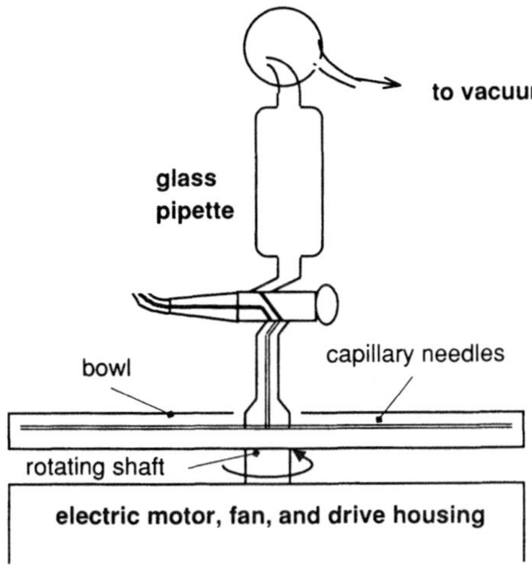

Figure 2.17. Diagram of pipette centrifuge.

Andreasen pipette technique. The modified expression of Stokes' law to account for the centrifugal field can be written as

$$x_1 = \sqrt{\frac{18\mu \ln(R/S)}{(\rho_s - \rho)\omega^2 t_1}} \tag{2.31}$$

where x_1 is the largest particle present in the initial sample withdrawn at time t_1, when the measurement radius is R and the surface is at radius S, while ω is the angular velocity. Allen (1981) presents a procedure for particle size evaluation, analogous to that of the Andreasen pipette, using Eq. (2.31).

Centrifugal sedimentation equipment can also use X-rays as a detection source. After passing through the suspension, particles are detected by a scintiliation counter. The signal is then processed to generate the size distribution. The attenuation is proportional to the mass concentration, which has to be converted to the size distribution by algorithms from a software.

Gravity and centrifugal sedimentation can be combined for the same sample in order to directly determine the Stokes diameter for a wide range of particle sizes. In such a way conversions are avoided and mass distributions, applicable to processes where gravimetric efficiencies are relevant, can be properly derived. Ortega-Rivas and Svarovsky (1994) determined particle size distributions of fine powders using a combined Andreasen pipette–pipette centrifuge method. They derived relations useful to model hydrocyclone separations, which were later employed to describe apple juice clarification. For more information on sedimentation equipment, refer to Allen (1997), who discussed different commercial gravimetric and centrifugal sedimentation particle analyzers in current use. Recently, the European Standards Organization of the European Community has prepared standards for centrifugal and gravity sedimentation methods.

2.3.6.4. Stream Scanning

Instruments used in stream scanning techniques have recently undergone significant developments. Instrumentation in this category comprises a variety of different techniques with which to measure particle size. For example, during stream scanning of a liquid or gas, in which particles are suspended, the medium can either be subjected to a specific restriction or exposed to various light sources, obtaining a response that is a function of the concentration and number of particles present in the suspension being analyzed. In stream scanning, particles are examined one at a time and their interaction with an external field is taken as a measure of their size. Stream scanning methods utilize different principles, namely:

- laser beam diffraction caused by the particle;
- electrical resistance as the particles pass through a field (Coulter counter);
- amount of particle cut off as a particle passes through a beam;
- signal created from rotating scanning beam through a particle;
- time of flight between two laser beams; and
- interference pattern as a particle passes through the intersection of two laser beams (Doppler effect).

Instruments that utilize light as a means of detecting particles in suspension generally operate by making the stream flow through a cell across which a light beam is passed. As each particle passes through the beam, a portion of the beam is blocked by the particle's cross-section. The number of particles in preset sizes is thus easily recorded. Instruments using a light-blocking technique will, however, be problematic in measuring particles whose refractive indices are close to those of the carrier liquid. In general, instruments will have problems with particles smaller than 2 μm.

Laser diffraction is the most widely used technique for particle size analysis. Instruments employed in this technique are considered fast, reproducible, and easy to use and particularly attractive in their capability to analyze (over a broad size range) a variety of dispersion media such as liquids or air streams. The basic principle upon which these instruments work is shown in Fig. 2.18. In essence, the presence of particles in a light beam causes light diffraction, and the light intensity distribution for a single opaque spherical particle falls off rapidly as particle size is reduced. This action results in a set of light rings at various radii around the incident beam. The most common beams are produced from intense light of fixed wavelength He–Ne gas lasers ($\lambda = 0.63$ μm). When an array of sizes is

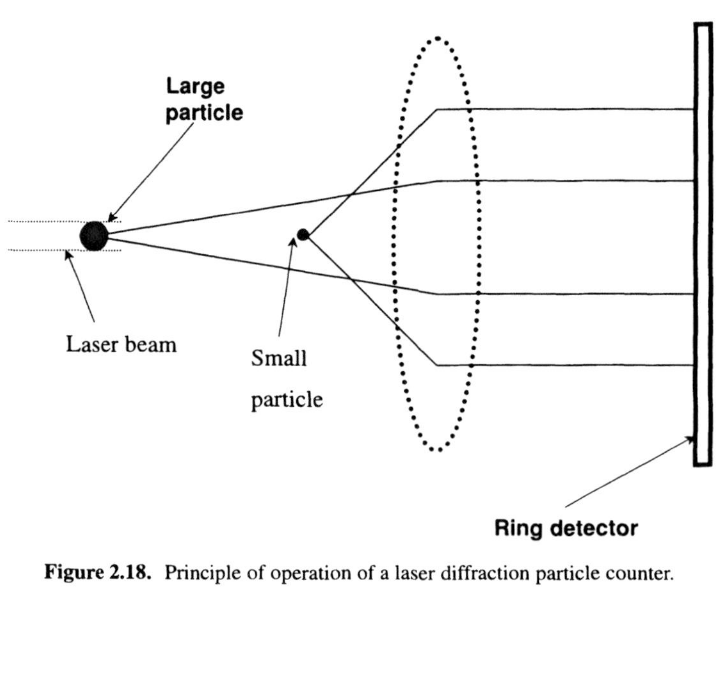

Figure 2.18. Principle of operation of a laser diffraction particle counter.

illuminated, a similar pattern emerges, but particles contribute to the intensities of more than one ring. Each set of diffraction rings is spaced radially at a distance fundamentally related to a specific particle diameter. This light scattering pattern must be deconvoluted in order to determine the size distribution from the scattered pattern measurement.

Representative samples can be as small as 4–10 g for dry powders and 1–2 g for liquid suspensions. During a test, the dry powder can be blown through the beam by means of pressure and sucked into a vacuum cleaner to prevent dust dispersion into the environment (may result in poorer dispersion than with liquid dispersing medium). Particles in suspension can be measured by recirculating the sample in front of the laser beam. This cloud or 'ensemble' of particles passes through a broadened beam of laser light and scatters the incident light onto a Fourier lens. The lens focuses the scattered light onto a detector array and, using an inversion algorithm, the particle size distribution is inferred from the collected diffracted light data. Sizing particles by this technique depends on accurate, reproducible, high resolution light scatter measurements and ensures full characterization of the sample. The size range covered by instruments employed is approximately 0.1–3,000 μm according to ISO 13320. The method rapidly produces a measurement in less than 1 min and, thus, is ideally suited for process control operations where results are required quickly with minimal operator attention. Recent developments in the use of laser diffraction techniques have enabled particles in the sub-micron range to be analyzed.

Many commercial instruments that apply this principle are available, such as those manufactured by Leeds & Northrup, Cilas, Coulter, Seishin, Shimadzu, Sympatec, Malvern, Beckman, Fritsch, Insitec, and Horiba & Nitto (Allen, 1997). Polarization Intensity Differential Scattering (PIDS) is a technique that overcomes the limitations of conventional laser diffraction in order to give high resolution submicrometer analysis. PIDS uses three different wavelengths of light (450, 600, and 900 nm) in two planes of polarization (vertical and horizontal) to irradiate the sample. The resultant scatter patterns of various sized submicrometer particles are easily differentiated from each other, providing well-resolved particle size distributions. Modern laser diffraction instruments use Mie Theory as the basis of their size calculations. As Mie Theory covers all light scatter from spherical particles, both PIDS data and laser diffraction data can be processed into a particle size distribution using one continuous algorithm.

In instrumental particle counters, such as the well-known *Coulter counter*, the stream containing the particles is forced through a flow restriction, which is then subjected to an electric field. The normal flow of electrical current between the poles is altered by the particles passing through the restriction. These electrical flow changes, as a function of the size of particles passing through, are registered as pulses, and then counted and grouped according to size. This type of counter will count and size particles in the range 0.5–800 μm, and since the basic response is directly related to particle volume, the instruments are for all practical purposes independent of problems associated with particle shape, color, or density, which affect most other methods.

2.3.6.5. On-line Measurement Techniques

Automation of process control has created a need for continuous monitoring of particle size of particulate matter in process streams. Some on-line particle size analysis instrumentation has been developed recently to meet this need. It can initiate regulatory or shutdown signals in control systems. The basic requirements for such instrumentation are that it must operate automatically and continuously under preset instructions, and the response time from observations to readout must be so short as to be nearly instantaneous. This relatively new range of techniques follows the general pattern found in the whole subject of particle size measurement. Some just give a measure of central tendency (mean diameter), while others give one or more points on the size distribution. They may be truly on-line in operating on the whole process stream, they may need a partial sample stream taken off the main one, or they may merely be automated rapid response batch techniques.

On-line measurement is an area that is undergoing fast growth. Equipment may be broadly divided into two categories: stream scanning and field scanning. Stream scanning includes different techniques following the basic principle described in the previous section adapted on-line to various processes. Field scanning usually applies to concentrated systems in which some size-dependent behavior of the bulk material is monitored and particle size deduced from theoretical or calibrated relationships. Ultrasonic attenuation, echo measurements, laser attenuation, on-line viscometry, electric noise correlation techniques, X-ray attenuation, and X-ray fluorescence are examples of such field scanning methods. Most relevant to solid–liquid separation problems are those field scanning instruments that deduce particle size from separation efficiency of some separational equipment using the concept of the analytical cut size. Within these methods, wet sieving machines and separation performance hydrocyclones can be mentioned, e.g., for starch production.

2.4. OTHER PRIMARY PROPERTIES

2.4.1. Surface Area

One of the most important characteristics of fine powders is the area of the surface of the solids, which is usually expressed as a specific surface area. This feature is a measure of the fineness of the powder as well as of its porosity, but it is unable to discriminate between a mono-disperse powder and one containing a wide range of particle sizes. Surface area is important in all applications where the process is surface dependent. Examples of such applications are mass and heat transfer, flow through packed beds, or fluidization. In food process engineering, combined heat and mass transfer is critical in quality control of many materials where moisture has to be removed to the lowest possible level, but the use of excessive heat may impair sensory attributes. Many food materials have to be finely divided in order to improve mass transfer, with a minimum of applied heat, in applications such as leaching of coffee beans and drying of different spices. In general terms, any surface-dependent phenomenon can be used for surface area measurement. Permeametry and gas adsorption are widely used in many industrial applications for this purpose. The specific surface determined by any method can be converted into an equivalent mean spherical diameter by the simple relation: $x_{sv} = 6/S_v$, where S_v is the volume specific surface. The equivalent mean diameter x_{sv} is the size of a spherical particle that, if the powder consisted of only such particles, would have the same specific surface area as the actual sample.

2.4.1.1. Permeametry

This method is based on measuring the permeability of a packed bed of powder to a laminar gas flow. Since permeability can be defined as the resistance of fluid flow through a compact bed of powder, evaluation of such resistance measures the surface area of a solid volume of powder. The surface area determined by permeametry is an external envelope area based on the concept that a packed bed of powder can be idealized as a bundle of capillaries. The surface area of the particles within the bed, which are presented to the moving fluid, is a function of the magnitude of the wall area of the capillaries/channels through the bed. The fluid mainly used is air, because with particles of less than 5 μm, liquid introduces errors into the surface area evaluated because of segregation, adsorption on, and aggregation of, fine particles. The surface area should technically be described as the effective permeability volume specific surface S_v. The permeability surface area per unit effective volume of particles can be transposed into a mass specific surface S_w by division of the solid particle density ($S_w = S_v/\rho_p$). The permeability process can be described by the Carman–Kozeny equation, which may be represented as follows:

$$q = \frac{\Delta P}{\mu L}\frac{\varepsilon^3 A}{k(1-\varepsilon^2)S_w^2} \tag{2.32}$$

where q is the rate of flow of incompressible fluid through a powder bed of viscosity μ, ΔP is the pressure drop across the bed, L is the bed depth, ε is the porosity or voidage, A is the cross-section area of powder bed, k is the Carman factor, and S_w is the mass specific surface described above. As stated earlier, the specific surface calculated involves only the walls of the pores of the bed, which are swept by the flow. The pores within the particles that do not contribute to the flow are not considered and, therefore, the surface measured is an envelope surface that can be very much smaller than the total surface measured by a different method, such as gas adsorption.

The effectiveness of Eq. (2.32) depends upon the Carman factor k, and also on the constancy of this factor with variation in porosity. The bed porosity is the ratio of the volume of inter-particle voids to the total bulk volume of the powder bed and can be calculated from the equation:

$$\varepsilon = 1 - \left[\frac{W}{LA\rho_p}\right] \tag{2.33}$$

where W is the mass of the powder bed and ρ_p is the particle density.

The Carman factor k is constant over a wide range of porosities. For example, for values of ε up to 0.8, it can be taken as 5. When ε begins to exceed 0.8, the values of k begin to rise rapidly. Such increase may be due to the fact that the Carman–Kozeny equation is based on channel flow (the pore spaces made by the solid particles are considered as a series of channels), thus as ε tends to 0.8 and higher values, there are fewer channels, the particles begin to become completely independent of each other and resistance to flow decreases. The Carman factor k can be considerable in error even for mono-size particles, and the error increases enormously for wide distribution of sizes. It consists of two terms:

$$k = k_0 k_2^2 \tag{2.34}$$

where k_0 is the shape factor of the channel and k_2 is the ratio of actual length to bed length.

For a uniform circular capillary, the value of k_0 equals 2; for other geometrical shapes, k_0 varies between 1.2 and 3.0. For most annular and elliptical shapes, k_0 lies between 2.0 and 2.5. The success of the Carman–Kozeny equation for granular beds can be attributed to the pore spaces which are continually breaking up, dividing and re-connecting along the length of the bed, so that the average shape lies in the range of 2.5. For log-normal capillary distributions, k_0 may be 2/3. Since most powders possess a large particle size range, there are no widely varying sizes of pore spaces. The ratio of actual length to bed length k_2 is known as the tortuosity factor and is considered to be independent of ε, although it is more probable that a slow variation occurs, since tortuosity approaches unity as porosity approaches unity.

The actual measurement of surface area is commonly performed in constant flow instruments (constant pressure drop), such as the Lea and Nurse apparatus. As shown in Fig. 2.19, in this equipment the sample is compressed to a known porosity in the permeability cell and maintained within it by means of porous paper discs at each end. The paper discs are supported by a rigid perforated plate. A standard cell size of cross section of 1.207 cm^2 is normally used. Dry air, drawn by an aspirator or

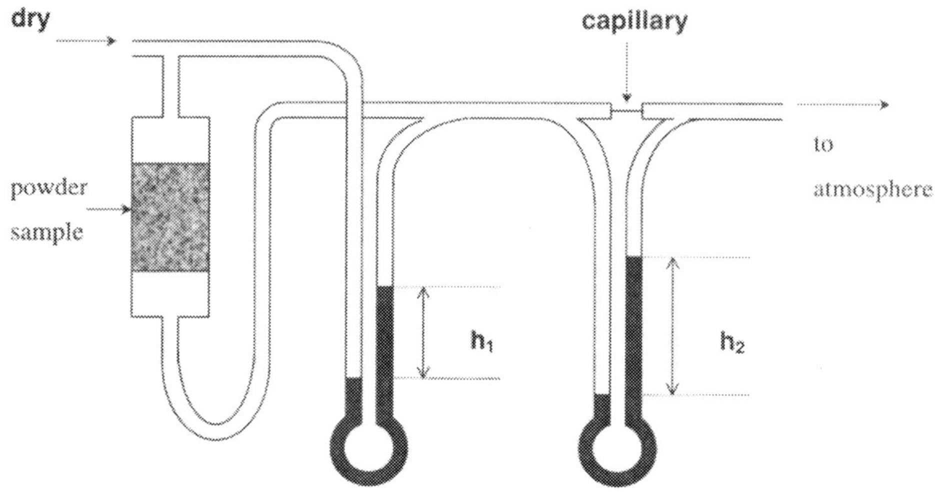

Figure 2.19. Diagram of the Lea and Nurse apparatus (adapted from Svarovsky, 1987).

a pump, flows through the bed at a constant rate and then passes through a capillary that serves as a flow meter. Static pressure drop across the powder bed is measured with a manometer as static head h_1, while the flow rate is measured by means of the capillary flow meter, giving a reading h_2 on the second manometer. Both pressure drops are small compared with atmospheric pressure and, thus, the compressibility of the gas can be neglected. The Carman–Kozeny equation can be simplified to include the static head measurements h_1 and h_2, taking the following form:

$$S_w = \frac{14}{\rho_s(1-\varepsilon)}\sqrt{\frac{\varepsilon^3 A h_1}{cLh_2}} \tag{2.35}$$

where A is the cross-sectional area of the bed, ρ_s is the solids density, and c is the flow meter conductance. The proposed ISO standard recommends that, in order to increase the accuracy of the permeability determination, the measurement be repeated at three different flow rates and an average value taken. It also recommends that the bed uniformity be tested by repeating measurements with different amounts of powder packed to the same porosity or under the same packing force. The porosity range should normally be between 0.45 and 0.7, and the optimum range should be checked for every new powder.

Another constant flow instrument is the Fisher sub-sieve sizer, now widely used in industry. This equipment is based on the Gooden and Smith method, which uses a single manometer to measure the pressure drop across the powder bed as well as the flow rate. Dry air is fed to the instrument at constant overpressure P_0 (controlled by a constant pressure regulator). It passes through the powder plug and is discharged into the atmosphere via capillaries of adjustable resistance. The overpressure P_0 is therefore the total resistance of the whole system and a manometer is used to determine how this pressure drop is divided between the powder plug and the flow meter resistance in series with it. Hence, the manometer reading of the absolute pressure after the bed P' gives the flow rate and the pressure drop across the powder bed is simply the difference between the feed pressure P_0 and the manometer reading P'. The commercial sizer uses a self-calculating chart that gives a direct reading of the bed porosity and the equivalent mean spherical diameter of the powder.

Permeametry is generally suitable for powders of average particle size between 0.2 and 50 μm, but it can also be used with coarser powders (up to 1,000 μm average particle size) using a suitably scaled-up test equipment. With very fine particles, the results are subject to systematic errors due to slip flow (a correction in the basic equation can be used to take this into account). With highly irregular particles, such as platelets or fibers, error is introduced because the Carman factor (5 for coarse granular solids) depends on particle shape.

There are also available instruments to measure surface area by permeametry that operate on a variable flow mode (constant volume). Examples of this type of instrument are the Griffin and George permeameter (oil suction), as well as the Reynolds and Branson apparatus (mercury suction). The constant volume methods do not use flow meters and therefore need frequent calibration against known reference materials. As previously mentioned, constant flow instruments are widely used in industry, and specific details of some other techniques appear in textbooks available on the subject (Parfitt and Sing, 1976).

2.4.1.2. Gas Adsorption

As the name suggests, gas adsorption methods measure the surface area of powders from the amount of gas adsorbed onto the powder surface. These methods measure both external and internal surfaces (including open pores in the particles) and can, therefore, yield physically meaningful average particle sizes only with non-porous materials. Nitrogen is most commonly used as operating gas, but krypton and argon, as well as organic adsorptives such as benzene and carbon tetrachloride, can also be employed. Physical adsorption is a relatively weak interaction between solids and gases and it can, therefore, be removed by evacuation. The quantity of physically adsorbed gas at a given pressure increases with decreasing temperature and, to maximize the quantities, the adsorption is usually carried out at temperatures close to the boiling point of nitrogen under atmospheric pressure. From the resulting plot of the volume adsorbed V against the relative pressure P/P_0 (the adsorption isotherm), it is possible to calculate the mono-layer capacity of the powder surface and hence its surface area, if the area corresponding to the nitrogen molecule is known. Normally, isotherms are reversible, but in some porous solids the adsorption and desorption curves form a hysteresis loop.

As a first approximation, the volume required for a mono-layer V_m can be taken as the first point where a change of slope occurs on the measured isotherm. A more precise calculation of the mono-layer capacity can be obtained by the application of the equation derived by Brunauer, Emmet and Teller, generally known as the BET equation:

$$\left[\frac{P}{V(P_0 - P)}\right] = \left(\frac{1}{V_m c}\right) + \left(\frac{c-1}{V_m c}\right)\left(\frac{P}{P_0}\right) \tag{2.36}$$

where V is the volume of gas adsorbed at pressure P, P_0 is the saturated vapor pressure of adsorbate at temperature of adsorption, and c is an equation parameter.

According to Eq. (2.36), the plot of $P/[V(P_0 - P)]$ against P/P_0, should be a straight line of slope $(c-1)/(V_m c)$ and intercept $1/(V_m c)$, from which V_m and c can be readily determined. Since the area occupied by each nitrogen molecule is 16.2 Å, the specific surface of the sample is given by $S_w = 4.38 V_m / W$, where W is the sample weight. If c is greater than 1, the small intercept on the y-axis can be neglected without serious loss of accuracy and V_m can then be calculated from $V(P_0 - P) = P_0 V_m$; this requires only one point of measurement (usually taken at $P/P_0 = 0.3$). This procedure is the basis of the single-point determinations and some instruments built for routine analyses. Such determinations are usually within about 7% of the values calculated from a plot of five or more points on the full isotherm using the full BET equation. The BET method is valid for relative pressures P/P_0 from 0.05 to 0.3 and, with nitrogen, it is not too precise for areas less than 2 m^2/g, i.e., particle size greater than 2 or 3 μm.

2.4.2. Moisture

Moisture content plays a vital role in all aspects of general handling of any type of powder. It is a critical aspect in food powders where it is associated with increased cohesiveness, mainly due to inter-particle liquid bridges. Inter-particle films or bridges are also said to be responsible for spontaneous agglomeration of particles. This feature is beneficially exploited in processes like instantizing dehydrated food powders, such as instant coffee and cocoa. Generally, however, the formation of liquid films or bridges at the surface of food powders may cause flow difficulties and, in extreme cases, severe caking problems.

Moisture or water content is in principle very simple to measure: it is the quantity of water in a unit mass of bulk powder, as a percentage or fraction by mass (or weight). Some confusion may arise as to whether the mass of bulk powder includes moisture or not. The moisture content may, therefore, be either expressed on a *dry basis* (i.e., per unit mass of dry powder) or on a *wet basis*, where it is regarded as a fraction or percentage of the total system (i.e., bulk powder plus water or moisture). In food process engineering, moisture content is commonly taken on a dry basis, so it would appear in calculations as units of water related to correspondent units of dry powder, regardless of the system of units employed, i.e., kg of water/kg of dry solid, lb of water/lb or dry solid, etc.

Moisture is generally determined by gravimetric means. A bulk powder sample is weighed, the moisture is removed by evaporation, and the sample is weighed again. There are two problems involved in this method: first, how much water has actually been driven off and, second, how much heat can be applied to the solid before it starts decomposing or liberating water through crystallization. Water between particles is sometimes referred to as "free" moisture, while the water within the pores of the particles is known as "inherent" moisture. At one extreme the sample can be dried under ambient conditions by natural air, and the result is the "air-dried moisture content" which, obviously, does not include the inherent moisture of the powder. Food powders that can be heated up to 105°C are measured using the oven drying method, where the sample is placed in a glass bottle or other air-tight container and weighed with the lid in place. For drying, the lid is removed and the sample placed in an oven and held at 105–110°C until it becomes dry. In order to prevent the sample from absorbing moisture from the atmosphere when taken out of the oven, sample in the container with the lid on is cooled in a desiccator before weighing it again to determine the moisture quantity by weight difference.

There are several alternatives to the standard laboratory oven: a microwave oven can be used providing that the powder is not allowed to exceed a given temperature (110°C) before all of the water is driven off. An infrared heater may also be used in place of an oven, for quick and less accurate determinations. There are some commercial balances available, the so-called "moisture balances," which incorporate such heaters directly above the pan of the balance. Apart from the gravimetric methods of moisture determination, there are sophisticated commercial instruments for measuring moisture, some suitable for on-line applications and monitoring. Those are based on back-scatter radiation, near infrared absorption, electrical conductance, or reflection photometry, and most require previous calibration.

As previously mentioned, moisture is critical in handling, storing, and processing of food powders. It practically affects any secondary property of a food powder such as bulk density, cohesiveness, adhesiveness, and so on. These bulk properties will be discussed in Chapter 3, and Chapter 12 brings a detailed description on related undesirable effects.

REFERENCES

Allen, T. (1981). *Particle Size Measurement*. Chapman & Hall, London.

Allen, T. (1997). *Particle Size Measurement*. Chapman & Hall, London.

Abrahamsen, A.R. and Geldart, D. (1980). Behaviour of gas-fluidized beds of fine powders, Part I. Homogeneous expansion. *Powder Technol.* **26**: 35–46.

Barbosa-Cánovas, G.V., Málave-López, J. and Peleg, M. (1985). Segregation in food powders. *Biotechnol. Prog.* **1**: 140–146.

Barbosa-Cánovas, G.V., Málave-López, J. and Peleg, M. (1987). Density and compressibility of selected food powders mixture. *J. Food Process Eng.* **10**: 1–19.

Beddow, J.K. and Meloy, J.K. (1980). *Testing and Characterization of Powders and Fine Particles*. Heyden and Son, London.

Beke, B. (1964). *Principles of Comminution*. Publishing House of the Hungarian Academy of Science, Budapest.

Bergstrom, B.H. (1966). Empirical modification of the Gaudin–Meloy equation. *AIME Trans.* **235**: 45–45.

Davies, R. (1984). Particle size measurement: experimental techniques. In *Handbook of Powder Science and Technology*, Fayen, M.E. and Otten L. (eds.). Van Nostrand Reinhold, New York.

Ehrlich, R. and Weinberg, B. (1970). An exact method for characterization of grain shape. *J. Sedim. Petrol.* **40**: 205–212.

Gotoh, K. and Finney, J.L. (1975). Representation of size and shape of a single particle. *Powder Technol.* **12**: 125–130.

Harris, C.C. (1968). The application of size distribution equations to multi-event comminution processes. *Trans. SME*, **241**: 343–358.

Harris, C.C. (1971/1972). A multi-purpose Alyavdin–Rosin–Rammler–Weibull chart. *Powder Technol.* **5**: 39–42.

Hatton, T.A. (1978). Representative sampling of particles with a spinning riffler. *Powder Technol.* **19**: 227–233.

Herdan, G. (1960). *Small Particle Statistics*. Butterworths, London.

Holt, C.B. (1981). Characterization of the shape of particles produced by crushing using harmonic series analysis. *Proc. Powtech.*, Birmingham, UK.

Jones, S. (1983). The problem of closure in the Zahn–Roskies method of shape description. *Powder Technol.* **34**: 93–94.

Kaye, B.H. (1981). *Small Characterization of Fine Particles*. John Wiley & Sons, New York.

Kaye, B.H. (1984). Multi fractal description of fine particle profiles. *Part. Charact.* **1**: 14–21.

Luerkens, D.W., Beddow, J.K. and Vetter A.F. (1982). Morphological Fourier descriptors. *Powder Technol.* **31**: 209–215.

Ma, L., Davis, D.C., Obaldo, L.G. and Barbosa-Cánovas, G.V. (1997). Mass and spatial characterization of biological materials. In *Engineering Properties of Foods and Other Biological Materials*. Washington State University Publisher, Pullman.

Mandelbrot, B.P. (1977). *Fractals, Form, Chance and Dimension*. Freeman, San Francisco.

Mandelbrot, B.P. (1982). *The Fractal Geometry of Nature*. Freeman, San Francisco.

Mohsenin, N.P. (1996). Physical characteristics. In *Physical Properties of Plant and Animal Materials. Structure, Physical Characteristics, and Mechanical Properties*. Gordon and Breach Science Publishers, Inc., Amsterdam, The Netherlands.

Okuyama, K. and Kousaka, Y. (1991). Particle density. In *Powder Technology Handbook*, Iinoya, K., Gotoh, K. and Higashitani, K. (eds.). Marcel Dekker, New York.

Ortega-Rivas, E., Meza-Velásquez, F. and Olivas-Vargas, R. (1997). Reduction of solids by liquid cyclones as an aid to clarification in apple juice processing. *Food Sci. Technol. Int.* **3**: 325–331.

Ortega-Rivas, E. and Svarovsky, L. (1994). A direct combined method for measuring particle size distribution in obtaining

Svarovsky, L. (1981). *Solid–Liquid Separation*. Butterworth, London.

Svarovsky, L. (1987). *Powder Testing Guide. Methods of Measuring the Physical Properties of Bulk Powders*. British Materials Handling Board, UK.

Webb, A.P. and Orr, C. (1997). *Analytical Methods in Fine Particle Technology*. Micrometrics Instrument Corp., Norcross, GA, USA.

Yan, H. and Barbosa-Cánovas, G.V. (1997). Size characterization of selected food powders by five particle size distribution functions. *Food Sci. Technol. Int.* **3**: 361–369.

Yu, A.B. and Standish, N. (1990). A study of particle size distribution. *Powder Technol.* **62**: 101–118.

CHAPTER 3

BULK PROPERTIES

Food powders can be classified into different categories according to their handling properties. The bulk properties of food powders are a function of physical and chemical properties of the material, the geometry, size and surface characteristics of the individual particles, as well as the history of the system as a whole. This chapter introduces some ways to evaluate food powder failure properties, such as angle of wall friction, effective angle of internal friction, failure function, cohesion, flowability, compressibility and other mechanical properties. These mechanical properties can be combined with environmental conditions such as moisture, temperature, particle size and chemical composition in order to condense physical and chemical issues related to the powder during manufacturing and distribution. Shear testers are used for bi-dimensional stress quantification that leads to the calculation of these properties. Other direct measurement methods for failure properties are also introduced. Handling properties such as angle of repose, angle of slide, conveying angle and angle of spatula are provided with some relevant reference methods, and new methods such as rapid methodologies are utilized for evaluating drainage, flow or conveying capabilities. Powders are classified according to handling in dispersion systems or according to their dynamic or static conditions during transport or storage.

Bulk density and porosity are of fundamental practical use when doing quality control in the food industry. These parameters can help to determine, for example, whether a raw material can be mixed or a final product can be packed in a predetermined container. Bulk density is classified depending on its measurement method. Particle size, voidage and other properties are studied with known equations, including bulk density as a variable to see the influence of other conditions. Food powder compressibility is useful to evaluate the extent of the intentional or unintentional compression that a powder suffers during transport or production. Examples of compressibility values are shown, as well as useful models are developed in order to explain compression mechanisms. Strength properties, especially for brittle and fragile powders such as agglomerated coffee or milk, explain attrition phenomena for these types of powders. Abrasion and friability illustrate how these degradating physical processes occur.

The last section of this chapter reviews instant properties in relation to powder solubility in water and rehydration properties. Different instant properties are defined, as well as standard and non-standard methods for their evaluation.

3.1. FLOW PROPERTIES

3.1.1. Introduction

Powder flow is defined as the relative movement of a bulk of particles among neighboring particles or along the container wall surface (Peleg, 1977). The practical objective of powder flowability investigations is to provide both qualitative and quantitative knowledge of powder behavior, which can be used in equipment design and in equipment performance prediction (Sutton, 1976). The flow characteristics of powders are of great importance in many handling and storage situations encountered in bulk material processes in the agricultural, ceramic, food, mineral, mining, and pharmaceutical industries because the ease of powder conveying, blending and packaging depends on them (Chen, 1994). The study of stresses developed as a result of storage in high bins, hoppers or silos plays a key role for bulk property evaluation. If a material has to be stored, removed from storage or transferred to some other place, no serious difficulties will be met if the material is a fluid. If, however, the material is a particulate solid, especially if it is made up of very small particles, cohesiveness, friction and interlocking between particles are factors that could complicate the handling of these solids. In order to ensure that a particulate material stored in any sort of container is recovered when required, it is important to study the powder flow properties under its environmental manufacturing conditions.

In the food processing industry, a specific size range characterizes particulate solid materials in order to be considered as "powders," according to the criteria defined in Chapter 2. For example, different cereal grains used as raw materials in different food processes would not be considered as powders. Many other particulate materials involved in food processes are finely divided and lie properly within the classification of fine and very fine, as defined in Table 2.2 of the previous chapter. For all fine and very fine powders, the attractive forces between particles are large when compared with the weight of individual particles, and so they are defined as cohesive. Cohesive powders normally present flow problems. Practically, cohesive materials can fail to flow out of a container with an opening about a thousand times larger than the particles' diameter. Flow problems occur with cohesive powders of any nature, but may be more serious with food powders because they are commonly related to releasing sticky substances (e.g., fat) or to the presence of hygroscopic behavior, temperature and time of consolidation.

To guarantee steady and reliable flow, it is crucial to accurately characterize the flow behavior of powders. The forces involved in powder flow are gravity, friction, cohesion (inter-particle attraction), and adhesion (particle–wall attraction). Furthermore, particle surface properties, particle shape and size distribution, and the geometry of the system are factors that affect the flowability of a given powder. It is, therefore, quite difficult to have a general theory applicable to the flow of all food powders in all possible conditions that might be developed in practice (Peleg, 1977). The first requirement is to identify the properties that characterize the flowability of a particular material and to specify procedures for measuring them. The way the shear strength varies with the consolidating stress, and the properties used to identify and quantify such interactions, are commonly known as the failure properties of a powder.

3.1.2. Failure Properties

3.1.2.1. Definitions

In order to flow, powders must fail and their strength must be less than the load put on them. Failure properties take into account the state of compaction of the powder, as this strongly affects its flowability unless the powder is non-cohesive, like dry sand, and it gains no strength on compression. These properties may also be strongly affected by humidity and, especially in the case of food and biological materials, by temperature. The time of consolidation can also have an effect on failure properties of powders. It is important, therefore, to test such properties under controlled conditions using sealed powder samples or air conditioned rooms or enclosures. Also, time consolidating samples must be tested to simulate storage conditions. The basic properties describing the failure condition are: (a) the angle of wall friction; (b) the effective angle of internal friction; (c) the failure function; (d) the cohesion; and (e) the ultimate tensile strength.

The angle of wall friction ϕ is equivalent to the angle of friction between two solid surfaces except that one of the two surfaces is a powder. It describes the friction between the powder and the material of construction used to confine the powder, e.g., a hopper wall. The wall friction causes some of the weight to be supported by the walls of a hopper. *The effective angle of internal friction* δ is a measure of the friction between particles and depends on their size, shape, roughness and hardness. *The failure function* is a graph showing the relationship between unconfined yield stress (or the strength of a free surface of the powder) and the maximum consolidating stress (also called major principal stress), and gives the strength of the cohesive material in the surface of an arch as a function of the stress under which the arch was formed. *The cohesion* C is, as mentioned earlier, a function of inter-particle attraction and is due to the effect of internal forces within the bulk, which tend to prevent planar sliding of one internal surface of particles upon another. *The ultimate tensile strength* of a compact powder is the most fundamental strength mechanism, representing the minimum force required to cause separation of the bulk structure without major complications of particle disturbances within the plane of failure.

There are several ways, direct or indirect, of testing the five failure properties defined above. Basically, all of them can be determined using a shear cell, but simplified or alternative procedures can be adopted when the aim is to monitor the flowability of the output from a process or to compare a number of materials.

3.1.2.2. Determinations Using Shear Cells

There are basically two types of shear cells available for powder testing: the Jenike shear cell, also known as the translational shear box, and the annular or ring shear cell, also called the rotational shear box.

Shear testers of various types have been used to determine the stress/strength relationships of bulk solids. To date, Jenike's direct shear cell tester and his proven procedure for design of bins for flow has become a benchmark in research and in industrial practice. The Jenike shear cell is circular in cross section with an internal diameter of 95 mm. A vertical cross section of the cell is shown in Fig. 3.1. It is an assembly of a base, a ring on top of the base and a cover. The ring and the base are initially attached with pins. The cell is filled with the powder gently, and the lid is placed in position. A vertical compacting load is applied to the powder sample by means of a constant weight

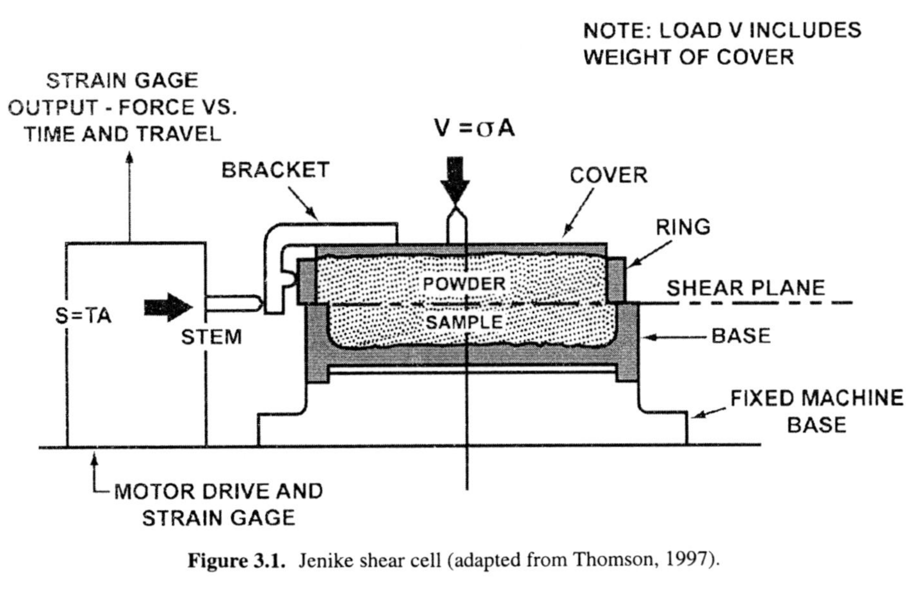

Figure 3.1. Jenike shear cell (adapted from Thomson, 1997).

Determining the Yield Locus with the Jenike Shear Cell. Test Procedure.

The Jenike test sequence simulates the change in stresses that act on an element of material as it flows through a bin. The test is accomplished in three steps. The first, called preconsolidation, is to ensure uniformity between samples. The second, called consolidation, reproduces flow with a given stress under steady state conditions. In the third step, the sample is sheared to measure shear stress at failure.

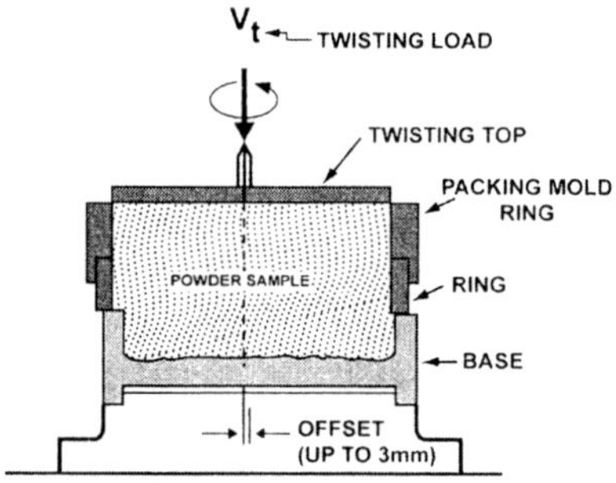

a)

a) Preconsolidation – With a packing ring in place, the cell is filled, a twisting top is placed on the sample, a force V_t is applied to the top while it is given a number of oscillating twists. The twisting top and force are removed and the powder surface scraped level with the shear ring.

b & c) Consolidation and preshear – A shear cover is placed over the powder sample and a selected normal force V is applied. A shear force is then continuously applied until it reaches a steady-state value indicating plastic flow. The shear force is then interrupted and the stem retracted. The measured steady state stress is point E on the yield locus (Fig. 3.3)

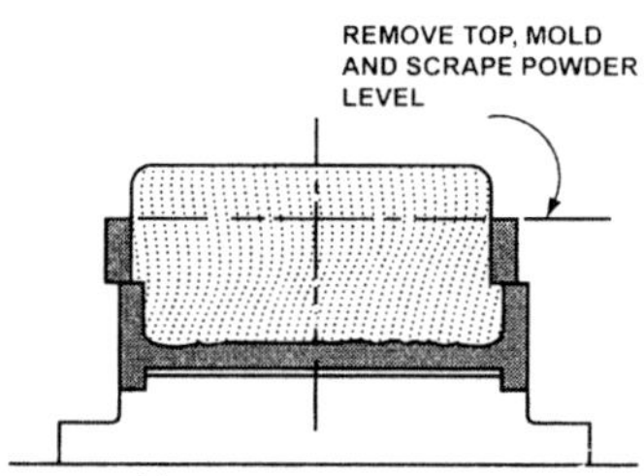

b)

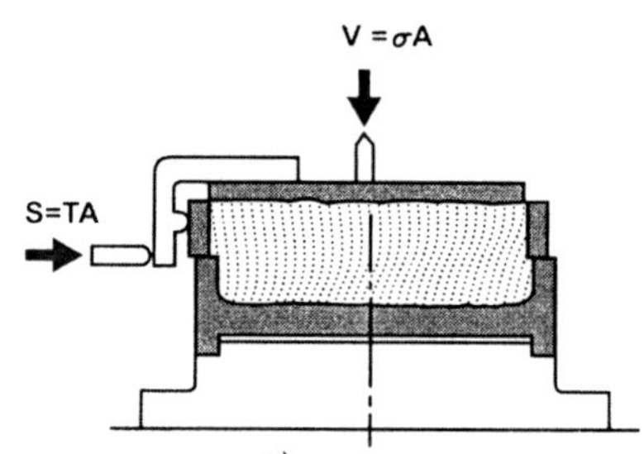

c)

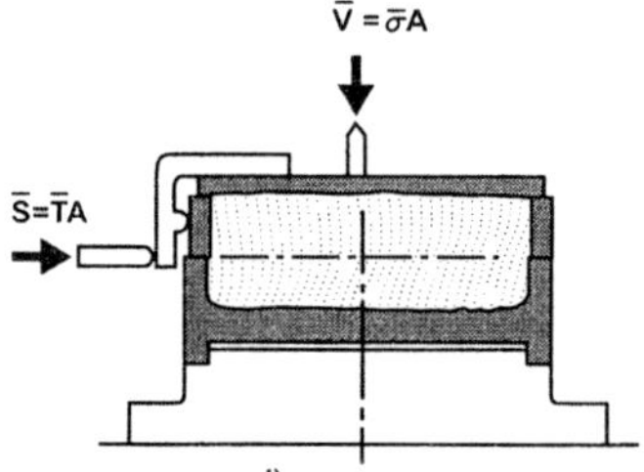

d)

d) Shear - The normal force V is replaced by a smaller force $\overline{V}$ and the shearing force is reapplied until the stress / strain peaks and falls off. Indicating failure plane in the sample and a point on the yield locus. This procedure is repeated several times with fresh samples, each consolidated as above but sheared with a progressively smaller normal force.

Figure 3.2. Test procedure for yield locus construction with Jenike shear cell (adapted from Thomson, 1997).

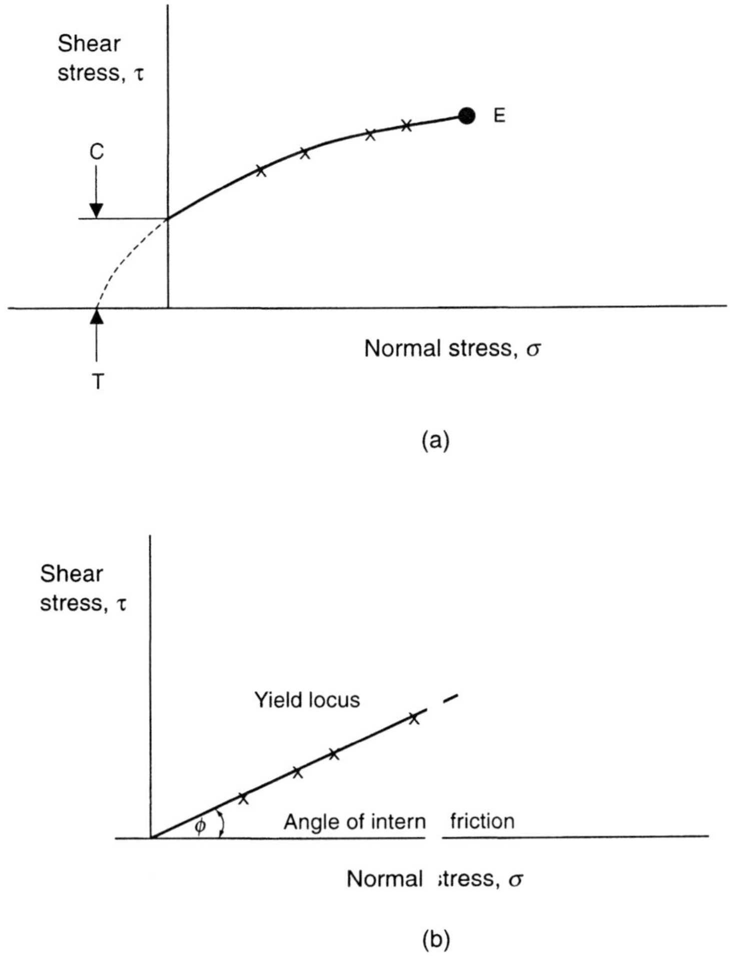

Figure 3.3. Different yield locus curves: (a) yield locus of a cohesive solid (e.g., brown sugar); (b) yield stress of free-flowing powder (e.g., sand).

or pressure at the center of the lid. Before the test starts, the pins are removed in order to let the ring slide on top of the base during the test. The test begins by moving the base by a motor driven stem at constant speed. A bracket, connected to the ring, measures the shear force necessary to slide the ring on top of the base at a constant vertical force. Shear force, displacement and time values are collected. A summary of the shear cell instructions is shown in Fig. 3.2.

The shear force can thus be obtained in function of time or horizontal displacement. Then, the peak or asymptotic yielding strength can be obtained and represented as one point in the yield locus. At this point, no more volume change is observed. Five or six different vertical loads are applied to a set of identical samples and the shear force needed to initiate flow is found in each case. The forces are divided over the cross sectional area of the cell to obtain the stress values, and the shear stress is plotted against the normal stress. The resulting graph is a yield locus, as shown in Fig. 3.3, and it is a line which gives the stress conditions needed to produce flow for the powder when compacted to a fixed bulk density.

Jenike was the first to use concepts of plastic failure for analyzing the flow of solids in bins and hoppers to develop the flow/no-flow criterion. This has produced an extremely useful quantitative

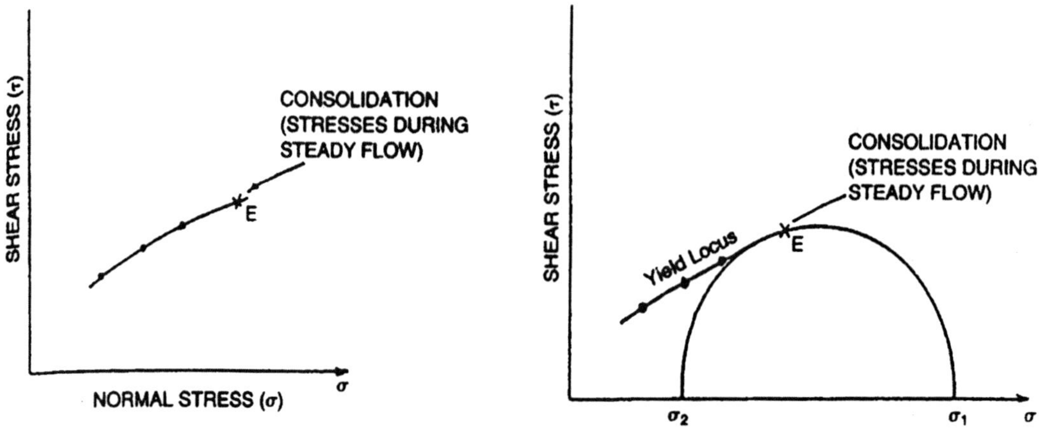

Figure 3.4. Yield locus and Mohr circle stress semicircle (σ_2 is the minor consolidation stress and σ_1 is the major consolidation stress (adapted from Thomson, 1997).

method for designing storage bins for gravity flow of solids. This method has been proven in the engineering practice. Furthermore, the Commission of the European Communities (1992) has created a standard for Jenike shear testing using limestone powder as a reference material. Furthermore, Jenike's method is described by the American Society for Testing and Materials in the ASTM D6128-00 for direct shear cells and in the ASTM D6773-02 for annular shear cells.

The yield locus for a cohesive solid is shown in Fig. 3.3a. However, the yield locus for a free-flowing powder, such as dry sand, will form a straight line, as shown in Fig. 3.3b. In his analysis, Jenike assumed that in the plastic region, solid properties at a point are the same in all directions (isotropic), and are frictional, cohesive and compressible. During incipient failure, the bulk solid expands; during steady flow, it can either expand or contract.

Mohr stress semicircles are used to identify the directional and strength properties of the sample from the yield locus, as shown in Fig. 3.4. A Mohr circle can represent the state of stress of any plane within the bulk solid. For any stress condition represented by a Mohr semicircle tangent to the yield locus, the bulk solids will be at yield, and the major principal stress σ_1 and minor principal stress σ_2 at this condition will be defined by the intersection of the semicircle with the σ axis. The yield locus terminates at the point of tangency of the Mohr semicircle through point E. This circle intersects the σ axis at the principal stresses σ_1 and σ_2. Major consolidation stress σ_1 will be used in Section 3.4 to define the flow factor.

If the material being tested is cohesive, the yield locus is not a straight line and does not pass through the origin. It can be shown that the graph when extrapolated downwards cuts the horizontal axis normally. As shown in Fig. 3.3, the intercept T is the tensile strength of the powder compacts tested, and the intercept C is called the cohesion of the powder; the yield locus ends at point E. A yield locus represents the results of a series of tests on samples that have the same initial bulk density. More yield loci can be obtained by changing the sample preparation procedure and, in this way, a family of yield loci can be obtained. This family of yield loci contains all the information needed to characterize the flowability of a particular material; however, it is not in a convenient form. More yield locus application for powder flowability is described in Section 3.2 of this chapter. For many powders, yield locus curves can be described by the empirical Warren–Spring equation (Chasseray, 1994):

$$\left(\frac{\tau}{C}\right)^n = \frac{\sigma}{T} + 1 \tag{3.1}$$

Table 3.1. Cohesion for some food powders.

Material	Moisture content (%)	Cohesion (g/cm^2)
Corn starch	<11.0	4–6
Corn starch	18.5	13
Gelatin	10.0	1
Grapefruit juice	1.8	8
Grapefruit juice	2.6	10–11
Milk	1.0	7
Milk	4.4	10
Onion	<3.0	<7
Onion	3.6	8–15
Soy flour	8.0	1

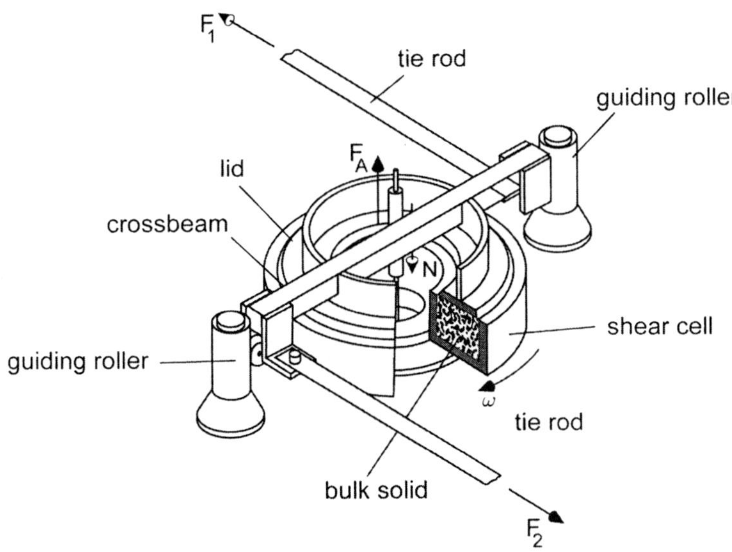

Figure 3.5. Annular shear cell (adapted from ASTM D6773-02).

where τ is the shear stress, C is the material's cohesion, σ is the normal stress, T is the tensile stress, and n is the shear index ($1 < n < 2$). Table 3.1 lists cohesion values for several food powders.

In annular shear cells, the shear stress is applied by rotating the top portion of an annular shear, as represented in Fig. 3.5. These devices allow much larger shear distances to be covered, both in sample preparation and testing, allowing a study of flow properties after testing. However, cell's geometry creates some problems. The distribution of stress is not uniform in the radial direction, but, for the ratio of the inner and outer radii of the annuli greater than 0.8, the geometrical effects are often considered negligible. The annular shear cells tend to give lower values for yield strength than the Jenike shear cell tester. The annular shear cell is also known as the ring shear or ring shear tester, because the cell is in the form of a full ring and is rotated. It has been reported to have the advantages of nearly unlimited shear deformation, possibility of measurements at very low consolidation stresses, ease of operation, and possibility of time consolidation measurements using a consolidation bench (Schulze, 1996). The results obtained using a ring shear tester are in reasonable agreement with those obtained with the Jenike shear cell.

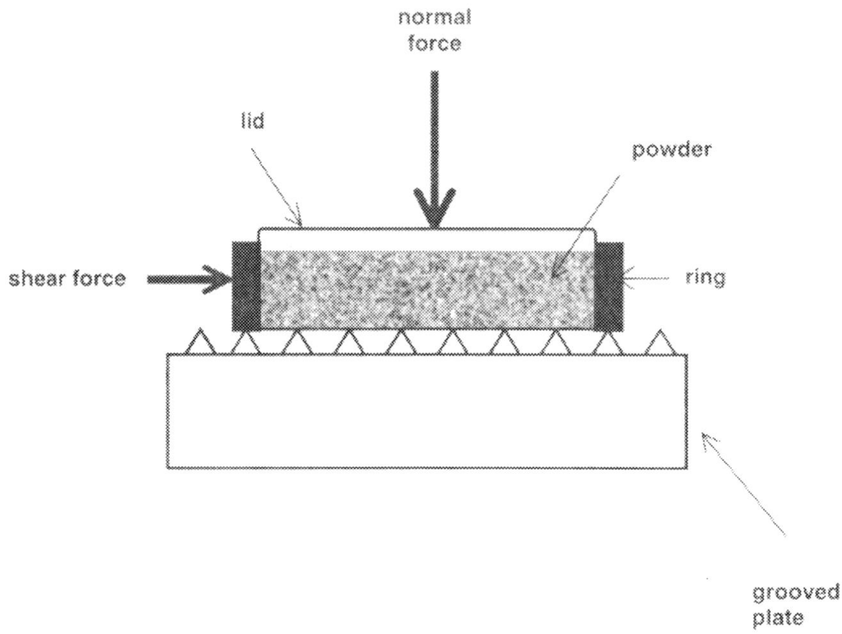

Figure 3.6. Grooved plate for the determination of the angle of internal friction (adapted from Svarovsky, 1987).

Two important properties defined above can be obtained from the yield loci: the first is the *effective angle of internal friction* δ, while the second is the *failure function* of the powder. The angle *of wall friction* ϕ' is measured by replacing the base of a Jenike shear cell with a plate of the material of which the hopper (or any sort of container) is made. The ring from the shear cell is placed on the plate and filled with powder and the lid is put in position. The shear force needed to maintain uniform displacement of the ring is found for different vertical loads on the lid. The slope of the graph of shear force against normal force gives the angle of friction between the particles and the wall, or angle of wall friction. This measure would complete the testing of a particulate material using only a Jenike shear cell.

3.1.2.3. Direct measurement of Failure Properties

The angle of internal friction ϕ can be measured directly by the "grooved plate" method. The base of the Jenike shear cell is replaced by a metal plate in which a number of saw-toothed grooves are cut (Fig. 3.6). These grooves are filled with the powder to be tested. The ring from the Jenike cell is then placed on the plate and filled with the powder and the lid is placed into position. A load is placed on the lid and the ring is pushed across the grooves until the shear force settles out at a constant value that is measured, and this action is repeated for a number of different vertical loads. The graph of shear force against normal force will be a straight line, with its slope being the angle of internal friction of the powder.

For direct measurement of the failure function, a split cylindrical die is used. The bore of the cylinder may conveniently be about 50 mm and its height should be just more than twice the bore. The cylinder is clamped so that the two halves cannot separate and it is filled with the powder to be tested, which is then scraped off level with the top face. By means of a plunger, the specimen is subjected to a known consolidating stress. The plunger is then removed and the two halves of the split die are separated, leaving a free-standing cylinder of the compacted powder. A plate is then placed on top of the specimen and an increasing vertical load is applied to it until the column collapses.

The stress at which this occurs is the unconfined yield stress f_c, i.e., the stress that has to be applied to the free vertical surface on the column to cause failure. If this is repeated for a number of different compacting loads and the unconfined yield stress is plotted against the corresponding consolidating stress, the failure function of the powder will be obtained. Although the results of this method can be used for monitoring or for comparison, the failure function obtained will not be the same as that given by shear cell tests, due to the effect of die wall friction when forming the compact. A method of correcting for friction has been described elsewhere (Williams et al., 1971).

Two methods can be used for direct measurement of tensile strength. In the first method, a mold of the same diameter as the Jenike cell is split across a diameter. The base of the cell is roughened by sticking sandpaper to the contacting surface. The two halves are clamped together, the cell is filled with the powder, and a lid is placed in position. The specimen is compacted by the application of a known vertical force to the lid and this, along with the clamp, are removed. The two halves of the cell containing the specimen rest on a base plate in which slots have been cut to form an air bearing. Air is introduced so that the cell can move horizontally without friction and the force needed to pull the two halves of the specimen apart is determined. Knowing the cross sectional area of the specimen, the tensile strength can be found. Measurements are made for a number of compacting loads, and tensile strength is plotted against compacting stress. This method is quite difficult to perform properly, requiring careful attention to details.

The second method is easier to use and gives results with less scatter. In this case a mold of the same diameter as the Jenike shear cell and a lid that just fits inside it are used. The base of the cell and the lower face of the lid are covered with sticking tape on which glue is spread. The cell is filled with the powder, which is scraped level with the top of the cell and the lid is placed in position. A compacting load is applied to the lid by means of a weight hanger and left in position until the glue has hardened. The lid is then attached through a tensile load cell to an electric motor, by which the lid is slowly lifted. The stress required to break the specimen is thus obtained. After failure, the lid and the base of the cell are examined; the result is accepted only if both are completely covered with powder, showing that tensile failure has occurred within the powder specimen and not at the surface. Figure 3.7 presents a diagram of these two methods for direct measuring of tensile strength.

The two methods described above for measuring the tensile strength of a lightly compacted powder do not give the same results, providing evidence that a bed of a particulate material will generally not be isotropic (i.e., its properties, including tensile strength, can change in different directions). Since the second method is easier to use, while the compacting stress and the tensile strength are applied in the same direction, it is most often recommended for general characterization purposes.

3.1.3. Other Handling Properties

As discussed previously, since most food powders of importance in process engineering are cohesive, the use of characterization variables which do not take into account the state of consolidation have to be used with extreme care. However, as has also been mentioned, there are a number of particulate systems relevant to the food industry that are free flowing, such as cereal grains and pulses. For these types of free flowing materials, some characteristic angles can be used for calculations

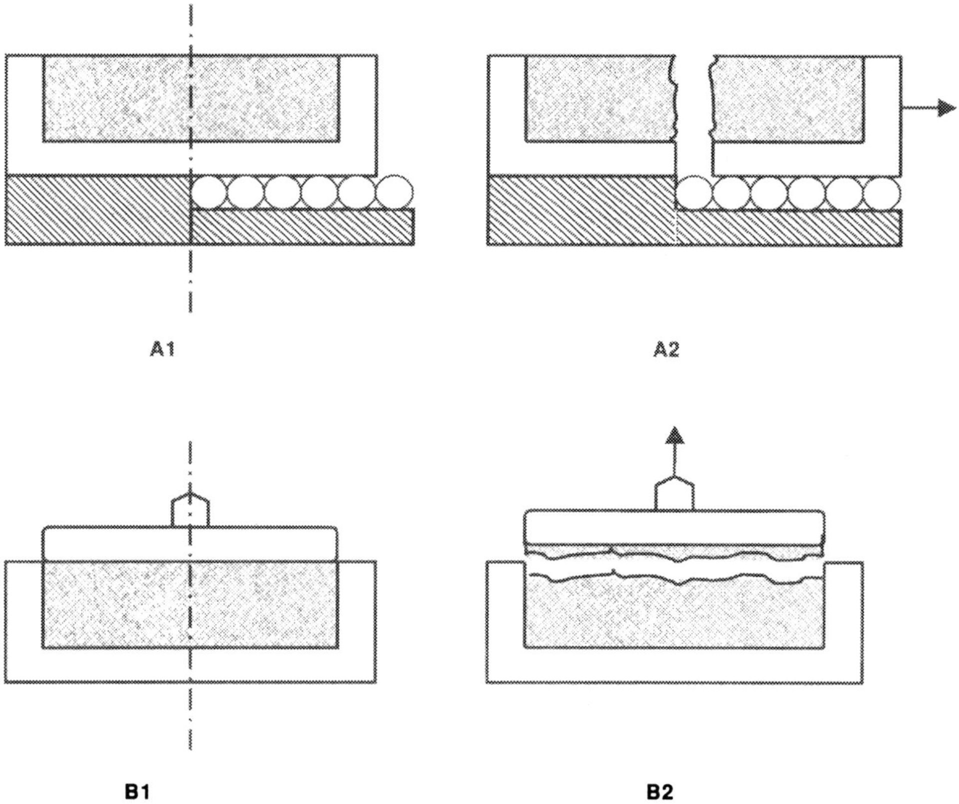

Figure 3.7. Tensile strength measurement (adapted from Peleg, 1978).

concerning aspects of storage and transport. Some of these angles are the angle of repose, the angle of slide, the conveying angle, and the angle of spatula. The description of some of these tests can be found in the ASTM D6393-99.

3.1.3.1. Angle of Repose

The *static angle of repose* is defined as the angle at which a material will rest on a stationary heap; it is the angle θ formed by the heap slope and the horizontal when the powder is dropped on a platform. Bulk solids such as cereals, milk, flour, salts, sugars, etc., when transported treated or stocked, can flow like liquids, but can also form a stable heap or pile due to internal forces. The most obvious characteristic of this heap is the angle of repose. Depending on the conditions under which the pile has been poured and how the angle is measured, different values of the angle can be obtained for the same powder, not being an intrinsic characteristic of the product. The measuring methods used, as well as experimental parameters, influence the results. Thus, published values of angle of repose are not always comparable. Several types of angle of repose are used to assess food powder flowability in an empirical manner. Teunou et al. (1995) reviewed a list of common measuring methods for the angle of repose which have been published in the literature, as shown in Fig. 3.8.

The most commonly employed are the drained (Fig. 3.8a) and the poured (Fig. 3.8b) angles of repose. The former is obtained from the discharge methods where the powder flows through an orifice situated at the base of a transparent and graduated cylinder. At the end of the flow, L/I relation

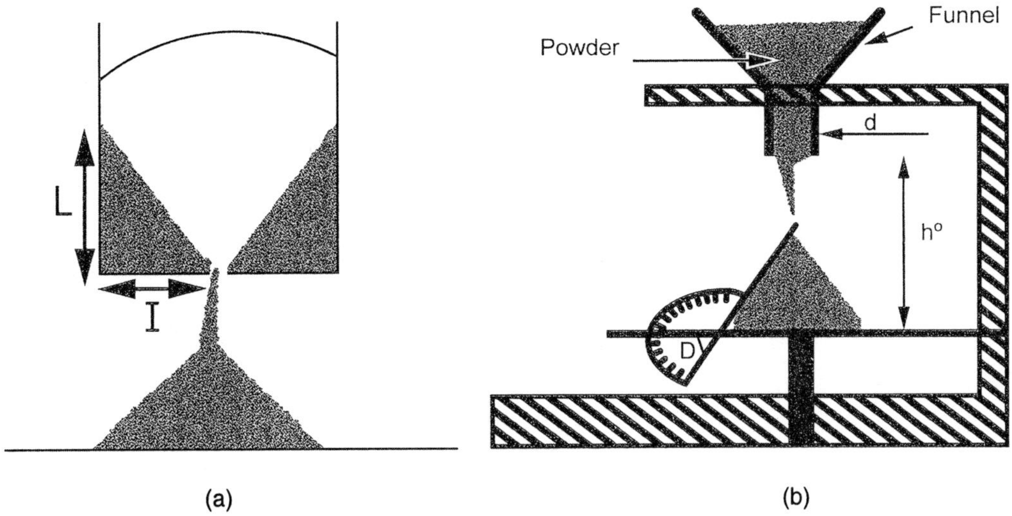

Figure 3.8. Methods to measure angle of repose (adapted from Teunou et al., 1995).

determines the drained angle (see Fig. 3.8a). The discharge method cannot be applied for cohesive materials because they can cake in the cylinder. The poured angle is measured on a pile poured freely onto a flat surface, while the latter is measured on the conical surface of powder in a flat-bottomed container when the powder has been allowed to flow out through an orifice in the base. This method is appropriate to all types of powders because cohesive and non-cohesive powders can be poured in a circular plate. The drained angle is always greater than the poured angle because, in the first case, the powder sliding or rolling down the slope is separating, while in the second case it is converging.

Although, as has been already discussed, the angle of repose could be considered meaningless in many practical applications, it can be used as a rough flowability indicator. In fact, it is the actual measurement applied by food industry quality control in order to evaluate flowability. According to Carr (1976), angles of up to 35° indicate free flowability, 35°–45° some cohesiveness, 45°–55° cohesiveness (loss of free flowability), and 55° and above very high cohesiveness and, therefore, very limited (or no) flowability. This characterization method can provide a rough flow indication on small quantities of powders which have not undergone any consolidation. The common belief that the flowability of a powder is a direct function of the angle of repose, especially when studying silo design (powder under high stresses), is misleading and should be avoided because, as stated earlier, most industrial powders show different grades of cohesiveness. The angle of repose of a cohesive material is indeterminate, being dependent on the previous history of a given sample. Therefore, for cohesive materials the angle of repose is irrelevant to the flow behavior of the material in any particular situation. The behavior of the material depends on the manner in which its strength varies with its state of compaction. For example, a slightly cohesive powder may be able to flow and be poured through a hole (forming a pile with a slope equal to the angle of repose), but after being compressed by simple tapping, the flow would not develop, due to the formation of a stable cohesive arch over the hole. A very small change in the bulk density of the powder would, therefore, have caused a dramatic change in its flowability.

Teunou et al. (1995) studied the effect of particle size, shape, moisture and its relation between flow properties in sugar particles of different sizes, agglomerated and fine milk and glass particles of varied sizes. It was found that the angle of repose decreases with increasing particle size, mainly because smaller particles tend to adhere much more strongly to each other. Particle shape determines internal force in powders and thus acts on the angle of repose. For example, a marked difference was noted between particles of spherical and angular forms, as in the case of glass particles, or crystalline forms and spangle forms, such as can be found as in sugar. Spherical glass particles, which have a low coefficient of friction, roll on a preformed conic heap, which remains very flat. Meanwhile, angular particles settle one on top of the other to increase the preformed heap leading to an increase in angle of repose. Moisture increases the cohesion of alimentary powders by increasing system plasticity, liquid bridges or solid bridges after re-crystallization or drying. Because of that, the angle of repose increases regularly with the powder's moisture content.

Nevertheless, the angle of repose can also be used for industrial design in the range of low stresses for the following determinations: heap basic height and diameter (valuable for grains and cereals), silo height, hopper angle determination for a good final sewage, conveyor belt dimensioning and forces on a horizontal silo's walls. All this information is very well summarized by Teunou et al. (1995). Its approximation to the angle of internal friction for free-flowing powder can be justified theoretically by the Coulomb model of the yield locus, but is not useful for powder flowability determination under high stresses. It is necessary to always consider the compression effect of the product or the range of stresses in the process before drawing any conclusion.

3.1.3.2. Angle of Slide

The *angle of slide* is closely related to the drained angle of repose (Wilkinson et al., 1983). It is the minimum angle to the horizontal of a flat inclined surface that will allow a bulk solid to flow from rest under its own weight. This angle is supposed to be useful in designing stationary chutes, but its measurement has not been described as standardized. The measured value is expected to be highly influenced by the material of construction of the chute, the amount of material on it, and by humidity. It has been shown (Augenstein and Hogg, 1978) that the motion of individual particles depends strongly on the nature of the surface over which they flow; highly roughened surfaces cause shear within the flowing stream, while smooth ones cause slip at the surface.

3.1.3.3. Conveying Angle

The *conveying angle*, also called the maximum angle of rise, is the angle at which a smooth belt conveyor can operate without the bulk powder running back down. There is a need for a standard test method to be developed for determination of this angle, to allow designers of belt conveying systems to have a reliable design criterion. A rough guide given in the literature (Wilkinson et al., 1983) is that the angle of rise used in such systems should not exceed about half the poured angle of repose.

3.1.3.4. Angle of Spatula

The *angle of spatula* is measured by spooning out powder with a 22-mm wide flat-blade spatula, which is inserted into the bulk powder parallel to the bottom of the container. It is then lifted straight up and out of the powder, keeping it parallel to a horizontal plane. The spatula is vibrated or tapped in a specified manner and the angle of the tangent to the powder surface at the edge of the spatula is measured, to the horizontal plane. The average angle of several measures is taken and, except for very free-flowing materials, its value is always greater than that of the poured angle of repose, since it is taken from a consolidated material. The angle of spatula is considered useful as a very simple and rough estimate of flowability; a value less than 40° indicates that the material is free flowing.

3.2. CLASSIFICATION OF POWDERS ACCORDING TO HANDLING

When considering powders as dispersion systems, they can be classified into different categories according to their particular behavior in dynamic or static situations. For this reason, powders or dry particulate materials can be categorized according to their handling and storage properties. The most important categories according to the handling in question can be the handling and storage of de-aerated powders, as well as the handling of aerated powders. The term "aerated" refers to particles separated from each other by a film of air, not being in direct contact with each other (e.g., gas fluidization).

Powders in the de-aerated state can be classified on the basis of direct tests results using a shear cell tester or a compression tackiness tester. There are several designs of shear cells that basically measure different shearing forces for corresponding normal forces on consolidated powder samples. Yield locus has been defined in Section 3.1.2.2. Mohr circle can be used to determine solid characteristics derived from the yield locus, which is the curve that represents the relation between the maximum shear and normal stresses. For each yield locus, with the use of Mohr circles, the unconfined yield stress and major consolidation stress σ_1 (i.e., the maximum normal stress undergone by a certain powder at a given state of compaction) can be obtained. Unconfined yield stress f_c is the normal stress necessary to make the powder yield at zero shear stress (i.e., no shear forces in the planes perpendicular to the normal force). This can be the case when the minimum principal stress σ_2 (Fig. 3.4) is equal to 0. Therefore, a Mohr circle through the origin, tangent to the yield locus, defines the largest stress f_c (Fig. 3.9) that solids can withstand at a free, unsupported surface.

The ratio of the major consolidation stress σ_1 found from the yield locus to the unconfined yield strength f_c is called the flow factor ff (ff $= \sigma_1/f_c$). Jenike (1964) proposed a classification according to the position of one point of the failure function at a fixed value of the unconfined yield stress with respect to the flow factor line. Figure 3.10 shows a schematic representation of classification of powders following this criterion. As can be seen, at a fixed value of the unconfined yield strength

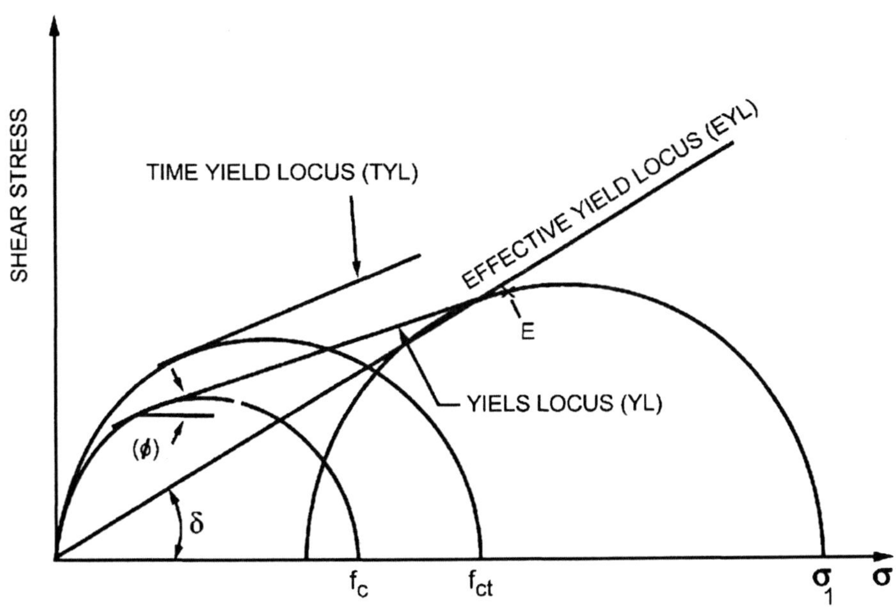

Figure 3.9. Mohr circle with powder flow characteristics (adapted from Thomson, 1997).

Table 3.2. Classification of powders according to Jenike (1964).

Type of powder	ff value or failure function slope value
Very cohesive	Less than 2
Cohesive	Between 2 and 4
Easy flow	Between 4 and 10
Free flow	More than 10

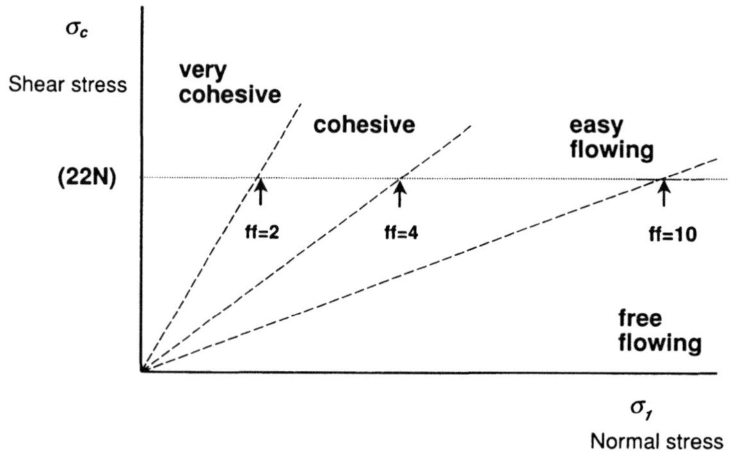

Figure 3.10. Diagramatic representation of Jenike's classification of powders (adapted from Svarovsky, 1987).

(e.g., 22 N), the straight lines through the origin at a slope of 1/ff would represent the categories of very cohesive, cohesive, easy flowing and free flowing, as represented in the mentioned figure, as well as in Table 3.2. In addition to Jenike's classification, Williams et al. (1971) also defined simple powders as those whose failure function is a straight line through the origin, providing that the slope of the line is less than 1/ff.

In Jenike's method, the failure function (or flow function) can be determined by shearing the powder at different initial levels of compaction set by different initial preconsolidation stresses (Fig. 3.3). The maximum or asymptotic shear stress obtained from a shear test using a certain normal stress, can be plotted vs. the applied normal stress at which the cell is run, and one point in the yield locus will be obtained. Using different normal stresses at same initial preconsolidation stresses, a yield locus can be formed. By the use of Mohr circles, the unconfined yield stress f_c and the major principal stress σ_1 for this specific initial consolidating stress can be obtained. Repeating this procedure for different initial consolidation stresses (e.g., according to pressure conditions at which the powder will be stored in a hopper), different pairs of f_c and σ_1 can be obtained. By plotting f_c vs. σ_1 the failure function can be obtained. Finally, the slope value of this curve is used as a classification method by using the same criteria defined in Table 3.2. The principles behind this method find direct application in hopper design as will be described in Chapter 4.

Another way of classifying powders is on the basis of the results obtained with a compression tackiness tester. This tester is a simplified version of the uniaxial compression test on a cylindrical briquette of powder, where the compression force at failure is a direct measure of powder tackiness (or adhesiveness through deformation). The consolidation stresses used to form a briquette or compacted particle mass are generally greater than those used with the Jenike shear test. The total compression force is supposed to simulate pressures greater than that of the bottom of a 6-ft high pile of bags

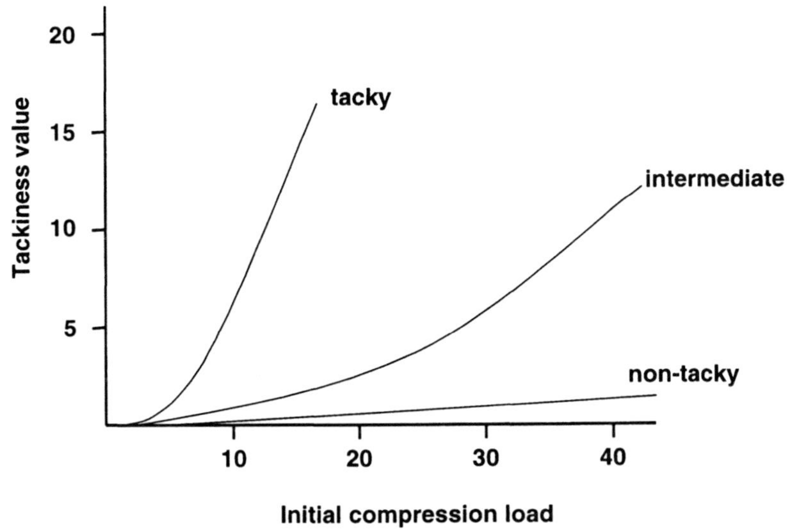

Figure 3.11. Classification of powders according to "tackiness" (adapted from Svarovsky, 1987).

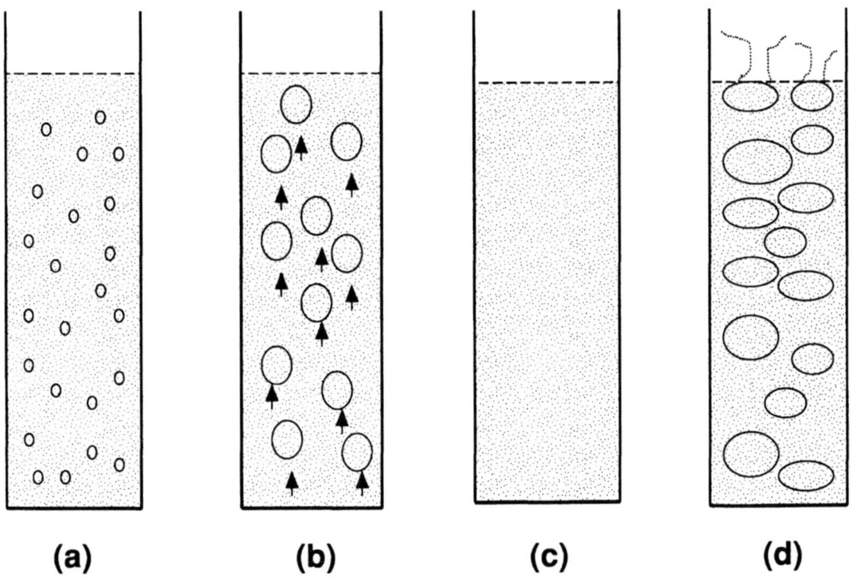

Figure 3.12. Geldart's classification of powders fluidized by air.

in storage. Tackiness values above 27 N usually indicate powders that can cause problems due to their sticky characteristics. The shape of a curve of tackiness value against initial compression load (Fig. 3.11) is also an indication of tackiness; non-tacky powders give essentially a horizontal line, while tacky products show the effect of increasing compression.

With regard to the aerated state, a widely accepted classification of powders is the one proposed by Geldart (1973), which takes the two most important particle properties into account: particle size and particle density. Geldart's classification is derived from the behavior of powders when fluidizing by air at ambient conditions. Figure 3.12 illustrates the four general manners in which different powders behave when subjected to these described conditions of fluidization. According to this

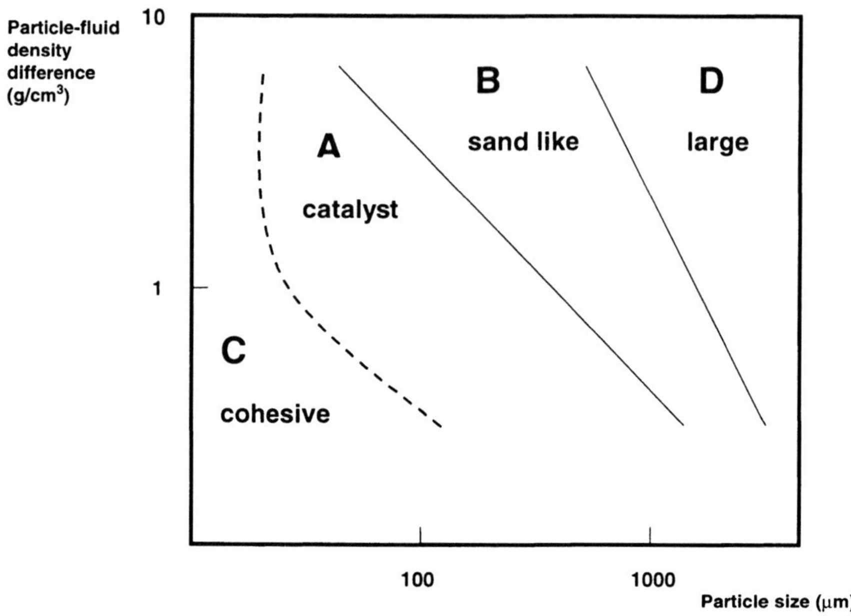

Figure 3.13. Graphic representation of classification of powders according to Geldart.

figure, powders can be termed as A, B, C, or D type and may be described as follows. Powders are sometimes referred to as slightly cohesive or catalyst type and exhibit large bed expansion after minimum fluidization and before initiation of bubbling (powder A). B powders are known as sand-like and present bubbling at the minimum fluidization velocity with a small bed expansion. C powders are called cohesive and are difficult to fluidize at all, while D powders are simply known as large and can form stable spouted beds if air is admitted only through a centrally positioned hole. Geldart's classification can also be represented in a plot of particle fluid density difference versus particle size, as shown in Fig. 3.13.

The above-mentioned classification of powders may be useful even when fluidization as such is not of interest. For example, in the filling of containers and in the residence time needed in hoppers to avoid flooding out when the discharge valve is opened, the bed expansion and the rate of de-aeration are of concern. There is a striking difference between A, B, and C powders in the way they behave in de-aeration. A type powders collapse at a constant rate, B type powders de-aerate almost instantaneously, while C type powders collapse initially much faster than those in group A but may remain in a slightly aerated state for a considerable period, with the pressure at the bottom of the bed decaying very slowly. The de-aeration rate can, therefore, also be used to characterize powders, but Geldart and Wong (1985) found this to be insufficiently sensitive and developed a standardized de-aeration time test. In the context of gas–solid suspensions, a cohesive powder may be defined as that in which the inter-particle forces become so great that they exceed the aerodynamic drag that can be exerted by the gas.

3.3. PACKING PROPERTIES

As already stated, a powder is a complex form of solid material made up of a very large number of individuals, each different from its neighbor. Every time a particular powder sample is poured into a receptacle, the individual particles are located in different places than before. Individual or inherent properties have already been discussed in the previous section under the common term of primary properties. Clearly, it is not possible to predict quantitatively how a powder will behave based on knowledge of the measured properties of individual particles; so direct measurement of bulk properties is necessary. As each repeated measurement of a sample will be based upon a rearrangement of its population, there will be an inevitable scatter of readings; consequently, the powder should be handled in as identical a manner as reasonably possible each time a measurement procedure is performed. Moreover, it would seem logical that there would be a development of standardized testing and characterization methods, but this has not happened. The main forms and procedures to evaluate secondary properties of powders will be discussed in this section.

3.3.1. Bulk Density and Porosity

Measurement of bulk density is of fundamental use by the industry to adjust storage, processing, packaging and distribution conditions. Particularly, bulk density is one of the properties used as part of the specifications for a certain final product derived from grinding or drying. For example, a common deviation that can occur when packaging ground coffee, is that the product's bulk density exceeds the specification range. Then, the powder will occupy a smaller portion of the intended volume in the package. Although the net weight is correct, the package will look as lacking coffee. On the other hand, if the bulk density lies below the bulk density allowed by the specification, the product volume will be larger than the package. As a result, the package volume will be completed with coffee particles, but it will have a lower net weight of coffee than what is declared in the label.

When a powder just fills a vessel of known volume V and the mass of the powder is m, then the bulk density of the powder is m/V. If the vessel is tapped (i.e., particles are packed by vertical vibration), it will be found in most cases that the powder will settle, and more powder needs to be added to achieve the initial volume once again because the bulk density has increased. Particles were rearranged by vibration, filling the voids in the bulk. The bulk voidage or porosity is directly related to bulk density. Bulk density is the mass of particles that occupies a unit volume of a bed, while porosity is defined as the volume of the voids within the bed divided by the total volume of the bed. The following mass balance relates these two properties:

$$\rho_b = \rho_s(1 - \varepsilon) + \rho_a\varepsilon \tag{3.2}$$

where ρ_b is the powder bulk density, ρ_s is the particle density, ε is the porosity, and ρ_a is the air density. Since the air density is small relative to the powder density, it can be neglected, and the porosity can thus be calculated simply as:

$$\varepsilon = \frac{\rho_s - \rho_b}{\rho_s} \tag{3.3}$$

Equation (3.3) gives the porosity or voidage of the powder; whether or not this includes the pores within the particles depends on the definition of particle density used in the evaluation, such as those discussed in Section 2.2.

Over the years, in order of increasing values, three classes of bulk density have become conventional: *aerated*, *poured*, and *tap*. Each of these depends on the treatment to which the sample is subjected. There is still some confusion in the current literature as to how these terms are interpreted. Some consider the *poured* bulk density as *loose* bulk density, while others refer to it as *apparent* density. *Aerated* density can also be considered to be quite a confusing term. Strictly speaking, *aerated* should mean that the particles are separated from each other by a film of air and not being in direct contact with each other. Some authors interpret the term as meaning the bulk density after the powder has been aerated. In this case, we refer to the most loosely packed bulk density.

Bulk density could be simply regarded as *loose* bulk density. This approach is implied in many investigations when dealing with cohesive powders. For many food powders, which are more likely cohesive in behavior, the terms more commonly used to express bulk density are *loose poured*, and *tapped* bulk density, after vibration. Another way to express bulk density is in the form of a fraction of its particles' solid density, which is sometimes referred as "theoretical density." This expression, as well as the use of porosity instead of density, enables and facilitates the unified treatment and meaningful comparisons of powders having considerably different particle densities.

3.3.1.1. Measurements of Bulk Density

Aerated bulk density

Aerated bulk density is, in practical terms, the density at which the powder is in its most loosely packed form. Such a form can be achieved by dropping a well-dispersed "cloud" of individual particles into a measuring vessel. Alternatively, gas fluidization can sometimes be used with the gas turned slowly off and the particles allowed to settle quietly into place. The structure within the vessel is held by the cohesive forces between the particles, and can be extremely fragile. Leveling off the surface of the powder at the top of the vessel is difficult to achieve without causing particle movement leading to error, as some structure collapses. Determination of aerated bulk density can be performed using an apparatus like the one illustrated in Fig. 3.14. As shown in the figure, an assembly of screen cover, screen, a spacer ring, and a stationary chute are attached. They go on top of a chute that can vibrate at variable amplitude. Once the stationary chute is aligned with the center of a pre-weighed 100-ml cup, the powder is poured through the vibrating sieve and allowed to fall from a fixed height of approximately 25 cm through the chutes into the cylindrical cup. The amplitude of the vibration is set so that the powder will fill the cup in 20–30 s. The excess powder is skimmed from the top of the cup using the sharp edge of a knife or ruler, without disturbing or compacting the loosely settled powder. The weight of the powder and volume retained in the cup are determined, and the loose bulk density is calculated from both values.

Poured bulk density

Poured density is widely used, but the measurement is adapted to the conditions of a particular company or industry. Certain precautions that should be taken are clearly stated: the measuring vessel should be fat rather than slim, the powder should always be poured from the same height and the possibility of bias in the filling should be as small as possible. Although measuring of poured bulk density is far from standardized, many industries use a sawed-off funnel with a trap door or stop to pour the powder into the measuring container. A standard volume, usually a measuring cylinder, can also be used for the volume measurement, varying from 50 to 1,000 ml, depending on the product and its size; 1 liter is now accepted as the preferred standard. It is also better to use a density cylinder with a 2:1 length to diameter ratio, rather than the much slimmer measuring cylinder. Rarely, the height of fall is standardized as well, realizing that the height will affect the compaction state of the powder. Finally, it is also the practice in some industries to allow the powder, after pouring it into the container, to settle to its stable density for about 10 minutes before scraping off the top.

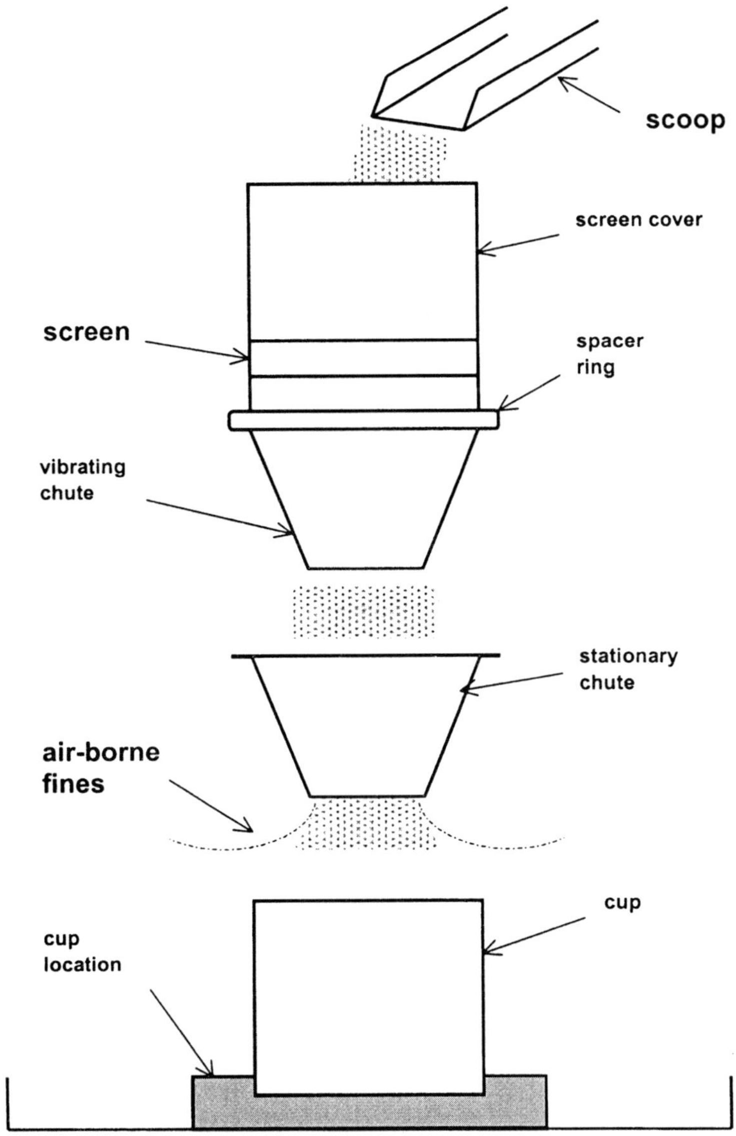

Figure 3.14. Aerated bulk density determination.

Tap bulk density

The tap bulk density, as implied by its name, is the bulk density of a powder that has been settled into closer packing than existed in the poured state by tapping, jolting, or vibrating the measuring vessel. As with poured bulk density, the volume of a particular mass of powder may be observed, but it is generally better to measure the mass of powder in a fixed volume. Although many in the industry measure the tap density by tapping the sample manually, it is best to use a mechanical tapping device so that the conditions of sample preparation are more reproducible. An instrument useful to achieve such reproducibility is the Hosokawa powder characteristic tester, which has a standard cup (100 ml) and a cam-operated tapping device which moves the cup upward and drops it

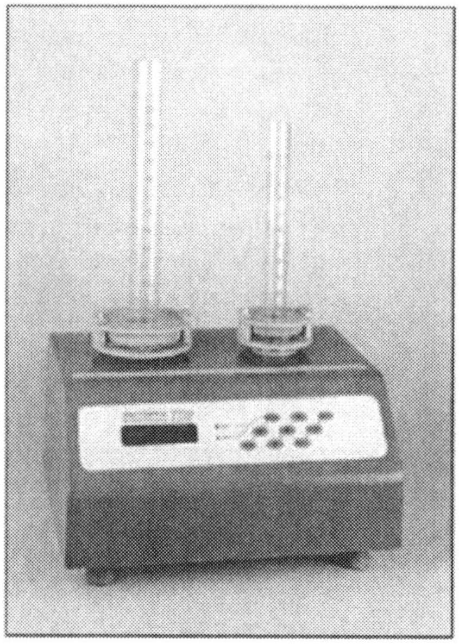

Figure 3.15. Tap density tester.

periodically (once every 1.2 seconds). A cup extension piece has to be fitted and powder added during the sample preparation so that at no time the powder packs below the rim of the cup. After the tapping, excess powder is scraped from the rim of the cup and the bulk density is determined by weighing the cup.

The tap density tester is described by norms from ASTM (American Society for Testing and Materials Standards), which gives standardized and repeatable results for measuring tapped or packed volumes of powders and granulated or flaked materials. This can have digital LED displays and user-selectable counter or timer operations, including dual non-rotating platform drive units and two graduated funnel top cylinders, which generally are of 100 and 250 ml, as described in Fig. 3.15.

Ultimate bulk density

Yan and Barbosa-Cánovas (2001) first proposed the concept of "ultimate bulk density" while studying density changes in agglomerated food powders due to high hydrostatic pressure. When the hydrostatic pressure was higher than a critical value (around 200 MPa for the powders used), the agglomerated food powders were compressed so densely that all of the agglomerates and primary particles were crushed and compressed together, leaving almost no open or closed pores. Since the final compressed bulk density is usually higher than the commonly used apparent "solid density," which is measured by a pycnometer based on gas- or liquid-displacement principle, the bulk porosity (i.e., 1 – (bulk density/solid density)) will be a negative value without physical meaning. It is believed that the "ultimate bulk density" is dependent on the product formulation, physical properties of product ingredients, and production conditions.

3.3.1.2. Hausner Ratio

A useful definition related to bulk densities is the so-called Hausner ratio, which is not a separate test, but one derived from bulk density tests; it can be defined as the ratio of the tap bulk density to the loose bulk density.

Malavé-López et al. (1985) defined as the ratio of asymptotic over initial bulk density by the relationship:

$$H_R = \frac{\rho_\infty}{\rho_0} \tag{3.4}$$

where H_R is the Hausner ratio, ρ_∞ is the asymptotic constant density after certain amounts of taps and ρ_0 is the initial bulk density. A more practical equation widely used to evaluate flow properties can be given by the following equation which calculates powder volume changes in a graduated cylinder after certain period of time or number of taps (Hayes, 1987):

$$H_R = \frac{\rho_n}{\rho_0} = \frac{V_0}{V_n} \tag{3.5}$$

where n is the number of taps provided to the sample, ρ_n and ρ_0 are the tapped and loose bulk density and V_0 and V_n are the loose and tapped volume, respectively.

The usefulness of this ratio arises because it deals with properties relevant to moving, rather than static, powder. For example, in the classification of the fluidization behavior of powders, Geldart (1973) was able to classify powders into four groups, but a good prediction of fluidization behavior was needed to avoid using relatively large-scale plant trials. This was achieved by measuring the Hausner ratio under controlled conditions (Geldart et al., 1984), and the results indicated that powders with Hausner ratios smaller than 1.25 are easily fluidized, while powders with Hausner ratios greater than 1.4 present fluidization problems. In another study, Kostelnik and Beddow (1970) found the Hausner ratio to be extremely sensitive to particle shape, and proposed that it could be used as an index of powder shape. Investigations using the Hausner ratio have also been carried out to understand the influence of relative humidity upon process operations where powders are lightly compacted, and can be classified as transitional between cohesive and free flowing. It has been found (Harnby et al., 1987) that it is possible to predict zones of relatively stable operation for processes in terms of particle size and relative humidity. There is no doubt that the Hausner ratio, as a factor so simple to measure, has great potential as a general fingerprinting criterion of powder handling behavior. More research is needed to relate it to other powder properties.

3.3.1.3. Factors Affecting Bulk Density

Approximate values of loose bulk densities of different food powders are given in Table 3.3. As can be seen, with very few exceptions, food powders have apparent densities in the range of 300–800 kg/m^3. As previously mentioned, the solid density of most food powders is about 1,400 kg/m^3, so these values are an indication that food powders have high porosity which can be internal, external, or both. There are many published theoretical and experimental studies of porosity as a function of particle size, distribution, and shape. Most of these studies pertain to free-flowing powders or models where porosity can be treated as primarily due to geometrical and statistical factors only (Gray, 1968; McGeary, 1967). Although in these cases porosity can vary considerably, depending on factors such as the concentration of fines, it is still evident that the exceedingly low density of food powders cannot be explained by geometrical considerations only. Most food powders are known to be cohesive and, therefore, an open bed structure supported by inter-particle forces very likely exists (Moreyra and Peleg, 1981; Scoville and Peleg, 1980; Dobbs et al., 1982). Since the bulk density

Table 3.3. Approximate bulk density and moisture of different food powders.

Powder	Bulk density (kg/m^3)	Moisture content (%)
Baby formula	400	2.5
Cocoa	480	3–5
Coffee (ground and roasted)	330	7
Coffee (instant)	470	2.5
Coffee creamer	660	3
Corn meal	560	12
Corn starch	340	12
Egg (whole)	680	2–4
Gelatin (ground)	680	12
Microcrystalline cellulose	610	6
Milk	430	2–4
Oatmeal	510	8
Onion (powdered)	960	1–4
Salt (granulated)	950	0.2
Salt (powdered)	280	0.2
Soy protein (precipitated)	800	2–3
Sugar (granulated)	480	0.5
Sugar (powdered)	480	0.5
Wheat flour	800	12
Wheat (whole)	560	12
Whey	520	4.5
Yeast (active dry baker's)	820	8
Yeast (active dry wine)		8

of food powders depends on the combined effect of interrelated factors, such as the intensity of attractive inter-particle forces, particle size, and number of contact points (Rumpf, 1961), it is clear that a change in any of the powder characteristics may result in a significant change in the powder bulk density. Furthermore, the magnitude of such change cannot always be anticipated. There is an intricate relationship between the factors affecting food powder bulk density, as well as surface activity and cohesion.

As with the moisture factor, also included in Table 3.3, moisture sorption is generally associated with increased cohesiveness, mainly due to inter-particle bridges. Many food powders are highly hygroscopic and, therefore, high moisture contents would result in lower loose bulk densities. However, this decrease would only be detected in freshly sieved or flowing powders, where the same inter-particle forces do not allow caking of the mass. Although sugar and salt are examples of powders that lower their densities as a result of increasing humidity, finer powders (e.g., baby formula and coffee creamer) that are very cohesive even in their dry form do not present such a trend. For these powders, it appears that the bed array has reached maximum voidage at low moisture contents, and further lowering of the density becomes impossible. It is also worth remembering that excessive moisture levels, especially in powders containing soluble crystalline compounds, may result in liquefaction of the powder with a consequent increase in density.

Anti-caking agents, also known as flow conditioners, are supposed to reduce inter-particle forces and cohesiveness, and as such, they are expected to increase the bulk density of powders (Peleg and Mannheim, 1973). It has been observed that there may be an optimal concentration beyond which the anti-caking effect will diminish bulk density, or it will be practically unaffected by the agent concentration (Hollenbach et al., 1982). It can also be observed that for a noticeable effect on the bulk density increment (i.e., an increase on the order of 10% or more), the agent and host particles must have surface affinity. If this is not the case, the conditioner particles

Table 3.4. Effect of moisture content on mechanical characteristics of some food powders.

Powder	Moisture (%)	Poured bulk density (kg/m^3)	Compressibility	Cohesion (g/cm^2)
Glass beads (175 μm)	Dry	1.72	~0	~0
	1.0	1.20	0.23	15
Powdered salt (100/200 mesh)	Dry	1.26	0.02	~0
	0.6	0.78	0.12	50
Powdered sucrose (60/80 mesh)	Dry	0.62	0.152	~10
	0.1	0.50	0.185	~14
Starch	Dry	0.81	0.12	~6
	18.5	0.69	0.15	~13
Powdered onion (80/120 mesh)	Dry	0.51	0.03	5
	5.2	0.51	0.05	15
Baby formula (commercial)	Dry	0.52	0.08	37
	2.7	0.41	0.08	Too cohesive
Coffee creamer (commercial)	Dry	0.46	0.08	49
	7.0	0.45	0.19	32
Active dry	5.2	0.52	0.05	~0
Baker's yeast	8.4	0.52	0.08	14
	13.0	0.49	0.26	Too cohesive

may segregate or fill the voids between particles randomly and, instead of reducing inter-particle forces and liquid bridge formation, will only fill inter-particle space, not varying the bulk density.

For example, the addition of GRAS anti-caking agents such as silicon oxide, sodium aluminum silicate, tricalcium phosphate and calcium stearate (0.1–2%) have been found to provide an increase in loose bulk density, depending on the agent and concentration (Hollenbach et al., 1982). Molina et al. (1990) studied the effect of the addition of Hubersorb®-600 (0.5%) on the density of ground coffee. Unlike in crystalline powders (e.g., ground sucrose or salt), the admixture of the conditioner at 0.5% concentration did not drastically alter the coffee's density. Therefore, a new research opportunity was opened for the addition of selected conditioners in coffee packed in bags or stored in bulk in order to protect its flavor as well as its physical stability. It seems, however, that there is very little information on the exact nature of these surface interactions, as well as the mechanism by which they affect the bed structure. Examples of effects of moisture and anti-caking agents on the bulk properties of selected food powders are given in Tables 3.4 and 3.5. More information about anti-caking agents can be found in Chapter 12.

3.3.2. Compressibility

Food powders can be compacted by tapping or by mechanical compression. These processes can occur either unintentionally as a result of handling or transporting, or intentionally as when tableting or agglomerating. In the food industry, unintentional compression is normally undesirable, while operations aimed at obtaining defined shapes are usually required in some processes. Unintentional compression will be discussed in this section, whereas the latter will be included under the general scope of processing operations in a subsequent part of this text.

The theoretical and empirical considerations of vibratory compaction have been mainly focused on nonfood powders (Hausner et al., 1976). Sone (1972) reported the following relationship for food

Table 3.5. Effect of anti-caking agents on the bulk density and compressibility of selected food powders.

Powder	Agent	Concentration	Poured bulk density (kg/m^3)	Compressibility
Sucrose (powdered)	None	—	700	0.066
	Calcium stearate	0.5	870	0.039
	Silicon oxide	0.5	750	0.052
	Tricalcium phosphate	0.5	760	0.044
Salt (powdered)	None	—	1,010	0.080
	Calcium stearate	0.1	1,140	0.032
	Silicon oxide	0.1	1,100	0.045
	Tricalcium phosphate	0.1	1,160	0.025
Soup mix	None	—	700	0.27
	Aluminum silicate	2.0	750	0.15
	Calcium stearate	2.0	630	0.27
Gelatin (powdered)	None	—	680	~0
	Aluminum silicate	1.0	700	0.016
Microcrystalline cellulose	None	—	350	0.017
	Aluminum silicate	1.0	360	0.030
Corn starch	None	—	620	0.109
	Calcium stearate	1.0	590	0.099
	Silicon oxide	1.0	670	0.077
	Tricalcium phosphate	1.0	610	0.062
Soy protein	None	—	270	0.040
	Calcium stearate	1.0	270	0.041
	Silicon oxide	1.0	270	0.036
	Tricalcium phosphate	1.0	310	0.024

powders:

$$\gamma_n = \frac{V_0 - V_n}{V_0} = \frac{abn}{1 + bn} \tag{3.6}$$

where γ_n is the volume reduction fraction, V_0 is the initial volume, V_n is the volume after n taps, and a and b are constants.

The applicability of Eq. (3.6) was tested through its fitting to the following linear form:

$$\frac{n}{\gamma_n} = \frac{1}{ab} + \frac{n}{a} \tag{3.7}$$

The constant a in Eqs. (3.6) and (3.7) represents the asymptotic level of the volume change or, in other words, the level obtained after a large number of tapings or a long time in vibration. The constant b is representative of the rate at which this compaction is achieved, i.e., $1/b$ is the number of vibrations necessary to reach half of the asymptotic change. In general, this form of data presentation is very convenient for systems comparisons, since it only involves two constants.

A very common undesirable aspect of compressibility is its negative influence on flowing capacity. In powder technology, great attention has been paid to the general behavior of powders under compressive stress (Peleg, 1977). Compression tests have been used widely in pharmaceutics, ceramics, metallurgy, civil engineering, as well as in the food powder field, as a simple and convenient technique to measure such physical properties as powder compressibility and flowability. In order to get the pressure–density relationship for a given powder, a set of compression cells (usually a piston in a cylinder) is used. The tested powder is poured into the cylinder and compressed with the piston attached to the crosshead of, for example, a TA-XT2 Texture Analyzer (Stable Micro Systems, England) or Instron Universal Testing Machine. Normally, the instrument will record a force–distance relationship during a compression test. It is relatively easy to change this relationship into a pressure–density relationship to get the compressibility after data treatment, when the cross section area of the cell and the initial powder weight are known. The compression process takes place in two stages: filling voids with particles of the same or smaller size than the voids by particle movement, and filling smaller voids by the particle's elastic, and/or plastic deformation, or fragmentation.

The pressure–density for powders in a compression test at a low-pressure range can be described by the following equation (Barbosa-Cánovas et al., 1987):

$$\frac{\rho(\sigma) - \rho_0}{\rho_0} = a + b \log \sigma \tag{3.8}$$

where ρ (σ) is the bulk density under the applied normal stress σ, ρ_0 the initial bulk density, and a and b are constants. The constant b represents, specifically, the compressibility of a given powder. Compression tests are useful in characterizing the flowability of powders because the interparticle forces enabling non-flowing open structures stand still in powder beds are crushed under relatively low pressures. As shown in Eq. (3.8), the constant b, representing the change in bulk density by applied stress, is referred to as powder compressibility. It has been found that b can be correlated with cohesion of a variety of powders and, therefore, could be a simple parameter to indicate flowability changes (Peleg, 1977). Generally, the higher the compressibility the poorer the flowability, but if quantitative information about flowability is required, shear tests are necessary (Schubert, 1987).

One of the standard methods for evaluating the flowability of a particulate system is to calculate the Hausner ratio after tapping. As described before, the Hausner ratio is defined as the ratio of a powder system's initial (loose) bulk density to its tapped bulk density (i.e., the ratio of loose volume to tapped volume). It is easy to calculate the Hausner ratio and evaluate flowability when the loose and tapped volumes of the test material are known. Hayes (1987) has defined different ranges for Hausner ratio (H_R) to characterize flowability:

- $1.0 < H_R < 1.1$, for a free flowing free flowing powder;
- $1.1 < H_R < 1.25$, for a medium flowing powder;
- $1.25 < H_R < 1.4$, for a difficult flowing powder;
- $H_R > 1.4$, for a very difficult flowing powder.

Adding a small amount of fine powders, such as anti-caking agents, is often used to improve the flow properties of powdered materials in the chemical, pharmaceutical, and food industries. In this case, fine particles coat the coarser particles of the main constituent and prevent them from sticking together. Damp or sticky solids, which are difficult to handle, may be converted into free flowing after the added fine powders absorb small quantities of liquids. Another way to improve flowability of food powders is by using the agglomeration process, which is accomplished by wetting the fine particles in an atmosphere of water or suitable solvent droplets, causing them to collide and stick together, and then drying the agglomerated material in an air stream. Apart from improving flowability, agglomerated powders may show better wettability and dispersibility in liquids, and tend to be dust-free (Hoseney, 1994). Agglomeration will be discussed further in a subsequent chapter of this text.

3.4. STRENGTH PROPERTIES

There are a number of properties of particulate materials that determine particle breakage and attrition. Many solid food materials, especially when dry, are brittle and fragile, showing a tendency to break down or disintegrate. Mechanical attrition of food powders usually occurs during handling or processing, when the particles are subjected to impact and frictional forces. Attrition represents a serious problem in most of the food processes where dry handling is involved, since it may cause undesirable results such as dust formation, health hazard, equipment damage, and material loss. Dust formation may be considered the worst of these aspects, as it may develop into a dust explosion hazard. The topics of attrition and dust explosion are included in Sections 12.1 and 12.4 of this book, respectively. Some procedures to assess strength properties will be discussed as follows.

3.4.1. Abrasion

Abrasiveness of bulk solids, i.e., their ability to abrade or wear surfaces with which they come into contact, is considered a property closely related to the hardness of the material. The hardness of powders or granules is defined, in direct analogy with the definition of hardness of solid materials, as the degree of resistance of the surface of a particle to penetration by another body. Hardness is often considered a relative rather than an absolute property and may be determined by using the well-known Mohs' hardness scale shown in Table 3.6. In this scale, the ten selected minerals are listed in order of increasing hardness, so that a material of a given Mohs' number cannot scratch any substance of a higher number, but will scratch those of lower numbers. In a qualitative manner, materials different from those included in the scale are referred to as having an equivalent number of hardness of the ten listed.

Likewise, the abrasiveness of food powders can be assessed in different ways. It can be implied from the relative hardness of the particles and the surface with which they are in contact, using the Mohs' hardness scale. It can also be described by an *abrasion index*, which combines the effects of particle hardness, shape, size distribution, and bulk density into one factor, independent of the nature of the contacting surface. The best way to assess abrasiveness is to use the actual bulk material and the contact surfaces in question. There have been some developed tests proposed for specific materials. For example, a test used for coke and coal, consisting of measuring the wear on a standard surface when it is brought into moving and intimate contact with the material under specific conditions, can be adapted to many different materials, including food powders. Abrasiveness and hardness are two major factors that govern the choice and design of different types of equipment, such as size reduction

Table 3.6. Mohs' scale of hardness.

Hardness number	Material	Notes
1	Talc, graphite	Can mark paper powdered by finger
2	Gypsum rock salt	Can scratch lead
3	Calcite	Can scratch finger nail
4	Fluorospar	Can scratch copper coin
5	Apatite	
6	Feldspar	Can scratch window glass
7	Quartz	Can scratch a knife blade
8	Topaz	
9	Sapphire, corundum	
10	Diamond	

machines, air classifiers, mixers, dryers, etc. Hardness, rather than abrasiveness figures and values, are normally found in the literature; from a practical standpoint hardness can be taken as the chief property of this kind when making a decision on design and operating aspects of processes involving equipment units like those mentioned above. As a rule of thumb, considering Mohs' hardness scale, materials can be generally rated as soft, medium hard, or hard, when they show values between 1 and 3, 3.5 and 5, and 5 and 10, respectively. Many food materials are normally soft according to this criterion and, thus, the problems related to strength of materials normally faced in the food industry have to do with attrition and friability, rather than hardness and abrasion.

3.4.2. Friability

This property is defined as the tendency of particles to break down during storage and handling, while attrition is the actual, unwanted breakdown of particles. The above two definitions imply total breakdown, but attrition usually means particles getting smaller due to their corners or surface irregularities being knocked off. Attrition is a serious, yet little understood problem in handling of food materials, which may be considered responsible for economical losses in the food industry. Friability can be commonly determined using impact, vibration, shear, and tumbler tests.

Breakdown of particles on impact can be tested either on single particles or on a quantity of the bulk solid, and the result is a measure of the material's friability. The most common types of multi-particle impact test are the drop and shatter tests in which a specified quantity of the material is dropped through a specified height onto a hard surface or into a container. Drop shatter tests are generally used on coarser solids than those within the scope of this text; the fine fraction is in fact removed from the bulk material before the test. With harder materials, like aggregates, an impact test machine can be used that employs a 14-kg hammer which drops from a height of 380 mm, a specified quantity of the sample in a cup. The amount of fines produced by the impact is an indication of shatter resistance. Friability of tablets or granules is tested quite commonly by vibration in a container or on a sieve. A similar procedure also has been used in testing finer materials like catalysts, bone char, or fertilizers, but no standard exists.

Shear cells, as used in testing yield strength of solids, may also be used for testing friability. Since a particle large strain is required in order to produce significant attrition, the annular shear cells, which permit infinite strain, are normally used. Tumbler tests, and, more specifically, drum tests, have probably been the most popular. They involve rotation within a drum with internal flights and analysis of the material for particle size distribution. Friability indices can be derived in terms of percentages of materials retained on specific screens after given numbers of rotations.

3.5. RECONSTITUTION PROPERTIES

Many powdered products produced by spray drying or grinding are difficult to rehydrate. In the context of food drying, reconstitutability is the term used to describe the rate at which dried foods pick up and absorb water, reverting to a condition which resembles the undried material when put in contact with an excessive amount of this liquid (Masters, 1976). Especially in rehydration operations, when water aided by capillary forces penetrates into the narrow spaces between fine particles (i.e., particle size less than 100 μm), the particles will start to dissolve and form a thick, gel-like mass that resists further water penetration. Thus, lumps containing dry particles in the middle will be formed requiring strong mechanical stirring to be homogeneously dispersed or dissolved in the liquid (APV, 1989).

In the case of powdered dried biological materials, a number of properties such as wettability, sinkability, and, dispersability may influence the overall reconstitution characteristics. Food powders obtained from drying processes are normally reconstituted for consumption. The selected drying method and adjustment of drying conditions can result in a product with good rehydration properties. For example, reconstitution characteristics will not be the same if drying methods such as freeze-drying or osmotic dehydration are used.

The most efficient method to improve the rehydration characteristics of dried food powders is probably the use of agglomeration (Barletta and Barbosa-Cánovas, 1993). In order to agglomerate particles, the powder is treated with steam or warm, humid air such that condensation occurs on the particle surface. Inter-particle contact is promoted, often by swirling the wetted powder in a vortex. By agglomerating fine powders of about 100 μm in size into particles with the size of several millimeters, the wetting behavior of the particles is improved and lump formation can be avoided (Schubert, 1987).

3.5.1. Instantizing Processes

The term "instant" is usually used in industries such as food, pharmaceuticals, animal feed, chemicals, and pigments to describe the dispersing and dissolving properties of powders. Some popular commercially available instant powders are milk, coffee, cocoa, baby foods, soups, sauces, soft drinks, sugar mixtures, as well as vitamins and medicated powders. Even though some powders are "naturally" instant, it is a common practice to apply a special treatment, a so-called instantizing process, to powdered materials. This treatment will provide food powders with the "instant" attribute so that they can be dissolved or dispersed more readily in aqueous liquids than when they are in their original powdered forms (Schubert, 1980).

There are two main groups of instantizing processes: agglomeration and non-agglomeration. Agglomeration processes include straight-through agglomeration (e.g., spray drying and agglomeration), rewetting agglomeration, spray-bed dryer agglomeration, and press agglomeration. The size enlargement of powders by agglomeration is a technique often used in a wide range of industries and it has had increased demand in recent years. Depending on its application or the area in which it is used in industry, the agglomeration process is also sometimes referred to as granulation or instantizing (APV, 1989). These methods will be covered in detail in Chapter 7. To better understand the instant properties of agglomerated food powders, it is important to have a fundamental knowledge of the inter-particle forces or the binding mechanisms that are involved in agglomeration. For all the particles in an agglomerated state, it is well known that the forces causing primary particles to stick together are solid bridging, liquid bridging, inter-particle attraction forces, and mechanical interlocking (Schubert, 1980).

Solid bridging forms as a result of sintering, solid diffusion, condensation, or chemical reaction. All of these are more likely to happen at an elevated temperature, but ex-solution of soluble material can form solid bridges at room temperature. *Liquid bridging* results from the presence of a bulk liquid between individual particles. Once a liquid bridge is established, any evaporation of liquid reduces the curvature radii of liquid–gas interfaces and thus increases the forces holding the particles together so that they approach each other more closely. *Inter-particle attraction forces* can be either electrostatic or Van der Waals forces (short ranged attraction forces between solid surfaces). Electrostatic forces arise through charging by contact with charged particles or friction. Van der Waal forces arise from electron motion within an atom, which protrudes beyond the surface of a particle. *Mechanical interlocking* occurs in agglomerates formed by particle interlocking and only if these are fibrous or plate-shaped particles. These mechanisms are explained in more detail in Section 7.2 when describing particle aggregation fundamentals.

Agglomeration processes can be accomplished between two particles if they are brought together (with or without pressure) and/or at least one of them has a sticky surface. What happens in the

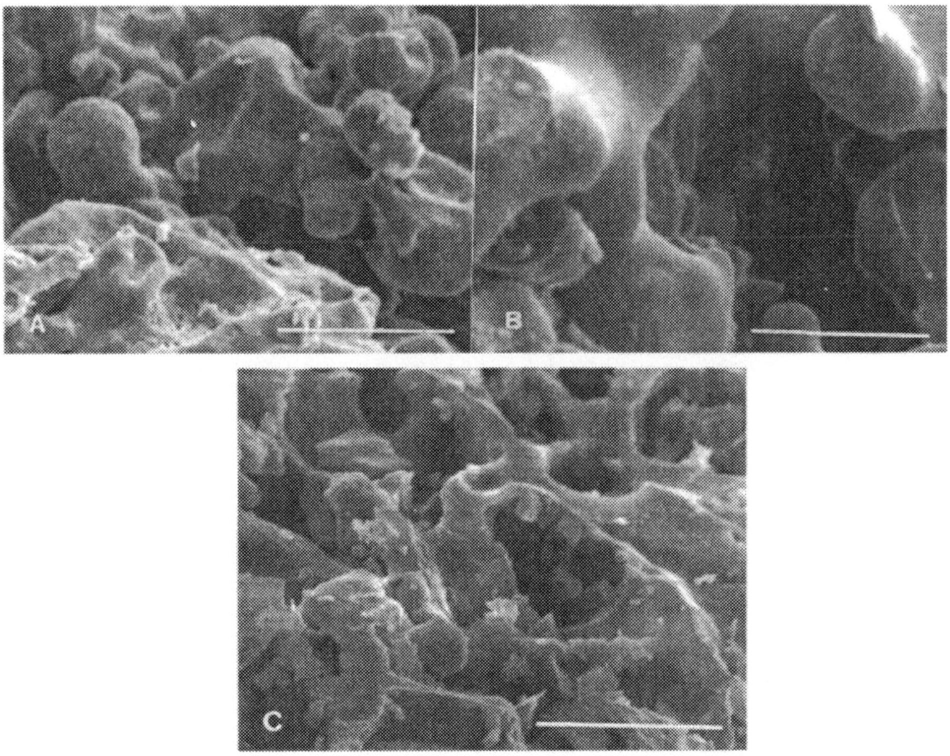

Figure 3.16. Microstructures of some typical agglomerated food powders observed under scanning electron microscopy: (A) spray-dried nonfat milk; (B) spray-dried coffee; and (C) freeze-dried coffee. Bar length = 60 μm in all cases.

rehydration of an agglomerated powder is that the large passages between the primary fine particles can assist in quickly displacing the air and allowing the water to penetrate before an impenetrable layer is formed. Therefore, the powder can disperse into the liquid and have complete dissolution (APV, 1989). Microstructures of some typical agglomerated food powders observed under scanning electron microscopy (SEM) are shown in Fig. 3.16.

Satisfactory instant properties can also be achieved using non-agglomeration techniques, such as freeze drying, osmotic drying, and drum drying, adding additives like lecithin (in dried whole milk), removing certain components like fat (low fat dried milk), and applying thermal treatment to amorphous materials (APV, 1989; Pietsch, 1999; Schubert, 1980; Schubert, 1981).

Freeze-drying consists of the production of ice crystals and their sublimation at very low pressures (Heldman and Singh, 1981). This procedure results in food particles with an open pore structure, which absorb water easily when they are reconstituted. Another alternative is the use of the so-called combined methods, such as osmotic dehydration followed by conventional drying. In osmotic dehydration, food particles are immersed in a concentrated solution. By osmotic pressure, the water inside the particles tends to migrate to the solution in order to equate water activities on both sides of the cellular wall (Monsalve-González et al., 1993). This partial dehydration will aid in the final stage of drying, and textural damage of the biological materials will be minimized. In this sense, biological materials dehydrated by combined methods will also have an open pore structure and, similar to freeze dried materials, will present good reconstitution properties. Beltran-Reyes et al. (1996) developed an apple powdered ingredient by grinding dried apples obtained by osmotic dehydration followed by conventional heated air drying. They determined that the firmness of the rehydrated mash, measured as an extrusion force in a texture analyzer, was a direct function of the particle size. For the same ingredient, Ortega-Rivas and Beltran-Reyes (1997) reported that rehydration improved as particle size decreased.

3.5.2. Instant Properties

Parameters that determine the properties of agglomerates include those related to primary particles and agglomerates. Among the parameters related to primary particles are particle size, size distribution, shape, and surface area. Agglomerates are related to the following parameters: particle size, size distribution and shape of the agglomerate, the apparent and bulk densities, porosity of the agglomerate, pore sizes and their distribution in the agglomerate, and the agglomerate strength (Pietsch, 1999). Compared with a finely dispersed particulate system, agglomerated powders have better flowability, lower tendency to stick together or to demix, improved packing density and homogeneity, and better instantaneous properties that are important for fast foods and beverages.

The instantaneous properties of agglomerates are the most desirable properties of agglomeration processes and they can be measured by the following four dissolution properties when agglomerates are spread on the surface of a liquid (Schubert, 1987):

- *Wettability*: liquid penetration into a porous agglomerate system due to capillary action or the ability of agglomerates to be penetrated by the liquid;
- *Sinkability*: the sinking of agglomerates below the liquid surface;
- *Dispersibility*: the dispersion of agglomerates with little stirring; and
- *Solubility*: dissolving of soluble agglomerates in the liquid.

The corresponding four phases of dissolution, i.e., wetting, sinking, dispersing and solution, can proceed individually, but some overlapping may occur, depending upon the amount of material involved. Wetting is a time controlling first step and it refers to a simple solid wetting assisted by the capillary suction of the pores in the agglomerates (Pietsch, 1999). The term "wettability" describes the capacity of the particles to absorb water on their surface, thus initiating reconstitution. This property depends largely on particle size. Small particles have a large value of specific area (i.e., the ratio of surface area to mass) and may not be wetted individually. Increasing particle size and/or agglomerating particles can reduce the incidence of clumping. The nature of the particle surface can also affect wettability. For example, the presence of free fat in the surface reduces wettability. The selective use of surface-active agents, such as lecithin, can sometimes improve wettability in dried powders containing fat.

Related to the second dissolution step, sinking or submergence is controlled by the mass of the agglomerates and is not critical if the wetting takes place easily (Pietsch, 1999). "Sinkability" describes the ability of the particles to sink quickly into the water. It depends mainly on the particle size and density, since larger and denser particles usually sink faster than finer and lighter ones. Particles with a lot of included air may be relatively large in size but exhibit poor sinkability because of their low density.

Finally, "dispersability" describes the ease with which the powder may be distributed as single particles over the surface and throughout the bulk of the reconstituting water, while solubility refers to the rate and extent to which the components of the powder particles dissolve in the water. Dispersability is reduced by clump formation and is improved when the sinkability is high, whereas solubility depends mainly on the chemical composition of the powder and its physical state.

For a dried product to exhibit good reconstitution characteristics there needs to be a correct balance between the individual properties discussed above. In many cases, alteration of one or two of these properties can markedly change the rehydrating behavior. Several measures can be taken

Table 3.7. Reconstitutability and some physical properties of different types of skim milk powder (adapted from APV, 1989).

	Ordinary spray dried powder	Integrated fluid bed agglomeration	Re-wetting agglomerated powder
Wettability (s)	>1,000	<20	<10
Dispersibility (%)	60–80	92–98	92–98
Insolubility index	<0.10	<0.10	<0.20
Average particle size (μm)	<100	>250	>400
Bulk density (kg/m^3)	641–689	449–545	449–497

in order to improve reconstitutability of dried food products. Reconstitutability and some physical properties of skim milk powder produced by different drying methods are summarized in Table 3.7, which shows that agglomeration processes greatly improve the powder wettability and dispersibility and also decrease the bulk density.

3.5.3. Instant Property Evaluation

Instant properties are a function of time of wetting and each industry has a more or less defined procedure to set a maximum allowable dissolution time when evaluating instant properties for quality control. A standard procedure to measure the instant properties must define the specific solvent temperature, the liquid surface area, amount of material to dissolve, the method of depositing a certain amount of material onto the liquid surface, unassisted or predetermined mixing steps, and the timing procedure (Pietsch, 1999). As classified by Schubert (1987), if agglomerates have good instant properties, the processes used to measure the four instant properties mentioned above should be completed satisfactorily in a few seconds if the layer thickness of the agglomerates spread on the liquid surface is about 10 mm.

Many objective methods to determine the instant properties are available, and satisfactory methods have been developed especially for the most important instant properties, e.g., wettability and dispersibility (Schubert, 1987). A simple test for measuring liquid penetration speed, a standard method to measure the dynamic wetting, a standard method to measure the dispersibility, and a standard method for dispersibility and wettability of dairy products is explained in detail below.

3.5.3.1. Penetration Speed Test

As shown in Fig. 3.17, the test cell contains a screen on which a layer of agglomerates with H thickness is retained and a plexiglass cylinder is added to avoid wetting from above. Without the cylinder, the powder bed may lift up and swim on the liquid surface at the beginning of the measurement, resulting in free particle sedimentation rather than penetration. When the test cell is put into the liquid, the time is measured until the entire material bed is submerged. These test results are closely related to the wetting behavior of agglomerates because not only the large pores between the agglomerates must be filled, but also the pores of agglomerates must be penetrated by liquid during wetting. It is also an effective method to determine the optimum agglomerate size for wetting if there are different sized fractions of an experimental batch of agglomerates to be evaluated (Pietsch, 1999).

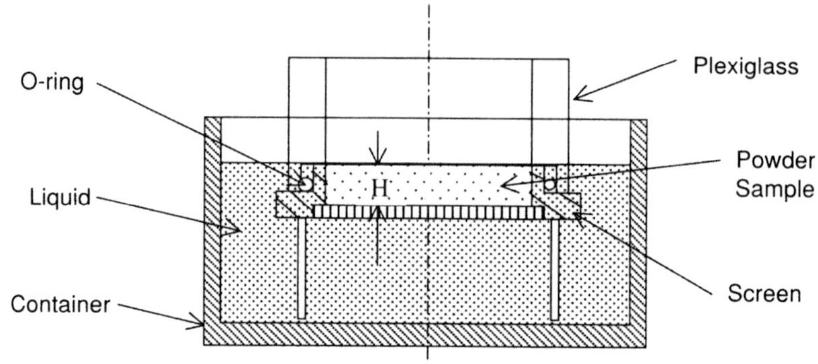

Figure 3.17. A simple test to measure the liquid penetration speed into a powder mass (adapted from Pietsch, 1999).

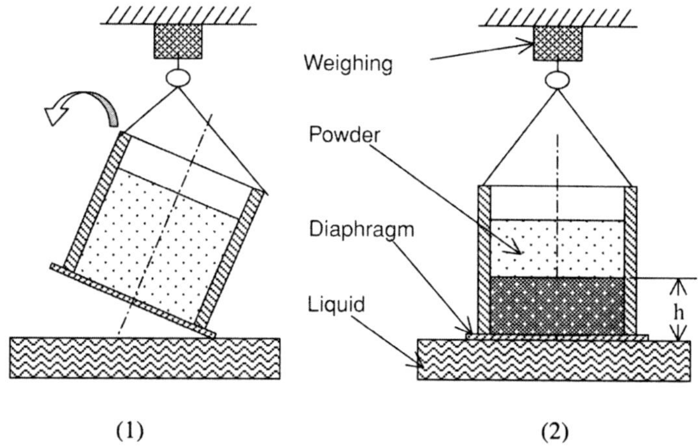

Figure 3.18. A schematic diagram of dynamic wetting test: (1) represents the cell position before the test and (2) for the cell position during the force–time measurement (adapted from Pietsch, 1999).

3.5.3.2. A Dynamic Wetting Test

The dynamic wetting behavior of agglomerates can be measured by using the device developed by Schubert (1980). As shown in Fig. 3.18, the material to be studied is put into a measuring cell that is fixed to a weighing cell and in contact with the liquid by tilting the cell onto the liquid surface. In this way air bubbles, which might disturb the penetration process, can be prevented from forming between the liquid surface and the porous plate at the cell bottom. The force measured by the weighing cell is proportional to the liquid volume absorbed due to capillary pressure and it is plotted against time. Thus, the penetration behavior of the agglomerate can be completely described by the force–time plot until all the material in the cell is wetted (Pietsch, 1999; Schubert, 1980).

3.5.3.3. Dispersibility Measuring Test

In this test, it is essential to assume that following a short period of dispersing, soluble particles are completely dissolved and suspended particles can remain in the liquid for a long time. All dissolved and suspended particles are regarded as dispersed. Other particles, mostly non-wetted and deposited or floating at the surface, are regarded as residual material. If the dispersed total sample is divided into the dispersed mass and residual mass according to the standard set by the International

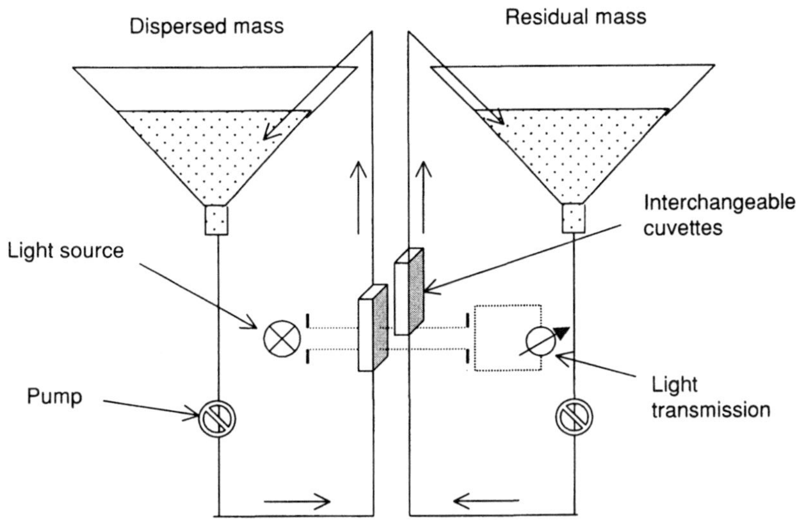

Figure 3.19. A rapid photometric method for measuring the dispersibility of instant food powders by light transmission (adapted from Schubert, 1987).

Dairy Federation (IDF), the following dispersibility value D can be calculated (Schubert, 1987):

$$D = \frac{125M_C(1 + M_C)}{99M_C + 151} \tag{3.9}$$

and

$$M_C = \frac{C_d}{C_r} \tag{3.10}$$

where C_d is the mass concentration of the dispersed proportion and C_r is the mass concentration of the residual material.

The mass concentrations C_d and C_r can be determined gravimetrically or from the sample weight, but the procedure is time consuming. A rapid photometric method, which is based on the relationship between the concentration C and the loss in light transmission T according to the Lambert–Beer law (i.e., $\ln T \propto C$), has been developed. The instrument for measuring the dispersibility of instant food powders by light transmission is shown in Fig. 3.19. In this method, the sample is dispersed and divided into dispersed and residual portions that are then homogenized and filled into a corresponding funnel, separately. Each portion is in turn pumped continuously through a cuvette to measure the light transmission value. The photometric dispersion measure M_T which corresponds to the C_d and C_r can be expressed as:

$$M_T = \frac{\ln T_d}{\ln T_r} \tag{3.11}$$

where T_d is the transmission value for the dispersed portion and T_r is the transmission value for the residual portion. M_T is the same as the M_C within the limits of measuring accuracy and can be directly used in Eq. (3.9) instead of M_C.

This dispersion test is only suitable for agglomerated food powders, but not for all instant powders because of the assumptions mentioned at the beginning. Another instrument using the same light transmission mechanism is described by Pietsch (1999). It can measure wettability, dispersibility, and solubility in the same test and provide more information on the progress of wetting, dispersion, and dissolution. The tests can be carried out under different specified dispersion/dissolution conditions, such as dispersion/dissolution without stirring, with different levels of stirring, or with the application of ultrasound (Pietsch, 1999).

3.5.3.4. IDF Standard Method

This method is completely described in the International IDF Standard 87 (IDF, 1979). It is specifically designed to determine the dispersibility of instant dried milk and is also a rapid routine method to determine wettability (i.e., wetting time). A certain amount of dried milk powder (26 g for instant dried skim milk and 34 g for instant dried whole milk), held in a glass tube with a glass plate on the top of a beaker, is scattered on the water surface by withdrawing the glass plate underneath the glass tubing. The water in the beaker weighs about 250 g and its temperature is adjusted to around 25°C. After a defined procedure for stirring, sampling, and solid content measuring, the dispersibility D is calculated as (IDF, 1979):

$$D = \frac{962C_S}{100 - (W + C_S)}, \quad \text{for instant dried skim milk} \tag{3.12}$$

$$D = \frac{735C_S}{100 - (W + C_S)}, \quad \text{for instant dried whole milk} \tag{3.13}$$

where C_S is the total solid content of the liquid in % (m/m), and W is the water content of the pre-treated test sample in % (m/m). According to this IDF standard, a dried milk powder is regarded as failing to possess sufficient instant characteristics if its D value is less than 85% for dried whole milk or less than 90% for dried skim milk.

The apparatus and procedure to measure the wettability of instant dried milk is almost the same as that for the dispersibility measurement, except that the following stirring, sampling and solid content measuring steps are omitted. After the test sample is evenly spread on the water surface, the time required for all the sample particles to sink below the water surface or remain on the surface but have a typical "wet" appearance is recorded. The wettability θ_{wet} is calculated as:

$$\theta_{wet} = \theta - 60 \tag{3.14}$$

where θ is the recorded time in seconds.

REFERENCES

Akers, R.J. (1992). The Certification of Limestone Powder for Jenike Shear Testing, CRM116. Commission of the European Communities (ECSC-EEC-EAEC), Luxembourg.

APV (1989). *Dryer Handbook (DRH-889)*. APV Crepaco Inc., Rosemont, IL.

ASTM D6128-00. *Standard Shear Testing Method for Bulk Solids Using the Jenike Shear Cell.* American Society for Testing and Materials.

ASTM D6393-99. Standard Test Method for Bulk Solids Chracterization by Carr Indices. American Society for Testing and Materials.

ASTM D6773-02. *Standard Shear Test Method for Bulk Solids Using the Schulze Ring Shear Tester.* American Society for Testing and Materials.

Augenstein, D.A. and Hogg, R. (1978). An experimental study of the flow of dry powders over inclined surfaces. *Powder Technol.* **19**: 205–215.

Barbosa-Cánovas, G.V., Málave-López, J. and Peleg, M. (1987). Density and compressibility of selected food powders mixture. *J. Food Proc. Eng.* **10**: 1–19.

Barletta, B.J. and Barbosa-Cánovas, G.V. (1993). An Attrition index to assess fines formation and particle size reduction in tapped agglomerated food powders. *Powder Technol.* **77**: 89–93.

Beltran-Reyes, B., Ortega-Rivas, E. and Anzaldua-Morales, A. (1996). Characterization of reconstituted apple paste in terms of rehydration and firmness. *Food Sci. Technol. Intl.* **2**: 307–313.

Carr, R.L. (1976). Powder and granule properties and mechanics. In *Gas–Solids Handling in the Processing Industries*, Marchello, J.M. and Gomezplata, A. (eds.). Marcel Dekker, New York.

Chasseray, P. (1994). Physical characteristics of grains and their byproducts. In *Primary Cereal Processing*, Godon, B. and Willm, C. (eds.). VCH Publishers, New York.

Chen, X.D. (1994). Mathematical analysis of powder discharge through longitudinal slits in a slowly rotating drum: objective measurements of powder flowability. *Powder Technol.* **21**: 421–437.

Dobbs, A.J., Peleg, M., Mudget, R.E. and Rufner, R. (1982). Some physical characteristics of active dry yeast. *Powder Technol.* **32**: 75–81.

Geldart, D. (1973). Types of gas fluidization. *Powder Technol.* **7**: 285–292.

Geldart, D., Harnby, N. and Wong, A.C.Y. (1984). Fluidization of cohesive powders. *Powder Technol.* **37**: 25–37.

Geldart, D. and Wong, A.C.Y. (1985). Fluidization of powders showing degrees of cohesiveness-II. Experiments on rates of de-aeration. *Chem. Eng. Sci.* **40**: 653–661.

Gray, W.A. (1968). *The Packing of Solid Particles*. Chapman and Hall, London.

Harnby, N., Hawkins, A.E. and Vandame, D. (1987). The use of bulk density determination as a means of typifying the flow characteristics of loosely compacted powders under conditions of variable relative humidity. *Chem. Eng. Sci.* **42**: 879–888.

Hausner, H.H., Roll, K.H. and Johnson, P.K. (1976). *Vibratory Compaction—Principles and Methods*. Plenum Press, New York.

Hayes, G.D. (1987). *Food Engineering Data Handbook*. John Wiley & Sons, New York.

Heldman, D.R. and Singh, R.P. (1981). *Food Process Engineering*. Van Nostrand Reinhold, New York.

Hollenbach, A.M., Peleg, M. and Rufner, R. (1982). Effect of four anticaking agents on the bulk characteristics of ground sugar. *J. Food Sci.* **47**: 538–544.

Hoseney, R.C. (1994). *Principles of Cereal Science and Technology*. American Association of Cereal Chemists, St. Paul, MN.

IDF (1979). *International IDF Standard 87: 1979*. International Dairy Federation, Brussels, Belgium.

Jenike, A.W. (1964). *Storage and Flow of Solids*. Bulletin No. 123, Utah Engineering Experiment Station, Salt Lake City, UT.

Kostelnik, M.C. and Beddow, J.K. (1970). New techniques for tap density. In *Modern Developments in Powder Metallurgy*, Hausner, H.H. (ed.). Plenum Press, New York.

Malave-Lopez, J., Barbosa-Cánovas, G.V. and Peleg, M. (1985). Comparison characteristics of agglomerated coffee during tapping. *J. Texture Studies* **24**: 253–258.

Masters, K. (1976). *Spray Drying Handbook*. John Wiley & Sons, New York.

McGeary, R.K. (1967). Mechanical packing of spherical particles. In *Vibratory Compacting*, Hausner, H.H., Roll, K.H. and Johnson, P.K. (eds.). Plenum Press, New York.

Molina, M., Nussinovitch, A., Normand, M.D and Peleg, M. (1990). Selected physical characteristics of ground roasted coffees. *J. Food Process. Preserv.* **14**: 325–333.

Monsalve-González, A., Barbosa-Cánovas, G.V. and Cavalieri, R.P. (1993). Mass transfer and textural changes during processing of apples by combined methods. *J. Food Sci.* **58**: 1118–1124.

Moreyra, R. and Peleg, M. (1981). Effect of equilibrium water activity on the bulk properties of selected food powders. *J. Food Sci.* **46**: 1918–1922.

Ortega-Rivas, E. and Beltran-Reyes, B. (1997). Rehydration properties of apple powders and particulates. *Powder Handl. Process.* **9**: 245–248.

Peleg, M. (1978). Flowability of food powders and methods for its evaluation—a review. *J. Food Proc. Eng.* **1**: 303–328.

Peleg, M. and Mannheim, C.H. (1973). Effect of conditions on the flow properties of powdered sucrose. *Powder Tecnhol.* **7**: 45–50.

Pietsch, W. (1999). Readily engineer agglomerates with special properties from micro- and nanosized particles. *Chem. Eng. Progress* (Aug.): **95**(8): 67–81.

Rumpf, H. (1961). The strength of granules and agglomerates. In *Agglomeration*, Knepper, W.A. (ed.). Industrial Publishers, New York.

Schubert, H. (1980). Processing and properties of instant powdered food. In *Food Process Engineering*, Vol. 1, Linko, P., Mälkki, Y., Olkku, J. and Larinkari, J. (eds.), pp. 675–684. Applied Science Publishers, London.

Schubert, H. (1981). Principles of agglomeration. *Intl. Chem. Eng.* **21**(3): 363–377.

Schubert, H. (1987). Food particle technology. Part 1: Properties of particles and particulate food systems. *J. Food Eng.* **6**: 1–32.

Schulze, D. (1996). Flowability and time consolidation measurements using a ring shear tester. *Powder Handl. Process.* **8**: 221–226.

Scoville, E. and Peleg, M. (1980). Evaluation of the effect of liquid bridges on the bulk properties of model powders. *J. Food Sci.* **46**: 174–177.

Sone, T. (1972). *Consistency of Foodstuffs*. D. Reidel, Dordrecht, The Netherlands.

Sutton, H.M. (1976). Flow properties of powders and the role of surface character. In *Characterization of Powder Surfaces*, Parfitt, G.D. and Sing, K.S.W. (eds.). Academic Press, New York.

Svarovsky, L. (1987). *Powder Testing Guide. Methods of Measuring the Physical Properties of Bulk Powders*. British Materials Handling Board, UK.

Teunou, E., Vasseur, J. and Krawczyk, M. (1995). Measurement and interpretation of bulk solids angle of repose for industrial process design. *Powder Handl. Process.* **7**(3): 203–227.

Thomson, F.M. (1997). Storage and flow of particlulate solids. In Fayed, M.E. and Otten, L. *Handbook of Powder Science & Technology*. Chapman and Hall, New York, pp. 389–436.

Wilkinson, H.N., Duffell, C.H., Reed, A.R. and Bunting, J. (1983). *Bulk Solids Physical Property Guide*. British Materials Handling Board, Ascot, UK.

Williams, J.C., Birks, A.H. and Bhattacharya, D. (1971). The direct measurement of the failure function of a cohesive powder. *Powder Technol.* **4**: 328–337.

Yan, H. and Barbosa-Cánovas, G.V. (2001). Density changes in selected agglomerated food powders due to high hydrostatic pressure. *Lebensmittel-Wissenschaft und-Technologie*, **34**(8): 495–501.

PART II

PRODUCTION, HANDLING, AND PROCESSING

CHAPTER 4

STORAGE

4.1. ALTERNATIVES FOR STORAGE OF BULK SOLIDS

4.1.1. Outdoors and Structured Storage

In the food and related industries there are some particulate systems, such as grains or pulses, which may be stored outdoors in large piles unprotected from the weather. When being taken for processing, such materials may be removed by dragline or tractor shovel and delivered to a conveyor. Space allowance can be done by estimating the volume of the pile through aerial or ground surveys and multiplying by the bulk density of the material. Since bulk food materials are able to deteriorate with outside ambient conditions such as moisture, there is a need to provide protection for these piled particulate systems. Several alternatives have been used such as domes and cylindrical structures with conical tops, which are normally known as silos. Some of these structured forms of storage have been termed according to a particular application, such as corrugated-wall structures known as grain silos. The dome silo storage system has been successfully used to store salt and different kinds of grains. The reclaiming systems for these types of silos are similar to those used in outdoors storage, e.g., draglines, scraper reclaimers, bucket wheel reclaimers, etc. Outdoor and structured storage systems are useful when huge amounts of materials need to be in inventory, but direct connection to feed processing lines is somewhat difficult.

4.1.2. Storage in Containers

Open-yard or structured storage is only a part of a bulk solids handling facility. Due to processing requirements, there is normally the need for an intermediate storage system, which may be directly connected to a processing line. In some other cases, the storage capacity does not involve large tonnages and, therefore, confined storage would be an obvious choice. Furthermore, considering food bulk solids which are very likely to undergo bio-deterioration when exposed to outside ambient conditions, piling is not a feasible manner of storage and, therefore, all these particulate systems may be most properly kept in containers or vessels also known as silos, but alternatively termed bins and hoppers. Silos, bins, and hoppers used to store materials vary in capacity from a few kilograms to multi-ton capacity vessels (Knowlton et al., 1994). A bin can be defined as any upright container for storing bulk solids, while a silo is a tall bin where the height is 1.5 times its diameter. A hopper is the converging sloping wall section attached to the bottom of a silo or a small bin with a sloping bottom, for temporary storage of solids prior to fed into a particular process. Generally, all these containers are loaded from the top and discharged from the bottom. A major problem in bin design is related to proper and controllable discharge.

Bin storage of bulk materials was studied, until recently, in a very empirical manner and success was assured sometimes by pure chance when free-flowing materials were involved. Prior to the 1960s, bulk solid storage was somewhat crude and the old silos were constructed with completely flat bottoms and multiple discharge holes, for withdrawal to complex screw systems for material handling. After the crude flat bottom, several alternatives were experienced, such as the central inverted cone design, which used an inverted cone made of concrete to reduce the exposed floor area and aid in the complete withdrawal of the stored material. A commonly found geometry of storing vessel for bulk solids is the cylindrical one, with the base in a conical form. Such a geometry would promote flow in an axial symmetric way being, thus, the mostly recommended to ensure uniform flow in discharge.

4.2. PRINCIPLES INVOLVED IN STORAGE BIN DESIGN

4.2.1. Basic Concepts of Bulk Solids Plant

The food and related industries handle considerable amounts of powders and particulate materials every year. Startup delays and ongoing inefficiencies are common in solids processing plants. An important cause of these problems is the improper design of bulk solids handling equipment. A 6-year study of 40 solids processing plants in the U.S. and Canada (Merrow, 1988) revealed that 80% of these plants experienced solids handling problems. The study also found that these plants were slow in smoothing operation, with an average startup time for some plants averaging 18 months. Once startup began, performance around 40–50% of the intended design was commonly observed.

Many powders and bulk solids do not flow reliably or uniformly through bins, hoppers, feeders, or chutes. The result is lost production, plant downtime, extra labor, poor quality control, spoiled material, and inefficient use of capital. Common flow problems in hoppers and silos can be summarized as: no flow (rathole and arching formation), segregation, and flooding. Lack of discharge in the no-flow situation can be attributed to the formation of a stable arch over the outlet, or a stable cavity called a "rathole" or "pipe" (Marinelli and Carson, 1992; Thomson, 1997). Irregular flow occurs if arches and ratholes are formed and collapse.

4.2.1.1. Ratholes

Piping or ratholing occurs when only the bulk solid above the outlet is flowing out, and the remaining bulk solid (at the dead zones) stays in place and forms a pipe (Fig. 4.1). After some material discharges as the gate opens or the feeder starts, the material's cohesive strength causes a narrow flow channel to form above the outlet. The rathole decreases the bin's usable (called live) capacity and can cause the material to cake or degrade, promote segregation, or cause the bin's structural failure. Unconfined yield strength of the bulk solid is the main property that can be associated with this situation. If the bulk solid consolidates increasingly with an increasing period of storage at rest, the risk of piping increases. For many cohesive bulk solids, and for the normal consolidation heads occurring in practice, bulk ratholes or pipes measuring several meters high are often observed. This makes control of the product discharge rate quite difficult. For example, cleaning of a funnel-flow type of bin with a rathole is often uncertain, because solids in the stagnant or dead region may pack and cake. In order to avoid the formation of ratholes, the stress imposed on the material will exceed the yield strength, and any rathole that tends to form will continually collapse.

4.2.1.2. Arching

An arch is a stable obstruction that forms over the point of narrowest cross-section of the storage vessel (usually the discharge outlet). The arch supports the rest of the bin's contents, preventing

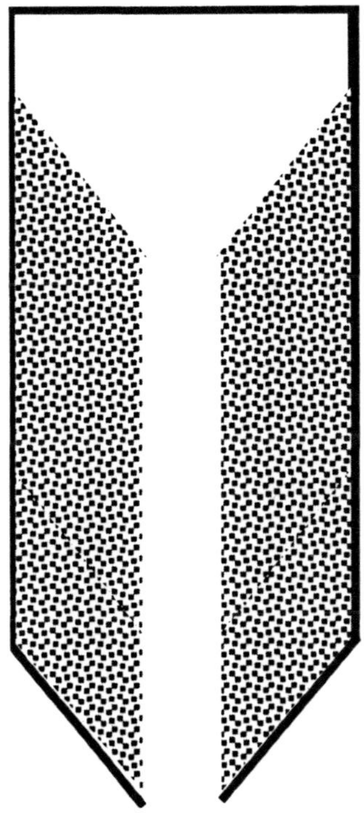

Figure 4.1. Stable rathole.

the discharge. In the case of fine powders, arching can also be explained by the unconfined yield strength, which is caused by the adhesion forces acting among the particles (cohesive arching). In the case of coarse bulk solids, arching is caused by the blocking or interlocking of single particles (interlocking arching). There is usually friction between the wall of the bin and the solid particles, and the interlocking of these particles causes a frictional effect throughout the bulk solid mass. In an extreme case, such frictional force causes the mass of bulk solids to arch, or bridge, so that it will not fall even if the material below is discharged. To break an arch and initiate flow a sledgehammer, a vibrator, or an air blaster can be used to apply a force greater than gravity to the compacted material.

Figure 4.2 depicts an element in an arch that has formed near the outlet of a hopper. The major principal stress on the arch (σ_1) acts perpendicular to the adjacent hopper wall. Since there is nothing acting against the arch from the underside, the minor principal stress σ_2 is zero. The individual arch element will rupture and fail, and the arch will collapse if σ_1 exceeds the material's cohesive strength. Cohesive arch profiles can be measured over a range of bulk solid conditions and hopper geometries using the technique of laser arching (Berry et al., 2000).

4.2.1.3. Erratic Flow

When an arch and rathole occur in the same bin, the combination can cause erratic flow. Firstly, the rathole is formed after initiating flow in the bin. Then, when force is used to collapse the rathole, the collapsing material forms an arch impacting the outlet. This series of events can become cyclic making the flow erratic.

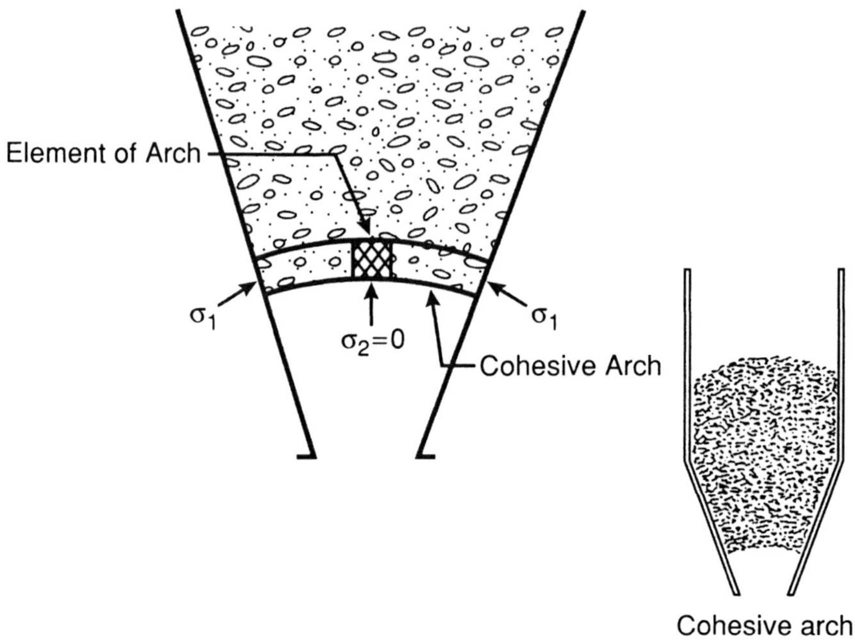

Figure 4.2. Element of a cohesive arch blocking the outlet of a hopper (from Bell, 1993).

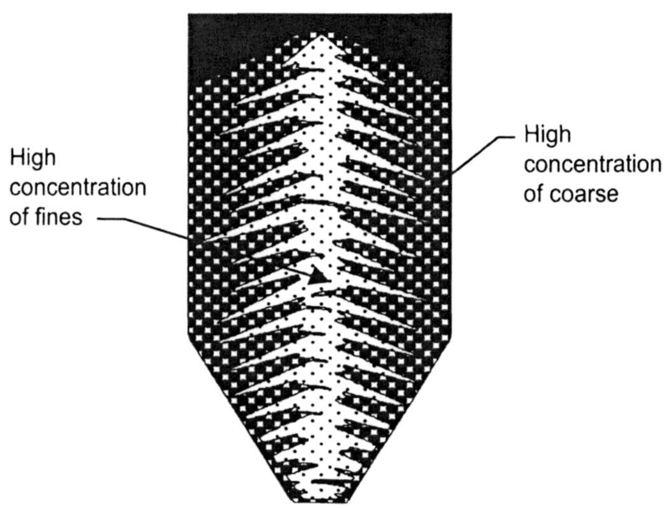

Figure 4.3. Sifting segregation during filling (adapted from Knowlton et al. with permission copyright © 1994 AICHE all rights reserved).

4.2.1.4. Segregation and Flooding

Mixtures of solid particles can separate or segregate while they are being handled. This often results in costly quality-control problems—as well as wasted raw materials, lost production, increased maintenance, and capital costs. Many materials experience separation of fine and coarse particles (Carson et al., 1986) and such separation can seriously compromise the quality of the final product and the efficiency of the process. Segregation occurs most frequently in free-flowing granular materials that have a wide size distribution, and seldom in fine powders where particle size is 70 μm or less; cohesive materials usually do not segregate during handling (Thomson, 1997).

The most common mechanism for particle segregation is called sifting (Fig. 4.3). Segregation can be minimized by changing the characteristics of the bulk solid (for example, moisture or particle size distribution), altering the process by which the bulk solids are handled, or redesigning the equipment used to handle the bulk solid. An example of an obvious segregation problem is a drink mix that varies in tartness due to fluctuations in the citric acid content. A less obvious example is packages that are routinely overfilled to ensure they meet the weights stated on the label. Addressing the non-uniformity of the feed would reduce the need for overfilling.

Bulk solids can become fluidized when falling downwards to the outlet opening, so that they flow out of the silo like a fluid. This behavior is called flooding or flushing (Schulze, 2000). Flooding can be caused by the collapse of a rathole in a bin containing fine powder, resulting in uncontrollable flow of material, loss of product and clouds of dust (Royal and Carson, 1993), among other problems. Pertaining structural failure, each year over 1,000 silos, bins and hoppers fail in North America alone. Most of these failures could have been prevented with proper and careful design, in which the loads imposed by the bulk solid being stored had been well considered.

The design of bins, hoppers and silos has never been given the attention it deserves by food or chemical engineers. For a long time the only engineers interested were civil engineers, because they are responsible for the structural stability of silos. On the other hand, mechanical, chemical and food engineers were in charge of flow inside silos and of the continuous uninterrupted discharge from the bottom opening of the silo. Since the pressure applied to walls, inserts and the bottom area are governed by the flow properties of stored bulk solid and flow pattern inside the silo, civil engineers and process engineers need to work closely together in this area. Approaches using properties such as angle of repose or angle of spatula in design considerations are ineffective, because the resulting values bear no relation to most of the design parameters needed to ensure reliable flow, mainly because particulate solids tend to compact or consolidate when stored.

The attempt of trying to model bulk solids as fluids also leads to a bottleneck, due to the fact that flowing bulk solids generate shear stresses and are able to maintain these stresses even when their flow rate is changed dramatically. It is also improper to consider bulk solids as having viscosity since almost all bulk solids exhibit flow properties that are flow-rate independent. Furthermore, a bulk solid is not able to carry a significant one-dimensional load without lateral support (Schwedes, 2001).

The systematic approach for designing powder handling and processing plants started in the mid-1950s by the pioneering work of Andrew W. Jenike. His concept was to model bulk solids using the principles of continuum mechanics. The resulting comprehensive theory (Jenike, 1964) describing the flow of bulk solids has been applied and perfected over the years, but is generally recognized worldwide as the only scientific guide to bulk solids flow.

The procedures for the design of a bulk solids handling plant are well established and follow four basic steps: (a) determination of the strength and flow properties of the bulk solids for the worst likely flow conditions expected to occur in practice; (b) calculation of the bin, stockpile, feeder or chute geometry to give the desired capacity to provide a flow pattern with acceptable characteristics, in order to ensure that discharge is reliable and predictable; (c) estimation of the loadings on the bin and hopper walls and on the feeders and chutes under operating conditions; (d) design and detailing of the handling plant including the structure and equipment.

In a storage tank, pressure increases linearly with depth and does not depend on the direction. Bulk solids transmit static shear forces and the pressures on wall and bottom do not increase linearly with depth. Instead, the pressure on wall quickly reaches a maximum value as depth increases, because part of the bulk solids weight is transmitted to the walls via friction forces. Furthermore, the wall pressure depends on the direction and varies for filling, storing, and discharging.

As mentioned before, a silo usually consists of a vertical part, called the bin, and a converging part, called the hopper. In the hopper part the vertical pressure is decreasing, because in addition to the friction force between bulk solid and wall, the vertical component of the stress normal to the wall carries an additional part of the weight of the bulk solid. During filling, the maximum consolidating stress—due to the action of gravity—is directed vertically in the axis of symmetry and the minor one horizontally. If bulk solid is discharged through the bottom opening, the material in the hopper is compressed horizontally—due to the converging action—while it expands vertically due to the open outlet. Now the horizontal stress becomes the major principal stress while the vertical is the minor one.

4.2.2. Elements of Bulk Solids Gravity Flow

Only fluids can flow; bulk solids under gravity forces can fall, slide or roll, but against gravity, they must be lifted by mechanical means. Solids cannot be pumped by centrifugal or reciprocating pumps; for pumping, bulk solids ought to be suspended in liquids or gases. There is no satisfactory term to describe "flow" of bulk solids as they do not follow strict definitions of fluid behavior, since a fluid is considered to be a continuum in which there are no voids. For a fluid, when the rate of shear is linearly proportional to the shear stress, it is said to be Newtonian and the coefficient of proportionality is called the absolute viscosity. Any deviation from this definition makes the fluid non-Newtonian. Solids in suspension can be referred to as non-Newtonian mixtures to differentiate them from a number of non-Newtonian fluids, which are a continuum or are perfectly homogeneous liquids. For all these reasons, bulk solids in suspension are occasionally referred to as "imperfect fluids".

The gravity flow of bulk solids occurs under the pressure corresponding to the equivalent of a "static head" of the material. Such head would be caused by the height of a solid column in a bin, but in practice it is often not available to produce the flow due to "arching" or "bridging". The velocity head at the discharge from the bin is usually a small fraction of the head, with the major part being consumed by the friction of the moving solids against the walls of the bin, as well as against solids. Friction is the resistance which one body offers to the motion of a second body when the latter slides over the former. The friction force is tangent to the surfaces of contact of the two bodies and always opposes motion. The coefficient of static friction μ for any two surfaces is the ratio of the limiting friction to the corresponding normal pressure, i.e.:

$$\mu = \frac{F}{N} \tag{4.1}$$

where F is the maximum friction of impeding motion and N is the normal pressure. If a body rests on an inclined plane and if the angle of inclination of the plane to the horizontal, α, is such that motion of the body impends, this angle α is defined as the angle of repose, so it follows that:

$$\mu = \tan \alpha. \tag{4.2}$$

When two surfaces move relative to each other, the ratio of the friction developed to the normal pressure is called the coefficient of kinetic friction and is independent of the normal pressure. The coefficient of kinetic friction is also less than the coefficient of static friction and independent of the relative velocity of the rubbing surfaces. There is experimental evidence that supports that the value of the kinetic friction coefficient increases as the velocity is decreased, and passes without discontinuity into that of static friction. All these principles would hold under conditions of a particular test, but must be modified in order to apply them to different conditions. Herein lies the main difficulty in applying existing test data on series of new tests because of the great variety of flowing conditions of bulk solids. The problem is particularly complicated since the properties of the flowing material depend on the time and method of storage operation.

When granular solids are stored in an enclosed container, the lateral pressure exerted on the walls at any point is less than predicted from the head of the material above such point. There is usually friction between the wall and the solid particles, and the interlocking of these particles causes a frictional effect throughout the bulk solid mass. The frictional force at the wall tends to offset the weight of the solid and reduces the effect of the head of solids on the floor of the container. In an extreme case, such frictional force causes the mass of bulk solids to arch, or bridge, so that it would not fall even if the material below were discharged. For many granular solids, when the height of the solid bed reaches about three times the diameter of the bin, the additional head of material shows virtually no effect on the pressure at the bin floor.

Solids tend to flow out of any opening near the bottom of a bin, but are best discharged through an opening in the floor. The pressure at a side outlet is smaller than the vertical pressure at the same level and removal of solids from one side of a bin considerably increases the lateral pressure on the opposite side while the solids flow. When an outlet in the bottom of a bin containing free-flowing solids is opened, the material immediately above the opening begins to flow.

A central column of solids moves downward without disturbing the material at the sides. Eventually, lateral flow begins, starting from the top layer of solids, and a conical depression forms in the surface of the mass of bulk solids being discharged. The material slides laterally into the central column moving at an angle approaching the angle of internal friction of the solids, as defined in Chapter 3, and the solids at the bin floor are the last to leave. If additional material is added at the top of the bin at the same rate as the material leaving through the bottom outlet, the solids near the bin walls remain stagnant and do not discharge as long as flow persists. The rate of flow of granular solids by gravity through a circular opening in the bottom of a bin is dependent on the diameter of the opening as well as on the properties of the solid (i.e., particle size and surface) and is independent, within wide limits, on the head or height of the solids.

4.3. FLOW PATTERNS IN STORAGE BINS

Since the early 1960s, mostly from the pioneering work of Andrew W. Jenike previously mentioned (Jenike, 1964), certain terms describing flow patterns became common in silo, bin and hopper design. The general theory pertaining to gravity flow of bulk solids has been documented through the years (Arnold et al., 1982; Roberts, 1988) and from the standpoint of flow patterns, there are three basic types of flow in symmetrical geometry: mass-flow, funnel-flow and expanded flow (Fig. 4.4).

4.3.1. Mass-Flow

In this mode of flow, the hopper is sufficiently steep and smooth to cause flow of all the solids in the bin without stagnant or steady regions during discharge. The bulk solid is in motion at every point within the bin whenever material is drawn from the outlet. In mass-flow bins, the flow is uniform and the bulk density of the feed is practically independent of the head of solids in the bin. Mass-flow guarantees complete discharge of the bin contents at predictable flow rates. When properly designed, a mass-flow bin can remix the bulk of the solid during discharge, minimizing segregation.

Mass-flow bins are classified according to the hopper shape and associated flow pattern. The main types of hoppers are: conical, wedge-shaped and chisel-shaped hoppers. Figure 4.5 shows these common designs of hopper geometry. Only conical hoppers operate with axi-symmetrical flow, while the remaining types tend to present a plane-flow pattern. In a plane-flow bin, the slope to the vertical of the hopper is, on average, 8°–10° larger than the corresponding value for axi-symmetrical bins with conical hoppers. Therefore, they offer larger storage capacity for the same head room than the axi-symmetrical bin, but this advantage may be somewhat offset by the long slotted opening needed, which can cause feeding problems.

The first-in, first-out sequence of flow makes mass-flow bins useful for storing solids that degrade with time. Mass-flow bins are generally recommended for cohesive materials, for materials that degrade with time, for fine powders and for particulate systems that need to be prevented from segregation. As mentioned in Chapter 3, normally food powders are highly cohesive (Table 3.1) and, therefore, the use of mass-flow bins represents a preferred alternative for their storage.

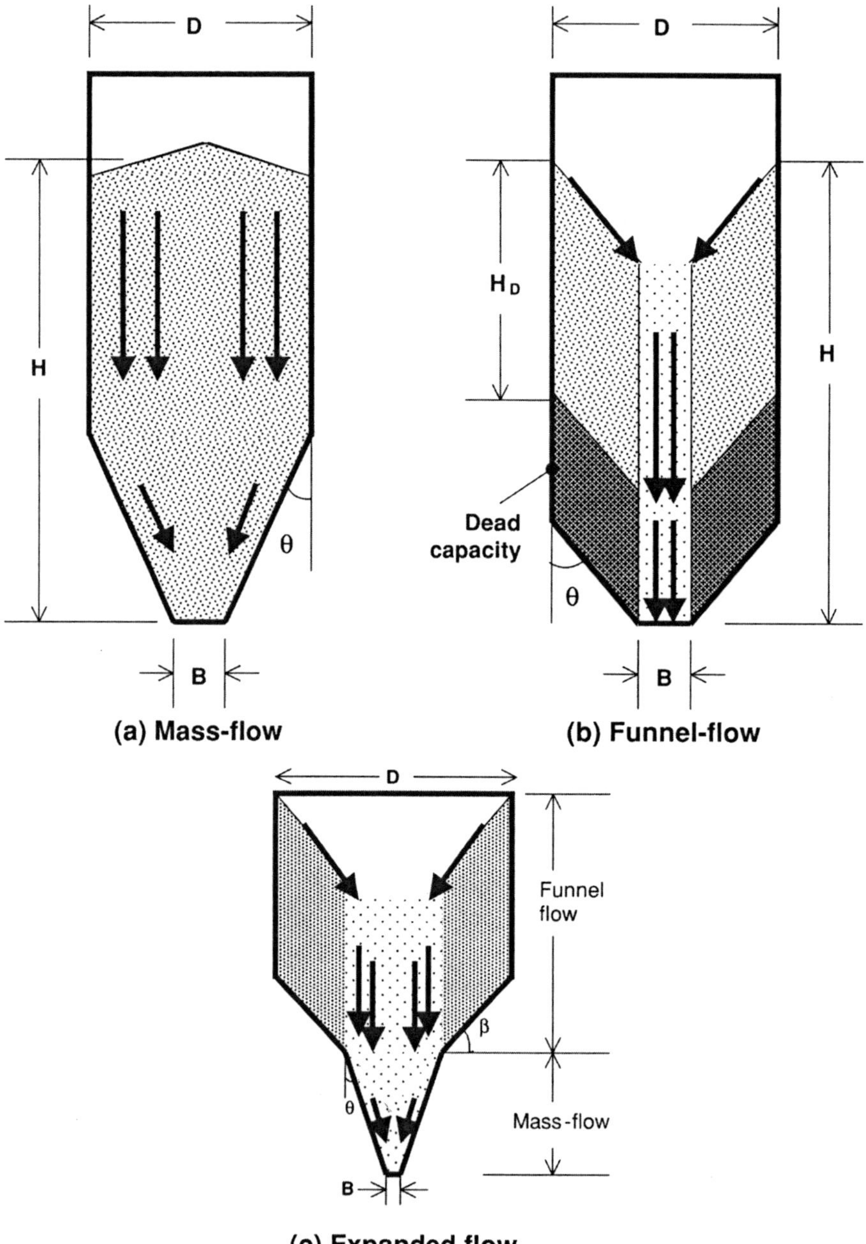

Figure 4.4. Types of flow patterns in hoppers.

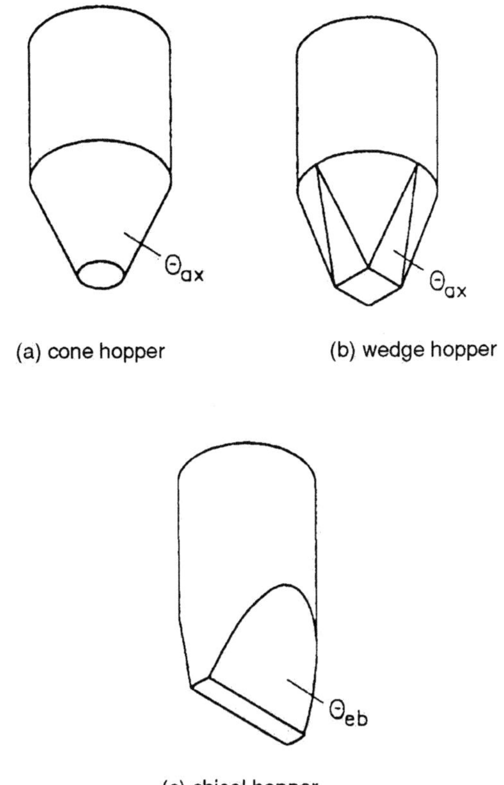

Figure 4.5. Common hopper geometries.

4.3.2. Funnel-Flow

Funnel-flow, suitable for coarse free-flowing materials, occurs when the hopper is not sufficiently steep or smooth enough to force the bulk solid to slide along the walls. It is also observed when the outlet of the bin is not fully effective, due to poor feeder or gate design. In a funnel-flow bin the stored material flows toward the outlet through a vertical channel forming within stagnant solids. The diameter of the channel approximates the largest dimension of the effective outlet. Flow out of this type of bin is generally erratic and gives rise to segregation problems. However, flow will continue until the level of the bulk solids in the bin drops an amount equal to the draw-down. At this level, the bulk strength of the contained material is sufficient to sustain a stable rathole, as illustrated in Fig. 4.4(b). Once the level defined by H_D in Fig. 4.4(b) is reached, there is no further flow and the material below this level represents dead storage. For complete discharge the bin opening needs to be at least equal to the critical rathole dimension, determined at the bottom of the bin corresponding to the bulk strength at this level. For many cohesive bulk solids, and for the normal consolidation heads occurring in practice, ratholes measuring several meters high are often observed. This makes control of the product discharge rate quite difficult and funnel-flow somewhat impractical.

The bin's major benefits are its low fabrication costs, because it is typically constructed of carbon steel rather than stainless steel, and its low headroom requirements. Funnel-flow has the advantage of providing wear protection of the bin walls as the material flows against stationary material. However, it is a "first-in last-out" flow pattern that is unsatisfactory for fine bulk solids of low permeability. Such materials can aerate during discharge through the flow channel, and this can give rise to flooding problems or uncontrolled discharge. Also, funnel-flow bins are more prone to cause arching of cohesive solids than mass-flow bins, so they usually require larger outlets for dependable flow. Furthermore, this type of bin may cause segregation of solids and is unsuitable for solids that degrade with time in the stagnant regions. Cleanout of a funnel-flow bin is often uncertain because solids in the stagnant region may pack and cake. For all these reasons, funnel-flow bins are only suitable for coarse, free-flowing or slightly cohesive, non-degrading solids when segregation is unimportant. Pertaining to food systems, funnel-flow bins may be used for grains, pulses, oilseeds, and so on, mainly for the application of directly feeding such materials to processing, such as in cereals extrusion or cereal milling.

4.3.3. Expanded Flow

This type of hopper combines characteristics of mass-flow and funnel-flow. The higher part of the hopper operates in funnel-flow, while the lower part operates in mass-flow. The mass-flow outlet usually requires a smaller feeder than would be the case for funnel-flow. The mass-flow hopper should expand the flow channel to a diagonal or diameter equal to or greater than the critical rathole diameter, thus eliminating the likelihood of ratholing. Funnel-flow bins provide the wall protection of funnel-flow, along with the reliable discharge of mass-flow. Expanded flow is ideal where large tonnages of bulk solids need to be stored, and is particularly suitable for storing large quantities of bulk solids while maintaining acceptable head heights. The concept of expanded flow may be used to advantage in the case of bins or bunkers with multiple outlets.

Expanded flow bins are recommended for the storage of large quantities of non-degrading solids. This design is also useful as a modification of existing funnel-flow bins to correct erratic flow caused by arching, ratholing or flushing.

4.3.4. Symmetrical and Non-symmetrical Flow

As previously mentioned, the above discussed flow patterns are axi-symmetrical in relation to the bin or silo walls since the outlet is located in the center line of the bin. This symmetry has many advantages in terms of flow and structural design. However, due to reasons of layout, capital cost or the designer's erroneous practices, many non-symmetrical bins and hoppers, or sometimes symmetrical bins with offset outlets, are built. This non-symmetrical configuration may cause eccentric withdrawal from multiple outlets or side outlets, and eccentric loading spouts. The resulting flow patterns can develop severe problems in the storage facilities. Quite often, eccentric withdrawal patterns can also be developed in symmetrical bins when the feeder is not properly designed or selected, or when the cutoff gate is left partially closed. The off-center vertical flow patterns caused by feeders or gates will have the same structural effects on the bin as a non-symmetrical bin.

Figure 4.6 shows various types of off-center flow channels. This type of funnel-flow enhances the problems of arching, creates a difficult rathole problem, allows discharge of segregated particle distribution, and causes non-uniform loading on the bin walls.

4.4. WALL STRESSES IN AXI-SYMMETRICAL BINS

4.4.1. Distribution of Bin Wall Loads

The prediction of wall loads in bins is an important piece of information for their design. It is necessary to estimate the pressures at the wall, which are generated when the bin is operated, in order

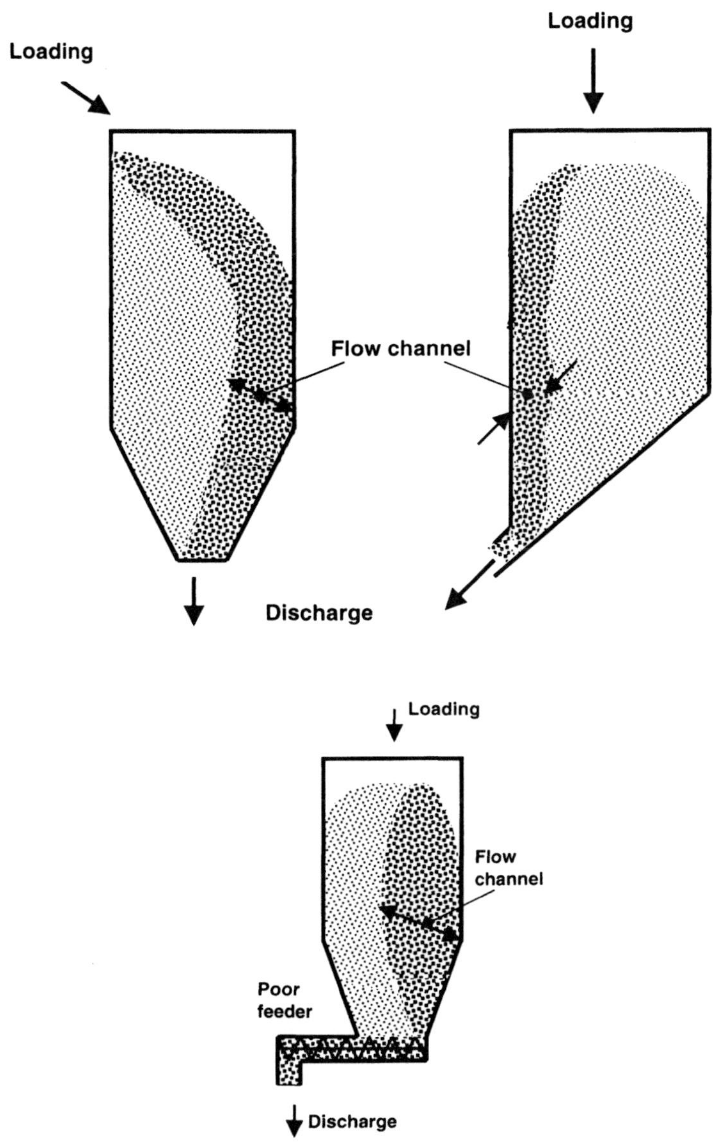

Figure 4.6. Different eccentric withdrawal patterns.

to design the bin structure efficiently and economically. Approaches to the study of bin wall loads are varied and involve analytical and numerical techniques, such as finite element analysis. Despite these varied approaches, it is clear that the loads are directly related to the flow pattern developed in the bin. The flow pattern in mass-flow bins is reasonably easy to predict, but in funnel-flow bins such prediction becomes quite a difficult task. For this reason, unless there are compelling causes to do otherwise, bin shapes should be kept simple and symmetric.

Research relating to wall stresses dates back to the 1800s when Janssen (1895) published his now famous theory. More recent investigations include those reported by Walker (1966), Walker and Blanchard (1967), Jenike and Johnson (1968, 1969), Walters (1973), Clague (1973), Arnold et al. (1982), Roberts (1988) and Thompson et al. (1997). Examination of these papers shows that the solution of the problem of stress distributions in bins is extremely complex. However, most researchers agree that the loads acting on a bin wall are different during the initial stage of filling and during the stage of flowing in discharge.

When bulk solids are charged into an empty bin, with the gate closed or the feeder at rest, the bulk solids settle as the solids head rises. In this settlement, the solids contract vertically in the cylindrical section and partially vertically in the hopper section. The major principal stress tends to align with the direction of contraction of the solids, forming what is termed as an active or peaked stress field. It is assumed that the solids are charged into the bin without significant impact to cause packing, and that the powders are charged at a sufficiently low rate so that they de-aerate. It is also assumed that the bin and feeder have been designed correctly for the solids to flow without obstruction. When the gate is fully opened, or the feeder operates so the solids start flowing out of the outlet, there is vertical expansion of the solids within the forming flow channel and the flowing mass of solids contract laterally. The major principal stresses within the flow channel tend to align with the lateral contractions, and the stress field is said to be passive or arched.

The region of switch from active to passive stress field originates at the outlet of the bin when the gate is opened or the feeder started, and rapidly moves upward into the bin as the solids are withdrawn. At the switch level a fairly large overpressure may be present, and it is assumed to travel upward with the switch at least to the level at which the channel intersects the vertical section of the bin. For a typical bin consisting of a hopper plus a cylindrical section above it, five stress fields have been recognized during the fill and discharge sequences: (a) in the cylindrical section during initial filling, where the state of stress is peaked or active; (b) in the cylindrical section during emptying, where the state of stress is either peaked or changes to arched, depending on whether the switch level is assumed to be caught at the transition; (c) in the converging hopper section during filling, where the state of stress is assumed to be peaked; (d) in the converging hopper section during emptying, where the state of stress is assumed to be arched; (e) the switch field, the region in the bin where the peaked stress field established during initial filling is transformed into the arched stress field. This switch starts at the outlet of the hopper, if newly filled from completely empty, and then travels up very quickly as emptying continues, generally to become caught in the transition.

Most of the researchers mentioned agree upon a wall pressure or stress distribution as shown in Fig. 4.7, which illustrates the approximate distribution of stresses on a single particle within the solids bulk as it flows along the wall of a mass-flow bin. At dynamic discharge conditions, the major consolidating stress, σ_1, increases exponentially with depth, abruptly increases at the transition, and then decreases toward zero at the vertex (area of radial stress).

4.4.2. Calculation of Loads in Bins

The above-mentioned Janssen theory (Janssen, 1895) includes possibly the oldest reported attempt to calculate pressures in silos. Janssen derived an equation for the calculation of vertical and horizontal pressures and wall shear stresses. He assumed a vertical force balance at a slice element spanning the full cross section of a silo being filled with bulk solids (see Fig. 4.8), and determined the wall friction coefficient with a shear tester as well as the horizontal pressure ratio from pressure measurements in a model bin. He also assumed a constant vertical pressure across the cross section of the slice element and restricted his evaluation to vertical silo walls. The Janssen equation for the vertical pressure p_v on dependence of the depth z below the bulk solids top level reads as follows for a cylindrical silo:

$$p_v = \frac{g\rho_b D}{4\mu' K'}\left[1 - \exp\left(\frac{4\mu' K' z}{D}\right)\right] \tag{4.3}$$

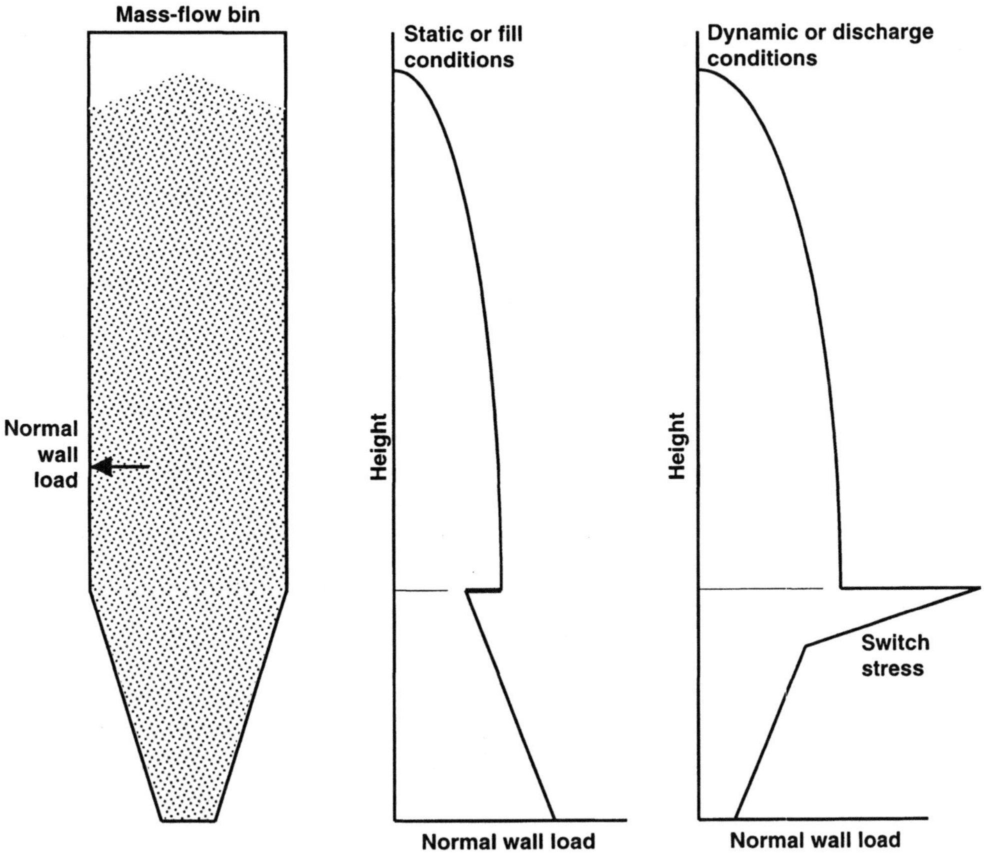

Figure 4.7. Wall load distribution in silo.

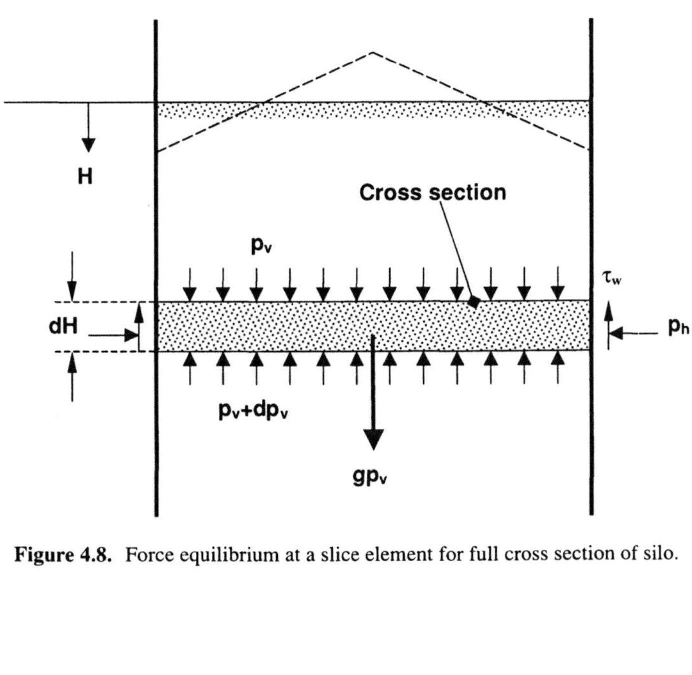

Figure 4.8. Force equilibrium at a slice element for full cross section of silo.

where g is the acceleration due to gravity, ρ_b is the bulk density of solids, D is the silo diameter, μ' is the sliding friction coefficient along the wall, and K' is the ratio of the horizontal to the vertical pressure, which can be expressed as:

$$K' = \frac{1 - \sin \delta}{1 + \sin \delta} \tag{4.4}$$

where δ is the angle of internal friction of solids.

The advantage of Janssen's equation is the simplicity of an analytical equation and its general in good agreement with pressure measurements in silos for the state of filling. The disadvantages are its non-validity for the hopper section, its assumption of a constant vertical stress across the cross section, and its assumption of plastic equilibrium throughout the stress field in the silo. The increase in horizontal pressure occurring when discharge initiates cannot be explained by the Janssen equation. This disadvantage can only be overcome by using finite element methods, more sophisticated yield criteria and a very high degree of computational effort (Häußler and Eibl, 1984).

Simplicity and analytical solution made the Janssen equation the basis for the first standard for the calculation of loads in silos more than 35 years ago. This equation is still the most widely used analytical solution for the calculation of pressures in silos. However, the structural design of silos requires the incorporation of experience, measurement results in model bins and full-scale silos, as well as accepted safety margins for uncertainties. Major factors contributing to the loads in silos are flow profile, flow behavior, the interaction between wall material and bulk solids, and the performance of feeders and discharge aids (Jenike, 1964). The loads in silos are influenced by many factors. Some of these are related to the bin structure, its material of construction and size. Many other factors, however, depend on the bulk solids flow properties, design of outlet size, type of feeder, discharge aids, and operating conditions. These factors are especially important for non-free flowing, cohesive bulk solids.

4.5. NATURAL DISCHARGE FROM BINS

The amount of solids discharged through an opening at the bottom of a bin, and the failure to restart the flow after intermission, depend on the bin design, shape, and the location of the opening, apart from the flowing properties of the solid. The flowing properties of granular material include grain nature, size, moisture content, temperature, adhesion, cohesion and, above all, time of consolidation at rest. There are very few solids that are free flowing and that will start flowing again after an extended period at rest. Examples include inert materials such as graded gravel and dry sand. In contrast, as has been stated, food powders are mostly cohesive, and their flow is very difficult even without pre-consolidation time. Pressure distribution within a bin affects its design for strength, but does not enter into calculation of the solids flow from the hopper. It has been mentioned that the volume of bulk solids discharged is independent of the head above the orifice, due to the arching effect. Therefore, the design of bins or silos in terms of their ability to initiate flow without any aid is based on solid mechanics theories considering only the hopper of the container.

The mass-flow and funnel-flow limits in silos are well known and have been used extensively in proper design. The limits for conical hoppers and plane hoppers depend on the hopper half-angle θ, the effective angle of internal friction δ and the wall friction angle ϕ. Once the wall friction angle and effective angle of internal friction have been determined by experimental means, the hopper half angle θ may be determined. In function form it can be expressed as:

$$\theta = f(\phi, \delta) \tag{4.5}$$

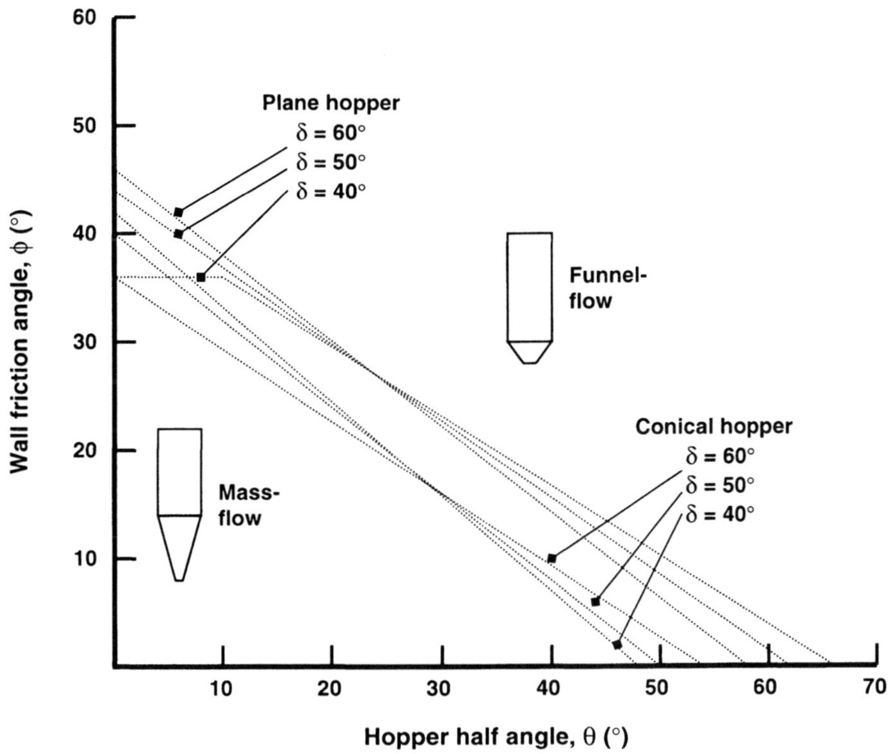

Figure 4.9. Mass-flow limit for plane-flow and conical hoppers.

The bounds for conical and plane-flow hoppers are plotted for three values of δ in Fig. 4.9. For the case of conical hoppers, it is recommended that the half-angle θ be chosen to be 3° less than the limiting value. For plane hoppers the bounds between mass-flow and funnel-flow are much less critical than for conical hoppers. In plane hoppers much larger hopper half-angles are possible, which means that the discharging bulk solids will undergo a significant change in direction as it moves from the cylindrical part to the hopper bottom.

4.5.1. Hopper Opening for Coarse Bulk Solids

Coarse bulk solids are those in which the particle size range is such that the air permeability is sufficiently high to allow air to percolate easily through the stored solids. According to the diagram shown in Fig. 4.10, the accelerated flow of bulk solids in the region of the outlet of the hopper can be considered when the air pressure gradient Δp_r equals zero. Analyzing the forces, it may be shown that:

$$\overline{\sigma}_1 = \frac{\rho_b g B}{H(\theta)} \left[1 - \frac{a}{g} \right] \tag{4.6}$$

where $\overline{\sigma}_1$ is the stress acting in arch at angle 45°, ρ_b is the bulk density of solids, g is the acceleration due to gravity, B is the hopper opening, a is the acceleration of discharging bulk solids and $H(\theta)$ accounts for variation in arch thickness, hopper half-angle and hopper type, whether conical or plane. $H(\theta)$ has an approximate value of 2.2 for plane-flow hoppers and 2.4 for conical hoppers.

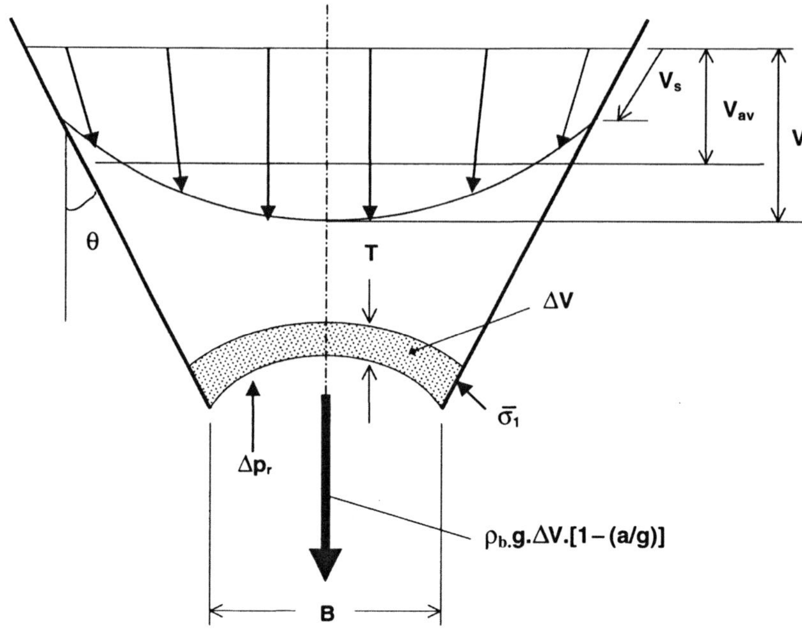

Figure 4.10. Accelerated flow of bulk solid in the outlet region of a hopper.

The minimum hopper opening to prevent a cohesive arch from forming occurs when static equilibrium prevails, i.e., when the acceleration of discharging bulk solids approaches zero, thus substituting $a = 0$ into Eq. (4.6) and transposing:

$$B_{\text{min}} = \frac{\overline{\sigma}_1 H(\theta)}{\rho_b g} \tag{4.7}$$

The hopper half-angle is chosen from the mass-flow limits as shown in Fig. 4.9, while the condition for $\overline{\sigma}_1 = \sigma_c$ is obtained from the intersection point of the flow factor ff line and failure function FF, as illustrated in Fig. 4.11. The stress σ_c acts at the abutment of any arch that tends to form, and is proportional to the span B. It will vary increasingly from the hopper slot upwards, reaching a steady value at the vertical walls. The flow/no-flow criterion states that an arch will form in a hopper when the unconfined yield strength f_c exceeds the stress σ_c, tending to break it.

As discussed in Chapter 3, the instantaneous failure function ff (relation of the major principal stress σ_1 and the unconfined yield stress f_c) is a bulk solid parameter and represents bulk strength, while the flow factor is a flow channel parameter. The flow factor and function $H(\theta)$ are given as design curves (Jenike, 1964; Arnold et al., 1982; Roberts, 1988). The limiting hopper slope δ_c for conical flow, or δ_p for plane flow, and the flow factor FF for mass-flow are determined by entering the measured angle of wall friction ϕ, as can be seen in the chart (Fig. 4.9), and moving right to the intersect on the boundary on the proper chart. At the intersection, the flow factor is read, and then required hopper slope (or half angle) is found by moving straight down in the diagram. The flow factor FF is a constant, and plots as a linear function through zero. When it is super-imposed on the flow function, the critical stress value σ_1 for determining the minimum hopper discharge opening (Eq. (4.7)) is found from the point of intersection.

If the FF and ff do not intersect, and FF lies completely below ff, the minimum hopper opening is very small and cannot be determined for this flow analysis. Opening size will be limited only by

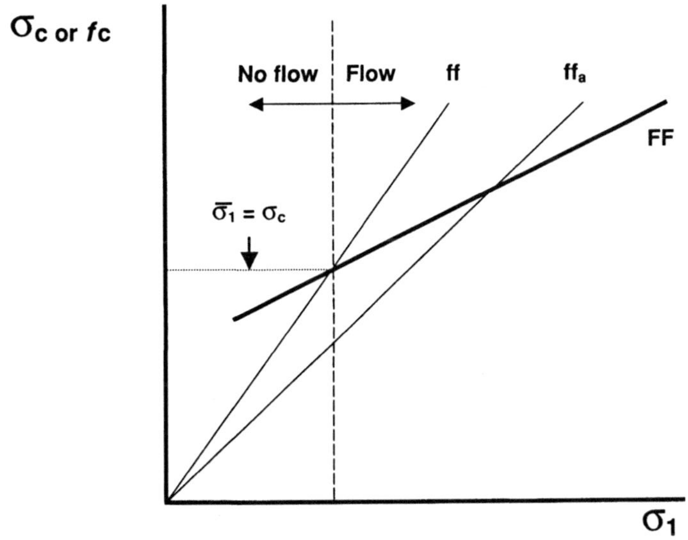

Figure 4.11. Graphical determination of arch stress at the hopper outlet.

the possibility of mechanical interlocking of particles, or by the required solids discharge rate. If the instantaneous FF lies below ff, and the time FF lies above, then it is usually possible to use vibration or other means to start flow after time consolidation, and thus return the solids to the instantaneous flow function. If there is no intersection, and the FF lies above ff, unassisted gravity flow is not feasible, and mechanical flow aids must be considered.

4.5.1.1. Mass-Flow Rate Calculation

Following the work of Johanson (1965), it may be shown that the acceleration in Eq. (4.7) may be expressed as:

$$a = g\left[1 - \frac{\text{FF}}{\text{FF}_\text{a}}\right] \tag{4.8}$$

where FF is the critical flow factor based on the minimum arching dimension and FF_a is the actual flow factor based on the actual opening dimension. This actual flow factor may be represented as the relationship between stresses, according to Fig. 4.11, by:

$$\text{FF}_\text{a} = \frac{\sigma_1}{\sigma_\text{c}} \tag{4.9}$$

where σ_1 is the major consolidation pressure at outlet corresponding to dimension B.

The acceleration has two components, i.e.:

$$a = a_\text{c} + a_\text{v} \tag{4.10}$$

where a_c is the convergence component due to flow channel and a_v is the component due to velocity increase as flow is initiated. This last component may be represented as a function by:

$$a_\text{v} = g\left[1 - \frac{FF}{FF_\text{a}}\right] - \frac{2V^2(m+1)\tan\theta}{B} \tag{4.11}$$

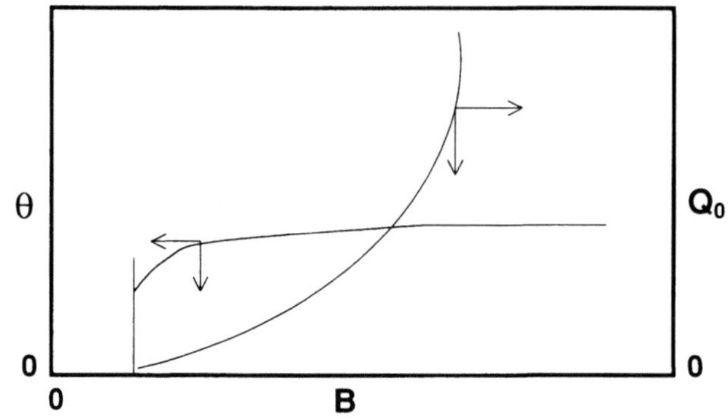

Figure 4.12. Hopper slope and discharge flow rate as a function of outlet opening.

where V is the discharge velocity and m is a constant which takes the value of zero for plane-flow hoppers and unity for axi-symmetrical or conical hoppers.

Equation (4.11) shows that as the discharge velocity V increases as flow is initiated, a_v tends to zero. Thus, an average terminal discharge velocity V_a is reached when $a_v = 0$:

$$V_a = \sqrt{\frac{Bg}{2(m+1)\tan\theta}\left[1 - \frac{FF}{FF_a}\right]} \tag{4.12}$$

The flow rate Q_0 may be calculated by the following equation:

$$Q_0 = \rho_b B^{(1+m)} L^{(1-m)} \left(\frac{\pi}{4}\right)^m V_a \tag{4.13}$$

where L is the length of slot for a rectangular opening in plane-flow hoppers, in which case B would represent the width of slot. It can be easily seen that for conical hoppers B would represent the diameter of the circular opening and, thus, Eq. (4.13) above is applicable for both cases of plane-flow hoppers, either with square or slot opening, and axi-symmetrical or conical hoppers.

For design purposes, B and Q_0 can be plotted as a function of θ, as presented in Fig. 4.12. In such a way, several options of hopper geometry can be explored. In many cases, the flow rate determined above the unimpeded discharge will be well in excess of plant requirements. For this reason, feeders may be used to control the discharge rate to any required value.

4.5.2. Hopper Opening for Fine Bulk Solids

For the case of fine powders, discharge from a hopper will be difficult due to the low permeability of the powder to the airflow. Flow rates very much lower than those calculated by the method described above can be common in practice. Analysis of the flow process is much more complex and involves a two-phase flow theory. The flow of fine powders has been studied over the years by several researchers such as McLean (1979), Arnold et al. (1982) and Arnold and Gu (1990). Larger hopper openings are required for the equivalent coarse bulk solids and, sometimes, air permeation is needed to bias the pore pressure in the hopper in order to assist the discharge. Fine powders are prone to flooding and uncontrollable discharge when allowed to aerate. For this reason, extreme care should be taken in designing and installing any air permeation system. Care must also be exercised in ensuring that the interface between the hopper and feeder is correctly designed to prevent problems due to flooding.

4.5.3. Velocity Distribution in the Hopper

Velocity distribution in a converging channel has been investigated by Johanson and Royal (1982), who were interested in computing the sliding velocity at a hopper wall in relation to the wear of the hopper linings. This concept has also been approached by Roberts et al. (1990), who examined wear in chutes, apart from hopper wall wear. The average velocity along the converging hopper (Fig. 4.7) is calculated from the flow rate as follows:

$$V_{av} = \frac{Q_m}{\rho_b A_z} \tag{4.14}$$

where Q_m is the mass-flow rate at section and A_z is the cross-sectional area.

The velocity profile in the hopper is as represented in Fig. 4.10. The velocity of sliding at wall is:

$$V_s = K_v V_{av} \tag{4.15}$$

The parameter K_v for conical and plane-flow hoppers may be obtained from Johanson and Royal (1982), but for a series of practical applications can be taken as 0.54. It may be noted that the velocity ratio V_0/V_s (see Fig. 4.10) and K_v, have a direct bearing on the blending characteristics of hoppers (Arnold et al., 1982).

4.5.4. Factors Influencing Bin Geometry for Mass-Flow

With few exceptions, in food processing symmetric bin shapes and mass-flow patterns will be the preferred alternatives for bulk solids storage facilities. Apart from grains, pulses and oilseeds, which are not powders, but particulate systems, most food materials are highly cohesive and hygroscopic and, thus, undisturbed storage time and variation in moisture content can significantly influence their unconfined yield strength.

Food powders are complex due to their composition (Schubert, 1987), their large distribution in particle size and the presence of solid–liquid–gas phases in the particle. Moisture has a great influence on flowability, and its presence and proportion within the food powder depend on the relative humidity of the surrounding atmosphere. The remaining factors that affect most powders flowability, i.e., failure properties, particle and bulk density, and so on, also affect food powder flowability and, therefore, have a direct influence on the design of bin geometry for mass-flow. Teunou et al. (1999) characterized representative food powders for their flowability and design of hoppers for mass-flow. Their main findings are summarized in Table 4.1, where it is surprising to find that flour, which has the poorest flowability, presented the best design dimensions. This may be due to the fact that hopper design calculations depend strongly on the values of the angles of effective internal friction and wall friction, and both were lower for the flour when compared with

Table 4.1. Physical properties and hopper dimensions of food powders (adapted from Teunou et al., 1999)*.

Powder	X (%)	ff	δ(g°)	ϕ (°)	θ(g°)	B (mm)
Flour	12.6	2.71	32	12.6	37	110
Skim-milk	4.6	11.04	50	13.0	32	270
Tea	6.6	4.22	43	15.0	31	130
Whey-permeate	3.8	5.85	49	15.0	30	180

*X is the water content in wet basis while the remaining variables are as defined in the text.

the other powders. Thus, in hopper design it is important to characterize as completely as possible any powder and to consider that failure properties are critical in the whole procedure. Moisture content, as important as it is, is a factor that is normally kept low in food powders in order to prevent spoilage. In such a way, this moisture control will aid the design in terms of keeping a moderate cohesiveness of the powder.

Furthermore, selected materials are also critical for the design. In mass-flow design, a low solid-wall friction ϕ is preferred for the hopper section. This permits larger hopper slope angles and thereby reduces the overall height of the hopper. Materials must be selected so that they resist abrasion from particles. Material abrasiveness is discussed in Chapter 3. On the other hand, walls in silos should not be smooth. As the solids-wall friction in the vertical part of the silo is reduced, more of the consolidating stress from the stored material is transmitted directly to the material in the converging hopper below. This could cause arching across the silo at the transition between cylinder and hopper. The vertical wall flow factor can be superimposed on the flow function as before to determine if the material will gain sufficient strength to arch across the bin at the transition. If the time flow function ff_t (obtained at predefined consolidation or storage times) continues to rise steeply at high pressures, even though it lies below the flow function (ff) (and indicates a small or zero arching diameter), arching may still be possible at the cylinder–cone transition. Then, a rougher surface should be specified.

4.5.5. Effect of the Gas Phase

During loading and subsequent settling, the gas entrapped within the solid can have a significant influence on wall pressures and flow behavior, a fact that is recognized in silo design codes. Entrapped gas in fine powders can be retained for an appreciable time. As mass-flow bins have come into widespread use, it has been found that the discharge of powders from these bins can become flow rate limited. Interstitial gas pressures within a powder bed change during flow, and this influences the rate of discharge from a mass-flow hopper.

The maximum flow rate of a fine powder through the outlet of a mass-flow bin is low compared to that of a coarse, granular solid. For fine materials, the expansion and contraction of voids during flow can create an upward air pressure gradient at the outlet of a mass-flow bin. During discharge, this upward gradient acts against gravity, reducing the discharge rate. Such gradients do not usually form with coarser particle materials. Because coarse materials are more permeable than fine materials, air is allowed to flow freely into and out of the voids as they expand and contract.

This phenomenon can be analyzed by considering how gas flows through a bed of powder when a pressure differential occurs across the bed. When the gas velocity is low, flow through the bed is laminar. Darcy's law can be used to relate gas velocities to gas-pressure gradients within or across the bed and can be expressed as follows:

$$u = -K\left(\frac{\mathrm{d}p/\mathrm{d}x}{\rho}\right) \tag{4.16}$$

where K is the permeability factor of the bulk solid, u is the superficial relative gas velocity through the bed of solids, ρ is the bulk density of the solid in the bed, and $(\mathrm{d}p/\mathrm{d}x)$ is the gas-pressure gradient acting at the point in the bed of solids where the velocity is being calculated.

The permeability factor K has units of velocity and is inversely proportional to the viscosity of the gas. A permeability test is run by passing air (or another suitable gas) through a representative column of solids. The pressure across the bed is regulated, and the rate at which the gas flows is measured.

This approach allows the permeability of the bulk solid to be determined as a function of its bulk density. Because mass-flow bins have stable flow patterns that mimic the shape of the bin, permeability values can be used to calculate critical-steady discharge rates from mass-flow hoppers. Permeability values can also be used to calculate the time required for fine powders to settle in bins and silos and to design solids-processing vessels to purge, heat, dry, or condition bulk solids.

The permeability of powders is a defining parameter that influences the rate of discharge. As an element of powder moves through a mass-flow silo to the outlet, the consolidating pressure on the element changes as described earlier. Initially, as the element is compressed, the voidage is reduced and interstitial air is squeezed out through the top surface. As it moves through the hopper, the consolidating pressures on the element decrease, the element expands, and the voidage increases. If the powder has a low permeability to air flow, the interstitial pressure in the lower region of the hopper can decrease to below atmospheric pressure. The resulting pressure gradient will cause an influx of air from the hopper outlet that will retard the solids flow.

In general, permeability is affected by particle size and shape, moisture content, and temperature. It decreases as particle size decreases. The better the fit between individual particles, the lower the bulk solid's permeability. As moisture content increases, materials tend to stick together, increasing permeability and therefore increasing discharge and settling rates. Since the permeability factor K is inversely proportional to the viscosity of the air or gases in the void spaces, heating causes the gas to become more viscous, which makes the bulk solid less permeable.

4.6. ASSISTED DISCHARGE

The above-mentioned procedures to determine optimum hopper outlet diameter might be taken only as an estimate because in real applications, problems due to flow blockages such as arching and ratholing may occur. In order to ensure flow from bins, even after the hopper geometry has been determined by following careful calculations such as those previously discussed, flow promotion may be necessary. Classification of flow promotion may be termed as passive or active involving energy. A third class of flow promotion may consider the use of feeders, which are useful not only to promote flow but also to control the flow rate.

4.6.1. Passive Devices

These types of devices for flow promotion do not require energy and are normally known as inserts. An insert is usually placed within the hopper section of a bin, with the purpose of expanding the size of the active flow channel in a funnel-flow bin to approach mass-flow. Another aim of an insert is to relieve pressure at the outlet region. Inverted cones and pyramids have been used for years in this regard, but with limited success. The typical position of a cone insert is illustrated in Fig. 4.13. Apart from the typical inverted cone or pyramid insert, the cone-in-cone insert, or Binsert® (Johanson, 1982) has also been tested as a flow promoter. As shown in Fig. 4.14, the Binsert® is actually a sort of funnel inside an expanded flow hopper, or a hopper within another hopper. In this insert, the material flows through the inner hopper, as well as through the annulus between the inner and the outer hopper. It has proved very effective and has the added advantage of promoting not only flow but also mixing or blending capability, provided the bin geometry and insert are correctly selected. In such a way, a completely uniform velocity with an absolute minimum degree of particle segregation can be achieved. This is particularly useful for food powders in which segregation would represent a serious problem, for example when preparing formulations for baking, instant drinks, etc.

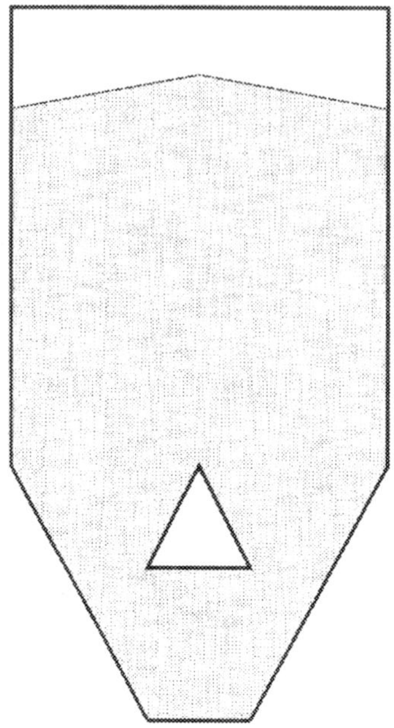

Figure 4.13. Inverted cone insert and its typical position within a hopper.

4.6.2. Active Devices

Vibrating hoppers powered by the use of electrical motors, pneumatic knockers, eccentric drives, or electromagnetic units, are some of the most important and versatile flow assisters. Figure 4.15 shows the normal configuration of a hopper equipped with any sort of vibrating device. The application of mechanical vibrators to promote gravity flow of bulk solids has been studied in detail by Roberts et al. (1986). They developed a dynamic shear test apparatus that permits shear strength to be determined in the presence of vibrations that may be applied over a range of amplitudes and frequencies. It was found that dynamic shear strength decays exponentially with increase in vibration, and that such decaying is similar to that in shear strength with voidage on the shear plane. These findings seem to suggest that vibration velocity is directly related to voidage. Reliable gravity flow in bins depends, to a very significant extent, on the magnitude of the friction angle between the flowing bulk solids and the hopper wall.

Vibrations applied to the hopper wall can reduce wall friction to a marked extent. Dynamic wall shear tests may be readily performed using the dynamic shear apparatus. The influence of mechanical vibration may induce flow by: (a) improving the mass-flow performance of an existing mass-flow bin by reducing the wall friction angle and critical arching dimension; (b) improving the funnel-flow performance of an existing funnel-flow bin by reducing the critical rathole dimension; and (c) lowering both the bulk strength and wall friction angle to convert an existing funnel-flow bin to a mass-flow bin. Effective flow promotion depends on the ability of the stored bulk mass to transmit vibration energy to the region of the flow blockage. It is usual to install the vibrator on the hopper wall to provide an immediate benefit through the resultant reduction in wall friction. If the flow blockage is in the form of an arch, vibration applied to the hopper wall at or near the outlet may cause the arch to fail and flow to occur. On the other hand, for stable ratholes, the vibration energy needs to be transmitted through the bulk mass in order to assist flow. The dynamic shear test

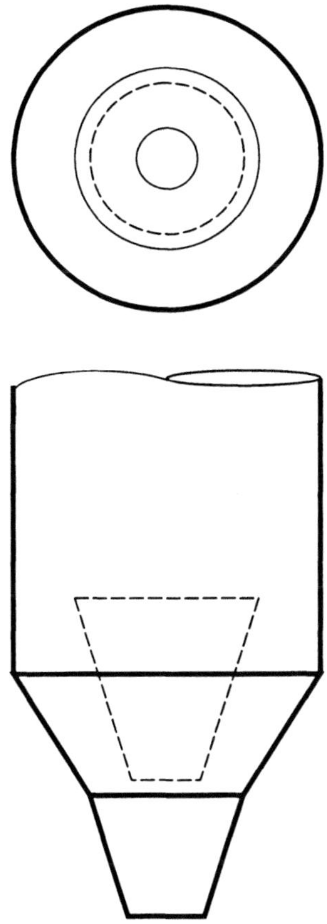

Figure 4.14. The Binsert® insert system.

apparatus, coupled with the theory of failure, provides information on the frequency and amplitude necessary to be applied at the zone of flow blockage (Roberts et al., 1986).

As shown in Fig. 4.16, air cannons or air blasters are also commonly used to promote gravity flow in bins. Terziovski and Arnold (1990) indicated that the sizing and placing of air blasters are normally carried out using techniques based on experience and guesswork. Based on their own research on the use of air blasters, they provided a procedure for the correct selection of these devices for efficient performance using the measured flow properties of the bulk solid and the bin geometry. In general, air blasters are used in a retro-fit situation to correct an inadequate design where flow blockages occur. However, in a new design where there is some uncertainty in the variations in bulk strength of a bulk solid for the full range of operating conditions, it is good practice to make provisions in the design for future installation of air blasters.

4.6.3. Use of Feeders to Control Discharge

In designing hoppers for silos, the procedure described so far consists of making calculations to determine optimum hopper slope and outlet opening in order to ensure flow. In case flow does not

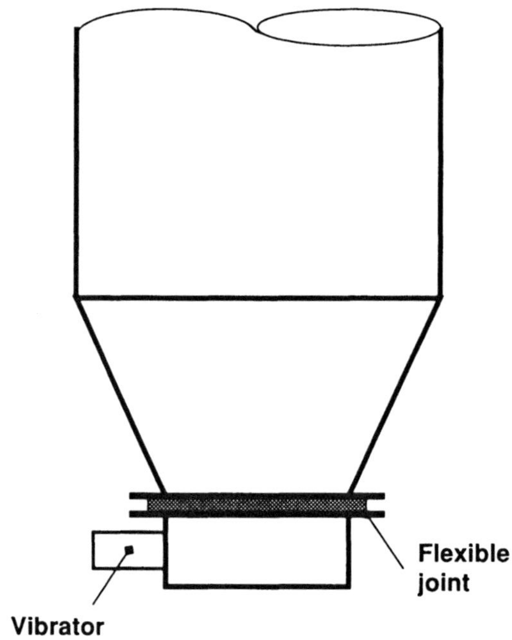

Figure 4.15. Vibrating hopper.

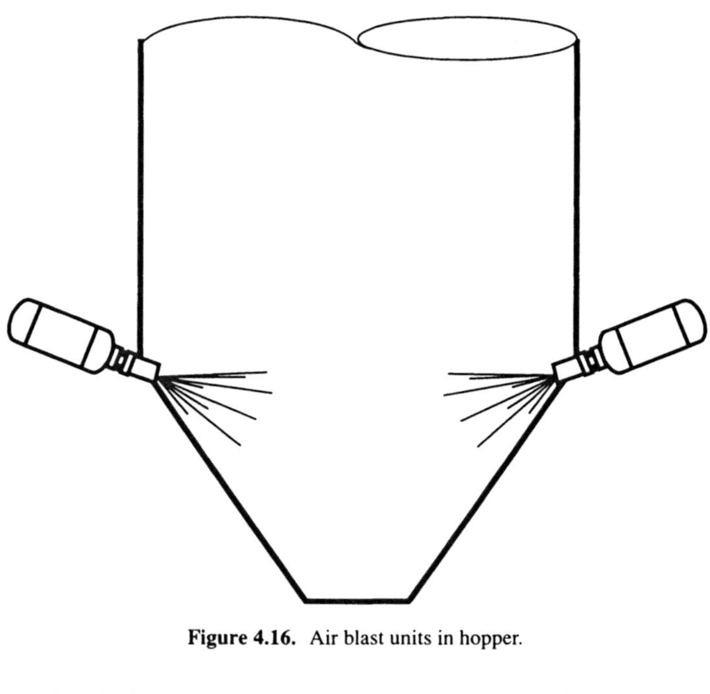

Figure 4.16. Air blast units in hopper.

occur, a flow promotion device can be selected after careful study of conditions and factors. Once flow out of a bin is guaranteed, the next step to complete the proper design of the bulk storage plant consist of controlling the flow rate to provide adequate feed to any given food powder process. In order to do so, the use of a feeder will become necessary.

Table 4.2. Guide for selection of feeders.

Bulk solid characteristics	Type of feeder
Fine, free flowing solids	Apron, vibratory, screw, star
Non-abrasive, granular materials	Apron, vibratory, screw
Difficult-to-handle (abrasive, hot, etc.) materials	Apron, vibratory
Heavy, lumpy or highly abrasive materials	Apron, vibratory

A feeder is a device used to control the flow of bulk solids from a bin. A feeder must be selected to suit a particular bulk solid and the range of feed rates required. It is particularly important to design the hopper and feeder as an integral unit in order to ensure that the flow from the hopper is fully developed with uniform draw of material from the entire hopper outlet. Feeders are usually rated by manufacturers on the basis of volume capacity. If the feeder selected on this basis has an inlet smaller than the minimum required hopper opening size, it is unacceptable. Too small an opening could result in bridging, ratholing, and erratic flow. To work in unison with the bin, the feeder must suit the material's flow properties, withdraw the material uniformly across the outlet's cross-sectional area, minimize the loads the material applies to the feeder, and accurately control the discharge rate (Marinelli, 1996).

The feeder can be volumetric or gravimetric. There are several types of volumetric feeders, but the most common are the belt or apron feeder, the screw feeder, the vibratory feeder, the star feeder, and the louvered feeder. Figure 4.17 presents schematic diagrams of the five types of feeders to be described next. Careful considerations, such as those described above, should be taken in selecting a feeder for a particular application. Table 4.2 shows a preliminary guide for choosing a volumetric feeder. A gravimetric feeder weights material to achieve a required feed-rate or batch size. Gravimetric feeders can operate in continuous or batch mode. The continuous feeder controls the weight of material fed per unit time (lb/h or kg/h). The batch feeder simply controls the weight of the material that is fed to the process. Common gravimetric feeders are weighbelt feeders, loss-in-weight systems, and gain-in-weight systems.

4.6.3.1. Volumetric Feeders

An *apron feeder* consists of a conveyor belt positioned just below a tapered opening of a hopper. Due to the geometry of the conveyor, apron feeders are particularly suitable for slot opening hoppers. Care must be taken that dead spots are not produced in the flow channel above the feeder belt. The capacity of a belt feeder can be increased by tapering the outlet in the horizontal and vertical planes. The feeder can handle friable, coarse, fibrous, elastic, sticky, and very cohesive materials; however, it is not well suited to floodable materials. In order to ensure proper flow of non-free flowing solids along the front bin wall, a sloping striker plate at the front of the hopper may be necessary. Belts have been used successfully under slot openings as long as 30 m with a constant slot width of 205 mm. Provisions should be made for field adjustment of the space between the skirt and the belt to provide uniform flow along the entire length. Since the minimum distance between the skirt and the belt should allow the largest particles to pass under, very long belt feeders are limited to the finer solids. The apron feeder is more accurate than the screw feeder, but it is subject to avalanching problems in some applications.

A screw adapted to a hopper bin is known as a *screw feeder*. Screw feeders are quite useful in producing uniform feed rates for a variety of bulk solids. An important factor in this kind of feeder is the need for a variable pitch screw to produce a uniform draw of material across the entire hopper

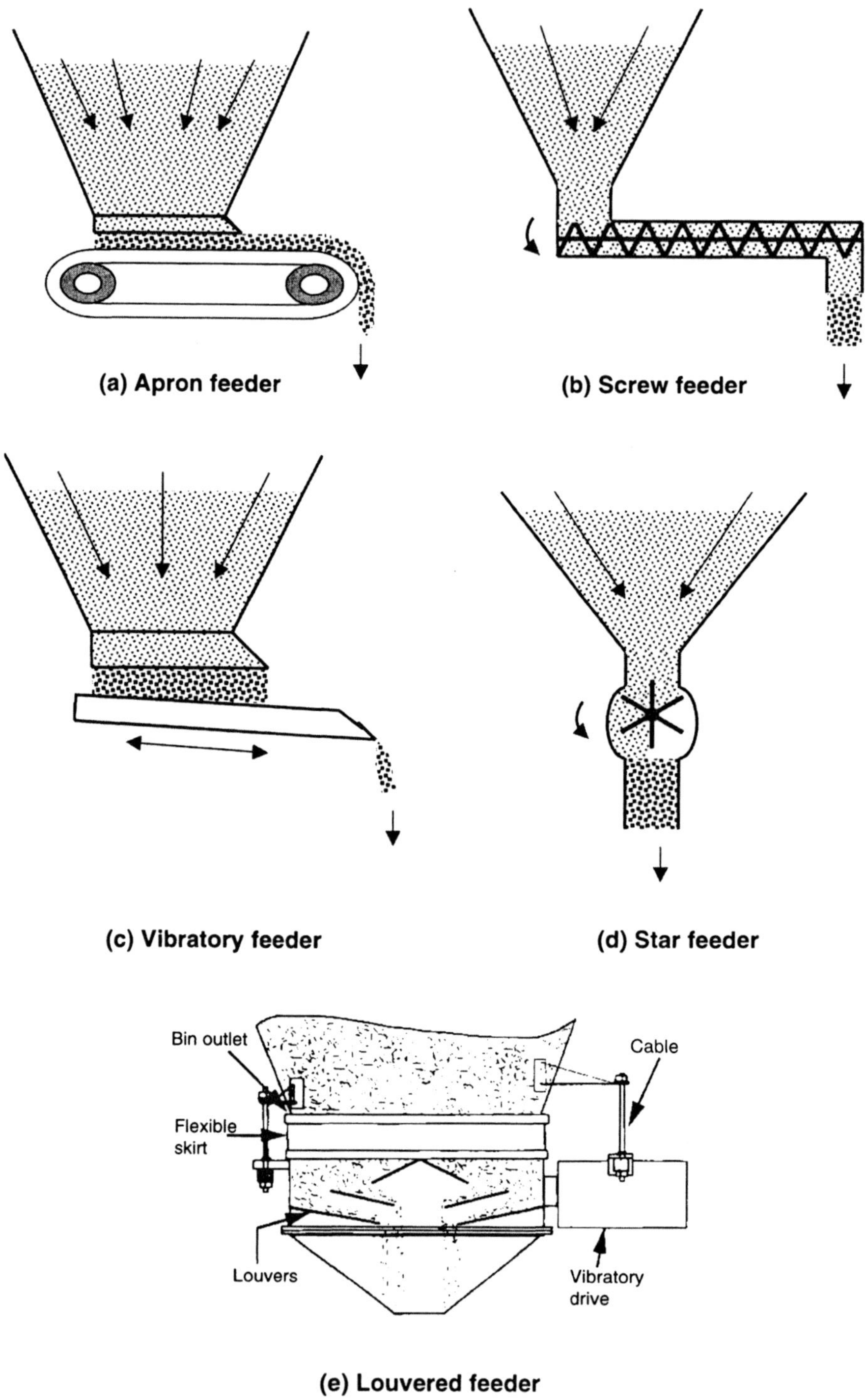

Figure 4.17. Different types of feeders to assist silo discharge.

opening. For uniform flow to occur, the screw feeder opening-to-diameter ratio should not exceed a value of 6. There are designs that adapt to round, square, or slotted outlets.

In a *vibratory feeder*, there is an inclined pan with different sorts of motion located below a hopper opening. Vibratory feeders provide uniform flow along a slot opening of limited length. The distance between the feeder pan and the hopper is increased in the feed direction. Slot length is limited by the motion of the feeder. Because in long slots the upward component of motion is not relieved by the front opening, solids tend to pack. This can cause flow problems with sticky bulk materials, as well as a large demand of power for free-flowing bulk solids. To overcome such difficulties, vibratory feeders and reciprocating-plate feeders are designed to feed across the slot. Although this type of feeder may require several drives to accommodate extreme width, the drives are small due to the short length of the feeder.

Star feeders may provide highly uniform withdrawal along a slot hopper opening. They normally comprise six paddles whose ends converse in a common point forming a radial arrangement, or a six-point star when looking at its cross section. When the star rotates, withdrawal of bulk solids is attained. A vertical section of at least one outlet width should be added above the feeder to ensure uniform discharge across the opening. The rotational speed of the paddles controls the discharge rate. Typically, the feeder dispenses to a pneumatic conveying system because the airtight seal of the feeder prevents countercurrent process air from flowing up through the material bed, thus, interrupting feeding. The rotary valve feeder can handle most materials, but typically is unsuitable for friable materials, which can be sheared by the rotating paddles, or materials with large particles, which can jam the paddles. The feeder is commonly used with a round or square bin outlet. A common problem with star feeders is that as the paddles rotate, they tend to preferentially pull material from one side of the bin outlet (Fig. 4.17(d)), which can destroy a mass-flow pattern. Furthermore, counter current air is filled back into the bin, reducing the material flow rate.

A *louvered feeder* consists of a cylindrical, square or rectangular housing enclosing a set of louvers. The feeder housing is suspended from cables and sealed to the bin outlet with a flexible skirt. A vibratory drive connected to the feeder provides the energy required for discharge. As the drive is energized, the louvers gently vibrate, overcoming the material's angle of repose and withdrawing the material uniformly across the bin outlet's entire cross-sectional area. The vibration frequency and amplitude control the feeder's output and discharge rate. Flow halts when vibration stops. The feeder provides moderate accuracy, comparable to that of the apron feeder.

4.6.3.2. Gravimetric Feeders

A *weighbelt feeder* consists of a conveyor belt supported on a weighing platform (or weighbridge). The weighing platform includes one or more load cells linked to a controller. A volumetric pre-feeder is located below the bin outlet and above the end of the conveyor belt. The discharge to the belt is located above a process or container. The pre-feeder withdraws material from the bin outlet and discharges it onto the belt. As the material travels over the weighting platform, the platform weighs the belt, conveyor drive and material, and sends the weight signal to the controller. In response to the weight signal, the controller multiplies the belt loading and speeds or slows the belt to meet the desired process specifications.

A *loss-in-weight system* (LIW) consists of an integral unit that is mounted below the bin. The integral unit includes a mass-flow weight hopper, a volumetric feeder mounted below the weight hopper, and either a platform scale or a suspended scale (Fig. 4.18(a)). A controller programmed with the desired feed rate is located outside the integral unit; the unit's feeder is linked to a refill device (a volumetric feeder such as a screw or rotary valve feeder) mounted below the material bin. After the refill device feeds material into the weight hopper to a preset level, the material flows

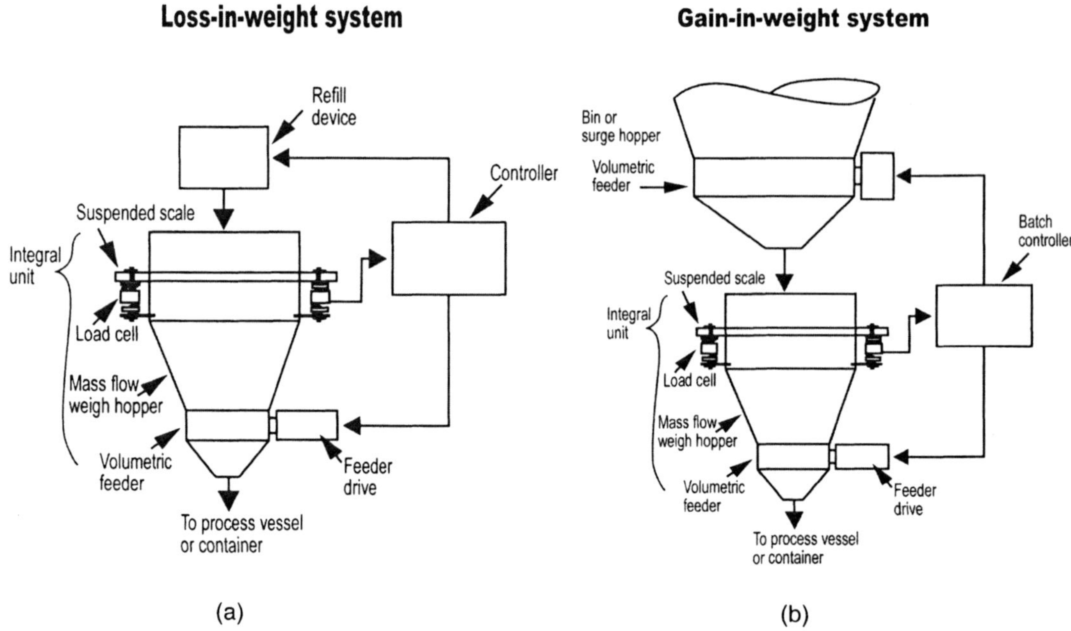

Figure 4.18. Gravimetric feeders: (a) loss-in-weight and (b) gain-in-weight systems.

from the weight hopper into the feeder, which dispenses the material at a controlled rate. The scale continuously sends readings to the controller, which calculates material weight loss in the hopper per unit time. It compares the calculated loss-in-weight with the controller and signals the feeder to either speed up or slow down to maintain an accurate feed rate. The LIW system can be used either in continuous or batch operations.

A *gain-in-weight system* (GIW) is only for batch operations. Like the LIW system, the GIW system consists of an integral unit that is mounted below the material bin and includes three devices: a mass-flow weight hopper, a volumetric feeder or slide-gate valve, and either a platform or suspended scale (Fig. 4.18(b)). A controller programmed to the desired batch weight links the scale to another volumetric feeder mounted below the material bin or surge hopper. The scale measures the weight of material entering the weight hopper (i.e., the gain-in-weight) and continuously sends weight readings to the controller. The controller compares the readings with the batch controller and signals the volumetric feeder below the material bin or surge hopper to feed quickly or slowly into the weight hopper until the weight matches the weight set in the controller. Then, the batch is discharged through the volumetric feeder or slide-gate valve into the process vessel or container. Both LIW and GIW are very accurate systems.

4.6.3.3. Loads on a Hopper Feeder

The estimation of feeder loads and drive powers requires knowledge of the stress fields generated in the hopper during the initial filling conditions, as well as during discharge. Under filling conditions, a peaked stress field is generated throughout the entire bin, as illustrated in Fig. 4.19. Once flow is initiated, an arched stress field is generated in the hopper and a much greater proportion of the bin load is supported by the walls of the hopper. Consequently, the load acting on the feeder is reduced considerably. This is also shown in Fig. 4.19. It is common for the load acting on the feeder under

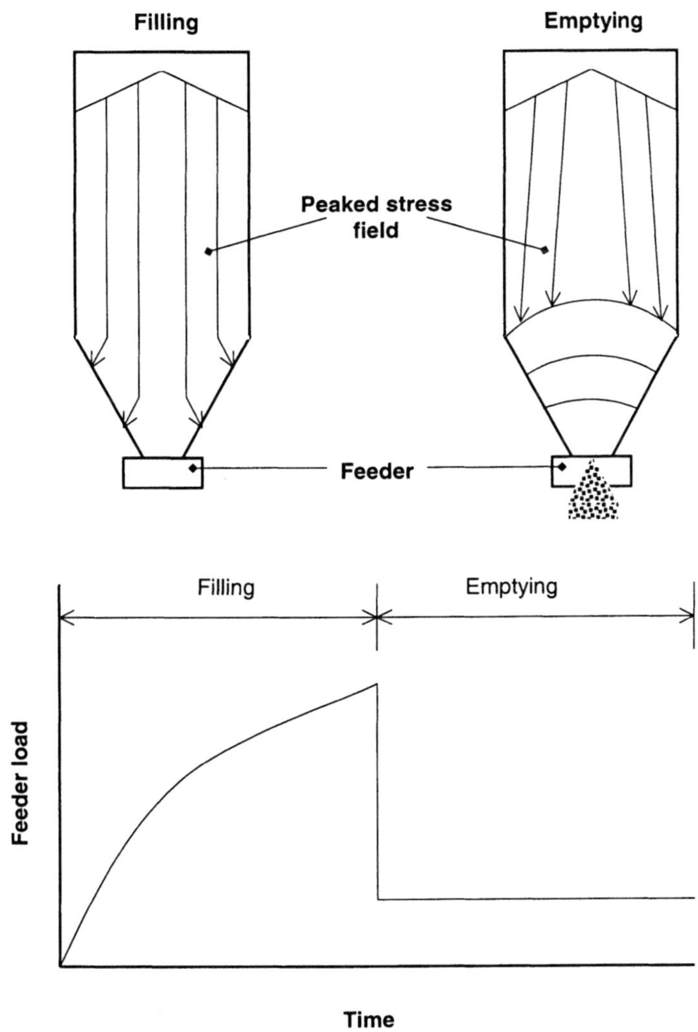

Figure 4.19. Loads on hopper feeder at filling and emptying conditions.

flow conditions to be in the order of 20% of the initial load. The arched stress field is stable and is maintained even if the flow is stopped. This means that once flow is initiated and then the feeder is stopped while the bin is still full, the arched stress field is retained and the load on the feeder remains at the reduced value. The load Q on the feeder is given by:

$$Q = q\rho_b g L^{(1-m)} B^{(2+m)} \tag{4.17}$$

where q is a non-dimensional surcharge factor, ρ_b is the bulk density of solids, g is the acceleration due to gravity, L is the length of the slotted opening, B is the width of slot or diameter of circular opening, and m is the symmetry factor whose values are 0 for plane flow hoppers and 1 for conical hoppers. Equations for q are available in the literature (Manjunath and Roberts, 1986a, 1986b).

The loads on feeders and the torque during start-up may be controlled by ensuring that an arched stress field fully or partially exists in the hopper prior to starting. This may be achieved by (a) cushioning in the hopper, i.e., leaving a quantity of material in the hopper as a buffer storage,

(b) starting the feeder under the empty hopper before initiating filling, and (c) raising the feeder up against the hopper bottom during filling and then lowering it to the operating conditions prior to starting; in this way an arched stress field may be partially established.

REFERENCES

Arnold, P.C. and Gu, Z.H. (1990). The effect of permeability on the flowrate of bulk solids from mass-flow bins. *Powder Handl. Process.* **2**: 229.

Arnold, P.C., McLean, A.G. and Roberts, A.W. (1982). *Bulk Solids: Storage, Flow and Handling*. The University of Newcastle Research Associates (TUNRA), Australia.

Bell, T.A. (1993). Measurement of Powder Flowability. In *Powders, Characterization and Quality Control*, Lawley, A. and Armour, S. (eds.). Proceedings of the 1993 International Conference & Exhibition on Powder Metallurgy & Particulate Materials, 169 pp.

Berry, R.J., Birks, A.H. and Bradley, M.S. (2000). Measurement of critical cohesive arches in silos using laser arching. In *From Powder to Bulk*. IMEChE, International Conference on Powder and Bulk Solids Handling. The Institutions of Mechanical Engineers, UK, 131 pp.

Carson, J.W., Royal, T.A. and Goodwill, D.J. (1986). Understanding and eliminating particle segregation problems. *Bulk Solids Handl.* **6**: 139.

Clague, K. (1973). *The Effects of Stresses in Bunkers*. Ph.D. Thesis. University of Nottingham, England.

Häußler, U. and Eibl, J. (1984). Numerical investigations of discharging silos. *J. Eng. Mech.* **100**: 957.

Janssen, H.A. (1895). Versuche über getreidedruck in silozellen. *Zeitschrift des Vereins Deutscher Ingenieure* **39**: 1045.

Jenike, A.W. (1964). *Storage and Flow of Solids*. Bulletin No. 123, Utah Engineering Experiment Station, Salt Lake City, UT.

Jenike, A.W. and Johanson, J.R. (1968). Bin loads. *J. Struct. Div.* **95**: 1011.

Jenike, A.W. and Johanson, J.R. (1969). On the theory of bin loads. *Trans. ASME Series B* **91**: 339.

Johanson, J.R. (1965). Method of calculating rate of discharge from hoppers and bins. *Trans. Min. Engrs. AIME* **232**: 69.

Johanson, J.R. (1982). Controlling flow patterns in bins by the use of an insert. *Bulk Solids Handl.* **2**: 495.

Johanson, J.R. and Royal, T.A. (1982). Measuring and use of wear properties for predicting life of bulk materials handling equipment. *Bulk Solids Handl.* **2**: 517.

Knowlton, T.M, Carson, J.W., Klinzing, G.E. and Yang, W.-C. (1994). The importance of storage, transfer, and collection. *Chem. Eng. Progress* **90**(4): 44–54.

Manjunath, K.S. and Roberts, A.W. (1986a). Wall pressure–feeder load interactions in mass-flow hopper/feeder combinations. Part I. *Bulk Solids Handl.* **6**: 769.

Manjunath, K.S. and Roberts, A.W. (1986b). Wall pressure–feeder load interactions in mass-flow hopper/feeder combinations. Part II. *Bulk Solids Handl.* **6**: 903.

Marinelli, J. (1996). Choosing a feeder that works in unison with your bin. *Powder Bulk Eng.* **10**(12): 43–57.

Marinelli, J. and Carson, J.W. (1992). Solve solids flow problems in bins, hoppers and feeders. *Chem. Eng. Prog.* **88**: 22.

McLean, A.G. (1979). *Flow Rates of Simple Bulk Solids from Mass-Flow Bins*. Ph.D. Thesis. University of Wollongong, Australia.

Merrow, E.W. (1988). Estimating startup times for solids-processing plants. *Chem Eng.* **89**: 89.

Roberts, A.W. (1988). *Modern Concepts in the Design and Engineering of Bulk Solids Handling Systems*. The University of Newcastle Research Associates (TUNRA), Australia.

Roberts, A.W., Ooms, M. and Scott, O.J. (1986). Influence of vibrations on the strength and boundary friction characteristics of bulk solids and the effect on bin design. *Bulk Solids Handl.* **6**: 161.

Roberts, A.W., Ooms, M. and Wiche, S.J. (1990). Concept of boundary, friction, adhesion and wear in bulk solids handling operations. *Bulk Solids Handl.* **10**: 189.

Royal, T.A. and Carson, J.W. (1993). How to avoid flooding in powder handling systems. *Powder Handl. Process.* **5**: 63.

Schubert, H. (1987). Food particle technology. Part I: Properties of particles and particulate food systems. *J. Food Eng.* **6**: 1.

Schulze, D. (2000). Silo design and bulk solid properties. In *From Powder to Bulk*. IMEChE, International Conference on Powder and Bulk Solids Handling. The Institutions of Mechanical Engineers, UK, 97 pp.

Schwedes, J. (2001). Flow properties of bulk solids and their use solving industrial problems. In *Handbook of Conveying and Handling of Particulate Solids*, Levy, A. and Kalman, H. (eds.). Elsevier, The Netherlands, 137 pp.

Terziovski, M. and Arnold, P.C. (1990). On the effective sizing and placement of air blasters. *Bulk Solids Handl.* **10**: 181.

Teunou, E., Fitzpatrick, J.J. and Synnott, E.C. (1999). Characterisation of food powder flowability. *J. Food Eng.* **39**: 31.

Thompson, S.A., Galili, N. and Williams, R.A. (1997). Lateral and vertical pressures in two different full-scale grain bins during loading. *Food Sci. Technol. Int.* **3**: 371.

Thomson, F.M. (1997). Storage and flow of particlulate solids. In *Handbook of Powder Science & Technology*, Fayed, M.E. and Otten, L. (eds.). Chapman and Hall, New York, pp. 389–486.

Walker, D.M. (1966). An approximate theory for pressure and arching in hoppers. *Chem. Eng. Sci.* **21**: 975.

Walker, D.M. and Blanchard, M.H. (1967). Pressures in experimental coal hoppers. *Chem. Eng. Sci.* **22**: 1713.

Walters, J.K. (1973). A theoretical analysis of stresses in silos with vertical walls. *Chem. Eng. Sci.* **28**: 13.

CHAPTER 5

CONVEYING

5.1. INTRODUCTION

Materials handling in the food and processing industries is concerned with movement of materials in different cases, such as from supply point to store or process, between stages during processes, or to packing and distribution. The movement of materials is a crucial activity that adds nothing to the value of the product, but can represent an added cost if not managed properly. For this reason, responsibility for materials handling is normally vested in specialist handling engineers, and many food manufacturers adopt this procedure. If a company does not have a specific department in charge of materials handling, the responsibility for efficient handling of materials falls on the production manager and his/her staff. It is important, therefore, for production executives to have a sound knowledge of the fundamentals of good handling practice.

Broadly speaking, the main types of handling equipment comprise: (a) conveyors, (b) elevators, (c) cranes and hoists, (d) trucks, and (e) pneumatic systems. Motion of the material can be caused either by gravity or manually, or by power. The method of motion provides a convenient basis for subdividing the types of conveyors into two main groups as follows: (i) gravity or manually moved: chute, roller, or skate-wheel conveyors; and (ii) powered: roller, belt, slat, chain, vibratory, magnetic, screw, flight and pneumatic conveyors.

The above categories include all the possibilities of materials handling within a processing plant, i.e., either in bulk or packed. For bulk particulate or powdered food materials, which fall within the scope of this book, a simpler and more convenient classification would comprise the following types of conveyors: belt, chain and screw, as well as pneumatic equipment. In the literature, all these forms of bulk material movement are known as conveyors and, thus, a proper categorization of the handling equipment for bulk particulate food solids would comprise the following groups:

(a) belt conveyors
(b) chain conveyors: scraper conveyors, apron conveyors and bucket elevators
(c) screw conveyors
(d) pneumatic conveyors: dense-phase systems and dilute-phase systems.

The above classification groups the different types of conveying systems by virtue of operating principle. For example, regardless of whether a bulk material is being moved horizontally, inclined, or vertically, chain conveying can perform the duty based on the same principle of fixing an element (a paddle or a bucket) to a system of chains externally powered.

Bulk solids conveyance represents one of those disciplines that have been neglected in terms of scientific research and development when compared with other operations in particle technology directly related to processing, such as attrition or agglomeration. Conveying systems and supplies are normally provided by a large, capable manufacturing industry, which holds much of the engineering information in the form of brochures, data sheets, and nomographs.

5.2. BELT CONVEYORS

The belt conveyor consists, essentially, of an endless belt operating between two or more pulleys, friction driven at one end and carried on an idler drum at the opposite end. Belt conveyors are used for movement of different types of bulk solids at distances varying from meters to kilometers and, thus, the belt and its load have to be supported on idlers on both conveying and return sections. Heavy-duty machines also have thick belts and these require larger diameter idler end drums. The construction of a belt conveying system requires a support structure, usually of steelwork. Belt conveyors have a high mechanical efficiency since, in larger installations, the entire load is carried on antifriction bearings. Damage to the product in movement is slight because there is little or no relative motion between the product being transported and the belt. The carrying capacity is high since relatively high speeds are possible when the whole system is operating. Bulk solid materials can be carried long distances mainly on the horizontal plane, although there is a possibility of inclined conveying with certain limits to the angle of elevation. Inclination is usually limited to 15 or 20° and, in order to avoid excess spillage or "runback," special belts with corrugated sidewalks and lateral ribs are used for steeper inclination, up to a nominal 45°. A properly designed and maintained belt system has a long service life, but the initial cost is usually high, so installation is only advisable when amortization of the high initial cost can be assured.

5.2.1. Components

The elements comprising a belt conveying system are the belt, drive, tension, idlers, and loading and discharge devices. Belts must be flexible enough to conform to the pulleys, wide enough to carry the quantity and type of material required, have strength enough to stand up under the expected load and operating tension, and possess a resistant surface. Stitched canvas, solid-woven balata and rubber belts may be commonly used. Stitched canvas and woven belts are normally impregnated with a waterproofing material. A rubber belt is made of canvas or woven material impregnated and vulcanized with rubber and covered with a rubber sheet. Balata belts are similar to rubber belts with regard to aging, but may be affected by temperatures over 50°C.

The drive should be at the discharge end of the belt and can be a conventional belt drive. The pulley must be large enough to provide proper contact surface with the belt to ensure a positive drive. Additional contact surface may be attained by using an idler pulley to provide more wrap contact of the belt. Pulley diameters must be large enough to keep from over-flexing the belt. The take-up that may be needed due to the stretch of the belt, as well as possible contraction and expansion as a function of changes in moisture and temperature, can be performed manually by adjusting screws, or automatically by attaching a dead weight. Such adjustment can be carried out on the foot end pulley or on an idler pulley.

As previously mentioned, the belt and its contents are supported by idler pulleys. Figure 5.1 illustrates the main types of idlers: troughed and flat. The straight pulley is used for the empty belt return and infrequently for carrying the load. It may be manufactured of plain wood or light steel. Troughed idlers are normally the support of the whole belt, to increase the conveyor capacity or to confine powdered and particulate materials that would not stay on a flat belt.

The material can be fed onto the belt by hand or by a mechanism providing continuous and steady flow. This may be a simple hopper with a gate valve or, if the material is not free flowing, coupled to an apron, screw, vibrating, or any other type of feeder. The material may be discharged over the end of the belt, by a diagonal scraper, by tilting one or more of the idler pulleys, or by a tripper. As shown in Fig. 5.2, a tripper consists of two idler pulleys mounted in such a way to cause the belt to take the shape of an S. The material is discharged over the top pulley and is caught by a chute

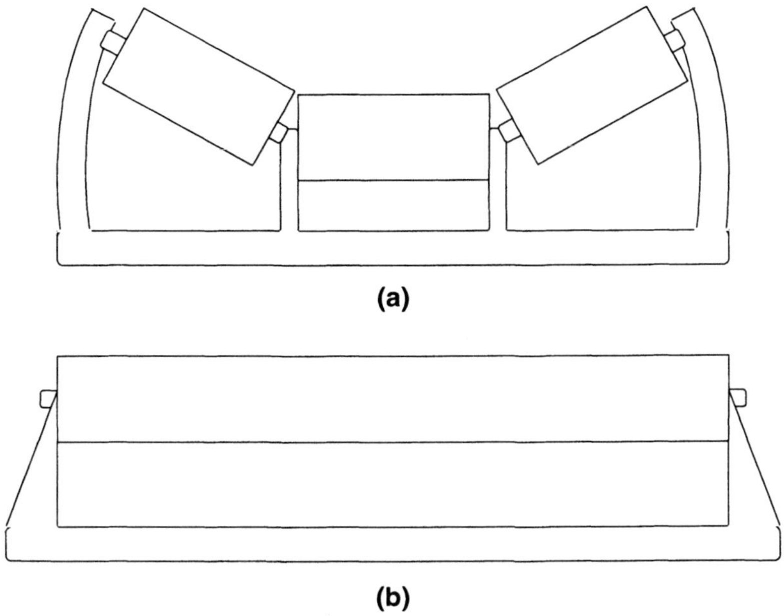

Figure 5.1. Idler pulleys for belt conveyors: (a) troughed; and (b) flat, generally used for empty belt return.

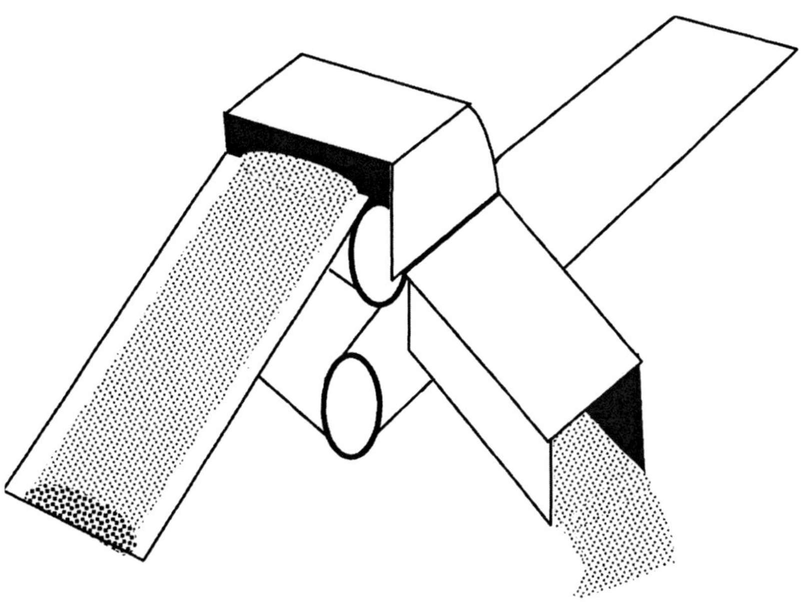

Figure 5.2. Diagram of the tripper, used for belt conveyor discharge.

that diverts it to either side of the belt, or to both. A short belt conveyor operating at right angles to the main belt may replace the chute when it is necessary to move the material to a considerable distance from the main belt. Discharge by tilting idlers may not be advisable, since the material is discharged over a considerable length of the belt and there is an additional strain placed upon the belt due to twisting. An angle scraper is the simplest discharging device and provides satisfactory discharge for many materials. The various discharge systems are designed so that they may be controlled remotely.

5.2.2. Design Aspects

A large belt conveyor system is an expensive and intricate facility that should be designed and installed under the supervision of specially trained conveyor engineers. During the last decades, knowledge about belt conveyor design has been enriched and new design procedures have been proposed based on the capabilities offered by modern computer systems (Lewis, 1985). Foote et al. (1988) examined the payoff of simulation and network analysis in conveyor system design. Roberts (1994) presented an overview of technical and economical factors involved in conveyor design. New techniques, based on intense exploitation of the existing design knowledge, have also been proposed (Dentsoras and Gavrielatos, 1993; Deepak, et al., 1994; Chattopadhyay et al., 1994). However, despite all these mentioned efforts, belt conveying design is far from standardized, and the following suggestions are only intended for preliminary approximations and calculations.

The width of the belt is determined by the size or amount of material to be conveyed and the type of service. The load cross section of a troughed belt is presented in Fig. 5.3, where the surcharge and the top profiles are shown for various slope angles. The surcharge is that portion of the load above the level indicated by the dotted line in Fig. 5.3 and may be considered as the load on a comparable flat belt. The surcharge angle is large for large lumped materials, particularly if mixed with fine material. The 20° arc shows the surface profile of most materials. Fine free-flowing materials can be blown or shaken off the conveyor at high belt speeds. Belt incline is limited to 15°–17° for grain, which is one of the main applications of belt conveyors for food materials.

The horsepower required for movement of material by belt conveyors can be calculated by conventional engineering methods by considering the lift, the frictional resistance of the belt and the frictional resistance of the different pulleys and trippers. However, the constants used in such a procedure vary with change in operating conditions. Also, flexing of the load and belt between supporting pulleys absorbs some energy. Power for standard installations, as well as some other operating variables, can be consulted in tabulated data such as those presented in Table 5.1 (Green and Maloney, 1999). Alternatively, power for belt conveying operation can be calculated from the following equations:

(1) Horsepower, HP, to drive empty conveyor:

$$\mathrm{HP} = \frac{F(L + L_0)(0.06Wv)}{270} \tag{5.1}$$

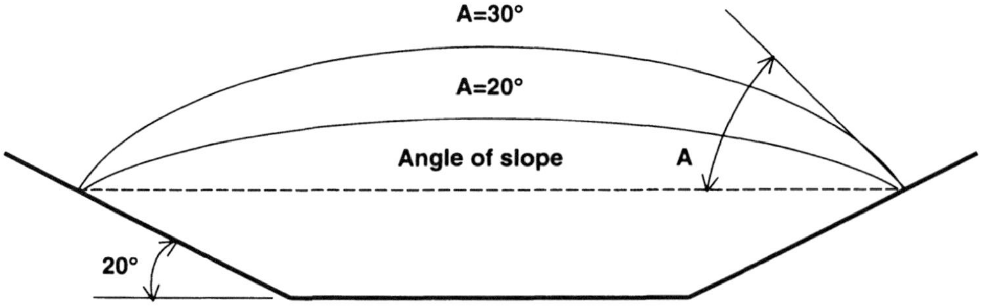

Figure 5.3. Cross section of a loaded belt showing the surcharge material and the top profile for a couple of slope angles A.

Table 5.1. Belt conveyor data for troughed anti-friction idlers (adapted from Green and Maloney, 1999).

Belt width (cm)	Cross-sectional area of load (m^2)	Maximum belt speed (m/min)	Maximum belt plies	Sized material, 80% under (mm)	Capacity for 1,600 kg/m^3 material (ton/h)	HP/30.5 m for 1,600 kg/m^3 material
35	0.010	91	5	51	96	1.32
40	0.013	91	5	64	132	1.68
45	0.017	107	6	76	190	2.42
50	0.020	107	6	89	230	2.90
60	0.030	122	7	114	392	4.04
75	0.049	137	8	178	710	6.74
90	0.072	183	9	203	1,380	9.52
105	0.101	183	10	254	1,980	13.68
120	0.136	183	12	305	2,640	18.20
135	0.177	183	14	356	3,420	23.60
150	0.223	183	16	406	4,320	29.90

(2) Horsepower, HP, to convey material on level:

$$\mathrm{HP} = \frac{F(L + L_0)T}{270} \tag{5.2}$$

(3) Horsepower, HP, to lift material:

$$\mathrm{HP} = \frac{\mathrm{TH}}{270} \tag{5.3}$$

The total power required is the sum of the powers calculated from Eqs. (5.1)–(5.3). The variables in the above equations are defined as follows: F is a friction factor whose main value is 0.05, L is the conveyor length in meters, L_0 is a constant whose main value is 30.5, W is the weight of all moving parts of the conveyor in kilograms per meter of overall length, v is the conveyor speed in m/min, T is the conveyor capacity in tons/h, and H is the height of lift in meters.

5.2.3. Applications

As stated earlier, belt conveyors require high initial capital, so their use may be well justified only for transportation at long distances, even in the range of kilometers. Also, due in part to this reason, they are not normally compatible with high standards of hygiene since they are typically installed outdoors. Consequently, their most common function is for handling heavy-duty minerals, high tonnage rates and/or long distances, e.g., coal, ore and similar materials not subject to problems of contamination. The main application in food systems has been in grain conveying. In fact, it has been mentioned (Wright et al., 1997) that belt conveyor drive power calculation had its origin in grain handling in the late 1700s in the United States.

5.3. CHAIN CONVEYORS

Compared with belt conveying systems, chain conveyors present a series of contrasting characteristics such as cheaper construction, noisy operation, slower movement, mechanic inefficiency, and no specialized skill required for their design. However, chain conveyors are very versatile in design, being more easily adaptable to different duties than belt conveyors. The main components of these types of conveying systems are the chains, the moving elements and the drives. The types of chain available for conveying are extremely varied and a simple complete classification would be difficult to provide. Types of chain commonly used in chain conveyors are the malleable detachable, malleable pintle, steel, roller and combination. The malleable detachable chain is, possibly, the most common and is normally used for light intermittent service. Pintle chain, characterized by a pin that connects the links, is used for more rigorous service such as vertical grain elevators. Steel chain is used where high strength or good wearing qualities, or both, are needed. Roller chain is fitted with rollers or wheels to minimize friction and reduce wear. Combination chain is made in such a way that different features of the above three types are combined to provide certain performance characteristics. Chains of special alloys are available for operation in the presence of extreme or adverse factors such as heat, chemicals, abrasive substances, etc. Some chain-link types are shown in Fig. 5.4. Depending upon how chain and moving elements are mounted for a specific duty, several types of chain conveyors can be obtained. Three different designs are commonly available: scraper conveyors, apron conveyors and bucket elevators.

5.3.1. Scraper Conveyors

These conveying systems are used for granular, nonabrasive materials and are simple, cheap, easily constructed and may operate at steep inclines. Power requirements are, however, high and wear may be excessive. Chain is made out of links, such as those illustrated in Fig. 5.4, with lugs to which flights of various types are fastened. The links may be fitted with rollers for carrying the load and to minimize frictional resistance. The simplest scraper conveyor is one made with "sawdust" chain, in which the links actually act as flights. As the name implies, it is used particularly for removing sawdust from sawing and milling operations, but it can be employed for many applications in food processes. For example, it can be used in dry and wet grain milling to remove hulls and husks where the quantities involved are not high.

Scrape conveyors with attached flights are designed in a variety of ways. The simplest is a single chain with flights operating in a wood or steel trough. Conveyors and elevators of this type are used extensively for moving the products of farming operations. Particular applications include moving raw products, such as grains and vegetables, into processing plants. Cross sections of several scraper conveyors are shown in Fig. 5.5. Materials of large granular size are conveyed on the top of the conveyor. For small granular material, the lower flights are enclosed and the material is conveyed at this point. This allows greater capacities because the trough can operate full and at a higher speed, compared with topside movement. Also, the enclosed feature permits high elevation angles. Discharge for both top and bottom-run conveying is usually at the head shaft. However, by putting gates in the lower trough, discharge for lower flight conveying can be made at any point of the conveyor. Similarly, top flight discharge can be facilitated by gates at the top trough and an open or skeleton lower trough.

Flight height, length and spacing depend upon the expected duty of the conveyor. For small grains and other particulate food materials, flat flights are recommended with a height of about 0.4 the flight length, and spaced at approximately one length. Flight speeds vary from 23 to 38 m/min. Low speeds should be used for materials of large granular size such as ear corn and walnuts. Small granular material, such as small grain and sesame seeds, can be moved at higher speeds, particularly if conveyed in the lower enclosed portion of the conveyor. High speeds can damage the product and so, where practicable, capacity should be provided by large-size flights rather than by high speeds. The capacity of a scraper conveyor operating on the level can be assumed as 115% of the rectangular

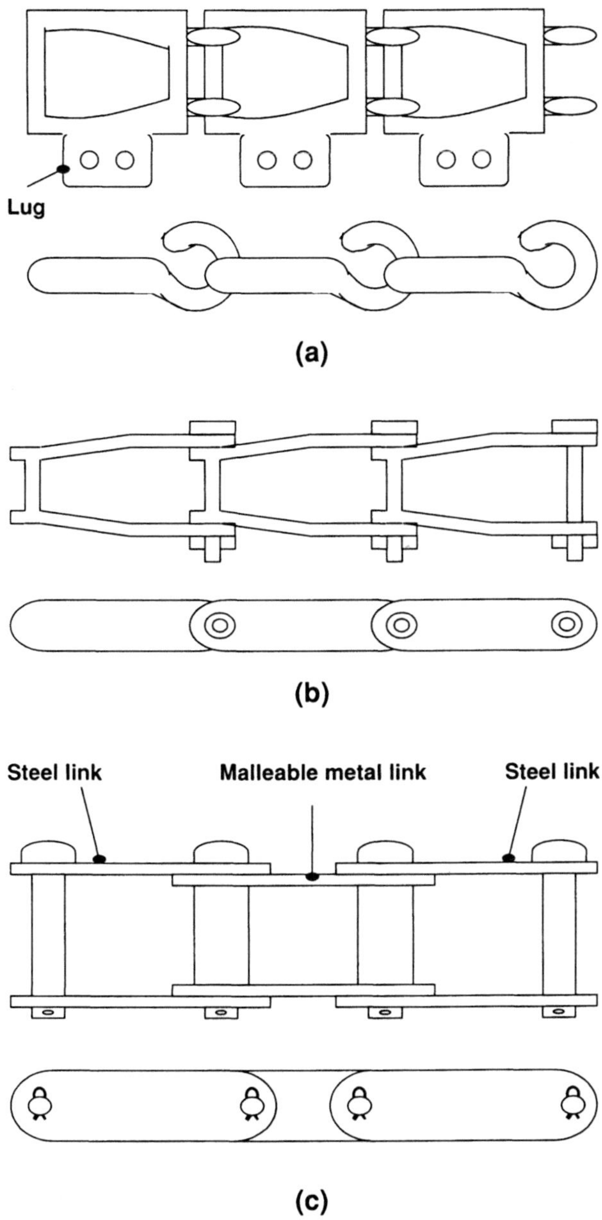

Figure 5.4. Different types of links for chain conveyors: (a) malleable metal equipped with lug; (b) steel; and (c) combined.

space between two flights when designed as suggested above. The capacity of a conveyor operating at an incline will be less than that of the conveyor on the level, according to the data given in Table 5.2.

The values defining the relative capacity of a scraper conveyor will vary considerably from material to material. For example, linted cotton seed will pile much higher on the conveyor than flax seed. This will affect the relative capacity when operating either on the level or at an angle. The

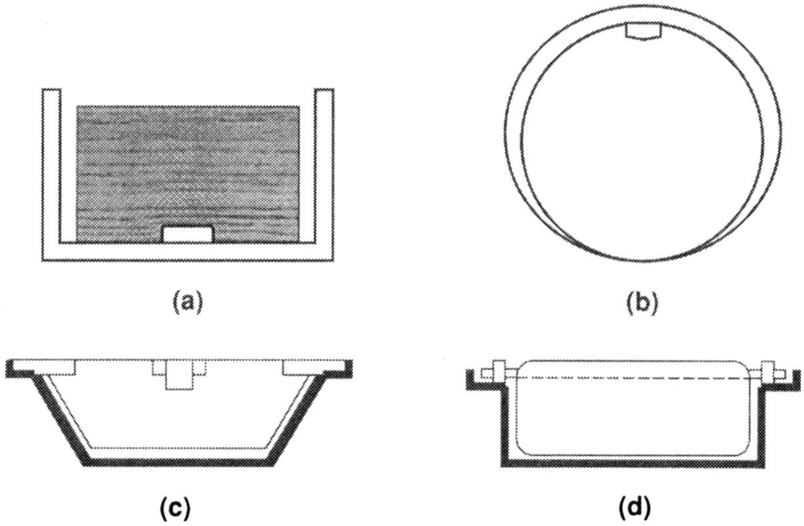

Figure 5.5. Cross sections of some scraper conveyors: (a) wooden; (b) cylindrical flight inside steel pipe; (c) flights supported on wearing plates; and (d) flights supported on rollers.

Table 5.2. Capacity of inclined conveyor compared with horizontal one.

Incline (°)	Approximate relative capacity
20	0.77
30	0.55
40	0.33

theoretical horsepower requirement, HP, for flight conveyors can be determined from the following equation:

$$\mathrm{HP} = \frac{(2vL_cWF_c) + C(LF_m + H)}{1400} \tag{5.4}$$

where v is the speed of the conveyor in m/min, L_c is the horizontal projected length of the conveyor in meters, W is the weight of the flights and chain in kilograms per meter of overall length, F_c is the coefficient of friction for the chain and flights, C is the capacity of material in kg/min, L is the horizontal projected length of the loaded conveyor in meters, F_m is the coefficient of friction for material, and H is the height of the lift in meters.

The calculated horsepower must be adjusted to compensate for expected maximum capacity, starting friction, loss in the driving mechanization, variation in friction coefficients, type of power unit, etc. The value of the coefficient F_c for chain dragging on metal can be taken as 0.33, while that for chain over wood can be approximated to 0.6. Friction coefficients for different materials are listed on Table 5.3. Some other typical design and capacity information is given in Table 5.4.

As previously mentioned, scraper conveyors have applications in the food industry mainly for conveying different sorts of grains. Flour and similar materials are also effectively conveyed at high capacity.

Table 5.3. Friction coefficients for sliding situation.

Material	Friction coefficient
Metal on oak	0.50–0.60
Oak on oak, parallel fibers	0.48
Oak on oak, cross fibers	0.32
Cast iron on mild steel	0.23
Mild steel on mild steel	0.57
Grain on rough board	0.30–0.45
Grain on smooth board	0.30–0.35
Grain on iron	0.35–0.40
Malleable roller chain on steel	0.35
Roller-bushed chain on steel	0.20

Table 5.4. Scraper-conveyor capacities (adapted from Green and Maloney, 1999).

Flight size (mm) and number of strands	Maximum size of lumps		Capacity (ton/h)* for various flight spacings, conveyor, horizontal (mm)		
	All lumps (mm)	10% lumps (mm)	460	610	915
255 × 100 and 1	38	76	32	25	16
305 × 130 and 1	45	89	46	35	23
380 × 130 and 1	51	102	66	50	33
380 × 155 and 2	89	178	87	67	44
405 × 205 and 2	102	203	110	82	55
460 × 205 and 2	127	229	124	93	62
510 × 255 and 2	152	254	—	141	94
610 × 255 and 2	203	305	—	176	116
765 × 255 and 2	254	355	—	—	250
305 × 130 and 1	45	89	56	42	28
380 × 180 and 1	64	114	78	58	39
460 × 205 and 1	76	127	124	93	62
305 × 130 and 2	51	102	56	4	28
380 × 155 and 2	76	127	76	57	38
460 × 180 and 2	102	203	96	72	48
610 × 205 and 2	203	305	—	124	83

*For 480 kg/m^3 bulk density material and conveyor velocity of 30.5 m/min.

5.3.2. Apron Conveyors

If the flights in the scraper conveyors are replaced with flat slats, steel plates or boards, a moving platform or apron is obtained and can be used for movement of sacked or large unit sized materials. The typical design is a series of pans mounted between two strands of roller chain, with pans overlapping to eliminate dribble, and often equipped with end plates for deeper loads. Apron conveyors are not suitable for conveying of powdered and particulate materials. For such materials, their main application is feeding at controlled rates, with lump sizes large enough to minimize dribbling. Apron-feeder applications range from fairly light duties with light-gauge steel pans, up to extremely heavy duties requiring reinforced manganese steel pans with center supports.

5.3.3. Bucket Elevators

These sort of conveying systems comprise high capacity units primarily intended for bulk elevation of relatively free-flowing materials, and may be considered a special adaptation of chain conveying. Bucket elevators are the simplest and most dependable equipment units for vertical lifting of different types of granular materials. They are available in a wide range of capacities and may be operated either entirely in the open or be totally enclosed. High efficiency in bucket elevators results from the absence of frictional loss from sliding of the material on the housing, and this feature distinguishes them from the vertical, or nearly vertical, scraper conveyor. The material-carrying element of this sort of conveyer is the bucket, which may be enclosed in a single housing called a leg, or two legs may be used. The return leg may be located some distance from the elevator leg. A single or double chain is used to attach the buckets. The most important considerations affecting the design and operation of bucket elevators are: (a) the physical properties of the conveyed material; (b) the shape and spacing of the buckets; (c) the speed at which the elevator is driven; (d) the method of loading the elevator; and (e) the method of discharging the elevator.

Important physical properties of the material being elevated are particle size, lump size, moisture content, angle of repose, flowability, abrasiveness, friability, etc. The design of the bucket has to do, principally, with capacity and ease of discharge. Buckets may be constructed out of malleable iron or steel and can be shaped with either sharp or round bottoms, as illustrated in Fig. 5.6.

Mounting and spacing of the buckets will conform to a specific elevator design. Some typical bucket elevators are shown schematically in Fig. 5.7. They may be fastened to the chain at the back (Fig. 5.7a) or at the side if mounted in two chains (Fig. 5.7b). Guides are sometimes used for two-chain installations, particularly in the up leg. Single-chain installations have no guides or supports between the head and foot wheels except, possibly, an idler or two placed at strategic points to eliminate whip. The center spacing of buckets varies with their size, shape, and speed, as well as

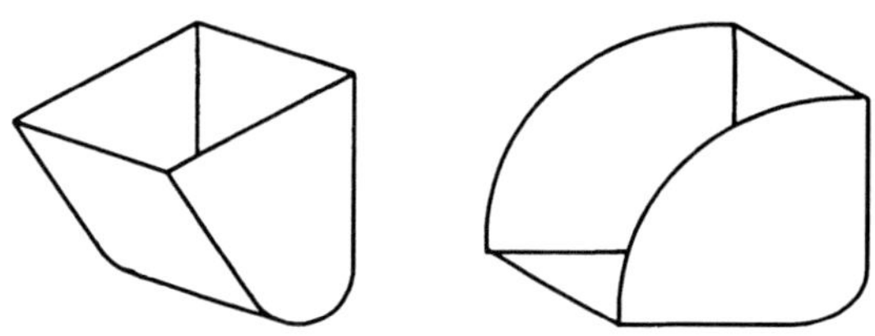

(a)

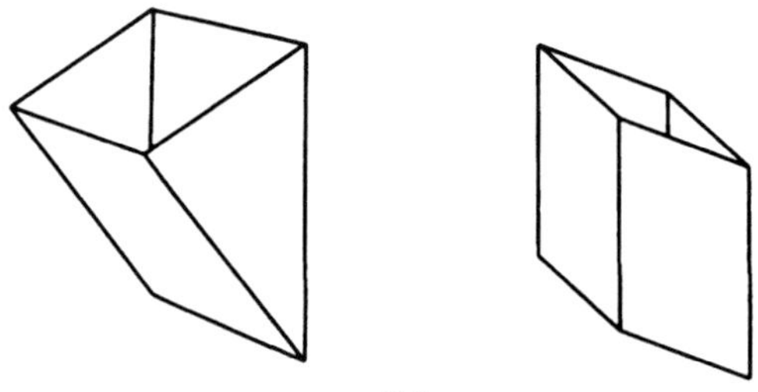

(b)

Figure 5.6. Buckets: (a) malleable iron, for spaced buckets elevators; and (b) steel, for continuous bucket elevators.

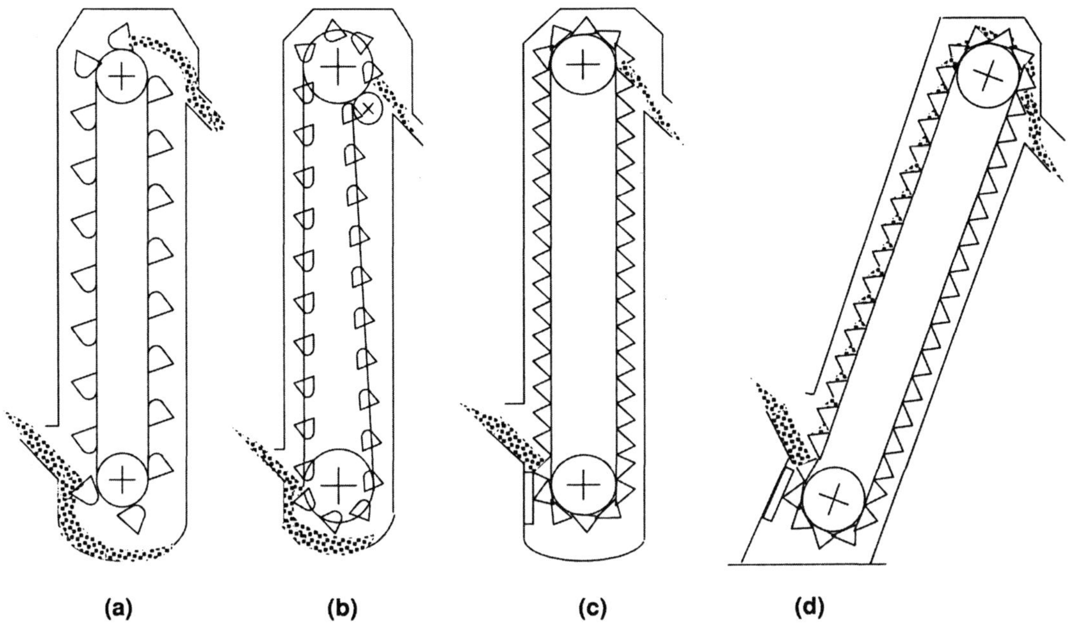

Figure 5.7. Bucket elevators: (a) centrifugal discharge spaced buckets; (b) positive-discharge spaced buckets; (c) continuous bucket; and (d) super capacity continuous bucket.

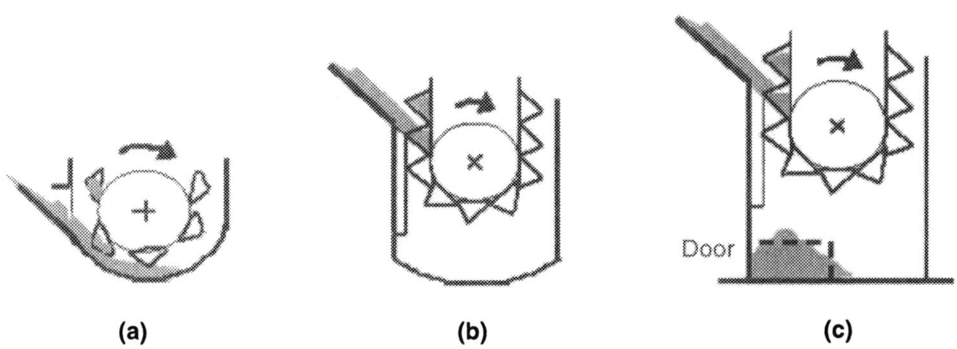

Figure 5.8. Modes of loading for bucket elevators: (a) partly direct and partly by scooping; (b) loading leg with feed spout above tail wheel; and (c) bottomless boot with cleanout door.

head and foot wheel diameter. The buckets must be placed so that the centrifugally discharging grain does not hit the bucket ahead of the one discharging. For general purposes, the spacing will be from 2 to 3 times the projected width.

The speed of the drive in bucket elevators, although depending much on the type of material, is mainly controlled by the rate and method of discharge. Three main types of discharge are generally recognized: centrifugal, positive and continuous, and they will be discussed later. A fourth type of discharge may sometimes be considered: gravity discharge, in which buckets are carried pivoted on two chains and are tipped mechanically to facilitate discharge.

As shown diagrammatically in Fig. 5.8, bucket elevators can be mainly loaded in three different ways. Spaced buckets receive part of the charge directly from a chute and part by scooping (Fig. 5.8a), continuous buckets are filled as they pass through a loading leg with a feed spout above the tail wheel (Fig. 5.8b) or, they can be loaded in a bottomless boot with a cleanout door (Fig. 5.8c).

Discharge can also be performed in three different ways, as illustrated in Fig. 5.7. Spaced bucket centrifugal discharge elevators (Fig. 5.7a) are the most common and are usually equipped with malleable iron round buckets (Fig. 5.6) spaced to prevent interference in loading or discharge. This type of elevator will handle almost any free flowing fine or small lump material. Speeds can be relatively high for fairly dense materials, but must be lowered considerably for aerated or low bulk-density materials to prevent fanning action. Spaced bucket positive discharge elevators (Fig. 5.7b) are essentially the same as centrifugal discharge units, except that the buckets are mounted on two strands of chain and are snubbed back under the head sprocket to invert them to allow positive discharge. These units are designed especially for materials that are sticky or tend to pack, and the slight impact of the chain seating on the snub sprocket, combined with complete bucket inversion, is generally sufficient to empty the buckets completely. Continuous bucket elevators (Fig. 5.7c) are generally used for larger lump materials or for materials too difficult to handle with centrifugal discharge elevators. Buckets used are of the steel type, such as those illustrated in Fig. 5.6. They are closely spaced and the back of the preceding one serves as a discharge chute for the bucket that is dumping as it rounds the head pulley. Close bucket spacing reduces the speed at which the elevator operates to maintain capacities comparable with the spaced bucket units. Gentle discharge is, therefore, promoted to help prevent excessive degradation, making this type of elevator suitable for handling finely pulverized or aerated materials. Super capacity continuous bucket elevators (Fig. 5.7d) are designed for high lifts and large lump material. They handle high tonnages and are usually operated at an incline to improve loading, as well as discharge conditions. Running speeds are low and, due to the heavy loads, the bucket supporting chain is usually guided on the elevating and return parts.

Except for overlapping buckets, which are not extensively used in processing, discharge depends upon centrifugal force in part or in full, or the ability of the material to be thrown into a chute as the buckets go over the head pulley. The characteristics of this feature and, in particular, the trajectory of the material after it leaves the bucket, are important to properly design and operate bucket elevators. Centrifugal discharge requires the speed of the chain to be held within close limits in order that the trajectory will fall within a specified region. Figure 5.9 shows a head wheel and a bucket in a series of positions. A unit mass of grain is subjected to two forces at the point the bucket starts to turn around the pulley. These forces are the weight of the unit volume W and the centrifugal force F_c acting radially, which is:

$$F_c = \frac{W v_t^2}{3600 g r} \tag{5.5}$$

where W is the weight of elemental mass, v_t is the tangential velocity, g is the acceleration due to gravity and r is the effective radius.

The resultant of these forces R, shown in Fig. 5.9, determines the point at which discharge takes place and its characteristics. The resultant for positions 1–4 in Fig. 5.9 is of such a direction that the material is held in the bucket, at position 5, F_c and W are opposing and R is zero, so there is no force on the material. Discharge begins at this point, the initial velocity and trajectory being that of the projected speed of the wheel at such point. Note that R in positions 6–8 is nearly in the same direction of motion of the bucket, thus forcing discharge. In order to produce this condition, F_c and W must be equal at a point near the top of the travel:

$$F_c = W = \frac{W v_t^2}{3600 g r} \tag{5.6}$$

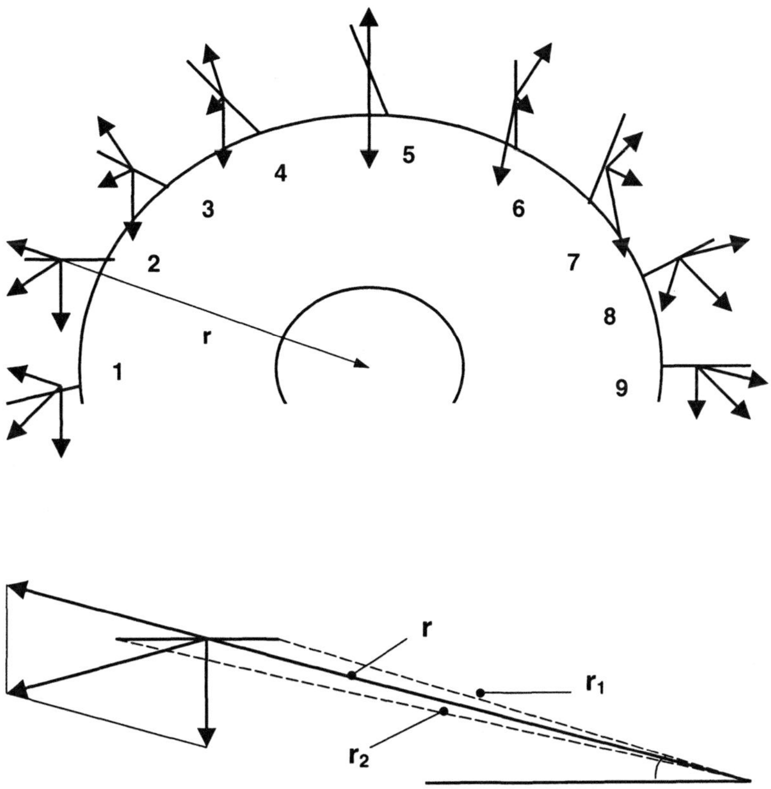

Figure 5.9. Force diagram of the loads in a head wheel bucket in a number of different positions. The effective radius of the head wheel bucket varies from r_1 to r_2.

so that:

$$v^2 = 3600gr \tag{5.7}$$

and since:

$$v = 2\pi r N \tag{5.8}$$

where N is revolutions per minute (rpm), then:

$$N = 54.19\left[\frac{1}{\sqrt{r}}\right] \tag{5.9}$$

Equation (5.9) shows the relationship between the effective head-wheel radius and its revolutions per minute for the most satisfactory discharge conditions. Discharge is not uniform or instantaneous because the effective radius varies from r_1 to r_2, as shown in Fig. 5.9. Thus, the material at the outer edge of the bucket discharges first.

Bucket elevator horsepower can be calculated quite easily using the following equations:

(1) Horsepower, HP, for spaced buckets and digging boots:

$$\text{HP} = \frac{TH}{152} \tag{5.10}$$

Table 5.5. Bucket elevator specifications for spaced buckets (adapted from Green and Maloney, 1999).

Size of bucket (mm)* and bucket spacing	Elevator centers (m)	Capacity (ton/h)**	Lump size (mm)	Head shaft (rpm)	HP required at shaft
152 × 102 × 108 and 305	7.6	14	19.0	43	1.0
	15.2	14	19.0	43	1.6
	22.8	14	19.0	43	2.1
203 × 127 × 140 and 356	7.6	27	25.4	43	1.6
	15.2	30	25.4	41	3.5
	22.8	30	25.4	41	4.8
254 × 152 × 159 and 406	7.6	45	32.0	43	3.0
	15.2	52	32.0	41	5.2
	22.8	52	32.0	41	7.2
305 × 178 × 184 and 457	7.6	75	38.1	41	4.7
	15.2	84	38.1	38	8.9
	22.8	84	38.1	38	11.7
355 × 179 × 184 and 457	7.6	100	44.5	38	7.3
	15.2	100	44.5	38	11.0
	22.8	100	44.5	38	14.3
406 × 203 × 216 and 457	7.6	150	50.8	38	8.5
	15.2	150	50.8	38	12.6
	22.8	150	50.8	38	16.7

*Bucket size given: width × projection × depth. Assumed bucket linear speed: 45.7 m/min.
**Capacities and horsepowers given for materials of 1,600 kg/m^3 bulk densities. For other densities, these will vary in direct proportion, e.g., a 800-kg/m^3 material will reduce the capacity and horsepower required by 50%.

(2) Horsepower, HP, for continuous buckets with loading leg:

$$\text{HP} = \frac{TH}{167} \tag{5.11}$$

In Eqs. (5.10) and (5.11), T is the bucket capacity in tons/h and H is lift in meters. Both equations include normal drive losses, as well as loading pickup losses, and are applicable for vertical or slightly inclined lifts. For estimating purposes, general bucket specifications are given in Tables 5.5 and 5.6 for spaced and continuous units, respectively.

As previously mentioned, bucket elevators are by far the most efficient way of elevating granular and particulate materials in a number of processing industries. In the food industry, they are employed extensively for elevating a variety of commodities such as sugar, beans, oilseeds, salt, and cereals.

5.4. SCREW CONVEYORS

5.4.1. Main Features

These systems are used to handle finely divided powders, damp materials, hot substances that may be chemically active, and granular materials of all types. Screw conveyors are also used for batch or continuous mixing, for feeding where a fairly accurate rate is required and for conventional conveying, as well as elevating jobs, particularly if the run is short. They operate on the principle of a rotating helical screw moving material in a trough or casing. Flights are made of stainless

Table 5.6. Bucket-elevator specifications for continuous buckets (adapted from Green and Maloney, 1999).

Size of bucket (mm)* and bucket spacing	Elevator centers (m)	Capacity (ton/h)**	Lump size (mm)	Head shaft (rpm)	HP required at shaft
203 × 140 × 197 and 203	7.6	35	25.4	28	1.8
	15.2	35	25.4	28	3.4
	22.8	35	25.4	28	5.0
254 × 178 × 298 and 305	7.6	60	38.1	23	3.0
	15.2	60	38.1	23	5.5
	22.8	60	38.1	23	8.0
305 × 178 × 298 and 305	7.6	70	38.1	23	3.5
	15.2	70	38.1	23	6.5
	22.8	70	38.1	23	9.5
356 × 178 × 298 and 305	7.6	80	44.5	23	4.0
	15.2	80	44.5	20	7.5
	22.8	80	44.5	20	11.0
356 × 203 × 298 and 305	7.6	100	50.8	20	5.0
	15.2	100	50.8	20	9.3
	22.8	100	50.8	20	13.3
406 × 203 × 298 and 305	7.6	115	50.8	20	6.0
	15.2	115	50.8	20	11.0
	22.8	115	50.8	20	16.0
457 × 203 × 298 and 305	7.6	130	50.8	20	7.0
	15.2	130	50.8	20	13.0
	22.8	130	50.8	20	20.0

*Bucket size given: width × projection × depth. Assumed bucket linear speed: 45.7 m/min.
**Capacities and horsepowers given for materials of 1,600 kg/m^3 bulk densities. For other densities, these will vary in direct proportion, e.g., a 800-kg/m^3 material, will reduce the capacity and horsepower required by 50%.

steel, copper, brass, aluminum or cast iron, principally. They may be hard surfaced with Stellite or similar materials to resist highly abrasive materials. Although screw conveyors are simple and relatively inexpensive, power requirements are high and single sections are limited in length. The standard pitch screw has a pitch approximately equal to the diameter and is used on most horizontal installations and on inclines up to 20°. Half standard pitch screws may be used for inclines greater than 20°. Double flight and triple flight, variable pitch and steeped diameter screws are available for moving difficult materials and controlling feed rate. Ribbon screws are used for wet or sticky substances, while special cut flight and ribbon screws are used for mixing. Figure 5.10 shows the main components of screw conveyors.

5.4.2. Operating Characteristics

As previously mentioned, horizontal screw conveying is the predominant way of operation. Horizontal screw conveyors usually run in a U-shaped trough, with or without cover, depending upon the type of service and the characteristics of the material being moved. The screw is supported by brackets at different standard spacing. For elevating at a steep incline, a cylindrical housing is required. The tube normally operates full with no brackets between the ends, since they would interfere with the movement of material. The material supports and guides the flight, so that satisfactory operation results if the conveyor is kept full.

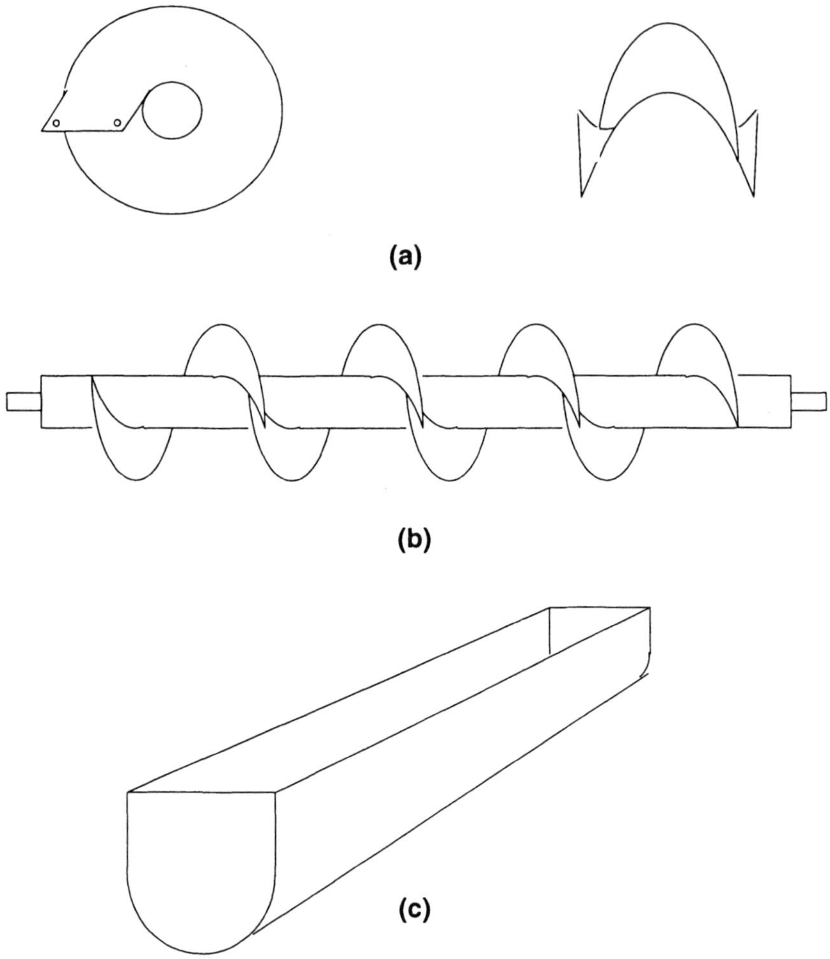

Figure 5.10. Screw conveyor components: (a) flight; (b) screw, formed by mounting flights on an axle; and (c) trough.

The use of hollow screws and pipes for circulating hot or cold fluids would permit screw conveyors to be used for heating, cooling, and drying operations. It is possible to seal a screw conveyor from the outside atmosphere in order to operate outdoors without special protection. The conveyor can be completely sealed to operate in its own atmosphere at positive or negative pressure. The casing can also be insulated to maintain internal temperatures in areas of extreme ambient temperatures. The casing can be designed with a drop bottom for easy cleaning to avoid contamination when different materials are to be conveyed through the same system.

Screw conveyors are usually made of standard sections coupled together, so special attention should be given to bending stresses in the couplings. Hanger bearings supporting the flights can obstruct the flow of material when the trough is loaded above its level. Thus, with difficult materials, the load in the trough must be kept below this level. Alternatively, special hanger bearings that minimize obstruction should be selected. Since screw conveyors operate at relatively low rotational speeds, the fact that the outer edge of the flight may be moving at a relatively high linear speed is often neglected. This may create a wear problem, and if wear is too severe it can be reduced by the use of hard-surfaced edges, detachable hardened flight segments, rubber covering, or high-carbon steels.

Screw conveyors are well suited as feeders or metering devices under bins or hoppers. The portion of the screw under the container is usually designed with half or third pitch, and the balance of the screw full pitch. Thus, the main portion of the screw operates half or third full, while the metering portion of the screw operates full.

5.4.3. Capacity and Power

Concise data and formulae are normally not available for individual design problems and consulting a specialized engineer when designing and installing large screw conveying systems is recommended. Data that could be available to assist in selection and design are normally empirical in nature. Roberts (1999, 2000) presented analytical data to predict the performance of screw conveyors. Recommended capacities of screw conveyors as a function of screw diameter and type of material in movement are given in Table 5.7.

The power requirement of a screw conveyor is a function of its length, elevation, type of hanger brackets, type of flights, the viscosity or internal resistance of the material, the coefficient of friction of the material on the flights and housing, and the weight of the material. Consideration must also

Table 5.7. Screw conveyor operating features (adapted from Green and Maloney, 1999).

Type of material	Screw diameter (mm)	Maximum lump size (mm)		Capacity (m^3/h)		Approx. area occupied by material
		25% lumps	100% lumps	At 1 rpm	At max. rpm	
From very fine to lumpy, different flowabilities, non-abrasive	152	19	13	0.064	10.6	45%
	229	38	19	0.226	34.0	
	305	51	25	0.546	76.5	
	356	64	32	0.872	113.3	
	406	76	38	1.320	158.6	
	457	76	51	1.872	215.2	
	508	89	51	2.690	283.2	
From very fine to lumpy, different flowabilities, mildly abrasive	152	19	13	1.042	2.1	30%
	229	38	19	0.158	7.9	
	305	51	25	0.377	18.8	
	356	64	32	0.597	29.9	
	406	76	38	0.889	44.5	
	457	76	51	1.286	64.3	
	508	89	51	1.758	87.9	
From very fine to lumpy, different flowabilities, very abrasive	152	19	13	0.021	0.7	15%
	229	38	19	0.079	2.5	
	305	51	25	0.190	5.7	
	356	64	32	0.297	8.5	
	406	76	38	0.444	12.0	
	457	76	51	0.643	16.7	
	508	89	51	0.881	22.1	

Table 5.8. Material factors for horizontal screw conveyors.

Type a ($F = 1.2$): light, fine, non-abrasive, free-flowing materials; ρ_b: 480–640 kg/m^3	Type b ($F = 1.4$–1.8): non-abrasive, granular or fines mixed with lumps; ρ_b: up to 830 kg/m^3	Type c ($F = 2.0$–2.5): non and mildly abrasive, granular or fines mixed with lumps; ρ_b: 640–1,200 kg/m^3	Type d ($F = 3.0$–4.0): mildly abrasive or abrasive, fine, granular or fines with lumps; ρ_b: 830–1,600 kg/m^3
Barley	Soy meal	Granular moist malt	Raw sugar
Granular dried malt	Cacao seeds	Cocoa	Bone meal
Corn flour	Coffee seeds	Dehydrated milk	
Cotton seed flour	Corn	Starch	
Wheat flour	Corn meal	Icing sugar	
Malt	Jelly granules		
Rice			
Wheat			

be given to additional power needed to start a full screw, to free a jammed screw or to operate with material which has a tendency to stick to the trough sides. The horsepower, HP, required to drive a screw conveyor depends upon the dimensions of the system and the characteristics of the material. A rough approximation for normal horizontal operation can be determined from the following relation:

$$\text{HP} = \frac{CL\rho_b F}{4500} \tag{5.12}$$

where C is the capacity in m^3/min, L is the conveyor length in meters, ρ_b is the apparent density of material in kg/m^3, and F is a factor depending on the type of material, as appearing in Table 5.8.

In Eq. (5.12), if horsepower is less than 1, it should be doubled; if it ranges from 1 to 2, it should be multiplied by 1.5; if it ranges from 2 to 4, it should be multiplied by 1.25; and if it ranges from 4 to 5, it should be multiplied by 1.1. No correction is necessary for values above 5 hp.

5.4.4. Main Applications

As previously stated, screw conveyors are versatile devices for the handling of a wide variety of materials horizontally, at an inclination and even vertically. They are suited for both dry bulk materials as well as semi-liquid non-abrasive materials. In the food industry, the applications are numerous, and they have been used for (a) conveying of different grains and oilseeds, such as barley, corn, rice, rye, wheat, cottonseed, and soy beans; (b) moving fine food powders such as flour, icing sugar, starch, and powdered milk; and (c) handling viscous food materials such as sugar beet pulp, peanut butter, and comminuted meat.

5.5. PNEUMATIC CONVEYING

5.5.1. Introduction

One of the most important bulk solids handling techniques in food processing is the movement of material suspended in a stream of air over horizontal, inclined or vertical surfaces, ranging from a few to several hundred meters. This type of conveying is one of the most versatile, handling materials that range from fine powders through 6.35 mm pellets and bulk densities of 16 to more than 3,200 kg/m^3. As compared with previously discussed methods, pneumatic conveying offers the containment and flexibility of pipeline transport for bulk solids that, otherwise, would be exposed to direct contact with moving mechanical parts. Most of the food powders and particulates handled in the food processing industries would present hygienic and contamination problems when conveyed in the opening; in such a way, pneumatic conveying represents an obvious choice for duties in which the integrity of handled products is paramount.

Pneumatic conveying has been used extensively for many years in many food processing operations. In fact, as reported by Reed and Bradley (1991), one of the earliest recorded uses was for unloading wheat from barges to flour mills at the end of the 19th century in London. Some other grains, as well as different cargo such as aluminum, cement, and plastic resins are still unloaded using the same basic methods. Other common applications include unloading trucks, railcars, and barges, transferring materials to and from storage vessels, injecting solids into reactors and combustion chambers, and collecting fugitive dust by vacuum.

Advantages in cleanliness and low contamination are reflected in the fact that if properly constructed and maintained, pneumatic systems can be virtually dust free. Vacuum systems offer the advantage that any leakage is in the pipeline, so that even damaged or leaky operations do not result in product loss. Sealed systems can prevent most forms of contamination, and contact with moving mechanical components is minimal. Pressure systems prevent inward leakage and can use dry, inert gas for conveying in order to exclude oxygen and moisture when handling sensitive biological materials.

Pneumatic conveying has some drawbacks, such as high power consumption, limitations in overall distance and capacity, as well as severe wear of equipment and attrition or degradation of the materials being conveyed. In terms of consumption, taken on the basis of cost per unit weight per unit distance conveyed, pneumatic conveying is by far the most expensive method of moving materials. In many practical cases, however, the higher cost is justified, because the alternatives are not practical due to layout limitations, containment and cleanliness, as well as low maintenance of this way of conveying. With regard to capacity and distance, pneumatic conveyors are generally restricted to about 300 tons/h and 1,000 m, although not simultaneously. Pertaining to wear and attrition, such problems normally arise only when systems are not properly designed and operated.

Despite the above-mentioned limitations, pneumatic conveying is an important and practical form of transporting bulk solids, with applications in almost every part of the food processing industries. The range of materials that can be handled is nearly unlimited, including powders and particulates of different types. The limitations on what can be conveyed depend more upon the physical nature of the material than on its generic classification. Particle size, hardness, resistance to damage, and cohesive properties are key factors in determining whether a material is suitable for this sort of conveying. Cohesive or sticky materials are often difficult to handle in a pneumatic conveyor. Moist substances that are wet enough to stick to the pipeline walls usually cannot be conveyed successfully. Materials with high oil or fat contents can also cause severe buildup in pipelines, making conveying impractical.

5.5.2. Theoretical Aspects

In contrast with those conveying methods previously discussed, pneumatic conveying can be perfectly identified as a case of two-phase flow, which is a topic well covered by fluid mechanics. Flow of gas in a pipeline is well understood with the conveying gas obeying the ideal gas law, and its density ρ_g being a function of pressure and temperature, as given by:

$$\rho_g = \frac{P}{RT} \tag{5.13}$$

where P is the absolute pressure, R is the universal gas constant, and T is the absolute temperature.

Mean gas velocity v in a pipeline is a function of mass flow rate of the gas and the density of the flow area:

$$v = \frac{\dot{m}}{\rho_g A} \tag{5.14}$$

where $\dot{m}$ is the mass flow rate of gas and A is the flow area.

By combining Eqs. (5.13) and (5.14), it follows that the mean gas velocity is a function of the gas pressure:

$$v = \frac{\dot{m}RT}{PA} \tag{5.15}$$

Assuming that the mass flow rate of the gas and the flow area are constant, as well as the gas temperature, the velocity at any two points in the line becomes proportional to the absolute gas pressure:

$$\frac{v_2}{v_1} = \frac{P_1}{P_2} \tag{5.16}$$

where P_1 and P_2 are absolute pressures.

The relationship between gas velocity and pressure drop in a straight pipe is found by the following simple relation:

$$\Delta P = f\left(\frac{L}{D}\right)\left(\frac{\rho_g u^2}{2}\right) \tag{5.17}$$

where f is the fanning friction factor, L is the pipe length, D is the pipe diameter, and u is the local gas velocity.

As Eq. (5.17) indicates, the pressure drop in a pipe is approximately proportional to the square of the gas velocity. The increase in velocity from one end of the pipe to the other results in a difference in pressure drop per unit length of more than 2. This illustrates the significance of density changes in the gas, as flow proceeds through the pipeline. Changes in the gas velocity also affect the suspension of solids in the gas stream. At low velocities, particles may be sliding on the bottom of the pipe, while at higher velocities particles will be fully suspended by the gas.

The above equations describe single-phase flow within a pipe and are considered the basic fundamentals for starting calculations in order to design pneumatic conveying systems. However, while single-phase flow in a pipe is well understood, adding solids into the moving stream in the so-called two-phase flow complicates matters significantly. As solid particles are introduced in a moving stream of gas, the pressure drop in the line begins to increase because momentum is transferred to the particles to accelerate them to the conveying velocity. The total pressure drop consists of two components: that due to the gas flow alone, and that required for transporting the particles. Apart from the gas velocity, some other variables, such as the amount of solids in the pipeline and the velocity of the solids relative to the gas, also affect the pressure drop.

The moving gas stream applies drag and lift to the particles. For particles to be conveyed in such a gas stream, the velocity of the gas must be sufficiently high to stop particles from settling out. In flow through horizontal pipes, the minimum air velocity to stop particles from settling to the bottom of the pipe is called the saltation velocity. The equivalent velocity for flow through vertical pipes is known as the choking velocity. The saltation velocity is a function of the density of the gas and the solids, as well as particle and pipeline diameter (Cabrejos and Klinzing, 1994). There is also a direct relationship between the saltation velocity and the solids loading ratio. Generally, saltation occurs at higher velocities when the solids loading ratio is also high.

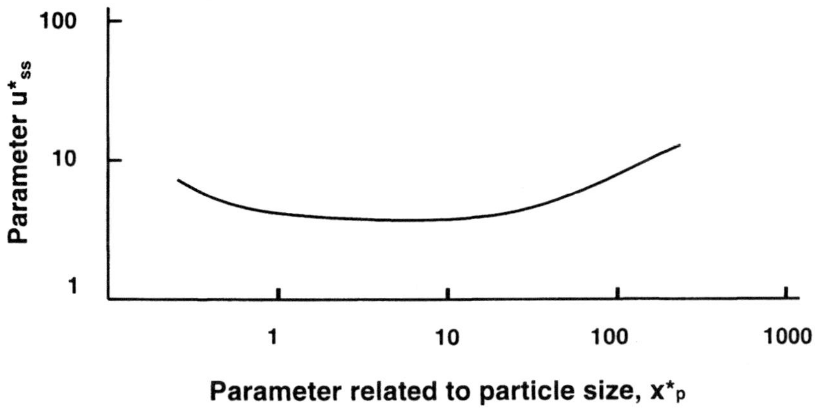

Figure 5.11. Minimum conveying velocity for a single particle for a pipe diameter D_t of 63.5 mm (adapted from Zenz, 1964).

In terms of designing equipment for pneumatic conveying, there is another type of velocity, i.e., the minimum conveying velocity, used to describe the correlation of gas velocity to the behavior of solid particles inside a pipeline. This velocity is the lowest necessary to prevent plugging the line in a given system for a given material. Some researchers have suggested using the saltation velocity with a safety factor, while others have developed empirical correlations. Some of these correlations, however, predict widely differing velocities for the same set of conditions (Wypych, 1999).

The saltation velocity can be calculated using the procedure presented by Jackson and Lamb (1981), as follows:

(1) Calculate the parameter x_p^* by the equation:

$$x_p^* = \frac{x_p}{[3(\mu_g)^2/4g\rho_g(\rho_s - \rho_g)]^{1/3}} \tag{5.18}$$

where x_p is the particle diameter, μ_g is the gas viscosity, and ρ_g and ρ_s are the gas and solid density, respectively. If there is a mixture of particle sizes, i.e., a particle size distribution which is quite common in most food powders, calculate x_p^* for the largest and smallest particle.

(2) From Fig. 5.11 calculate u_{ss}^* from x_p^* and use the following relation to determine the minimum conveying velocity u_{ss} for a single particle:

$$u_{ss} = 0.19(u_{ss}^*)\left[\frac{4g\mu_g(\rho_s - \rho_g)}{3(\rho_g)^2}\right]^{1/3}(D_t)^{0.4} \tag{5.19}$$

where D_t is the particle diameter in mm.

(3) Estimate n, the gradient of the curve in Fig. 5.11 at x_p^*. For a particle size distribution n is the slope of the line joining the values of u_{ss}^* for the largest and smallest particle.

(4) Transpose u_s, the saltation velocity from the following equations:

$$\frac{G_s}{\rho_s} = \frac{0.214(n)^{1.5}(u_s - u_{ss})}{u_{ss}} \tag{5.20}$$

for $n > 0.068$, or:

$$\frac{G_s}{\rho_s} = \frac{0.0032(u_s - u_{ss})}{u_{ss}} \tag{5.21}$$

for $-0.11 < n < 0.068$.

In both Eqs. (5.20) and (5.21), G_s is the flux of solids. For particle size distributions, the largest particle size should be used in these equations in order to avoid underestimation of the saltation velocity.

Pressure drop is another key variable involved in design and operation of pneumatic conveying systems. Pressure drop requirement is closely associated with gas flow and solids rate. The same amount of solids can be conveyed in a line using a number of velocity and pressure drop combinations. The study of the correlation of pressure drop and mean gas velocity, maintaining constant solids flow rate, gives origin to the most general classification of pneumatic conveyors, which divides them into dense-phase systems and dilute-phase systems.

5.5.3. Classification of Conveying Systems

Pneumatic conveying systems can be categorized in a number of ways depending on their function, as well as type and magnitude of operating pressure. Solids loading is a useful criterion for classifying pneumatic conveyors, which can run over a wide range of conditions, bounded on one end by gas alone with no entrained solids and at the other end by a completely full pipe where the solids are plugging the line. Most industrial conveying systems operate somewhere in between these two extremes, being ranked broadly as either dense-phase or dilute-phase systems, depending upon the relative solids loading and velocity of the system. This is best illustrated graphically in a general state diagram, which is a plot of pressure per unit length of pipe as a function of conveying gas velocity, with constant solids flow rate. As shown in Fig. 5.12, at higher velocities, particles are generally suspended in the gas with low solids loading ratio, typically below 15, and termed as dilute-phase conveying. If the gas velocity is slowly decreased, the pressure required to convey a constant amount of solids also drops. After reaching a minimum, further reduction in gas velocity results in an increase in pressure as particles begin to fall out of suspension and inter-particle collision increases. This region, with a solids loading ratio typically higher than 15 and the gas velocity below the saltation velocity, is that of dense-phase conveying. With many materials it is difficult to establish a definite boundary separating dense-phase and dilute-phase regions, and conveying can occur over

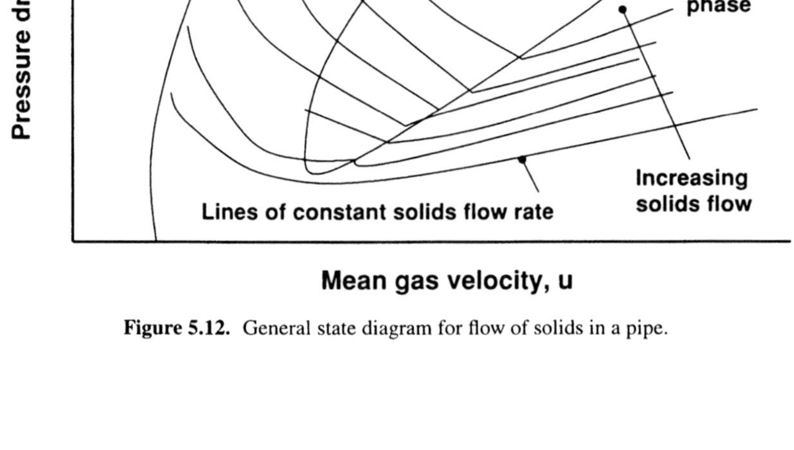

Figure 5.12. General state diagram for flow of solids in a pipe.

a continuous range from fully suspended to a slow moving bed. With other materials, very distinct regions are observed and the conveying progresses in either a very stable or unstable way.

Dense-phase conveying, also termed "non-suspension" conveying, is normally used to discharge particulate solids or to move materials over short distances. There are several types of equipment such as plug-phase conveyors, fluidized systems, blow tanks, and, more innovative, long-distance systems. Dilute-phase, or dispersed-phase conveyors, are more versatile in use and can be considered the typical pneumatic conveying system as described in the literature. The most accepted classification of dilute-phase conveyors comprises pressure, vacuum, combined, and closed-loop systems.

5.5.4. Dense-Phase Conveyors

The development of dense-phase conveying technology has significantly broadened the potential for using these systems to cover longer moving distances. Compared with conventional dilute-phase transport, these relatively high-pressure systems can be designed safely to operate at reduced conveying velocities which, in turn, leads to lower pipeline wear, smaller pipe sizes, and lower operation costs.

5.5.4.1. Plug-Phase Conveyors

Plug-phase conveying basically involves the transportation of a limited batch of material per conveying cycle. A typical arrangement for a plug-phase feeder is shown in Fig. 5.13. As can be seen, the device functions by introducing pressurized air on top of a head of material contained in the pressure vessel. If the material is free flowing, it will flow through a valve and move into a short conveying line. When using this system, care should be taken to avoid surges of air caused either by the tank emptying or by the air breaking through the product. This type of conveying system can be used to transport granular materials such as grains and other food particulates over relatively short distances, e.g., up to 200 m. The method of air injection can have a significant impact on the overall

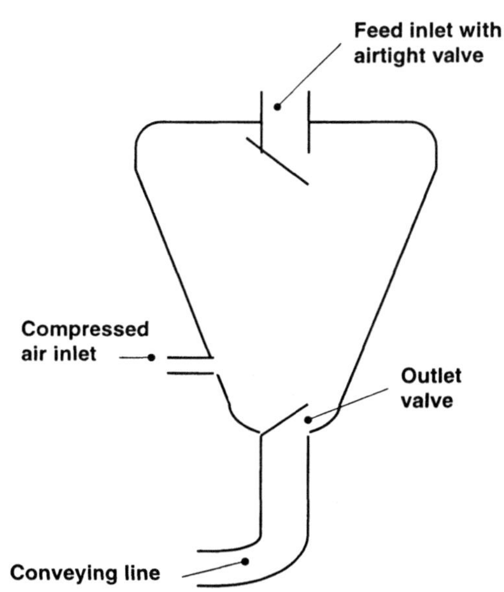

Figure 5.13. Diagram of a plug-phase feeder.

performance and efficiency of the system. Compared with conventional conveying systems, both dense-phase and dilute-phase, plug-phase conveying offers some advantages such as use of lower average conveying velocities, and minimization of system erosion.

5.5.4.2. Fluidized Systems

Particulate solids, contained in silos or bulk transportation vehicles, may be fluidized to assist discharge. Other applications are conveying of pre-fluidized, finely divided non-free-flowing materials over short distances, such as from storage to the entrance of a main conveying system. In fluidized systems, air is introduced in sufficient quantity and at sufficient pressure to produce gravity flow of the material. A diagram of a fluidized system is given in Fig. 5.14. Fluidizing is accomplished by means of a chamber in which air is passed through a bed upon which the material to be conveyed rests. The beds used consist of porous ceramic tiles, sintered metal or plastic, fine wire-mesh, and similar materials. As air passes through the porous bed, each particle is surrounded by a film of air. At the point of incipient fluidization, the material starts behaving as a free-flow powder or particulate. Fluidized solids may be forced through pipelines by air under pressure passed through a perforated flexible inner tube running the length of the pipeline. The fluidized material is then conveyed by establishing a pressure drop along the pipeline.

Fluidized systems are the most reliable and economical method of conveying certain powders and bulk solids over distances ranging from a few meters up to about one kilometer. They take advantage of the fluidization and air retention properties of the bulk solid (Mainwaring and Reed, 1987). The suitability of a given material to this moc of conveying, and the best type of feeder

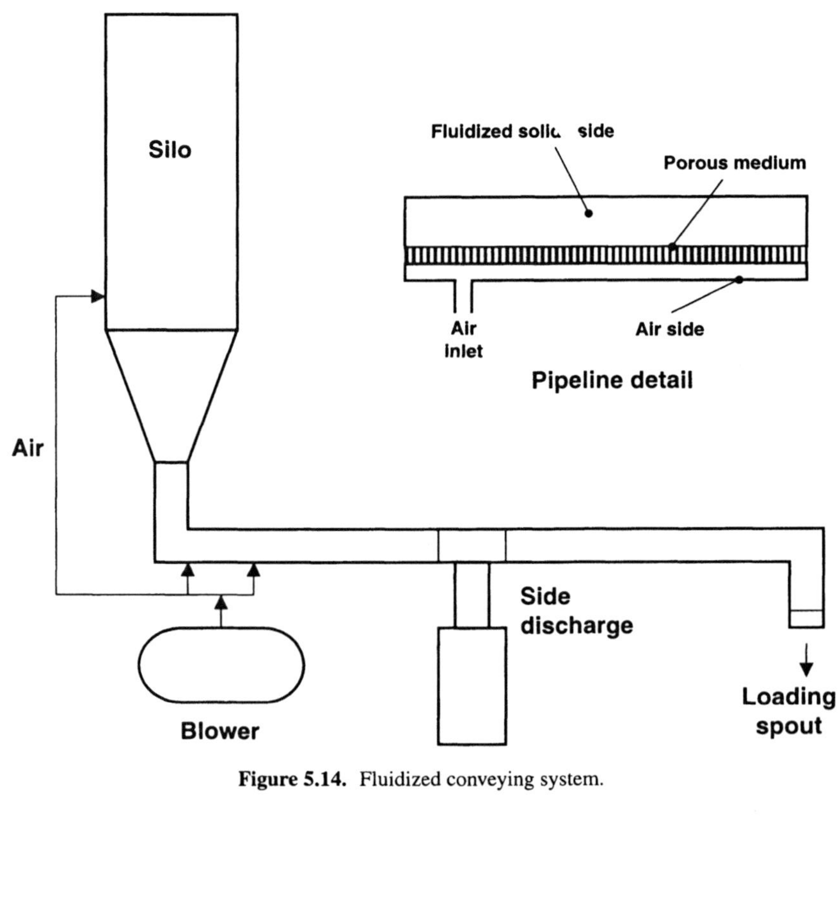

Figure 5.14. Fluidized conveying system.

for these systems, should be determined by test work. Typical materials discharged, using specially designed road and rail vehicles, include bulk loads of sugar, salt, and flour.

5.5.4.3. Blow Tanks

This type of slow motion, dense-phase pneumatic conveying allows friable products to be conveyed with extremely low levels of product degradation or damage, because solids conveying velocities usually occur in the range of 0.25–2 m/s. Such features are very important for a wide variety of food products, especially for those prone to attrition. Food materials conveyed successfully by this system include sugar, wheat, barley, spray-dried milk powder, peanuts, milled grain, as well as powdered and granulated coffee.

Low velocity systems operate at relatively high conveying pressures, e.g., 200–500 kPa, and consequently must have reliable hardware. Also, due to the extremely high levels of concentration that occur during conveying, the subsequent operation conditions depend strongly on the nature and physical properties of the material being transported. It is important, for these reasons, that full-scale tests be exercised prior to the selection or design of such equipment. A schematic diagram of a blow tank is presented in Fig. 5.15.

5.5.4.4. Long Distance Systems

One of the most interesting and innovative applications of dense-phase pneumatic conveying is long distance transportation. Conveying of fine and coarse materials over distances up to 3 or 4 km are possible with this system. This can be achieved, as illustrated in Fig. 5.16, by tandem arrangement of blow tanks. Long distance systems operate at high pressures and particular attention must be given to the control of solids feeding into the pipeline. This can be achieved by the cone dosing system shown in Fig. 5.16. In such a system, the flow of solids is controlled by the movement of a double cone located near the blow tank outlet. The cone moves continuously up and down inside the vessel, but the stroke is adjusted by a proportional-integral controller, based on a conveying line back-pressure measuring signal and set-point. The higher the conveying or operating pressure, the lower the cone set point inside the vessel. The flow of solids into the conveying pipeline are restricted in this way until a satisfactory conveying pressure is restored. Furthermore, the cone dosing system provides an additional and useful feature of mechanical agitation, as well as flow assistance of cohesive materials.

5.5.5. Dilute-Phase Conveyors

As stated earlier, dilute-phase conveying is the commonly employed method for transporting a wide variety of suspended solids using air flowing axially along a pipeline. The method is mainly characterized by the low solids-to-air ratio and by the fact that air and solids flow as a two-phase system inside a pipeline.

5.5.5.1. Types of Conveyors

Figure 5.17 shows the four main types of dilute-phase conveying systems. The pressure system, also called a positive pressure or push system, operates at super-atmospheric pressure and is used for delivery to several outlets from one inlet (Fig. 5.17a). Although most applications of these systems lie within the scope of dilute-phase conveying, under certain arrangements they can also operate as high pressure, dense-phase conveyors. In general, pressure systems can hold higher capacities and

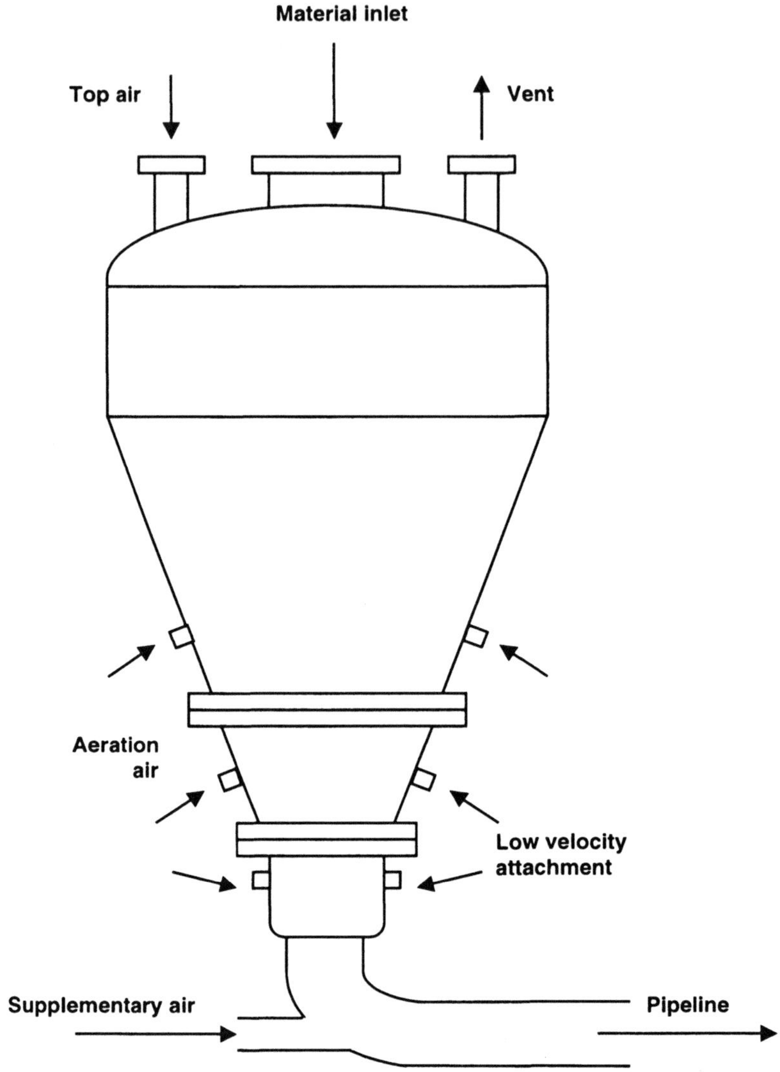

Figure 5.15. Low-velocity blow tank.

longer conveying distances than negative pressure systems. The vacuum, negative pressure or pull system works at sub-atmospheric pressure and is used for delivery to one outlet from several inlets (Fig. 5.17b). Vacuum systems are usually limited to shorter distances than positive ones and are more restrained to operate with dilute, low solids loading than pressure systems. When both features of pressure and vacuum systems are combined in a unit, the advantages of each can be exploited (Fig. 5.17c). These arrangements consist of two sections: a pull/push system with a negative pressure front end, followed by a positive pressure loop. The benefit is that they capitalize on the ease of feeding into a vacuum and combine this with the higher capacity and longer conveying distance when using positive pressure. Recirculation of the conveying air, as in the closed-loop system (Fig. 5.17d), reduces contamination of the product by air and limits product dehydration. However, such systems are often difficult to control and an intercooler may be required to prevent the pump from overheating the re-circulated air.

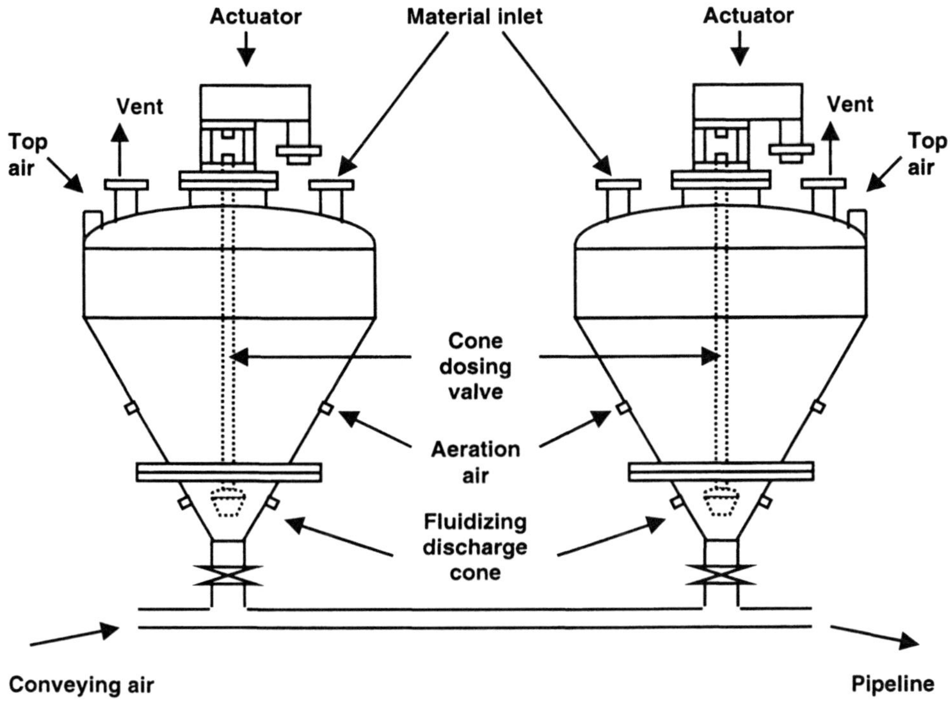

Figure 5.16. Tandem of blow tanks for long distance, dense-phase pneumatic conveying.

5.5.5.2. Operating Features

Pertaining to air velocity, most food materials may be conveyed satisfactorily at air speeds within the range of 15–25 m/s. Above this, abrasion of tube bends and product damage may present difficulties. At extremely low speeds, solids tend to settle out and block horizontal pipe runs. In terms of pressure drop, if air at high pressure is used, its corresponding high initial energy will enable more conveying to be accomplished, per kilogram of air, than if low pressure air is used. High pressure systems are, however, proportionately more expensive than low pressure ones. The maximum pressure recommended for general purpose, dilute-phase conveying of food materials is about 170 kPa. With regard to solids-to-air ratio, for maximum efficiency this should be as high as possible, but without invading the range of dense-phase conveying. For flour and salt, such a ratio may be up to 80 kg of solid/m^3 of air, while for wheat it may be limited to 30 kg of solid/m^3 of air. There is an upper limit for this ratio for specific materials; exceeding it will cause blockage of the system due to saltation. Finally, material properties such as size, shape, density, and surface properties need to be carefully considered in operation and selection of dilute-phase conveying systems. Other important properties are friability, hygroscopicity, as well as susceptibility to impact, abrasion damage or oxidation.

5.5.5.3. System Components

The main components of dense-phase pneumatic conveying systems include gas movers, solids feeders, pipelines, and solids separators. Their location within the system may vary from type to type, but their functions are the same. The gas mover provides an appropriate flow rate of gas required

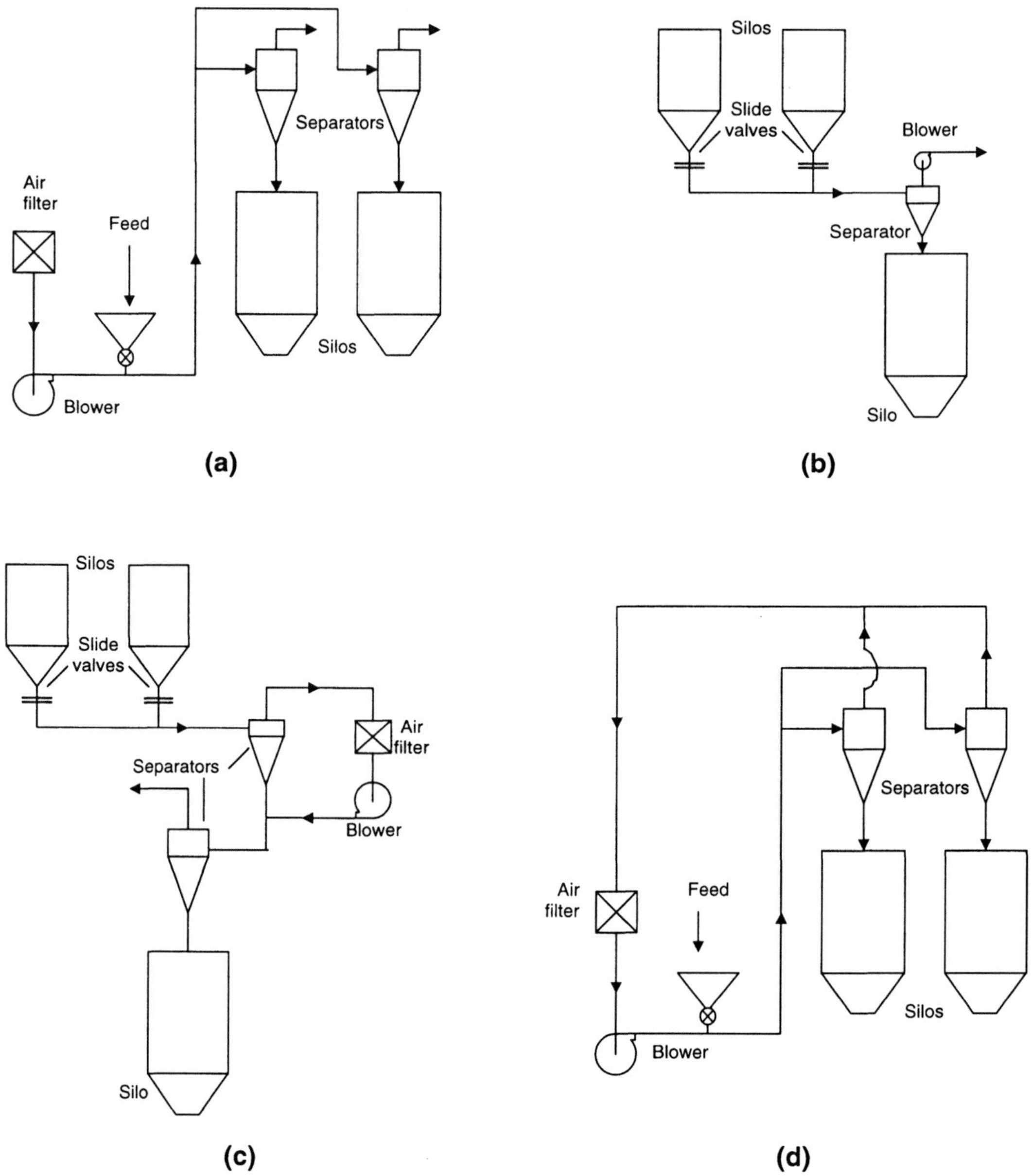

Figure 5.17. Dilute-phase pneumatic conveyors: (a) pressure system; (b) vacuum system; (c) combined system; and (d) closed-loop system.

for transport at the right velocity and pressure. The solids feeder introduces the solids particles at a controlled rate into the pipeline where they are mixed with the conveying gas. The pipeline consists of straight sections, both horizontal and vertical, connected together with bends. At the end of the trajectory, in any type of solid–gas separator, the solids are decelerated and recovered from the gas stream to be stored in a silo or fed into another unit. The gas, which is typically air, is generally released into the atmosphere. Secondary components comprise controls, safety equipment, and instrumentation.

By far, the most common device for moving gas in a pneumatic conveying system is the Roots-type rotary lobe blower. Such a blower is preferred since it provides an economical source of gas flow that meets the pressure or vacuum and flow requirements for the largest category of systems. An important feature of this type of blower is that it delivers a nearly constant volume over its operating pressure range. This is important because control of the gas in pneumatic conveying systems is critical for stable operation. Single-stage blowers generate up to 200 kPa, while double-stage blowers generate up to about 270 kPa. For low pressure conveyors, centrifugal fans can be used as gas movers. High speed, single-stage centrifugal fans deliver air at about 120 kPa and must be supplied with clean air if excessive abrasion of the light impellers is to be avoided. The slower running, multi-stage, heavy duty centrifugal fan can deliver up to 140 kPa and, being slower running, is relatively resistant to particle abrasion. Compressors deliver air at high pressures but require lubrication, a potential contaminant of the solids in movement. They provide pressures up to 800 kPa; with certain measures, they find applications in high pressure systems, as well as in dense-phase conveying.

For proper operation of a dilute-phase pneumatic conveying system, the solids fed into the pipeline must be carefully controlled. Two common types of feeders are the venturi feeder and the rotary valve or star feeder. The venturi feeder is only suitable for low pressure systems. The rotary valve feeder, also known as a star feeder, as described in Chapter 4, is widely used for feeding medium pressure conveyors. This feeder is efficient and simple in principle, but requires careful design in order to minimize air leakage. Excessive air loss from feeders wastes power, causes dust, and leads to system instability. Rotary feeders work well on free flowing, non-abrasive powders, and special types are available for more difficult materials.

Pertaining to pipelines, the use of bends represents the most critical design problem. The difficulties that may arise by the use of bends are abrasive wear, product degradation, product buildup, and pressure loss. The latter problem is primarily related to operation. In general terms, all bend associated drawbacks are minimized by the use of smooth, long radius bends instead of short radius and blind tees types of bends. The system layout is also important. It is often more troublesome to have many bends in short distances than few bends in long distances. In the former case, the contribution of each bend will be more significant than in the latter. However, for systems handling non-abrasive, non-degradable materials, the number and type of elbow is not so critical, and selection or design should probably be based on cost.

Separating the solids from the air stream when conveying concludes can be accomplished in a number of ways. For most cases, a single technique is insufficient to separate the conveyed material completely and satisfactorily. The most common approach consists of using a cyclone followed by a bag filter. Both separating techniques will be discussed in a later chapter. The selection of a gas–solids separating system should be based on the material characteristics, degree of separation required, environmental regulations, concentration of solids, and cost.

5.5.5.4. Selection and Design

Selecting the best system for a particular application depends on the process requirements and the characteristics of the material to be conveyed. Most of the criteria used in design of dilute-phase conveying systems are based upon theoretical aspects of two-phase flow. As in previously discussed conveying methods, relevant information in the form of brochures, design sheets, tabulated data, and nomographs is in the possession of conveyor manufacturers. A theoretical approach presented by Jackson and Lamb (1981) consists of determining pressure drop as a function of potential energy, kinetic energy, and frictional energy changes in horizontal and vertical sections, as well as in bends in the conveying line. The estimation of the overall pressure drop results from summing up all the individual contributions, depending on the specific layout of a pneumatic conveying system. The detailed procedure is as follows:

(1) In horizontal sections, the potential energy loss is zero, and the kinetic energy gain, E_k, is given by:

$$E_k = G_s u_p \tag{5.22}$$

where G_s is the flux of solids and u_p is the particle velocity.

(2) If particles have accelerated to maximum value, the particle velocity is defined by:

$$u_p = u_0 - u_t \tag{5.23}$$

where u_0 is the air velocity and u_t is the terminal settling velocity of the particles.

(3) The behavior of particles settling in a fluid is well understood, and the settling velocity depends on the flow regime:

$$u_t = \frac{x_p^2(\rho_s - \rho_g)g}{18\mu_g} \tag{5.24}$$

for $Re_p < 0.4$, which is the well-known Stokes' law. Also:

$$u_t = \left[\frac{4(\rho_s - \rho_g)^2 g^2}{225\rho_g \mu_g}\right]^{1/3} (x_p) \tag{5.25}$$

for $Re_p < 500$. And, finally:

$$u_t = \left[\frac{3.1 x_p(\rho_s - \rho_g)g}{\rho_g}\right]^{1/2} \tag{5.26}$$

for $Re_p < 2 \times 10^5$.

In Eqs. (5.24)–(5.26), Re_p refers to the particle Reynolds number, which is defined by:

$$Re_p = \frac{x_p \rho_g u_t}{\mu_g} \tag{5.27}$$

and all the variables have been previously defined.

(4) The frictional energy loss due to the gas flow E_{fg} is given by:

$$E_{fg} = \frac{2 f_g \rho_g u_0^2 L}{D} \tag{5.28}$$

where f_g is the gas friction factor, L is the pipe length, and D is the pipe diameter. The gas friction factor f_g, adopts values according to the system Reynolds number Re:

$$f_g = 0.0791(\text{Re})^{-0.25} \tag{5.29}$$

for $3 \times 10^3 < \text{Re} < 10^5$. And:

$$f_g = 0.0008 + 0.0552(\text{Re})^{-0.237} \tag{5.30}$$

for $10^5 < \text{Re} < 10^8$.

As indicated above, the Reynolds number in Eqs. (5.29) and (5.30) is that of the system:

$$\text{Re} = \frac{D \rho_g u_0}{\mu_g} \tag{5.31}$$

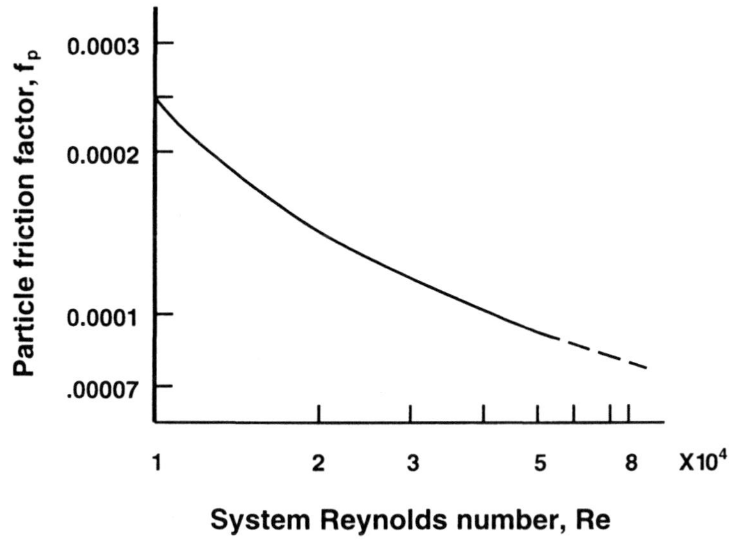

Figure 5.18. Particle friction factor as a function of system Reynolds number, Re (adapted from Rose and Barnacle, 1957).

(5) The frictional energy loss due to the particle flow E_{fp} can be estimated by:

$$E_{fp} = \left(\frac{\pi}{8}\right)\left(\frac{f_p}{f_g}\right)\left(\frac{\rho_s}{\rho_g}\right)^{1/2}\left(\frac{G_s}{G}\right) E_{fg} \tag{5.32}$$

where G is the flux of air and f_p is the particle friction factor, which can be determined graphically, as a function of Re, from Fig. 5.18.

(6) The frictional loss in flow round bends E_{fb} can be calculated from:

$$E_{fb} = 2f_b\left(\frac{G_s}{G}\right)\left(\frac{u_0^3}{u_p}\right)\rho_g \tag{5.33}$$

where $f_b = 0.375$, 0.188 or 0.125 for bend radius: pipe diameter ratios of 2, 4, 6, or more, respectively.

(7) For vertical sections, there will normally be no kinetic energy change. The frictional energy losses are calculated as for the horizontal sections. The potential energy gain E_p is given by:

$$E_p = \left(\frac{G_s}{G}\right)\left(\frac{u_0}{u_p}\right)\rho_s gh \tag{5.34}$$

where h is the height of the section.

As stated earlier, the sum of all the energy losses described above, applied to a particular system layout, will give the total pressure drop through the conveyor. A practical approach for preliminary calculations in selection and design of pneumatic conveying systems is the use of nomographs, such as those presented by Green and Maloney (1999).

5.5.6. Applications

As can be gathered from the discussion in this section, pneumatic conveying is, possibly, the most applicable type of technology for diverse conveying tasks in the food industry. All the many advantages previously mentioned pertaining to the various types of pneumatic conveyors described, suit food materials perfectly. The predominant features of many food powders and particulates, in the sense of being susceptible to damage when handled, makes pneumatic conveying an obvious alternative for food processing. It is worth pointing out, however, that for a number of examples already mentioned, such as sugar, spray-dried milk powder, as well as powdered and granulated coffee, this type of conveying has been found to be the most appropriate way of transportation with a minimum amount of damage. A final reference of the advantages of pneumatic conveying systems related to handling of food materials has to do with their self-cleaning capacity, virtual dustless operation, and generally sanitary conditions.

REFERENCES

Cabrejos, F.J. and Klinzing, G.E. (1994). Minimum conveying velocity in horizontal pneumatic transport and the pickup and saltation mechanisms of solid particles. *Bulk Solids Handl.* **14**: 541–550.

Chattopadhyay, A., Rao, K.R. and Parameswaran, M.A. (1994). Development of database system on bulk solids—a relational model approach. *Bulk Solids Handl.* **14**: 585–590.

Deepak, C.R., Chattopadhyay, A. and Rao, K.R. (1994). An integrated knowledge-based software for belt conveyor design. *Bulk Solids Handl.* **14**: 775–779.

Dentsoras, A. and Gavrielatos, T. (1993). Applying expert systems to belt conveyor design. *J. Syst. Eng.* **3**: 121–134.

Foote, B.L., Bobbie, L. Radidran, A. and Badiru, A.B. (1988). Simulation and network analysis payoff in conveyor system analysis. *Ind. Eng.* **20**: 48–53.

Green, D.W. and Maloney, J.O. (1999). *Perry's Chemical Engineers' Handbook*. McGraw-Hill, New York.

Jackson, A.T. and Lamb, J. (1981). *Calculations in Food and Chemical Engineering*. The MacMillam Press, London.

Lewis, R.K. (1985). The impact of technology on conveyor belt performance. *Bulk Solids Handl.* **5**: 25–29.

Mainwaring, N.J. and Reed, A.R. (1987). Permeability and air retention characteristics of bulk solid materials in relation to modes of dense-phase pneumatic conveyor performance. *Bulk Solids Handl.* **7**: 415–425.

Reed, A.R. and Bradley, M.S.A. (1991). Advances in the design of pneumatic conveying systems. A United Kingdom perspective. *Bulk Solids Handl.* **11**: 93–97.

Roberts, A.W. (1994). Advances in the design of mechanical conveyors. *Bulk Solids Handl.* **14**: 255–281.

Roberts, A.W. (1999). The influence of granular vortex motion in the volumetric performance of enclosed screw conveyors. *Powder Technol.* **104**: 56–67.

Roberts, A.W. (2000). Predicting the performance of enclosed screw conveyors. In *From Powder to Bulk Conference*. IMechE, London, June 2000.

Rose, H.E. and Barnacle, H.E. (1957). Flow of suspensions of non-cohesive spherical particles in pipes. *Engineer, London* **203**: 898, 939.

Wright, H., McElhinney, I. and Lemmon, L. (1997). Current UK drive power formulae for belt conveyors—fact or friction. *Bulk Solids Handl.* **17**: 201–204.

Wypych, P.W. (1999). The ins and outs of pneumatic conveying. In *International Symposium on Reliable Flow of Particulate Solids III*, Porsgrunn, Norway, August 11–13, 1999.

Zenz, F.A. (1964). Conveyability of materials of mixed particle size. *Ind. Eng. Chem. Fundam.* **3**: 65–69.

CHAPTER 6

SIZE REDUCTION

6.1. PRINCIPLES OF SIZE REDUCTION

6.1.1. Introductory Aspects

In many food processes it is frequently necessary to reduce the size of solid materials for different purposes. In this case, size reduction may aid other processes such as expression and extraction, or may shorten heat treatments such as blanching and cooking. Comminution is the generic term used for size reduction and includes different operations such as crushing, grinding, milling, mincing, and dicing. Most of these terms are related to a particular application, e.g., milling of cereals, mincing of beef, dicing of tubers, or grinding of spices. The reduction mechanism consists of deforming the food piece until it breaks or tears. Breaking of hard materials along cracks or defects in their structure is achieved by applying diverse forces.

The objective of comminution is to produce smaller particles from larger ones. Smaller particles are the desired product either because of their large surface or because of their shape, size, and number. The energy efficiency of the operation can be related to the new surface formed by the reduction in size. The geometric characteristics of particles, both alone and in mixtures, are important for product evaluation after comminution. In an actual process, a given unit does not yield a uniform product, whether the feed is uniformly sized or not. The product normally consists of a mixture of particles, which may contain a wide variety of sizes and even shapes. Some types of equipment are designed to control the magnitude of the largest particles in their products, but the fine sizes are not under such control. In some machines, fines are minimized, but they cannot be totally eliminated. In comminuted products, the term "diameter" is generally used to describe the characteristic dimension related to particle size. As described in Chapter 2, the shape of an individual particle is conveniently expressed in terms of the sphericity Φ_s, which is independent of particle size. For spherical particles Φ_s equals unity, while for many crushed materials its value lies between 0.6 and 0.7.

The different types of particle size distributions have been discussed extensively in Chapter 2. No single distribution applies equally well to all comminuted products, particularly in the range of coarser particle sizes. For finer particles, however, the most commonly found distribution follows a log-normal function (Herdan, 1960), which is the most useful among the different types of functions (Beddow and Meloy, 1980).

6.1.2. Forces Used in Size Reduction

As previously mentioned, in comminution of food products the reduction mechanism consists of deforming the food piece until it breaks or tears, and such breaking may be achieved by applying diverse forces. The types of forces commonly used in food processes are compressive, impact, attrition or shear and cutting. In a comminution operation, more than one type of force is usually acting. For example, crushing, grinding, and milling take place in powdered sugar, flour, mustard, and cocoa production. Table 6.1 summarizes these types of forces in some of the mills of common use in the food industry.

Table 6.1. Types of force used in size reduction equipment.

Force	Schematic diagram	Principle	Example of equipment
Compressive	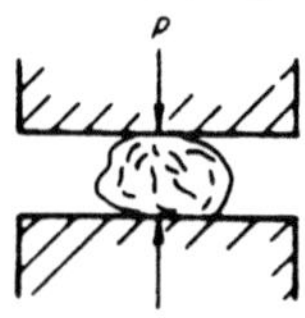	Nutcracker	Crushing rolls
Impact	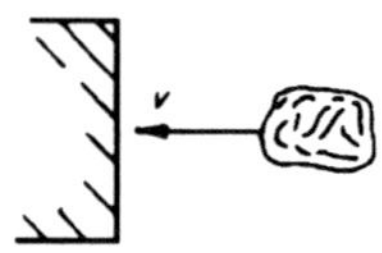	Hammer	Hammer mill
Attrition	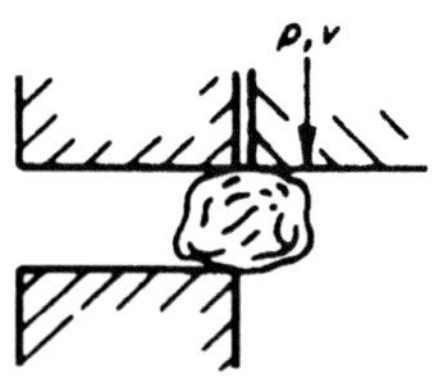	File	Disc attrition mill
Cut		Scissors	Rotary knife cutter

Compressive forces are used for coarse crushing of hard materials. Coarse crushing implies reduction to a size of about 3 mm. Impact forces can be regarded as general purpose forces and may be associated with coarse, medium and fine grinding of a variety of food materials, such as during the breakage of nuts. Shear or attrition forces are applied in fine pulverization, when the size of products can reach the micrometer range. Sometimes a term referred to as ultra-fine grinding is associated with processes in which the sub-micron range of particles is attained. Finally, cutting gives a definite particle size and may even produce a definite shape.

6.1.3. Mechanical Resistance Involved in Size Reduction

Mechanical resistance refers to all the properties that describe the behavior of a solid material as it deforms and breaks under the influence of an applied stress (Loncin and Merson, 1979). The deformation of a certain food material can be elastic when the applied stress remains below a limiting value, or inelastic. The material experiences elastic deformation when it returns to its original shape when the force is removed. If the stress exceeds the elastic limit, the material undergoes permanent (inelastic) deformation until it reaches the yield point when it begins to flow (region of ductility) under the action of the applied stress until it finally breaks. This process defines the *elastic stress limit, yield stress, breaking stress*, and the *region of ductility*.

The *breaking* stress or ultimate stress is a property of the material. Breaking occurs along cracks or defects in the piece structure. A large piece with many defects can be broken with a small stress with very little deformation. Smaller pieces have fewer defects remaining and will need a higher breaking strength. In the limit of very small particles, purely intermolecular forces must be overcome. This is why grinding is so difficult to achieve below a certain size. For example, fine grinding of roasted coffee (e.g., to less than 50 μm) is best recommended under cryogenic conditions (i.e., subzero temperatures) in order to accomplish the desired grinding efficiency.

The elastic state is described by Hooke's law ([stress] $= E \cdot$ [strain]). The most important characteristic is the modulus of elasticity E, which is the stress causing a unit change in length in the same direction as the applied force. The Poisson coefficient or bulk modulus permits prediction of the transverse contraction or expansion that occurs when a stress is applied longitudinally. Inelastic behavior is defined for stresses greater than the elastic limit but smaller than the breaking limit. The behavior is described by viscoelastic models (such as the Maxwell model or the Kelvin model), which combine elements of inelastic behavior and elements of viscous flow.

Stress can be applied to the particle in a variety of ways such as traction, compression, or shear (or combination of those methods). Although the stress limits are not the same for these different modes, there is clearly a relation among them. Consequently, it is often possible to use a compression test as an indication of the breaking load under tension. Since breaking occurs along cracks, in some materials the breaking point measured by compression is usually higher than when measured by traction; tension enhances the cracks, whereas compression tends to close them up.

Figure 6.1 represents the stress characteristics of materials. A food may be hard or soft; increased hardness is correlated with an increase in the modulus of elasticity. A strong material possesses a high elastic stress limit; and a weak material has a low elastic limit. Brittleness is a measure of the size of the region of ductility, and a brittle material breaks soon after the stress exceeds the

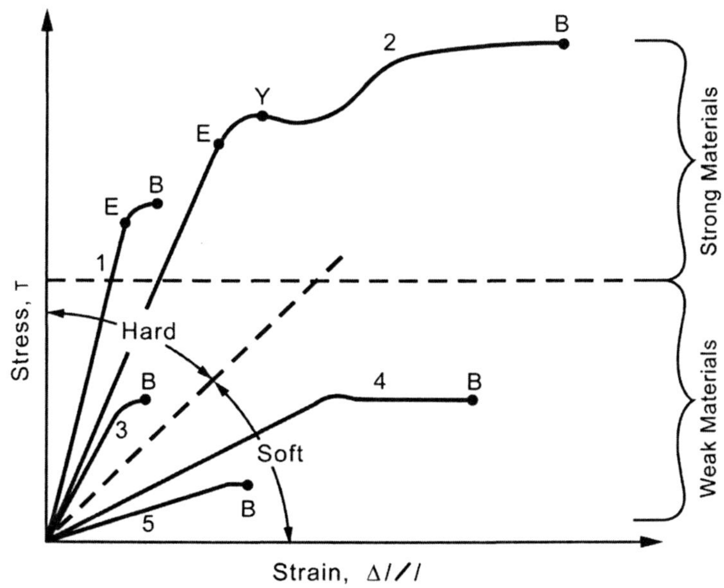

Figure 6.1. Stress–strain diagram for various types of solids. E is elastic limit, Y is yield point, B is breaking point, OE is elastic region, EY is elastic deformation, and YB is region of ductility. Different curves are represented for different types of material depending on their mechanical behavior: material (1): hard, strong, and brittle; material (2): hard, strong, and ductile; material (3): hard, weak, and brittle; material (4): soft, weak, and ductile; material (5): soft, weak, and brittle. From Loncin and Merson (1979).

yield stress. Conversely, a ductile material can deform considerably without breaking. A further property is toughness. A tough material has the ability to resist the propagation of cracks. Fibers impart toughness by relieving stress concentrations at the end of cracks. The opposite of toughness is fragility.

6.1.4. Properties of Comminuted Products

As stated earlier, the breakdown of solid material is performed through the application of mechanical forces that attack fissures present in its original structure. These stresses have been traditionally used to reduce the size of hard materials, either from inorganic origin (e.g., rocks and minerals) or from organic origin (e.g., grains and oilseeds). In both cases, comminuted particles obtained after any size reduction operation will resemble polyhedrons with nearly plane faces and sharp edges and corners. The number of major faces may vary, but will be usually between 4 and 7. As previously mentioned, a compact grain with several nearly equal faces can be considered as spherical, so the term diameter is normally used to describe the particle size of these comminuted products.

The predictable shape of the products described above has to do with molecular structure, since silicon and carbon, elements of the same group in the periodic chart, are generally key components of the crystal units which form the solid matrix. In this sense, a good number of food materials will present the hardness associated with the rigid structure of carbon derivatives and, as such, they will fragment following the same pattern of their relatives in the inorganic world whose structure is due to the presence of silicon components. An ideal size reduction pattern to achieve a high reduction ratio of hard brittle food materials, such as sugar crystals or dry grains, could be obtained first by compressing, then by using impact force, and finally by shearing or rubbing. Therefore, only these hard brittle food materials would produce powders when subjected to different forces in a comminution operation, whereas tough ductile food materials such as meat can only be reduced in size by applying cut forces. In fact, cutting is considered a process totally different to comminution because its operating principles are quite different from those governing the size reduction of hard materials.

In a comminution operation of food materials more than one type of the above-described forces is actually present. Regardless of the uniformity of the feed material, the product always consists of a mixture of particles covering a range of sizes. Some size reduction equipment is designed to control the size of the largest particles in its products, but the fine sizes are not under control. In spite of the hardness of the comminuted materials, the above-mentioned shape of produced particles would be subjected to attrition due to interparticle and particle—equipment contacts within the dynamics of the operation. Thus, particle angles will smooth gradually, with the consequent production of fines. In actual practice, any feed material will possess an original particle size distribution while the obtained product will end with a new particle size distribution having a whole range finer than the feed distribution.

A product specification will commonly require a finished product not to contain particles greater than (or smaller than, depending on the application) some specified size. In comminution practice, particle size is often referred to as screen aperture size (Chapter 10). The reduction ratio, defined as the relation between average size of feed and average size of product, can be used as an estimate of the performance of a comminution operation. The values for average size of feed and product depend on the method of measurement, but the true arithmetic mean, obtained from screen analyses on samples of the feed and product streams, is commonly used for this purpose. Reduction ratios depend on the specific type of equipment. As a general rule, the coarser the reduction, the smaller the ratio. For example, coarse crushers have size reduction ratios of below 8:1, while fine grinders may present ratios as high as 100:1. However, large reduction ratios, such as those obtained when dividing relatively large solid lumps to ultra-fine powders, are normally attained by several stages using diverse crushing and grinding machines. A good example of this is the overall milling of wheat grain into fine flour, in which crushing rolls in series of decreasing diameters are employed.

6.2. ENERGY REQUIREMENTS: COMMINUTION LAWS

In the breakdown of hard and brittle food solid materials, two stages of breakage are recognized: (a) initial fracture along existing fissures within the structure of the material; and (b) formation of new fissures or crack tips followed by fracture along these fissures. It is also accepted that only a small percentage of the energy supplied to the grinding equipment is actually used in the breakdown operation. Figures of less than 2% efficiency have been quoted (Coulson and Richardson, 1996); thus, grinding is a very inefficient process, perhaps the most inefficient of the traditional unit operations. Much of the input energy is lost in deforming the particles within their elastic limits and through interparticle friction. A large amount of this wasted energy is released as heat which, in turn, may be responsible for heat damage of biological materials.

Elastic and inelastic properties of a given food material often vary considerably with moisture content and the distribution of water in the material. Further complications arise because these properties are often strongly anisotropic, with various layers or parts having extremely different mechanical resistances. Furthermore, the properties of materials can vary with the rate with which the stress is applied; some materials are plastic and ductile if the stress is applied slowly, but can be elastic or brittle if the stress is applied by impact. Consequently, it is not possible, at present, to describe a food material or furnish the parameters needed to design a size reduction operation. These parameters must be determined experimentally.

The energy needed to cause rupture is the work needed to deform the material plus the energy needed to form the new surface. The latter is given by:

$$E = \Delta(\sigma A) \tag{6.1}$$

where σ is the interfacial energy of the surface and A is the surface area. The minimum work of distortion can be measured by placing a sample in tension (or compression) in a machine (such as an Instron testing machine) that simultaneously measures both the applied force and the elongation up to the breaking point. By plotting force vs. elongation and measuring the area under the curve between zero elongation and the elongation $(\Delta x)_{max}$ at rupture, one can evaluate the energy needed for breaking the piece:

$$E = \int_0^{(\Delta x)_{max}} F \, dx \tag{6.2}$$

In an actual grinding machine, the particles undergo many elastic or inelastic deformations that do not exceed the breaking stress and therefore do not cause breakage. These deformations require work, however, which is entirely wasted except for newly formed cracks that facilitate breakage on subsequent impacts. In fact, only 0.06–1% of the energy used in grinding is used to create new surface. The remaining energy appears as heat in the product and equipment, and high temperatures may result.

Theoretical considerations suggest that the energy required to produce a small change in the size of unit mass of material can be expressed as a power function of the size of the material:

$$\frac{dE}{dx} = -\frac{K}{x^n} \tag{6.3}$$

where dE is the change in energy, dx is the change in size, K is a constant, n is the power value and x is the particle size.

Equation (6.3) is often referred to as the general law of comminution and has been used by a number of workers to derive more specific laws depending on the application. It has been developed based on the energy needed for causing deformation, creating new surface, or enlarging cracks.

6.2.1. Rittinger's Law

Rittinger's assumption was based on the fact that particles do not deform before breaking, therefore being infinitely brittle. He considered that for the grinding of solids, the energy required should be proportional to the new surface produced, and gave to the power n the value of 2, thus obtaining the so-called Rittinger's law by integration of Eq. (6.3):

$$E = K\left[\frac{1}{x_2} - \frac{1}{x_1}\right] \tag{6.4}$$

where E is the energy per unit mass required for the production of a new surface by reduction, K is called Rittinger's constant and is determined for a particular equipment and material, x_1 is the average initial feed size, and x_2 is the average final product size. Rittinger's law has been found to hold better for fine grinding, where a large increase in surface results.

6.2.2. Kick's Law

Kick reckoned that the energy required for a given size reduction was proportional to the size reduction ratio, and took the value of the power n as 1. In such a way, by integration of Eq. (6.3), the following relation, known as Kick's law is obtained:

$$E = K\left[\ln\frac{x_1}{x_2}\right] \tag{6.5}$$

where x_1/x_2 is the size reduction ratio. Kick's law has been found to hold more accurately for coarser crushing, where most of the energy is used in causing fracture along existing cracks.

6.2.3. Bond's Law and Work Index

A third version of the comminution law is the one attributed to Bond (1963), who considered that the work necessary for reduction was inversely proportional to the square root of the size produced. In Bond's consideration n takes the value of $3/2$, giving the following version (Bond's law) also by integrating Eq. (6.3):

$$E = 2K\left[\frac{1}{\sqrt{x_2}} - \frac{1}{\sqrt{x_1}}\right] \tag{6.6}$$

where x_1 and x_2 are measured in micrometers and E in kWh/ton, $K = 5E_i$, where E_i is the Bond Work Index, defined as the energy requires to reduce a unit mass of material from an infinite particle size to a size such that 80% passes a 100 micrometer sieve. The Bond Work Index is obtained from laboratory crushing tests on the feed material. Bond's law holds reasonably well for a variety of materials undergoing coarse, medium and fine size reduction.

6.3. SIZE REDUCTION EQUIPMENT

6.3.1. Classification

Size reduction is a unit operation widely used in a number of processing industries. Many types of equipment are used in size reduction operations. In a broad sense, size reduction machines may be classified as crushers used mainly for coarse reduction, grinders employed principally in intermediate and fine reduction, ultra-fine grinders utilized in ultra-fine reduction, and cutting machines used for exact reduction (McCabe et al., 1992). Equipment is generally known as a crusher when it performs coarse reduction and as a mill when used for all other applications. The above-mentioned classification includes several categories of each type of machine; in total, approximately 20 different designs are recognized in comminution processes. In the food industry, not every piece of equipment has important applications. For example, larger types of coarse crushers, such as jaw and gyratory ones, are not normally encountered in the food industry. Table 6.2 lists the principal size reduction machines used for applications in food processing.

6.3.2. Features

Machines of various types and sizes are available for the comminution of materials in the food processing industry. The main characteristics of the most commonly employed units are discussed below.

6.3.2.1. Crushing Rolls

In this type of equipment, two or more heavy steel cylinders revolve towards each other (Fig. 6.2), so particles of feed are nipped and pulled through. The nipped particles are subjected to compressive force, causing the reduction in size. In some designs, differential speed is maintained to exert shearing forces on the particles as well. The roller surface can be smooth or can carry corrugations, breaker bars or teeth, as a manner of increasing friction and facilitating the trapping of particles between the rolls. Toothed-roll crushers can be mounted in pairs, like the smooth-roll crushers, or with only one roll working against a stationary curved breaker plate. Toothed-roll crushers are much more versatile than smooth-roll crushers, but have the limitation that they cannot handle very hard solids. They operate by compression, impact, and shear and not by compression alone, as do smooth-roll

Table 6.2. Size reduction machines used in food process engineering.

Range of reduction	Generic name of equipment	Type of equipment
Coarse and intermediate	Crushers	Crushing rolls
Intermediate and fine	Grinders	Hammer mills Disc attrition mills Tumbling mills (rod mills)
Fine and ultra-fine	Ultra-fine grinders	Hammer mills Tumbling mills (ball mills)

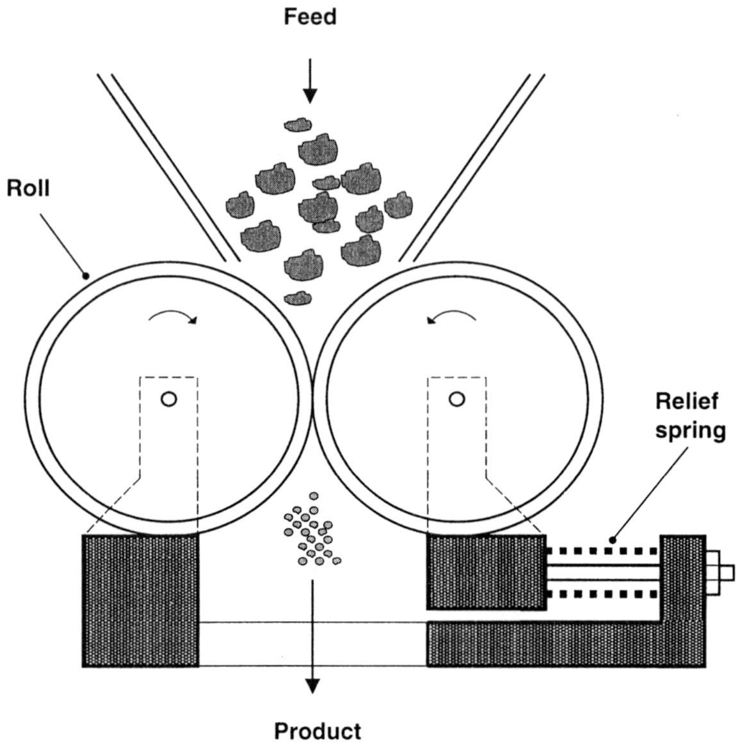

Figure 6.2. Diagram of crushing rolls.

crushers. Crushing rolls are widely applied in the milling of wheat, in roasted coffee grinding, and in the refining of chocolate (Clarke, 1987; Niediek, 1988).

6.3.2.2. Hammer Mills

Figure 6.3 shows a hammer mill, which contains a high-speed rotor turning inside a cylindrical case. The rotor carries a collar bearing a number of hammers around its periphery. By the rotating action, the hammers swing through a circular path inside the casing containing a toughened breaker plate. Feed passes into the action zone with the hammers driving the material against the breaker plate and forcing it to pass through a bottom mounted screen by gravity when the particles attain a proper size. Reduction is mainly due to impact forces, although under choke feeding conditions, attrition forces can also play a part in such reduction. The hammers may be replaced by knives or other devices to give the mill the possibility of handling tough, ductile or fibrous materials. The hammer mill is a versatile piece of equipment that gives high reduction ratios and may handle a wide variety of materials from hard and abrasive to fibrous and sticky. In the food industry, its applications are varied, with extensive use for grinding spices, dried milk, sugar agglomerate, cocoa press cake, tapioca, dry fruits, dry vegetables, and extracted bones.

6.3.2.3. Disc Attrition Mills

These types of mills, as illustrated in Fig. 6.4, make use of shear forces for size reduction, mainly in the fine size range of particles. There are several basic designs of attrition mills. The single

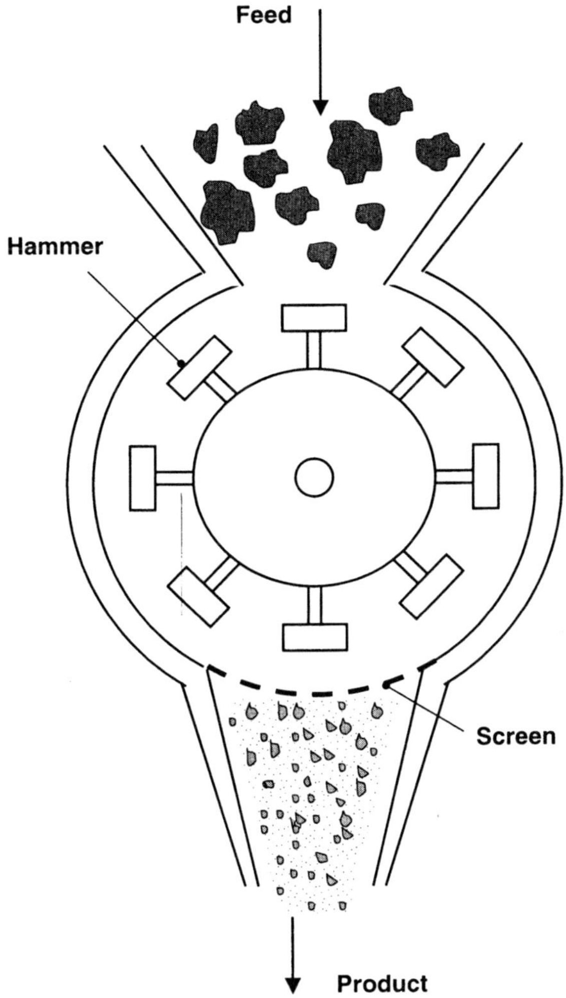

Figure 6.3. A hammer mill.

disc mill (Fig. 6.4a) has a high speed rotating grooved disc leaving a narrow gap with its stationary casing. Intense shearing action results in comminution of the feed. The gap is adjustable, depending on feed size and product requirements. In the double disc mill (Fig. 6.4b), the casing contains two rotating discs that rotate in opposite directions, giving a greater degree of shear compared with the single disc mill. The pin-disc mill carries pins or pegs on the rotating elements. In this case, impact forces also play an important role in particle size reduction. The Buhr mill (Fig. 6.4c), which is the older type of attrition mill originally used in flour milling, consists of two circular stones mounted on a vertical axis. The upper stone is normally fixed and has a feed entry port, while the lower stone rotates. The product is discharged over the edge of the lower stone. The applications of attrition mills in the food industry are extensive. They have been employed in dry milling of wheat, as well as wet milling of corn for the separation of starch gluten from the hulls. Other applications include breaking of cocoa kernels, preparation of cocoa powder, de-germination of corn, production of fish meal, manufacture of chocolate, and grinding of sugar, nutmeg, cloves, roasted nuts, peppers, etc.

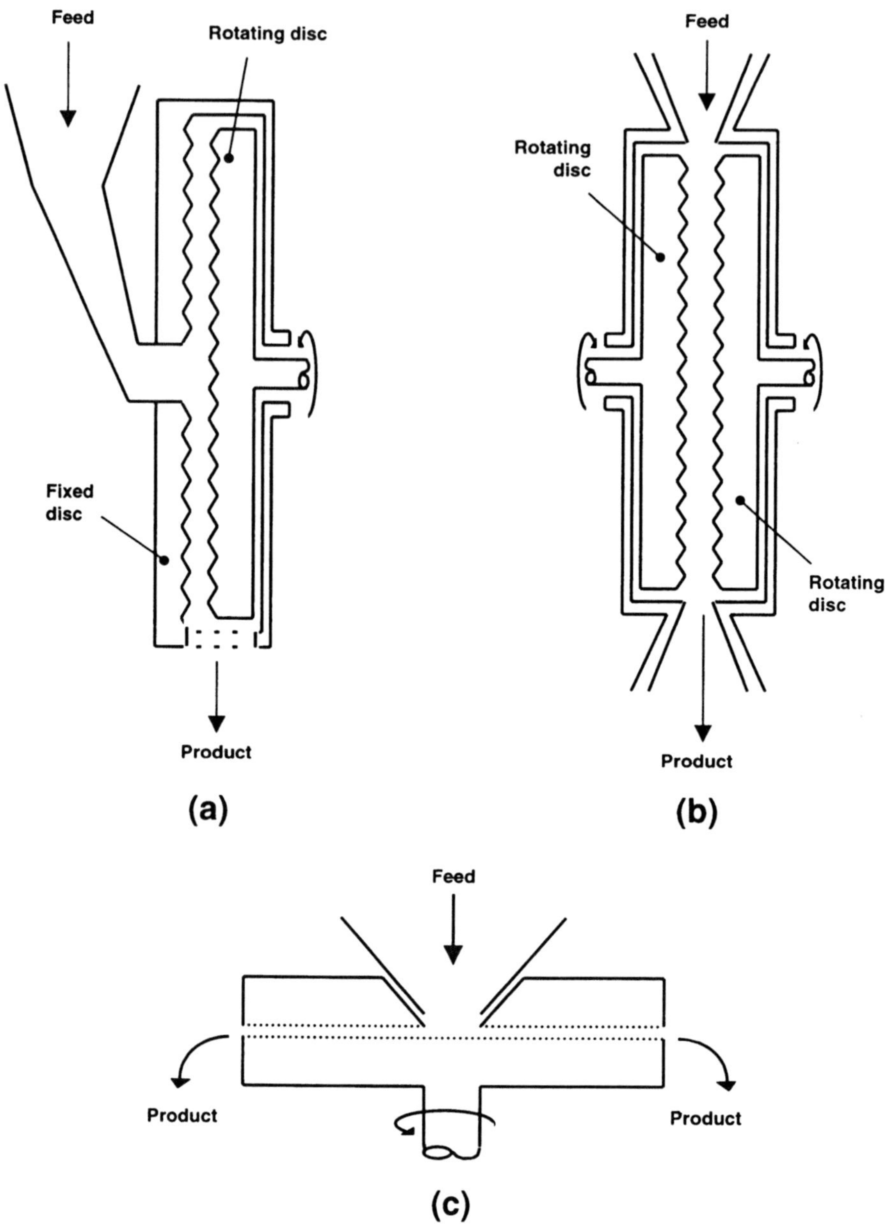

Figure 6.4. Disc attrition mills: (a) single disc mill; (b) double disc mill; and (c) Buhr mill.

6.3.2.4. Tumbling Mills

A tumbling mill is used in many industries for fine grinding. It consists of a horizontal slow speed rotating cylinder partially filled with either balls or rods. The cylinder shell is usually of steel, lined with carbon-steel plate, porcelain, silica rock, or rubber. The balls are normally made of steel or flint stones, while the rods are usually manufactured of high carbon steel. The reduction mechanism is carried out as follows: as the cylinder rotates, the grinding medium is lifted up the sides of the

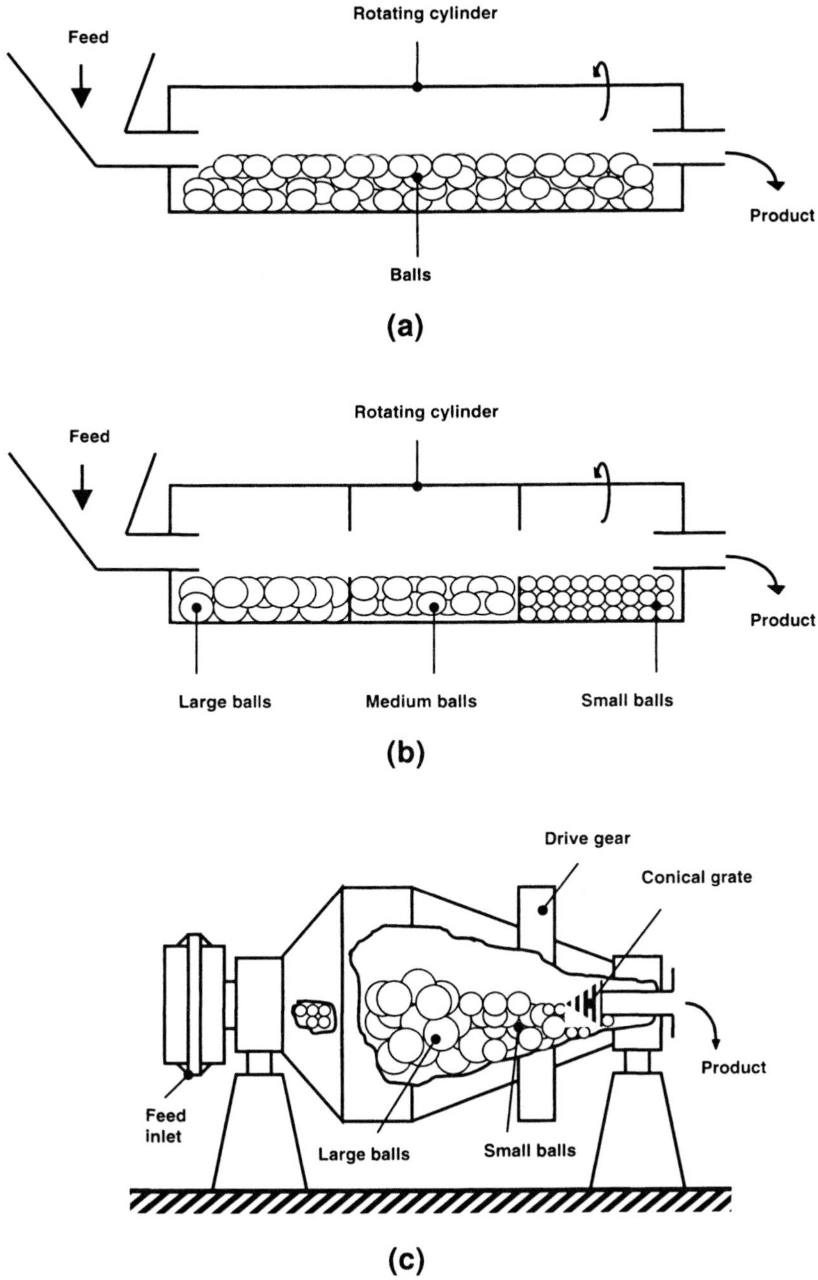

Figure 6.5. Tumbling mills: (a) trunnion overflow mill; (b) compartment mill; and (c) conical mill.

cylinder and dropped onto the material being comminuted, which fills the void spaces between the medium. The grinding medium components also tumble over each other, exerting a shearing action on the feed material. This combination of impact and shearing forces brings about a very effective size reduction. As a tumbling mill basically operates in a batch manner, different designs have been developed to make the process continuous. As illustrated in Fig. 6.5a, in a trunnion overflow mill, the raw material is fed in through a hollow trunnion at one end of the mill and the ground product overflows at the opposite end. Putting slotted transverse partitions in a tube mill converts it into a compartment mill (Fig. 6.5b). One compartment may contain large balls, another small balls, and a third pebbles, thus achieving a segregation of the grinding media with the consequent rationalization of energy. A very efficient way of segregating the grinding medium is the use of the conical ball mill shown in Fig. 6.5c. While the feed solid enters from the left into the primary grinding zone where the diameter of the shell is maximum, the comminuted product leaves through the cone at the right end where the diameter of the shell is minimum. As the shell rotates, the large balls move toward the point of maximum diameter, and the small balls migrate toward the discharge outlet. Therefore, the initial breaking of feed particles is performed by the largest balls dropping the greatest distance, whereas the final reduction of small particles is done by small balls dropping a smaller distance. In such an arrangement, the efficiency of the milling operation is greatly increased. Among the applications of tumbling mills in the food industry is the reduction of fluid cocoa mass.

6.3.3. Operation

The diversity of designs of the above-described machinery implies that their operating variables differ considerably. While energy requirements are generally governed by the comminution laws previously discussed, some other features such as capacity and rotational velocity are particular for each type of equipment and may be related, in some way, to the predominant force performing the reduction action.

In crushing rolls, the angle formed by the tangents to the roll faces at the point of contact between a particle and the rolls is called the angle of nip. It is an important variable for specifying the size of a pair of crushing rolls for a specific duty and is found as follows. Figure 6.6 shows a pair

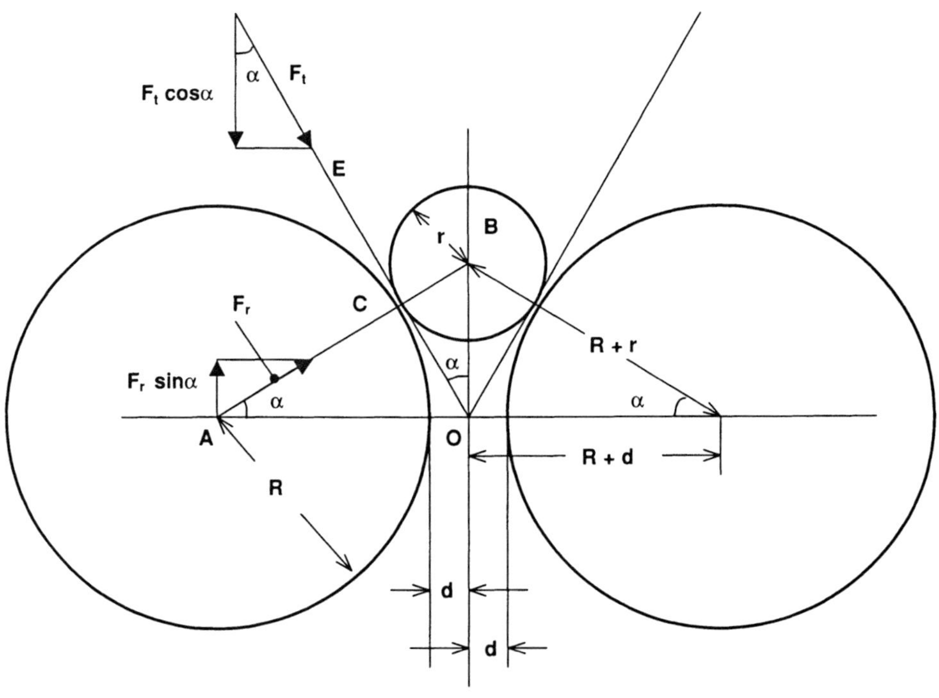

Figure 6.6. Diagram of forces and angle of nip in crushing rolls.

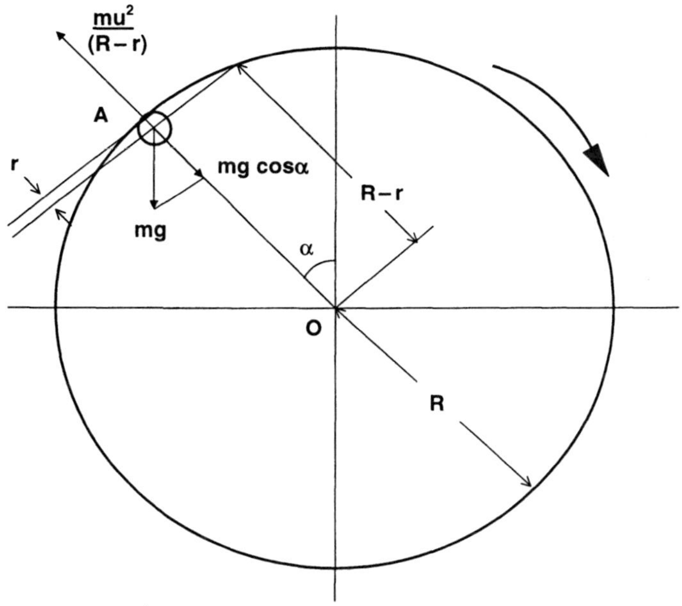

Figure 6.7. Diagram of forces on ball in a ball mill.

of rolls and a spherical particle being gripped between them. The radii of rolls and particle are R and r, respectively. The clearance between the rolls is $2d$. Line AB passes through the centers of the left roll and the particle, as well as through point C, which is the point of contact between the roll and the particle. As shown in Fig. 6.7, if α is the angle between line AB and the horizontal, line OE is a tangent to the roll at point C and it makes the same angle α with the vertical. Neglecting gravity, two forces act at point C: the tangential frictional force F_t, having a vertical component $F_t\cos\alpha$, and the radial force F_r, having a vertical component $F_r\sin\alpha$. Force F_t is related to force F_r through the coefficient of friction μ', so $F_t = \mu' F_r$. Force $F_r \sin\alpha$ tends to expel the particle from the rolls, while force $\mu' F_r \cos\alpha$ tends to pull it into the rolls. If the particle is to be crushed:

$$F_r\mu' \cos\alpha \geq F_r \sin\alpha \tag{6.7}$$

or:

$$\mu' \geq \tan\alpha \tag{6.8}$$

When $\mu' = \tan\alpha$, the angle α is half the angle of nip. A simple relationship exists between the radius of the rolls, the size of the feed, and the gap between the rolls. Thus, from Fig. 6.6:

$$\cos\alpha = \frac{R+d}{R+r} \tag{6.9}$$

The largest particles in the product have a diameter $2d$; Eq. (6.9) provides a relationship between the roll diameter and the size reduction that can be expected in the mill.

From Fig. 6.6 it follows that $2R$ will represent the diameter of the roll D_r, $2r$ will be the diameter of the feed D_f (when the feeding pieces have a proper sphericity Φ_s), and $2d$ will be considered the product diameter D_p. Taking these definitions, the theoretical volumetric capacity Q

of crushing rolls is the volume of the continuous ribbon of product discharged from the rolls and can be given by:

$$Q = \frac{ND_{\mathrm{r}}D_{\mathrm{p}}L}{60} \tag{6.10}$$

where N is the roll speed in rev/min, and L is the length of face in m.

Knowing the bulk density of the discharge stream, the approximate mass flow rate may be estimated. In practice, the actual capacity is found to lie between 0.1 and 0.3 of the theoretical capacity.

The load of balls in a tumbling mill should be such that when the mill is stopped, the balls occupy somewhat more than one-half the volume of the mill. In operation, the balls are picked up by the mill wall and carried nearly to the top where they fall to the bottom to repeat the process. Centrifugal force maintains the balls in contact with the wall and with each other during the upward trajectory. While they keep in contact with the wall, the balls exercise some grinding action by slipping and rolling over each other. Most of the grinding occurs, however, at the zone of impact where the free-falling balls strike the bottom part of the mill.

The faster the mill rotates, the farther the balls are carried up inside the wall and the greater the power consumption. The added power is profitably used because when the balls are carried at the higher point, they will have a greater impact on the bottom and perform a better reduction capacity. When the speed is too high, however, the balls are carried over and the mill is practically centrifuging the balls, impeding them from falling. The speed at which centrifuging occurs is called the critical speed, and little or no grinding is carried out when the mill operates at this, or higher, velocity. Operating speeds must be well calculated in order not to be considerably less than the critical speed, because little grinding action will occur, or considerably higher than the critical speed, because centrifuging will cancel the grinding capacity of the mill.

The speed at which the outermost balls lose contact with the wall of the mill depends on the balance between gravitational and centrifugal forces. Referring to the diagram in Fig. 6.7, considering the ball at point A on the periphery of the mill, the radii of the mill and the ball will be R and r, respectively. Thus, the center of the ball is $R - r$ from the axis of the mill and the radius AO form the angle α with the vertical. Two forces act on the ball: the force of gravity mg, where m is the mass of the ball, and the centrifugal force $mu^2/(R - r)$, where u is the peripheral speed of the center of the ball. The centripetal component of the force of gravity is $mg \cos \alpha$, which opposes the centrifugal force. As long as the centrifugal force exceeds the centripetal one, the particle will not lose contact with the wall. As the angle α decreases, however, the centripetal force increases. If the speed does not exceed the critical value, a point is reached where the opposing forces are equal and the particle is ready to fall. The angle at which this occurs is found by equating the centrifugal and centripetal forces:

$$mg \cos \alpha = \frac{mu^2}{R - r} \tag{6.11}$$

Transposing for $\cos \alpha$, Eq. (6.11) transforms to:

$$\cos \alpha = \frac{u^2}{(R - r)g} \tag{6.12}$$

The speed u is related to the speed of rotation by the equation:

$$u = 2\pi N(R - r) \tag{6.13}$$

Thus, Eq. (6.13) can be written as:

$$\cos \alpha = \frac{4\pi^2 N^2 (R - r)}{g} \tag{6.14}$$

At the critical speed, $\alpha = 0$, consequently $\cos\alpha = 1$, and N becomes the critical speed N_c. With all these considerations, Eq. (6.12) transforms into:

$$N_c = \frac{1}{2\pi}\sqrt{\frac{g}{R - r}} \tag{6.15}$$

Rod mills can produce 5–200 ton/h of material reduced to about 1 mm sizes, while ball mills can give 1–50 ton/h of fine powder with 70–90% sizes in the range of 70 μm. The total energy requirement for a typical rod mill grinding different hard materials is about 5 hp-h/ton, whereas for a ball mill it is approximately 20 hp-h/ton. Tube mills and compartment mills normally need somewhat more power than these figures. As the product becomes finer, the capacity of a given mill diminishes and the energy requirement increases.

6.4. CRITERIA FOR SELECTION OF COMMINUTION PROCESSES

6.4.1. General Considerations

In deciding how to crush or grind a food material, process engineers should consider factors such as the size distributions of feed and product, the hardness and mechanical structure of the feed, the moisture content, and temperature sensitivity of the feed. Regarding the size distributions of materials, each type of crusher or grinder is intended for a certain size of feed and product. It is usually possible to exercise some control over the size of feed, but sometimes it must be taken as it comes. As there is an upper limit on the size that can be accepted by a machine without jamming, for oversized material a guard screen is needed to keep large pieces out of the crusher or grinder. In the case of too much undersized material, pre-screening the feed can cut the amount that goes through the equipment. For small scale operation, such a cut is important, as it decreases the capacity required; in large scale equipment, though, the undersized particles simply pass through the throat, where there is always ample room, so removing them does not greatly affect capacity. A general guide for equipment selection, as a function of food material and reduction range, is presented in Table 6.3.

6.4.2. Hardness and Abrasiveness

One of the major factors that govern the choice and design of size reduction machines is the hardness of the material to be processed. As a general rule, hardness is defined in accordance with Mohs' scale, which is divided into 10 grades of hardness (Chapter 3, Table 3.6). As a rule of thumb, using Mohs' scale, any material is considered soft when it has a value between 1 and 3, medium-hard if presenting a value between 3.5 and 5, and hard when showing a value between 5 and 10. Many food materials, especially when dry, are brittle and fragile with hardness in the Mohs' scale on the order of 1–2. According to this, ball mills, hammer mills, roller mills, and attrition mills are very suitable for treating most solid foods of common use in the food industry. A knowledge of the mechanical structure of the feed material is useful to determine the most likely force to be used in its size reduction. As mentioned above, many food materials are brittle and fragile, so compressive forces may be employed. Some other food materials have a fibrous structure and are not easily disintegrated

Table 6.3. Application examples of size reduction machines.

Fineness range:	Crushing rolls	Hammer mills	Attrition mills	Tumbling mills
Coarse	•			
Intermediate	•	•	•	•
Fine and ultra-fine		•	•	•
Chocolate	•			•
Cocoa			•	•
Corn (wet)			•	
Dried fruits		•		
Dried milk		•		
Dried vegetables		•		
Grains	•		•	
Pepper		•	•	
Pulses			•	
Roasted nuts			•	
Salt		•		•
Spices		•		
Starch (wet)			•	
Sugar		•		•

by compressive or impact forces, so cutting may be required. For example, roasted whole beans require a cutting rather than a crushing action to provide a ground coffee with particles of suitable size and shape.

6.4.3. Mechanical Structure

A knowledge of the particular structure of the feed material can indicate the type of force most likely to be used in performing the size reduction. If the material is friable or has a crystalline structure, fracture may occur easily along cleavage planes, with larger particles fracturing more easily than smaller ones. In these cases, crushing using compressive forces would be recommended. When few cleavage planes are present, and new crack tips have to be formed, impact and shear forces may be more advisable. Many food materials have fibrous structures, so they are not easily reduced by compression or impact. In such cases, shredding or cutting may bring the force needed to perform the desired size reduction.

6.4.4. Moisture

The presence of moisture can be either beneficial or inconvenient in comminution processes. Safety problems caused by dust formation, arising during the dry milling of many solid materials, are well known and reviewed in Chapter 12 (Section 12.4). The presence of small quantities of water has been found useful in the suppression of dust and, in applications where the presence of moisture is acceptable, water sprays are often used to reduce dust formation. Some other applications allow for large quantities of water to be introduced in the size reduction process; wet milling of corn is a good example of this. On the other hand, in many cases feed moisture content in excess of 2–3% can lead to clogging of the mill with a consequent effect on throughput and efficiency. Agglomeration can also be caused by moisture, which is undesirable when a free flowing powder is needed to control feed rate.

6.4.5. Temperature Sensitivity

As stated, comminution is possibly the most inefficient unit operation in the food processing industry. The excessive friction presented in most size reduction machines releases heat that can lead to a considerable rise in temperature of the material being processed. Since food materials are normally heat sensitive, degradation reactions can occur. The release of sticky substances caused by the heat rise may also pose a problem. For example, in the grinding of sugar (Niediek, 1971), the energy released during fracture along fault lines causes each new surface formed to experience a short period (less than a microsecond) when the surface temperature is above 1,000–2,000°C. This causes the surfaces to pass the glass transition temperature and become amorphous. The amorphous sugar layer is hygroscopic until the sugar in these regions can re-crystallize by the formation of liquid and solid bridges through caking. For these reasons, some crushing and grinding machinery may be equipped with cooling devices such as jackets and coils.

6.5. APPLICATIONS

Size reduction is normally applied in an infinite variety of grinding characteristics. These range from readily grindable (sugar and salt), through tough-fibrous (dried vegetables) and very tough (gelatin), to those materials that tend to deposit (full-fat soy, full-fat milk powder). The fineness requirements may vary immensely from case to case. Many examples of applications of size reduction in food processes have been mentioned throughout this chapter. To summarize, the milling of wheat, the refining of chocolate, the grinding of spices and dried vegetables, the breaking of cocoa kernels, the preparation of cocoa powder, the de-germination of corn, the production of fish-meal, the manufacture of chocolate, the roasted coffee processing, etc., all require different methods for size reduction. Thus, for the food industry size reduction is, without doubt, one of the most fundamental processing steps.

REFERENCES

Beddow, J.K. and Meloy, J.K. (1980). *Testing and Characterization of Powders and Fine Particles*. Heyden and Son, London.

Bond, F.C. (1963). Some recent advances in grinding theory and practice. *Brit. Chem. Eng.* **8**: 631–634.

Clarke, J.K. (1987). Roasting and grinding. In *Coffee*, Vol. 2, Clarke, R.J. and Macrae, R. (eds.). Elsevier Applied Science, London, pp. 73–107.

Coulson, J.M. and Richardson, J.F. (1996). *Chemical Engineering*, Vol. 2. Butterworth-Heinemann, Stoneham, MA.

Herdan, G. (1960). *Small Particle Statistics*. Butterworths, London.

Loncin, M. and Merson, R.L. (1979). *Food Engineering. Principles and Applications*. Academic Press, New York, pp. 229–270.

McCabe, W.L., Smith, J.C. and Harriot, P. (1992). *Unit Operations in Chemical Engineering*. McGraw-Hill, New York.

Niediek, E.A. (1971). Studies on crushing of crystal sugar. *Z. Zuckerindus.* **21**(9): 432–439.

Niediek, E.A. (1988). Particle size reduction. In *Industrial Chocolate Manufacture and Use*, Beckett, S.T. (ed.). Blackie & Son Ltd. London.

CHAPTER 7

SIZE ENLARGEMENT

7.1. INTRODUCTION: SIZE ENLARGEMENT PROCESSES

The term "size enlargement" includes a number of processes that purposely combine small particles into large permanent masses in which the initial primary units are still identifiable. Applications in food processing are surprisingly numerous and are becoming increasingly important as more "structured" foods are developed. Size enlargement operations are used in the process industries with different aims such as improving handling and flowability, reducing dusting or material losses, producing structural useful forms, enhancing appearance, etc.

Size enlargement operations are known by many names, including compaction, granulation, tabletting, briquetting, pelletizing, encapsulation, sintering and agglomeration. While some of these operations could be considered similar, e.g., tabletting and pelletizing, some others are relevant to a specific type of industry, e.g., sintering in metallurgical processes. Briquetting and tabletting are important in the pharmaceutical and chemical industries where such food ingredients as dextrose, gelatin, glucose, sucrose, lactose, starch, and food gums are used as binders. In addition, pressure compaction with binders in roll plate presses or pellet machines is used for candies and dried soups.

In the food industry, the term "agglomeration" is applied to the process in which its main objective is to control porosity and density of materials in order to influence properties like dispersibility and solubility. In this case the operation is also often referred as *instantizing*, because rehydration and reconstitution are important functional properties in food processes. For example, instantizing is used for the production of milk powder, spray dried coffee, flours, starches, dry soups, cocoa products, dextrins, and dry pudding mixes. In this process, the surface of the dried particles is rewetted with steam or water mists; the particles are then mixed, usually in a turbulent gas stream, causing them to form clusters by collisions. Then the agglomerates are re-dried and sized. Agglomerates have a coarse, open structure form 0.1 to 3 mm. On the other hand, when size enlargement is used with the objective of obtaining definite shapes, the food industry takes advantage of a process that may shape and cook at the same time, known as *extrusion*. In a more general context, however, instantizing and extrusion of food processes are the two common categories of agglomeration tumble/growth and pressure agglomeration, and are referred as such in the literature.

Agglomeration has improved the dispersibility of powdered products by creating a form that is wetted simultaneously and uniformly when dropped into cold or hot water. In addition, agglomeration gives the product a psychological advantage over non-agglomerated dried powder forms. Color, particle shape, and appearance are more pleasing to the eye, giving the product more sales appeal.

7.2. AGGREGATION FUNDAMENTALS

Agglomeration can be defined as the process by which particles are joined or bind with one another, in a random way, ending with an aggregate of porous structure much larger in size than the original material. The term includes varied unit operations and processing techniques aimed at agglomerating particles (Green and Maloney, 1999). Powdered foods can be considered as a mixture of organic compounds, and as such they possess no precise melting point. At a given temperature, the food particles begin to plasticize and agglutinate. The point of agglutination is dependent upon the moisture content of the powder. To agglomerate powdered foods successfully requires selected treatment of the powder, at least for a short period, within or near the agglutination moisture—temperature zone. Powders can be brought into the zone by rewetting. Other methods utilize pressure or the addition of binders in order to form the agglomerates.

7.2.1. Mechanisms of Particle Bonding

Bonds are formed between powder particles so that they adhere together to form granules, and these bonds must be sufficiently strong to prevent breakdown of the final dried granules to powder in subsequent handling operations. The size of the particles, the structure of the granule, the moisture content, and the surface tension of the liquid determine the magnitude of these interparticle forces. Rumph (1962) identified five mechanisms responsible for interparticle forces operating during and after agglomeration, but stated that more than one apply to any particular system for agglomerate stability. These include solid bridges, liquid bridges, Van de Waals forces, electrostatic forces, and interlocking bonds (Parikh, 1997).

7.2.1.1. Solid Bridges

Solid bridges arise from the material deposited between the agglomerated particles. They may develop by diffusion of molecules from one particle to another through partial melting at points of contact where high pressures develop, especially at elevated temperatures. Solid bridges can also be built up by chemical reaction, crystallization of dissolved substances, hardening of binders, and solidification of melted components. The force of cohesion depends on the diameter of the contact area and the strength of the bridge material. Since solid bridges are often porous themselves, it is usually easiest to measure the strength of solid bridges experimentally.

7.2.1.2. Immobile or Freely Movable Liquid Bridges

In liquid bridges, the force of particle adhesion arises either from the surface tension of the liquid/air system (as in the case of a liquid droplet) or from capillary pressure. The availability of sufficient moisture to produce a thin, immobile adsorption layer can contribute to the bonding of fine particles by effectively decreasing the distance between particles and increasing the interparticle contact area. Thin, immobile films of highly viscous bonding media can form exceptionally strong bonds, the strength of which can exceed that produced by mobile liquid layers. On the other hand, when the liquid level on the surface increases beyond that of the film, mobile liquid forms bridges wherein capillary pressure and interfacial forces create strong bonds. Mobile liquid bridges are a prerequisite to the solid bridges formed by binders or other substances dissolved in the granulating fluid.

7.2.1.3. Attraction Forces Between Solid Particles

These are surface forces, which decrease particle size favoring the agglomerate stability by increasing the surface/mass ratio. Among these, Van de Waals forces result from dipole interactions on a molecular level and thus act over very short distances. Electrostatic forces are longer ranging forces that arise from surface changes on the particles. Their behavior depends on the particles'

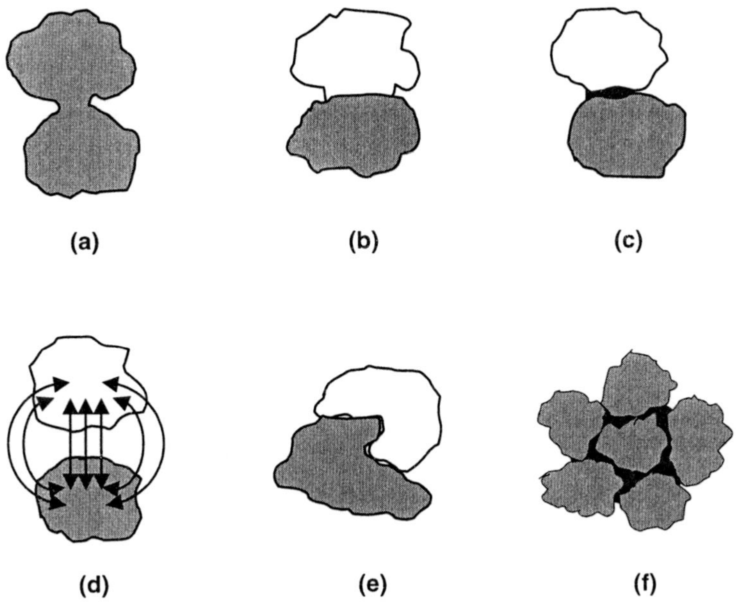

Figure 7.1. Different binding mechanisms in agglomeration: (a) partial melting sinter bridges; (b) chemical reaction hardening binders; (c) liquid bridges hardening binders; (d) molecular and like-type forces; (e) interlocking bonds; and (f) capillary forces (adapted from Pietsch, 1991).

composition, which defines their surface conductivity, and therefore whether the charges form points of high potential through the particle.

7.2.1.4. Form-Closed Bonds or Interlocking Bonds

Fibers, little platelets, or bulky particles can interlock or fold around each other resulting in "form-closed" bonds. Although mechanical interlocking of particles influences agglomerate strength, its contribution is generally considered to be small in comparison with the other binding mechanisms.

Agglomeration is used in food processes mainly to improve properties related to handling and reconstitution. Figure 7.1 shows some common binding mechanisms of agglomeration with bridges or force fields at the coordination points between particles (Pietsch, 1991). The two-dimensional structure represented in such a figure is in reality three-dimensional, containing a large number of particles. Each particle interacts with several others surrounding it and the points of interaction may be characterized by contact, or by a distance small enough for the development of binder bridges. Alternatively, sufficiently high attraction forces can be caused by one of the short-range force fields. The total number of interaction sites of one particle within the agglomerate structure is called the coordination number. Particles in an agglomerate could be quite numerous, making it difficult to estimate the coordination number. Indirect measurement of the coordination number can be made as a function of other properties of the agglomerate. In regular packs of mono-sized spherical particles, the coordination number k and the porosity or void volume ε, are related by:

$$k\varepsilon \approx \pi \tag{7.1}$$

Equation (7.1) gives good approximation of the coordination numbers of ideal agglomerate structures. Table 7.1 lists several values of coordination numbers calculated using Eq. (7.1) and compared with the ideal number for different structures, such as those illustrated in Fig. 7.2.

Table 7.1. Geometric arrangement, porosity, and coordination number of packings of mono-sized particles, as shown in Fig. 7.2.

Geometric arrangement	Porosity (ε)	Coordination number (π/ε)	k
Cubic	0.476	6.59	6
Orthorhombic	0.395	7.95	8
Tetragonal–spheroidal	0.302	10.39	10
Rhombohedral (pyramidal)	0.260	12.08	12
Rhombohedral (hexagonal)	0.260	12.08	12

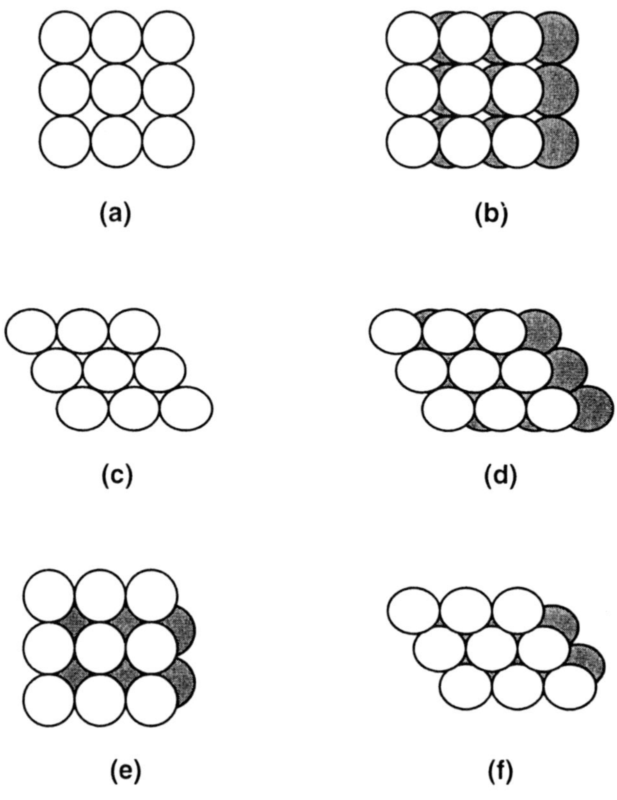

Figure 7.2. Packings of mono-sized spherical particles: (a) cubic; (b, c) orthorhombic; (d) tetragonal–spheroidal; (e) rhombohedral (pyramidal); and (f) rhombohedral (hexagonal).

7.2.2. Strength of Agglomerates

The strength of an agglomerate depends on the forces holding the agglomerate together, and there is often more than one force responsible for the agglomerate strength; therefore, it is often difficult to calculate the strength of agglomerates based on any one type of force. To gain better understanding of the granulation process, researchers have developed theoretical models to describe the strength of agglomerates. Because of the complexity of the granulation processes, simplifying assumptions have usually been made. For example, the mean tensile strength of a particle assembly in which bonds are localized at point contacts was estimated assuming the following:

- The particles may be represented by a large number of mono-disperse spheres that are distributed statistically in the agglomerate.
- There are a large number of bonds in the stressed cross section.
- The bonds are statistically distributed over the fracture section.
- The interparticle bond strength between individual particles is equivalent to the mean value of the entire agglomerate.

A general relation describing the tensile strength of agglomerates σ_t held together by binding mechanisms acting at the coordination points is:

$$\sigma_t = \frac{1-\varepsilon}{\pi} k \frac{\sum_{i=1}^{n} A_i(x,\ldots)}{x^2} \tag{7.2}$$

where A_i is the adhesion force caused by a particular binding mechanism and x is the representative size of the particles forming the agglomerate.

Substituting Eq. (7.1) into Eq. (7.2), the following relation is obtained:

$$\sigma_t = \frac{1-\varepsilon}{\varepsilon} \frac{\sum_{i=1}^{n} A_i(x,\ldots)}{x^2} \tag{7.3}$$

A further simplification results because many binding mechanisms are a function of the representative particle size x, and thus:

$$\sigma_t = \frac{1-\varepsilon}{\varepsilon} \frac{\sum_{i=1}^{n} A_i(x,\ldots)}{x} \tag{7.4}$$

The three dots in parentheses in Eqs. (7.2)–(7.4) indicate that A_i is also a function of other unknown parameters.

When liquid bridges have formed at the coordination points, A_i depends on the bridge volume and the wetting characteristics represented by the wetting angle. There are models available for predicting adhesion forces of various types (Pietsch, 1991), but A_i might be of different magnitude at each of the many coordination points, due to roughness or microscopic structure of particulates forming the agglomerates. All the forces defined above generally increase linearly with particle size. One exception to the rule is when the material is charge insulator, and in this case attraction forces vary with the square of the diameter. Van der Waals forces decrease very quickly with distance, however, and are practically zero for interparticle distances of 1 μm or more. As particles are pulled apart, liquid bridges, if present, become thinner and finally fail; before failure, however, the liquid bridge strength is not very sensitive to interparticle distance. For particles that are separated more than 1 μm, electrostatic forces are mainly available (when charges remain over the particles) to spontaneously bring them together into agglomerates. Once agglomerates are formed, the strength depends largely on liquid bridges if moisture is present; otherwise, the weaker van der Waals forces are important.

The representative particle size most appropriate to describe the agglomeration process is the surface equivalent diameter, x_{sv}, because porosity is surface dependent. As described in Chapter 2, such diameter is the size of a spherical particle which, if the powder consisted of only these particles, would have the same specific surface area as the actual sample. When determining the specific surface area, methods must be chosen which measure only the outer particle surface, excluding the accessible inner surface due to open particle porosity. One of these recommended methods is permeability, also discussed in Chapter 2.

From the previous discussion, it can be gathered that the strength of agglomerate structures held together by bonding mechanisms is highly dependent on porosity and particle size, or, more properly, specific surface area. The relationship would be inversely proportional in both cases, i.e., higher strength at lower porosities and lower surface areas.

Particle porosity will also determine how easily water penetrates the agglomerate structure. The porous surface of the agglomerate draws water rapidly into the interior by capillary action, causing rapid wetting of all surfaces of the individual grains. Consider the capillary pores of the agglomerates as tubes of a defined radius. The liquid will best penetrate if this radius is large (i.e., large grain size) and the capillary length is small (i.e., small agglomerate size) (Loncin and Merson, 1979). The liquid penetration time can be expressed using the mean radius of the pores in terms of the ratio of the surface area of the grains with respect to the volume:

$$t_\mathrm{p} = k\frac{l^2\eta}{\upsilon\cos\varphi}\frac{A_\mathrm{v}}{\varepsilon^{2.55}} \tag{7.5}$$

where A_v is the surface area of a particle per unit volume of particle (equal to $6/D$ for spheres), ε is the void fraction within an agglomerate of several particles, k is the experimentally determined constant, l is the liquid penetration depth through the capillary, η is the viscosity of the penetrating liquid in the pore, υ is the liquid surface tension, and φ is the angle of penetration of the liquid–air boundary through the capillary. This expression suggests reducing the area A_v, or increasing the agglomerate porosity in order to decrease the penetration time.

Agglomerates that are completely filled with liquid obtain strength from the negative capillary pressure in the structure. A relationship for this case is:

$$\sigma_\mathrm{t} = c\frac{1-\varepsilon}{\varepsilon}\alpha\frac{1}{x_\mathrm{sv}} \tag{7.6}$$

where c is a correction factor, α is the surface tension of the liquid and x_sv is the surface equivalent diameter of the particle. In order to apply Eq. (7.6), there must be a complete wetting of the solids by the liquid.

For high pressure agglomeration and the effect of matrix binders, general formulas have not been yet developed. It can be considered, however, that the effects of variables would follow the trend described before, with porosity, particle surface, contact area, and adhesion all playing an important role. For non-metallic powders, the following equation can be used to evaluate the needed applied pressure p to agglomerate:

$$\log p = mV_\mathrm{R} + b \tag{7.7}$$

where V_R represents the relationship V/V_s, V being the compacted volume at a given pressure and V_s is the volume of the solid material to be compacted, m and b are constants.

7.3. AGGLOMERATION METHODS

The principal methods of granulating powdered foods can be classified into three main categories: tumble/growth (or rewetting) agglomeration, pressure agglomeration, and other processes. Other processes include, among others, spray dryer agglomeration, its combination with tumble growth, and freeze-drying. Also, agglomerates can be obtained either by using binders or in a binderless manner. The tumble/growth method produces agglomerates of approximate spherical shape by buildup during tumbling of fine particulate solids; the resulting granules are at first weak and require binders to facilitate formation, and post-treatment is needed to reach final and permanent strength.

On the other hand, products from pressure agglomeration are made from particulate materials of diverse sizes. They are formed without the need of binders or post-treatment, and acquire immediate strength.

7.3.1. Tumbling of Powders (Rewetting Agglomeration)

The mechanism of tumble/growth agglomeration is illustrated in Fig. 7.3. As shown, the overall growth process is complex and involves both disintegration of weaker bonds and re-agglomeration by abrasion transfer and coalescence of larger units (Cardew and Oliver, 1985). Coalescence occurs at the contact point when, at impact, a binding mechanism develops that is stronger than the separating forces. Additional growth of the agglomerate may proceed by further coalescence, or by layering, or both. The most important and effective separation force counteracting the bonding mechanism is the weight of the solid particle. For particles below approximately 10 μm, natural attraction forces such as molecular, magnetic and electrostatic, become significantly larger than the separation forces due to particle mass and external influences. In such a way, natural agglomeration occurs.

The mechanism of tumble/growth agglomeration is similar to that of natural agglomeration. The particles to be agglomerated are larger, however, the particle-to-particle adhesion needs to be increased by the addition of binders, such as water or other more viscous liquids, depending on the properties of the particles being agglomerated and the required strength of the agglomerate structure. On the other hand, collision probability may be enhanced by providing higher particle concentration.

The conditions needed for tumble/growth agglomeration can be provided by inclined discs, rotating drums, any kind of powder mixer, and fluidized beds (Fig. 7.4). In general terms, any equipment or environment creating random movement is suitable for carrying out tumble/growth agglomeration. In food processing, most units use static and vibrating fluidized beds to mix the powder, promote interparticle collisions and dry the granules. Fluidized beds provide good random mixing of the powder, along with stable and uniform temperature profile. Furthermore, effective interphase contact between solid and liquid and controlled high rates of heat and mass transfer can be achieved (Coucoulas, 1992). In some cases, granules may be formed by shear processing in planetary mixers, ribbon blenders, Z-blade units and high speed intensive mixers—all batch equipment.

In certain applications, very simple tumbling motions, such as on the slope of storage piles or on other inclined surfaces, are sufficient for the formation of crude agglomerates. The most difficult task of tumble/growth agglomeration is to form stable nuclei due to the presence of few coordination points in small agglomerates. Also, since the mass of particles and nuclei are small, their kinetic energy is not high enough to cause the microscopic deformation at the contact points that enhances bonding. Recirculation of undersized fines provides nuclei to which feed particles adhere more easily to form agglomerates. In the whole process, tumble/growth agglomeration first renders weak agglomerates known as green products. These wet agglomerates are temporarily bonded by surface tension and capillary forces of the liquid binder.

This is the reason why, in most cases, tumble/growth agglomeration requires some sort of post-treatment. Drying and heating, cooling, screening, adjustment of product characteristics by crushing, rescreening, conditioning, and recirculation of undersize material are some processes that have been used as post-treatment in tumble/growth agglomeration. Sometimes, a large percentage of recycled fine product must be rewetted for agglomeration and needs to be processed again, causing economical burden to this technology (Pietsch, 1983).

For larger size or mass of the particles being agglomerated by tumble/growth methods, the forces that separate newly created bonds during growth become significant, until further size enlargement by tumbling is not possible, even if strong binders are added. There is, therefore, a definite limitation on the coarseness of a particle size distribution, in the range of x_{sv} between 200 and 300 μm.

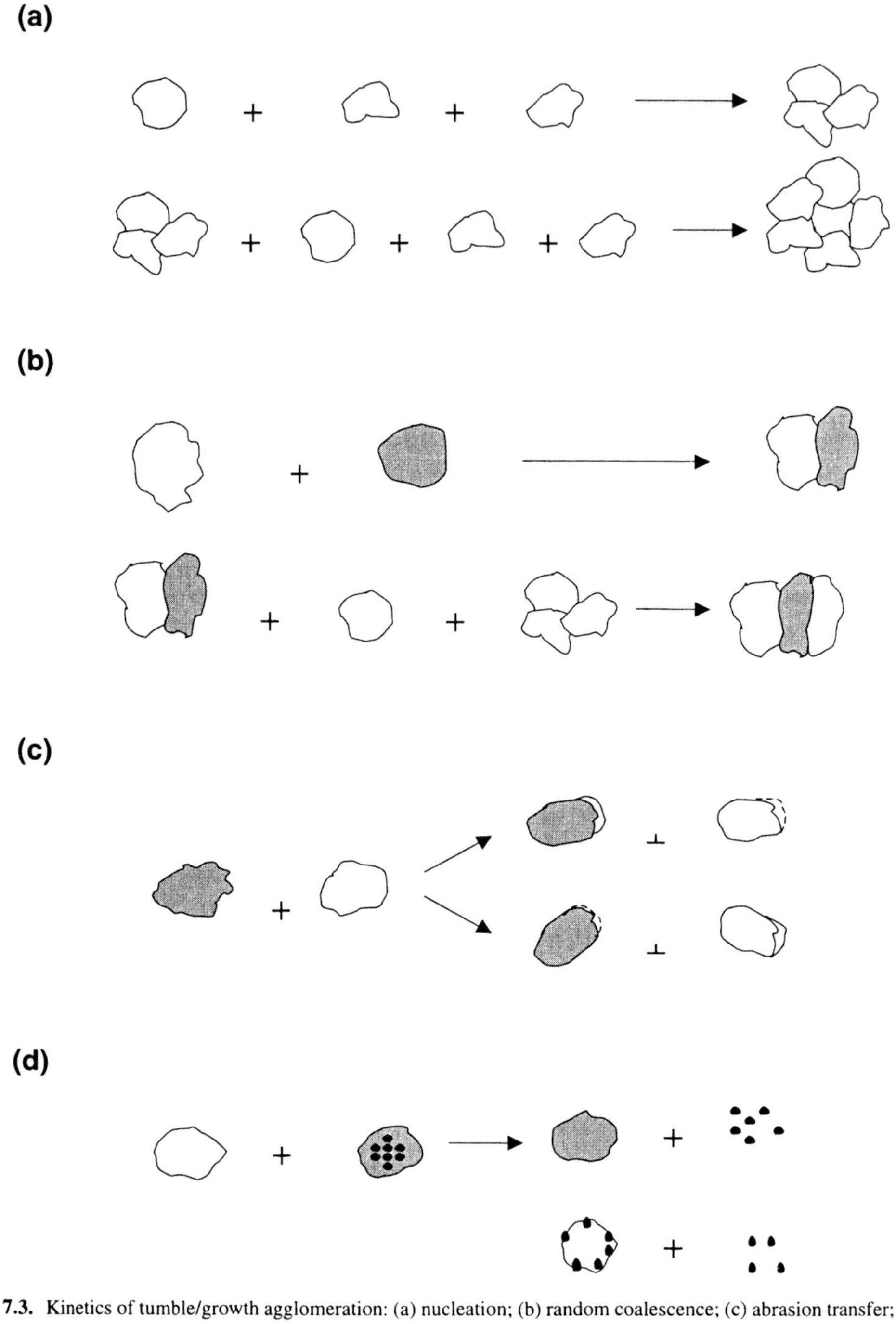

Figure 7.3. Kinetics of tumble/growth agglomeration: (a) nucleation; (b) random coalescence; (c) abrasion transfer; and (d) crushing and layering (adapted from Cardew and Oliver, 1985).

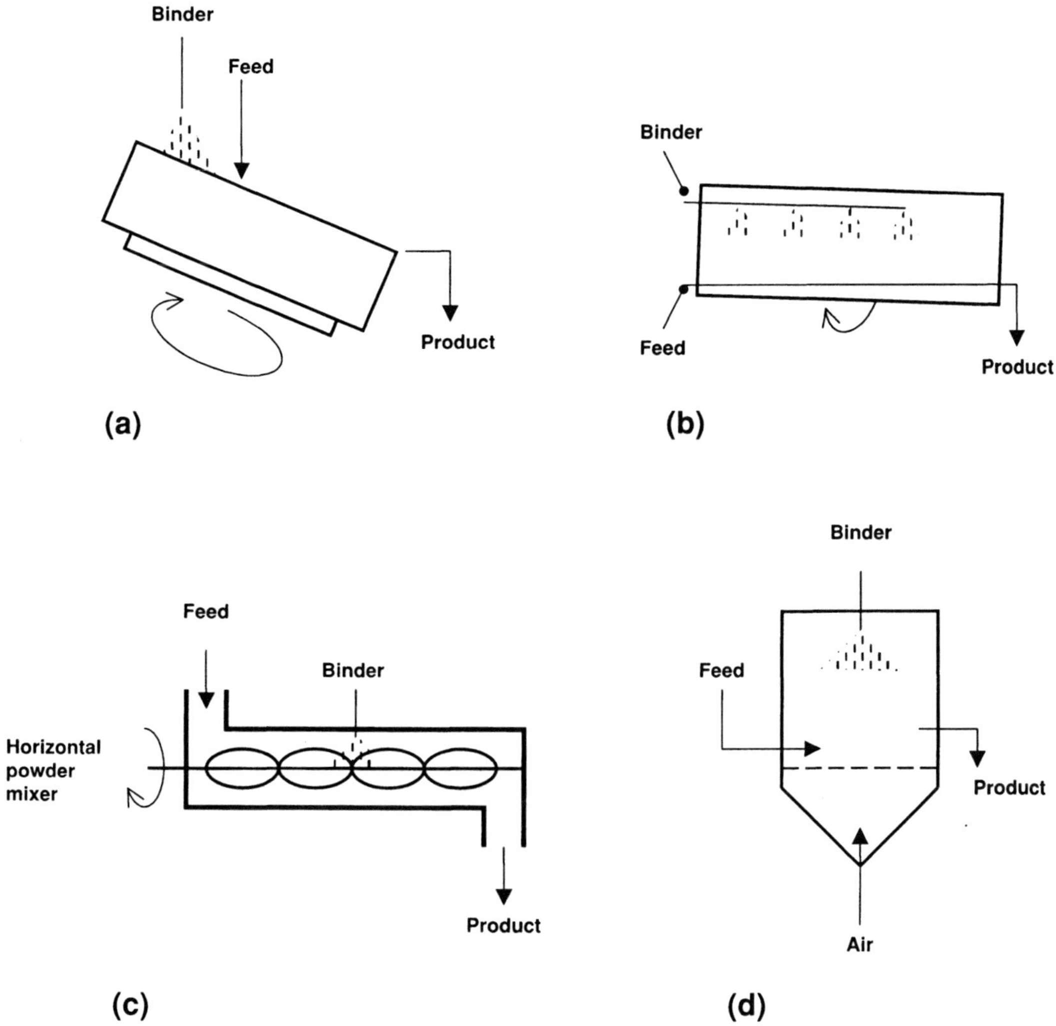

Figure 7.4. Equipment for tumble/growth agglomeration: (a) inclined rotating disc; (b) inclined rotating drum; (c) ribbon powder blender; and (d) fluidized bed.

In many cases, for instant product manufacture, a dried fine powder is kept in a fluidized state by either mechanical movement (using a horizontal or vertical mixer) or gas agitation (fluidized bed), as shown in Fig. 7.4c and d. Agglomeration is initiated by rewetting the powder with a liquid binder (e.g., water) that may be atomized pure liquids or solutions, steam, or combination of both. Figure 7.5 shows the agglomeration stages by rewetting. For example, granulated enzyme products can be manufactured by mixing the enzyme solution with a suitable filler to form a dough which is then pressed into fine granules. The granules are then sprayed with a suitable binder and further dried in a fluidized bed dryer (Härkönen et al., 1993).

Thus, liquid bridges are formed due to interfacial and capillary forces at the beginning, and solid bridges are formed by recrystallization of dissolved substances that are the soluble components of the powder itself or the easily soluble binder brought in during rewetting. Drying is promoted by passing warm air through the mixer or by using warm air for fluidization (APV, 1989; Pietsch, 1999). In most

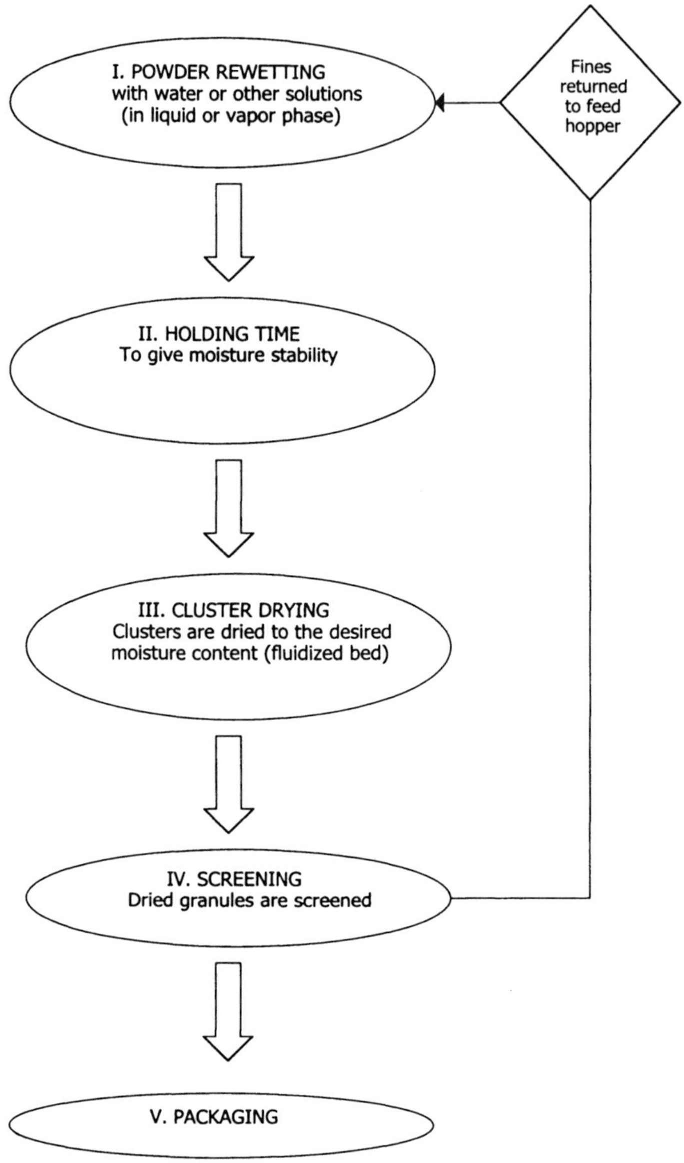

Figure 7.5. Agglomeration stages by rewetting.

cases, water is used as the rewetting media, but a special binder material, which must not compromise the final product integrity, is used in the case of relatively water insoluble materials. The solubility rate of the powder determines how the wetting agent should be applied. For optimum wetting conditions, the size of the wetting droplets must be equal to or smaller than the dried powder particle size. This condition is fulfilled when a wet stream is used as the wetting agent. Steam condensation on powder particles corresponds to wetting by very fine water droplets. When water extracts or liquid additives are used as wetting agents, atomizers producing very fine sprays (droplets no larger than 50 μm) are adopted. The frequently used binders for food powders include maltodextrines, sugar, starch, gelatin, arabic gum, and molasses (APV, 1989). Typically, this rewetting agglomeration method can produce agglomerates with superior redispersion characteristics.

Other rewetting agglomeration methods can be performed by surface techniques using a rotating disk or a vertical tube (Masters and Stoltze, 1973). A rotating disk receives the powder from a vibratory feeder, initially coming from a cyclone. Moistening of product during powder fall is carried out by injecting a wet steam or by spraying liquid into the powder stream. For products such as coffee, coffee substitute and other highly hygroscopic products, surface wetting readily renders the product sticky. The disk is rotated at high speed (500–1,200 rpm) to prevent product build-up on the disk, ensure formation of even-sized agglomerates, and distribute the agglomerates evenly within the agglomerating chamber. For less hygroscopic products, the disk is rotated at lower speed (20–50 rpm). For the vertical tube method, wetting contact with moist air or steam takes place in a vortex manner, and powder particle stickiness is achieved through condensation effects. Agglomerates leaving the agglomerating tube pass into a drying chamber and are contacted with drying air.

7.3.2. Pressure Agglomeration

In contrast to tumble/growth agglomeration where no external forces are applied, in pressure agglomeration pressure forces act on a confined mass of particulate solids, which is then shaped and densified (Engelleitner, 1994). Pressure agglomeration is normally carried out in two stages. The first stage comprises a forced rearrangement of particles due to applied pressure, while the second step consists of a steep pressure rise during which brittle particles break and malleable particles deform plastically (Pietsch, 1994). The mechanism of pressure agglomeration is illustrated in Fig. 7.6. There are two important phenomena that may limit the speed of compaction and, therefore, the capacity of the equipment: compressed air in the pores and elastic springback. Both can cause cracking and weakening which, in turn, may lead to destruction of the pressure agglomerated products. The effect of these two phenomena can be reduced if the maximum pressure is maintained for some time, known as dwell time, prior to its release.

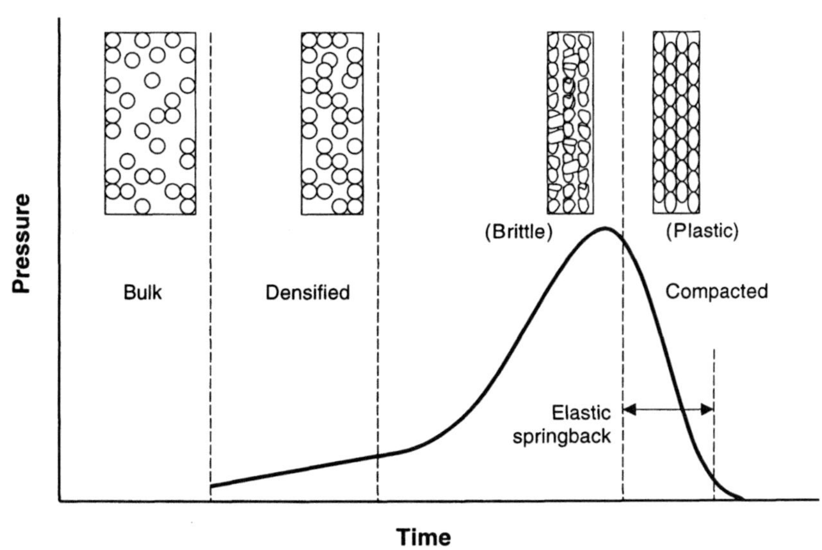

Figure 7.6. Mechanism of pressure agglomeration.

Pressure agglomeration can be performed employing a low, medium, or high pressure mode. When utilizing low or medium pressure agglomeration, relatively uniform agglomerates can be obtained. Under these conditions, the porosity of the feed material is changed, but no change in particle size or shape occurs. The feed mixture is often prepared with fine particles and binders, thus giving a sticky mass, which may be formed by forcing it through holes in differently, shaped screens or perforated dies. Agglomeration and shaping are, therefore, due to pressure forcing the material through the holes, as well as by frictional forces. High pressure agglomeration is characterized by a large degree of densification, resulting in low product porosity. Typically, the products from high pressure agglomeration feature high strength immediately after discharge from the equipment. To increase strength further, additions of small amounts of binders or use of post-treatment methods are possible. High pressure agglomeration is considered a versatile technique by which particulate material of any kind and size, from nanometers to centimeters, can be successfully processed under certain conditions.

Pressure agglomeration can be carried out in different types of equipment. Generally, low and medium pressure agglomeration is achieved in extruders, including the screen extruder, the screw extruder, and the intermeshing gears extruder. On the other hand, high pressure agglomeration is performed in presses such as the punch-and-die press, the compacting roller press and the briquetting roller press. Low and medium pressure agglomeration yield relatively uniform agglomerates of elongated spaghetti-like or cylindrical shape, whereas high pressure agglomeration produces pillow or almond-like shapes. Figure 7.7 presents equipment used for low and medium pressure agglomeration, while Fig. 7.8 illustrates some common machinery for high pressure agglomeration.

In press agglomeration techniques, instant agglomerates can also be produced by processes called compaction/granulation and extrusion/crumbling. One of the most common binding mechanisms in these processes is caused by the short-range molecular attraction forces, i.e., electrostatic and van der Waals forces, rather than solid bridging forces. Because these forces are reduced in liquids by a factor of around 10, the particles bonded by them can disperse easily in liquid, exhibiting expected instant properties. In the compact/granulation process, dry powdered mixtures are first compacted by high pressure, and then crushed and screened into a granular product. No drying or cooling operation is needed and the product is much denser. When this method is not feasible, the extrusion/crumbling process can be used, in which a powder mixture is blended with a binder liquid, additives, or dispersants extruded at low pressure and then dried, cooled, and crumbled to get a final instant product (Pietsch, 1999).

7.3.3. Specific Agglomeration Methods

7.3.3.1. Straight-Through Agglomeration

In the straight-through process, liquid concentrate is used. When powders are produced by spray drying, the agglomeration process can be accomplished in a fluidized bed connected directly to the spray dryer, where the operating conditions can be controlled so that the partially dried particles formed in the upper part of the dryer are still sticky. As shown in Fig. 7.9, fine particles, either from recycle or the drying chamber, are fed into an external fluidized bed to undergo cluster formation. Sometimes, steam or atomized water can be injected into the fluidized bed to assist in the agglomeration process. Final drying and cooling are also accomplished in the bed, and the agglomerated product is removed for storage or packaging. This process can produce agglomerates with instant properties sufficient for many applications (APV, 1989). This procedure is more adequate for coffee and baby foods (Masters and Stoltze, 1973). Such a system has also been used to agglomerate skim and whole milk, non-caking whey, milk replacer, and ice cream mix. In instant whole milk

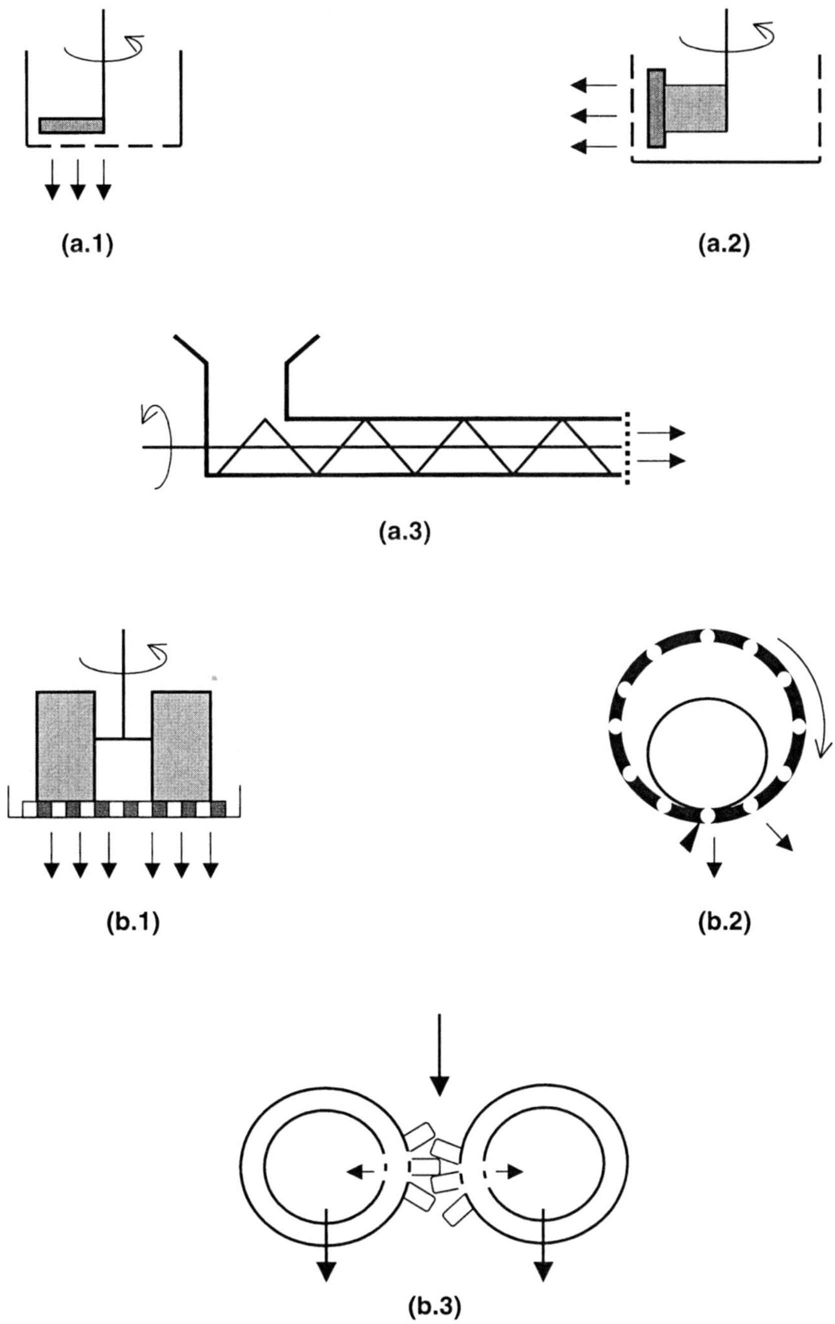

Figure 7.7. Equipment used for (a) low and (b) medium pressure agglomeration: (a.1) screen extruder; (a.2) basket extruder; (a.3) cylindrical die screw extruder; (b.1) flat-die extruder; (b.2) cylindrical die extruder; and (b.3) intermeshing gears extruder.

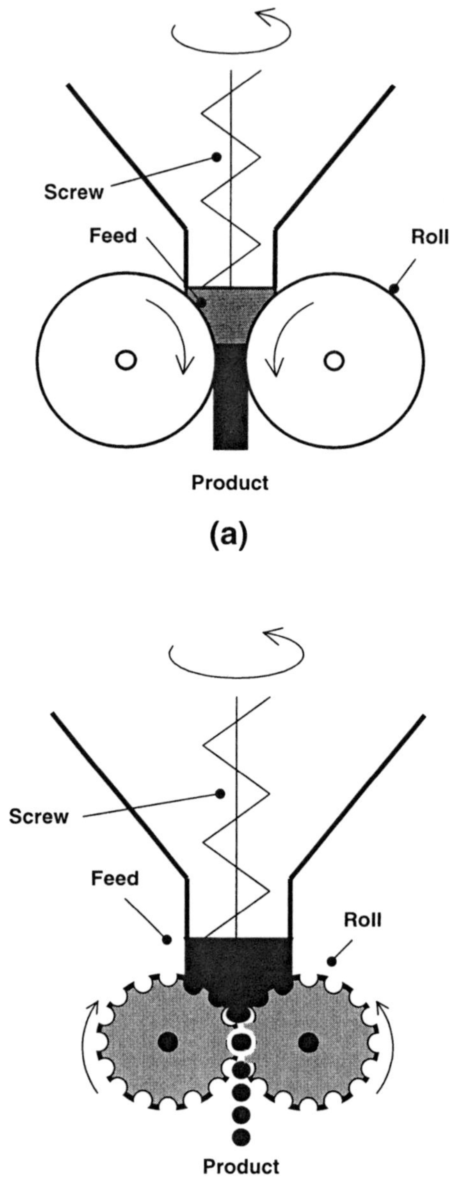

Figure 7.8. Equipment used for high-pressure agglomeration: (a) compacting roller press; and (b) briquetting roller press (adapted from Pietsch, 1997).

production, the straight-through process is followed by lecithin addition, improving wettability and dispersibility. For products that contain sugar or other carbohydrates, a twin atomizer arrangement is recommended. One atomizer is used for the food concentrate and the other for the sugar solution. Two types of droplets in the drying chamber causes multiple and effective droplet collisions that lead to strong, stable and high quality agglomerates.

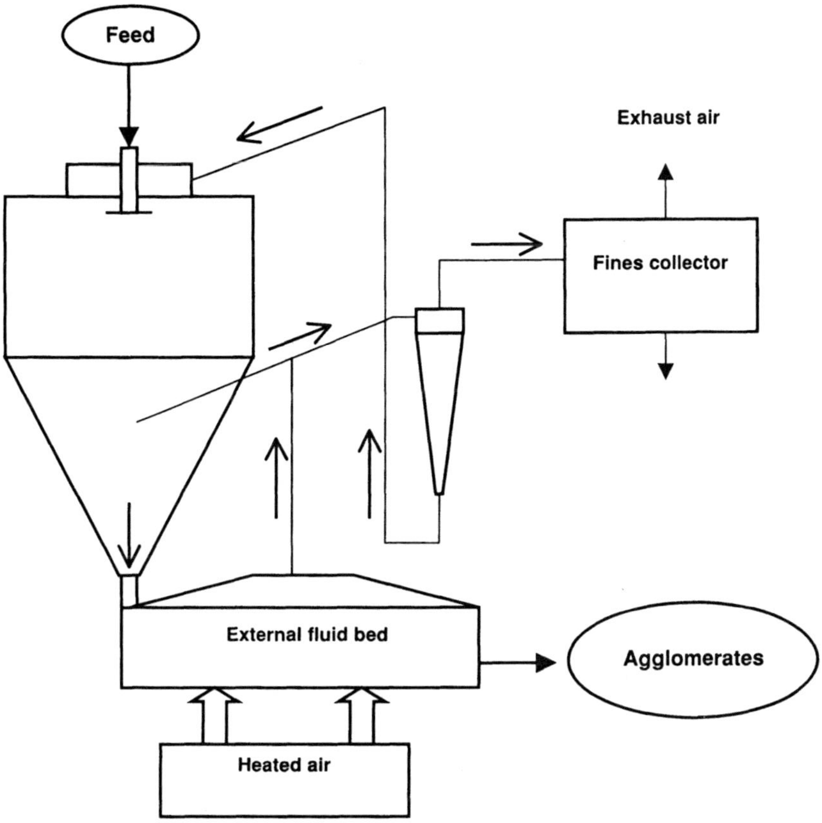

Figure 7.9. A schematic diagram of straight-through agglomeration process (adapted from APV, 1989).

7.3.3.2. Spray-Bed Dryer Agglomeration

This process has been employed for decades and found its earliest widespread acceptance in the dairy industry. Spray dryers to produce powdered milk, whey, and baby formulas are still one of the largest applications of the technology. In this new generation of spray dryers, a fluidized bed is integrated into a spray dryer chamber, combining spray drying with fluidized bed agglomeration. Particles formed in the spray drying zone enter the integrated fluidized bed at the bottom of the dryer with high moisture content, and become agglomerated in the bed where they are vigorously agitated by a high fluidization velocity. An external fluidized bed is connected to the integrated fluidized bed for final product drying and cooling. This type of dryer is most suitable for small to medium sized plants and can produce agglomerated powders with excellent properties (APV, 1989). A film-forming, easily soluble polymer can be dissolved in the continuous phase of emulsions or dispersions in the feeder, causing microencapsulation during drying. Such capsules, even if they are formed from emulsions, yield a dry, free-flowing powder, which in most cases possesses instant characteristics.

7.3.3.3. Atomizer Wheel

This is used when the powder cannot withstand a forceful agglomeration process, or where small agglomerates are desired. The layout closely resembles the agglomerating tube method. The only difference is that a rotary atomizer replaces the agglomerating tube. The powder falls around the rotating atomizing wheel and is sprayed with water or binder solutions. This system is used for certain baby foods, beverage whiteners, and cocoa/sugar mixtures (Masters and Stoltze, 1973).

7.3.3.4. Freeze-Drying

Freeze-drying processes are known as a useful alternative method to obtain dried food products with good aroma and flavor retention. In addition to these advantages, the porous structure resulting from ice sublimation allows good instant properties in freeze-dried products (Barbosa-Cánovas and Vega-Mercado, 1996). Freeze-dried coffee was first introduced into the market in 1964 with a great success and a large number of consumers who demonstrated willingness to pay a higher price for an improved soluble coffee (Nair and Sivetz, 1973). Other freeze-dried instant products include tea extract and fruit juices. The major steps in producing instant coffee by freeze-drying include extraction, extract freezing, granulation and classification, freeze-drying, and packaging. More details about freeze-drying can be found in Chapter 11.

7.3.4. Binders

Binders are adhesives that provide the cohesiveness essential for the bonding of solid particles. In the wet granulation process, binders promote size enlargement to provide granules and thereby improve flowability of the blend during the manufacturing process. During wet massing, the binder may be dissolved in the granulating solvent, which is then added to the powder, or mixed dry with the powder and the granulating solvent (generally water). In the second case, the binder is dissolved in the solvent *in situ*. Binders are classified as natural polymers, synthetic polymers, or sugars. The selection of a binder for a particular system is mostly empirical and depends on the previous experience of the formulator. Selection of the quantity of binder required in a particular system can be studied in optimization studies, using parameters such as granule friability, hardness or disintegration time. Some commonly used binders in wet granulation are listed in Table 7.2. Dry and wet mixing classification in the table refers to whether binders are added in solution or in its dry form.

Table 7.2. Commonly used binders in pharmaceutical and food technology (adapted from Kristensen, 1993).

Binder	Granulation technique
Natural polymers	
Starch	Wet mixing
Gelatin	Wet mixing
Acacia gum	Wet mixing
Alginic acid	Dry mixing
Sodium alginate	Wet mixing
Synthetic polymers	
Methylcellulose	Wet/Dry mixing
Na-carboxymethylcellulose	Wet mixing
Sugars	
Glucose	Wet mixing
Sucrose	Wet mixing
Sorbitol	Wet mixing

Binders differ in their bonding efficiency. For example, gelatin or acacia gum provide high hardness and slow disintegration to the agglomerate. Methylcellulose produces granulations that compress easily. Glucose or sucrose can be applied as syrups in concentrations above 50% in wet granulation processes exhibiting good bonding properties, although sucrose produces hard and brittle bridges. Among the factors influencing binder efficiency are concentration, viscosity, mechanical properties of the binder, interparticulate interactions between the binder and the substrate, and binder distribution.

During the wet granulation process, the binder forms an internal matrix and the granule strength increases as binder concentration in the formulation increases. The mechanical and film-forming properties of a binder, as measures of binder efficiency, determine the strength and deformation behavior of a binder matrix. Major determinants of granule strength are wettability of the substrate by binder, binder cohesion, and binder–substrate adhesion. Granule disintegration, dissolution, and flow properties are also useful to predict binder–particle interactions. Furthermore, the distribution of the binder in the granules influences its ability to produce strong and non-friable granules. For example, very viscous binder solutions, such as starch paste, may produce more friable granules.

7.4. SELECTION CRITERIA FOR AGGLOMERATION METHODS

There is a large variety of techniques and equipment available to carry out agglomeration duties in the food and processing industries. Some guidelines are given in the literature or can be provided by manufacturers. Table 7.3 summarizes some of the preliminary considerations to start a selection process of an agglomeration method by a practicing engineer. In general terms, features of the feed, the product, and the method are the most important considerations in the selection process (Pietsch, 1991).

7.4.1. Feed Characteristics

The first characteristics of feed to be considered are particle size and distribution. A limit in the range of a few hundred micrometers defines the applicability of methods using growth mechanisms based on coalescence in moving beds of particles. Larger particles, which may also refer to seed agglomerates, can only be incorporated if an adequate amount of binder or enough small particles are present. Since small particles embed in larger ones, the strength of the agglomerate is caused by the matrix of fine powder in this case. In general terms, it is difficult to agglomerate narrow particle-size distributions or mono-sized particles. Adding a binder can cause relatively large particles to agglomerate. It may be more economical, however, to crush larger particles in order to obtain material suitable for growth agglomeration. This is particularly crucial when a high porosity product is desirable.

Table 7.3. Guidelines for the selection of an agglomeration process.

Feed variables	Product variables	Method variables
Particle size and distribution	Agglomerate size and shape	Continuous or batch
Moisture content	Green strength	Capacity
Material characteristics	Cured strength	Wet or dry operation
Special features of material	Porosity and related features	Simultaneous processing
Bulk characteristics		Energy requirements
Binding characteristics		Costs

Pressure agglomeration is more suitable for larger feed size particles, e.g., sand-like material or particles up to 20–30 mm. Since the external forces acting upon the mass result in particle disintegration or deformation, the upper limit of feed particle size is determined more by restrictions of the feeder than in the ability to agglomerate. In most cases consolidation occurs in a short period, so a considerable amount of air must be removed during compaction in order to obtain sound agglomerates. There is an increasing resistance to flow with decreasing particle size due to the smaller pore radii. For this reason, fine bulk solids of about 150 μm or less, can only be agglomerated by pressure methods if certain preconditions, such as low speed and dwell time, are established.

Free moisture can play an important role in growth agglomeration by coalescence. In such a case, moisture provides the binder or prevailing binder mechanism. The maximum volume of liquid must not be more than 95% of anticipated agglomerate porosity. Wet agglomeration is sensitive to this limit because a small excess of moisture will cause the entire mass to achieve mud-like consistency. In the case of fluidized bed agglomerators, which may also act as driers, moisture content is less critical because it only has to be high enough to make the feed pumpable. In pressure agglomeration, moisture must be kept low, and it is often a precondition for a completely dry feed. Due to the extreme compression forces in high pressure agglomeration, crushing, rearrangement, and deformation of the solid take place and result in a considerable reduction of porosity. Excess water is either squeezed out or remains in the mass as an incompressible component, with the effect of resultant low strength of the agglomerate.

Material characteristics such as chemical properties, particle density, brittleness, elasticity, plasticity, wettability, and abrasivity play important roles in the choice of an agglomeration method. Particular chemical characteristics may be necessary to bring about the required chemical bonding, or may be incompatible with certain conditions of a method, such as addition of water or other liquids in most growth agglomeration techniques. Density of the feed particles determines particle weight and other field forces that may be counteractive to addition by coalescence. Brittleness, elasticity, plasticity, and abrasivity are most important for pressure agglomeration and of less concern for growth methods. Wettability, on the other hand, is paramount for all agglomeration methods using surface tension and capillary forces in the growth regime. Wetting of particle surfaces is a requirement for green strength.

Bulk properties, such as bulk density and flowability, as well as temperature, can be adjusted prior to agglomeration in order to improve size enlargement. In preparation to briquetting into ration-size agglomerates, vegetables, fruit juices, and food pulps may be frozen. High bulk density and unacceptable flowability are sometimes correlated by using two agglomeration methods in series. For example, fine feed food powders are pre-agglomerated to reduce the compaction stroke and improve the flow of feed into a die. This increases the speed of the turret for rotary table tabletting machines. At the same time, this technique avoids segregation of the feed mix by stabilizing the blend in a granular form.

Finally, the binding features of a given material must be considered for a possible agglomeration process. The binding characteristics should be assessed in order to decide whether agglomeration can be carried without adding binders, and only due to the presence of an inherent binder in the material, or with the addition of binders such as water or other liquids.

7.4.2. Agglomerated Powder Properties

Some properties like shape, dimension, and particle size distribution of the agglomerated product also influence proper selection of a suitable method. Agglomerated products are normally expected to improve in properties such as free flowing characteristics or dust-free features. Granular, free flowing, dust-free products can be manufactured using almost all methods of size enlargement. The task of narrowing the size distribution of the discharge is done by screening out under and oversized components. Since the fines are recirculated to the agglomerator, oversize particles are crushed and either rescreened or directly re-circulated with the fines. Granular products can also be obtained by crushing and screening large agglomerates using criteria such as product porosity, density, solubility, or reactivity.

The shape of the product is another important property. Spherical products are often desirable in an agglomeration operation, and such shape can be obtained using any growth agglomeration method. Contrastingly, spherical products cannot normally be obtained with high pressure agglomeration, unless extremely accurate feed control can be established. By using some types of pressure agglomeration equipment, such as tabletting machines, approximations to the spherical shape like pillow-, lens-, or almond-shaped compacts can be obtained.

Particle size is a fundamental parameter for the evaluation of final product quality and agglomeration process performance. For determination of agglomerate size or particle size distribution, or determining the fraction of fine material arising, for instance, as a result of attrition, the methods of particle size measurement described in Chapter 2. However, since many agglomerates are of the order of magnitude of 1 mm, sieving is often preferred since it is a comparatively simple method (Schubert, 1981).

Particle strength is relevant for the final product, but also plays a role during the size enlargement operation. In growth agglomeration, green agglomerates are formed first and then must be cured to obtain permanent bonding. A weak state could exist if the binding mechanism of the green agglomerate disappears before the permanent, cured bond sets in. Unless large amounts of matrix binders are used, or agglomerates are cured at high temperatures or by some chemical reaction, growth agglomeration products will normally be weaker than pressure agglomeration products.

Different strength levels develop primarily because agglomerates growing by coalescence feature higher porosity than those from pressure agglomeration. Materials that may disperse easily and are only agglomerated to improve handling of the intermediate product should have just enough strength to survive their short existence. In other cases a large, specific surface is more important than high density and strength. Normally, an increase in external forces acting on the particulate matter during size enlargement will cause porosity and related characteristics to decrease, while density and strength increase. In order to measure porosity of agglomerates, sectional microphotographs and pore size distributions can give an extensive description of the state of the agglomerate. Pore size distributions can be determined from sorption and capillary condensation measurements by the use of mercury porosimetry and from capillary pressure measurements with wetting liquids (Schubert, 1981). Interpretation of measured results requires much experience, particularly for the evaluation of pore models.

7.4.3. Alternative Methods

Agglomeration processes can be carried out in a batch or a continuous manner, depending on specific requirements and applications. Batch modes are generally low in capacity, but are characterized by a better control than that exercised in a continuous process. Most large-volume applications operate in a continuous form, but may be accompanied by significant variations in quality. In growth agglomeration, uncontrolled buildup must be removed, whereas in pressure agglomeration, worn parts must be replaced. Most of the growth methods are wet processes using binding liquids for forming green agglomerates, while high-pressure techniques are normally operated as dry processes. Agglomeration can sometimes be carried out simultaneously with other process. Simultaneous processing occurs in mixer-granulators, granulator-driers, or even mixer-granulator-driers. Mixers are often also granulators in which both processes occur in different zones. In fluidized bed granulators, however, agglomeration and drying can take place simultaneously.

Agglomeration shows potential for diverse applications such as recycling of wastes containing valuable ingredients or disposal of particulate wastes without value in an environmentally safe and acceptable way. These applications cannot always find economic justification, as they must be typically performed in compliance with legislation. Finely divided particulate material is often released to the environments in processes such as dry milling of cereals. Often, these materials are precipitated or removed by pollution control devices, but recontamination of the environment is an obvious concern and is normally regulated. In these cases, agglomeration methods can be employed to obtain a size-enlarged material in order to handle and dispose of it in a convenient way. Since in most cases one of the reasons for size enlargement is improvement of material handling, an agglomeration facility must be located near the particulate solid source. A suitable method must therefore include consideration of the availability and cost of utilities and ancillary devices such as binders or energy sources when required in wet granulation. Sometimes, since disposal may represent the main aim of the process, the same task may be accomplished using roller presses for dry compaction, and granulation by crushing and screening.

7.5. DESIGN ASPECTS OF AGGLOMERATION PROCESSES

Agglomeration processes consist of varied operations within a complete system aimed at obtaining a desired product. Mixing and screening are two common operations that are part of an agglomeration process. There are other varied post-treatments, such as those previously mentioned, which are used mainly to give strength to the agglomerate.

In tumble/growth agglomeration, if more than one feed powder is treated, the components must be metered and premixed. Homogenization may also be necessary due to the risk of selective agglomeration, as the particulates fed would involve many different sizes. During mixing, liquid or dry binders could be added. It is also possible to feed all or part of the recycle into the mixer. Aeration of the premixed material is an important factor, so particles will be loose and able to move randomly, in order to pick up the binder and to agglomerate upon impact. A metered addition to the agglomerator improves and accelerates agglomerate growth by seeding the charge. This is because recycle, despite representing an undersize product, consists largely of somewhat pre-agglomerated material. Control of the growth mechanism also requires addition of some of the liquid or dry binders in the agglomerator. Tumble/growth equipment produces green agglomerates that are better bonded by liquids. The agglomerate sizes and shapes are extremely varied, ranging within wide limits. Sometimes it is possible to screen the green agglomerates and feed only a narrow particle size distribution to the post-treatment stage. The moist recycle should be sent directly to the agglomeration unit. Green agglomerates are often weak and sticky, so they tend to blind screens quite easily (i.e., green agglomerates get attached to the screen openings). For this reason, separation of over or undersized material at this step must be avoided. The discharge from the post-treatment may also be screened to remove fines which may be formed by abrasion and breakage or, contrastingly, in order to retain oversize agglomerates that may have developed by secondary agglomeration of the still-moist and sticky green agglomerates. Oversized agglomerates could then be crushed to obtain a recycle that is normally dried, so it should be returned directly to the mixer and

Table 7.4. Selected patented processes for food powder size enlargement (extracted from Pintauro, 1972).

Product	Method name	Description
Non-fat dry milk*	Pebbles agglomeration process[RA]	A stream of air is introduced into a chamber where the powder is caused to commingle with continuously introduced warm water vapor and droplets of moisture. Moist and puffed material of varied particle sizes is removed from the chamber and conveyed to a vibrating screen that receives warm drying air upwardly.
	Straight-through process (with one or multiple nozzles)	Skim milk concentrate is supplied to a spray drying operation that converts the concentrate to the form of a moist divided material. The moist discrete material is collected by conventional separating cyclones. Then the material is supplied to a special conditioning operation. The moist discrete material is then dispersed into a hot moist atmosphere of steam and water droplets bringing sticky particles into random contact to form porous aggregates. Drying takes place in a series of shaker dryers.
	Spray drying process	A stream of gas carrying the pulverulent material is located in a concentrated countercurrent vapor stream inside a spray drying chamber. Powders are received dried in a conveyor belt.
	Single stage drying and agglomeration[RA]	Skim milk, sprayed through a nozzle and heated air, are brought together into a drying chamber. Then the mixture is subjected to a partial rehydration in which cool or unheated air is introduced into the mixture by directing it upward in the processing chamber.
	Freeze-drying method	Material is reduced into a discrete frozen solid, then vacuum dried by water sublimation, maintaining a vapor pressure of less than 4 mm Hg. At different time intervals, the material is fluidized by upwardly throwing a special dry gas (e.g., nitrogen at low pressure) where particle formation takes place.
Chocolate drink powders (blended cocoa, sugar, non-fat milk, stabilizer, salt and flavorings)	Pillsbury fluidized bed process (also for flour–sugar mixtures and angel cake)	A moving fluidized bed of powdered material is used. The bed is successively treated with vapor while injecting gas from below to form agglomerates through contacting and colliding particles.
	Agglomerating cocoa with fat	Granules consist of a sugar core onto which a coating of finely divided dry cocoa powder is agglomerated with an edible vegetable fat. An inclined pan granulator is used at about 45°C and rotated at 25–30 rpm.
	Binding with emulsifiers[RA]	Solid crystalline sugar is coated in its hot state with a nonionic emulsifying agent. The resulting product is then coated with cocoa in a mixer-cooler.
	Low temperature addition of moisture in cocoa and sugar mixtures	Water is added to the mixture at a temperature below 10°C. Mixing is accomplished by tumbling. The drying of the mixture is accomplished below 34°C (cocoa fat melting point) with dry air or silica gel with agitation.
	Defatted cocoa powder[PA]	Commercial cocoa powder is defatted by hexane extraction, then ground and mixed with sugar and vitamins. The mix is then extruded and dried.
	Co-crytallization by addition of crystallizing syrup: pelletizing or instantizing	Crystallizing sugar syrup is added to the powdery cocoa and mixed.

(*continued*)

Table 7.4. (*Continued*)

Product	Method name	Description
	Agglomerating nozzle (for cocoa-sugar powders)[RA]	Cylindrical enclosure (agglomerating tower) is equipped with devices for projecting jets of fluids on the inside. The pulverulent mixture, introduced at the top, is humidified while passing the fluid jets, agglomerates while descending the tower. Hot air currents dry the agglomerates at the end.
Natural sugars	Graining (granulated sugar)[RA]	Sugar syrup is concentrated and boiled in vacuum concentrators. Slow cooling occurs in vacuum pans (1–10°F/min) until reaching a moisture content of 1%. The transformed sugar is then milled and screened. Then it is agglomerated on an inclined surface by rewetting.
	Double agglomeration[RA]	Sugar of 1–2 μm is increased to 4–8 μm in the first stage by rewetting. The second stage increases particle size to more than 32 μm. Then a third stage agglomerate yields particles of 64–128 μm.
	All-purpose sugar process[RA]	Sugar is moistened by downwardly or horizontally directed jets of steam. Then, a separate turbulent agitation with directed air jets cause agglomeration.
	Brown sugars and fondant mixture	A mixture of fractured sugar crystals and syrup is formed so that the syrup coats the entire surface of the particles. Porous agglomerates are formed by adding moisture and agitating. Agglomeration takes place in a series of compartments where screening, drying, and cooling takes place.
Soluble coffee and soluble tea	Pebbles agglomeration process[RA]	Hydrous coffee concentrate is spray dried. The resulting finely divided powder is conveyed pneumatically from a supplied hopper into a chamber. Coffee is put in touch with finely atomized water. The agglomerates are dried, passing through a screen, which is in vertical motion. Warm dry air is delivered upwardly through the screen.
	Falling curtain method (with single or double steam zone)[RA]	A vertical plane of discretely arranged powder particles is formed. A jet of steam (or two) is introduced in a normal direction to the plane of powder particles to wet them sufficiently so that they adhere to one another. The wetter powder particles carried in the path of steam are caused to contact one another and agglomerate. Agglomerates are then dried on a conveyor belt with cold inlet air.
	Extract agglomerating fluid	A highly concentrated solution of coffee solubles in water is sprayed into fine instant coffee particles forming structurally strong agglomerates. No post-drying step is required.
	Freeze-dried particle control[PA]	Tea and coffee extracts are solidified by freezing. Frozen extract is ground and separated into two fractions. The coarser fraction is freeze-dried and the finer fraction is compressed in an extrusion device at low temperature to create a solid. A part of the solid is subdivided and then freeze-dried.
	Soluble tea process[RA]	Fine tea particles are fed onto a platform, which is vibrated with a circular motion. The platform is maintained in a humid atmosphere saturated with finely atomized water droplets onto the surface of the feed. The moist powder will cluster during vibration of the platform due to the

Table 7.4. (*Continued*)

Product	Method name	Description
		rotary motion of the particles. The fine particles will travel outwards of the platform and will be recycled back onto it. Agglomerates are removed from the center of the platform by suction and subsequently dried.
Flour	Diffusion wetting process[RA]	Moisture is adjusted to a desired degree by predrying. The flour particles are then subjected to a warm, moist atmosphere, and particles will collide and aggregate. Final drying takes place in an air inlet cone.
	High protein flour[RA]	Cohesive high protein flour is moisturized (to 20–35%) in a chamber and combined with agitation. Added moisture is carefully controlled. The material is removed from the chamber *en masse* while being simultaneously and continuously agitated in a determined conveyor. Then it is dried and screened.
	Agglomeration with shortening (flour and sugar mixes)[RA]	Melted shortening is atomized into a shower of particles of a homogeneous mixture of powdered ingredients in free or delayed gravitational fall. The mixture is passed through a vibrating trough that provokes the delayed fall of the mix into a granulator or any type of rotating container where at the same time the shortening is spayed. Cooling and then solidifying the shortening forms agglomerates.

RA: rewetting agglomeration; **PA:** press agglomeration.
Source: Pintauro (1972).
*Whole milk should be blended in its dried form in order to reach desirable wettability and dispersibility.

incorporated back to the process. In this way, tumble/growth agglomeration is an efficient process because recycling is occurring continuously and losses are minimal. The problem of attrition and fines production is normally presented in handling and distribution of agglomerated food products after being released from the agglomeration process. This problem is severe and will be discussed in Chapter 12.

With regard to pressure agglomeration, post-treatment is normally needed only for the case of low and medium-pressure agglomeration. These methods typically require liquid binders to ensure easy formability. High pressure agglomeration, on the other hand, does not include post-treatment and, in most applications, only dry additives are added. In contrast to tumble/growth agglomeration, which requires fine particulates as well as dispersion and aeration features, pressure agglomeration operates well using particles of wide size distributions and without aeration. In fact, aeration of the feed prior to agglomeration must be avoided to facilitate the operation and the maximum particle size that can be handled with increasing pressure. Large particles do not segregate and are easily incorporated during the formation of the agglomerate under pressure. When high forces are applied, brittle disintegration and plastic deformation occur. Also, a considerable volume reduction takes place with densification ratios as high as 1:5. Agglomerate strength increases with higher pressures during densification and forming. Knives could be used to cut extrudates and diverse types of separators may be used to break strings of briquettes into single units. Pressure agglomeration can also be used to obtain granulate products. In this case, a separator is used as a pre-breaker. The product is obtained between the two decks of double-deck screens. The oversized material is crushed and rescreened, while the undersized one is recirculated. Multiple step crushing and screening operations may be employed in order to improve yield and to obtain cleaner granular products.

7.6. APPLICATIONS OF AGGLOMERATION

Agglomeration has many applications in food processing. In the context of instantizing, tumble/growth agglomeration is used in the food industry to improve reconstitutability of a number of products including flours, cocoa powder, instant coffee, dried milk, sugar, sweeteners, fruit beverages powders, instant soups and spices. With regard to shaping, extrusion has been extensively used in grain process engineering to obtain an array of products from diverse cereals, principally ready-to-eat breakfast cereals. Pintauro (1972) developed an extensive selection of approximately 110 patented processes created by internationally recognized companies and researchers in the field. Table 7.4 provides a short summary of different processes and products described in this guide for size enlargement and instant product development. Interested readers can comprehend the processes in much greater detail in the mentioned reference. This table presents different possibilities for agglomeration with products of different composition and their mixes. General processes discussed in this chapter can be found in each description.

REFERENCES

APV (1989). *Dryer Handbook (DRH-889)*. APV Crepaco Inc., Rosemont, IL.

Barbosa-Cánovas, G.V., Vega-Mercado, H. (1996). *Dehydration of Foods*. Chapman & Hall, New York.

Cardew, P.T. and Oliver, R. (1985). *Kinetics and Mechanics in Multiphase Agglomeration Systems*. Notes of course on Agglomeration Fundamentals. In Proceedings of 4th International Symposium on Agglomeration. Toronto University, Waterloo, Ontario, Canada.

Coucoulas, L. (1992). Agglomeration food for thought. *Process Eng.* **8**: 29–31.

Engelleitner, W.H. (1994). *Method Comparison*. Notes of course on Briquetting, Pelletizing, Extrusion, and Fluid Bed/Spray Granulation. The Center for Professional Advancement, Chicago, IL.

Green, D.W. and Maloney, J.O. (1999). *Perry's Chemical Engineers' Handbook*. McGraw-Hill, New York.

Härkönen, H., Koskinen, M., Linko, P., Siika-aho, M. and Poutanen, K. (1993). Granulation of enzyme powders in a fluidized bed spray granulator. *Lebensm,-Wiss. u.-Technol.*, **26**: 223–241.

Kristensten, H.G. (1993). Binders. In *Encyclopedia of Pharmaceutical Technology*. Marcel Dekker, New York, pp. 452–465.

Loncin, M. and Merson, R.L. (1979). *Food Engineering. Principles and Applications*. Academic Press, New York, pp. 229–270.

Masters, K. and Stoltze, A. (1973). Agglomeration Advances. *Food Eng.* **45**(1): 64–67.

Nair, J.H. and Sivetz, M. (1973). Coffee and tea. In: *Food Dehydration*, Vol. 2, Arsdel, W.B.V., Copley, M.J., Morgan, A.I. (eds.). The AVI Publishing Company, Inc., Westport, CT, pp. 384–436.

Parikh, D.P. (1997). Handbook of pharmaceutical granulation technology. In *Drugs and Pharmaceutical Sciences*, Vol. 81. Marcel Dekker, New York.

Pietsch, W. (1983). Low-energy production of granular NPK fertilizers by compaction-granulation. In *Proceedings of Fertilizer'83*. British Sulphur Corp., London, UK, pp. 467–479.

Pietsch, W. (1991). *Size Enlargement by Agglomeration*. John Wiley & Sons, Chichester, UK.

Pietsch, W. (1994). Parameters to be considered during the selection, design, and operation of agglomeration systems. In Proceedings of *Preprints of 1st International Particle Technology Forum, Part I*. AIChE, New York, pp. 248–257.

Pietsch, W. (1997). Granulate dry particulate solids by compaction and retain key powder particle properties. *Chem. Eng. Progress* **4**: 24–46.

Pietsch, W. (1999). Readily engineer agglomerates with special properties from micro- and nanosized particles. *Chem. Eng. Progress* **8**: 67–81.

Pintauro, N. (1972) Agglomeration processes in food manufacture. *Food Processing Review* No. 25.

Rumph, H. (1962). The strength of granules and agglomerates. In *Agglomeration*, Knepper, W.A. (ed.). Wiley-Interscience, New York, pp. 379–418.

Schubert, H. (1981). Principles of agglomeration. *Int. Chem. Eng.* **21**(3): 363–377.

CHAPTER 8

ENCAPSULATION PROCESSES

8.1. INTRODUCTION

In the past, a wide range of food products were technically not feasible for manufacture but are possible today through the development and design of encapsulated ingredients. Such formulations derive from processes that totally envelop the active material in a coating or "capsule," thereby conferring distinct physico-chemical capabilities compared to the original non-encapsulated ingredients. Encapsulation can be defined as a process where a continuous thin coating is formed around solid particles, liquid droplets, or gas cells that are fully contained within the capsule wall (King, 1995). In particular, food processing encapsulation is directly related to the coating of minute particles of ingredients (e.g., acidulants, fats, and flavors), as well as whole ingredients (e.g., ground raisins, nuts, and confectionery products), by microencapsulation and macrocoating techniques, respectively (Shahidi and Han, 1993). Encapsulation is a topic of interest in a wide range of scientific and industrial areas, varying from pharmaceutics to agriculture and from pesticides to enzymes. Although the technology of encapsulation and controlled release is undoubtedly the most developed in the area of drug delivery systems, it has also revolutionized the food and fragrance industries (Greenblatt et al., 1993).

Encapsulation technology has been used in the food industry for more than 60 years as a way to provide liquid and solid ingredients as an effective barrier for environmental and/or chemical interactions until release is desired (Reineccius, 1994). In effect, nearly any material that needs to be protected, isolated, slowly released over time, or released at a certain time can be encapsulated. Those food ingredients that may benefit from encapsulation include flavors, acids, alkalis, buffers, lipids, enzymes, microorganisms, artificial sweeteners, vitamins, minerals, preservatives, antioxidants, cross-linking agents, leavening agents, colorants, and nutrients (Risch, 1995; Reineccius, 1994).

Encapsulation is especially widely used in dry flavor production, while the vast majority of flavor compounds used in industries are in the form of liquids at room temperature. For food products and beverages such as cake and soup mixes, jelly crystals, dry beverage mixes, and instant breakfast drinks, the use of liquid flavoring is not technologically acceptable. Therefore, it is necessary to present the flavoring components in the form of a dry, free-flowing powder. This can be achieved by encapsulation technology.

In addition to the great advantage of the physical state change from liquid to solid, other typical advantages of using encapsulation in food ingredients include (Versic, 1988b; Greenblatt et al., 1993; DeZarn, 1995):

(a) controlling the release of encapsulated ingredients (e.g., gradual release of flavors during microwaving, leavening agents in baking, and citric acid release during sausage manufacture);
(b) enhancing stability to temperature, moisture, oxidation and light (e.g., aspartame protection during baking, oxidation barrier for beta-carotene, protection during freeze and thaw cycles, and increased shelf life);

(c) masking undesirable flavors (e.g., taste-masking of potassium chloride for nutritional supplements);
(d) reducing negative interactions with other compounds (e.g., microencapsulation of such acidulants as citric acid, lactic acid, and ascorbic acid to maintain color, texture, nutrient content, and flavor of foods, and encapsulation of choline chloride to inhibit interaction with vitamins in premixes); and
(e) promoting easier handling of the core or interior material by preventing lumping, improving flowability, compression, and mixing properties, reducing core particle dustiness, and modifying particle density.

Microcapsule properties may be changed to suit specific ingredient applications, including composition, mechanism of release, particle size, final physical form, and cost. When designing encapsulation processes, it must be clearly established what type of functions encapsulated ingredients can provide to the final product in order to select the most suitable coating material. Furthermore, the different processing conditions that the product will go through before release are of essential consideration. Other important features to take into account are the optimum concentration of the active ingredient, the mechanism of release, the final particle size, density, and stability requirements of the encapsulated ingredient. Last but not least, the cost constraints of the encapsulated ingredients are fundamental for a profitable outcome of the final product. Based on the general technical and economical considerations, a microcapsule wall polymer should have enough mechanical strength to allow minimum wall thickness and hence maximum capsule size, and therefore maximum payload (Finch, 1993).

8.2. MICROCAPSULES

The final capsules from encapsulation processes change their denomination depending on their particle sizes. Encapsulated particles are called microcapsules when the size range is between 0.2 and 5,000 μm, macrocapsules when the range is larger than 5,000 μm, and nanocapsules when the range is smaller than 0.2 μm (King, 1995). Table 8.1 shows different encapsulation processes that produce particle sizes of varied ranges. Capsule structure is divided into the core and the coating material. The interior contents of capsules, i.e., the core, can also be called ingredients, substrate, fill, or active agents. The coating material, also denominated as the wall, shell, membrane, carrier, or coat, is the external layer or layers that cover the core material. It can be made from natural, semi-synthetic, or synthetic polymers.

Table 8.1. Particle size ranges for selected encapsulation methods (adapted from Vasishtha, 2002).

Encapsulation method	Size range (μm)
Spray drying	20–150
Centrifugal extrusion	125–3,000
Air suspension coating	50–10,000
Extrusion	700–6,000
Coacervation	1–500
Centrifugal suspension-separation	5–1,000

Table 8.2. A summary of typically used coating material in food industries (adapted from Greenblatt et al., 1993; Shahidi and Han, 1993).

Type	Example
Carbohydrates	Corn syrup, dextrins (malto and cyclodextrins), starch, sucrose
Gums	Agar, gum arabic, gum acacia, sodium alginate, butyrate phthalate, carrageenan
Lipids	Beeswax, diglycerides, fats, hardened oils, beeswax, monoglycerides, oils, paraffin, stearic acid, tristearin
Inorganic materials	Calcium sulfate, clay, silicates
Celluloses	Acetylcellulose, carboxymethylcellulose, nitrocellulose, methylcellulose, cellulose acetate-butyrate-phthalate, cellulose acetate-phthalate, ethylcellulose,
Proteins	Albumin, casein, gelatin, gluten, hemoglobin, peptides
Synthetic elastomers	Ethylenevinyl acetate, polyacrylamide, polyacrylate, polyethylene, polyvinyl alcohol, polyvinyl acetate
Synthetic polymers	Acrylonitrile, polybutadiene

8.2.1. Coating Material for Encapsulation

Coating substances that are basically film forming materials can be selected from a wide variety of natural or synthetic polymers, depending on the material to be coated and the characteristics desired in the final microcapsules. The coating composition is the main determinant of the functional properties of the microcapsule and of the method to be used to improve the performance of a particular ingredient. An effective coating material should have good rheological properties at high concentration and ease of manipulation during the process of encapsulation. It should be selected so that it produces a stable emulsion or dispersion with the active ingredient, and does not react or degrade the active material during processing and storage. Furthermore, it should meet specified or desired capsule solubility properties and active material release properties.

A summary of coating materials typically used in the food and fragrance industries is listed in Table 8.2. Coating materials for encapsulation of food ingredients can be subdivided into carbohydrates, cellulose, gums, lipids, and proteins. For example, encapsulation by spray drying and extrusion depends primarily on the carbohydrates used for the encapsulation matrix. Gums, usually used as texturing ingredients, stabilize emulsions, control crystallization, and inhibit syneresis (the release of water from fabricated foods), thereby improving coating properties. Lipids are generally used for encapsulation for water soluble ingredients. Protein ingredients are also effective in encapsulating food ingredients. In particular, gelatin is used in coacervation.

8.2.2. Types of Encapsulated Food Ingredients

The types of food ingredients that can be encapsulated are shown in Table 8.3 (Gibbs et al., 1999). Applications for encapsulation have been slow to expand since the technique was formerly thought to be too expensive and highly specific. However, since production volumes have increased and become more cost-effective, a wide variety of encapsulated foods can be found. Flavored oil encapsulated in food-grade hydrocolloid is an example of water soluble capsules commonly found.

Flavoring agents and spices are encapsulated by a variety of processes and offer numerous advantages to the food processor. Citrus oil and other flavors, for example, provide enhanced stability to oxidation, volatilization, and light, controlled release, resistance to clumping and caking, and substantially longer shelf life (Shahidi and Han, 1993). Encapsulated flavors are available as natural flavors, natural and artificial flavors, essential oils (menthol, peppermint, and spearmint), oleoresins, natural flavors with other natural flavors added, chips, and artificial flavors. Although encapsulated

Table 8.3. Various food ingredients that can be encapsulated (adapted from Gibbs et al., 1999).

Type of ingredient
Flavoring agents
Acids, alkalis, buffers
Lipids
Redox agents (bleaching, maturing)
Enzymes and microorganisms
Artificial sweeteners
Leavening agents
Preservatives
Colorants
Cross-linking and setting agents
Agents with undesirable flavors and odors
Essential oils, amino acids, vitamins, minerals

flavors may be used in many different applications, they are currently gaining considerable attention for their stability through high-temperature/short-time processes such as those utilized in preparing extruded foods and microwavable foods.

Acidulants are added to foods as flavor modifiers, preservatives, and processing aids. Unencapsulated food acids can react with food ingredients to produce many undesirable effects. These include deceased shelf life of citrus flavored foods and starch containing foods, loss of flavor, degradation of color, and separation of ingredients. Encapsulated food acids resolve these and other problems because they preclude oxidation and provide controlled release, with their coating formulated to dissolve or melt at specific temperatures. Furthermore, encapsulated acids reduce hygroscopicity, reduce dusting, and provide a high degree of flowability without clumping. Examples of encapsulated acidulants that are commercially available are adipic acid, ascorbic acid, citric acid, fumaric acid, lactic acid, and malic acid (Shahidi and Han, 1993). Encapsulated acidulants can be used as dough conditioners and in meat processing (e.g., in cured meat products). For example, uncoated lactic acid and citric acid cannot be used in the production of cured meats because they react almost instantaneously with the meat, rendering it unsuitable for further processing. However, an encapsulated acid that is formulated for delayed release at smoldering temperatures can be used, reproducing the same pH as that obtained with lactic acid bacteria, eliminating the need for fermentation. Thereby, production time can be reduced.

Microencapsulation also enables ingredients such as enzymes to maintain their viability for extended periods of time, avoiding their exposure to ions, protons, free radicals or other type of deleterious agent. Sweeteners are often subject to the effects of moisture and/or temperature. Encapsulation of sweeteners, namely sugars and other nutritive sweeteners, reduces their hygroscopicity, improves their flowability, and prolongs their sweetness perception. Sodium chloride, encapsulated with a variety of coatings, including partially hydrogenated vegetable oil, is used in formulations to control color degradation, rancidity, water absorption, and yeast growth. The encapsulated form also improves flowability and reduces clumping and caking. Typical product applications include ground meats, pretzel snacks, and yeast dough (Shahidi and Han, 1993).

Leavening agents such as sodium bicarbonate are used in baked goods to achieve volume and lightness of texture. Encapsulated sodium bicarbonate protects the base from premature reaction with acid or water, and delays the release of its contents until optimum baking conditions are present. This ensures that maximum leavening is achieved and proves to be economically attractive. Encapsulates are produced with a variety of fat and oil coatings which allow the food product developer more flexibility in optimizing the formulation.

Natural colors such as annatto, β-carotene, and turmeric present solubility problems during their use and may create dust clouds. Encapsulated colors are easier to handle, and offer better solubility, stability to oxidation, and control during dry blending. Their shelf life is extended at least 2 years compared to 6 months for the uncoated form (Gibbs et al., 1999).

Vitamins and minerals are added to nutritional dry mixes to fortify a variety of foods, including breakfast cereals, dairy products, baby formulas, and pet foods. Both fat- and water-soluble vitamins and minerals may be encapsulated with a variety of coatings to provide many advantages. Encapsulation increases stability and improves flow properties and reduces dusting when the nutrients are added to dry mixes (Gibbs et al., 1999).

8.2.3. Microcapsules: Their Structure and Release Mechanisms

Microcapsules can be divided into three main classifications in terms of their conformation: single particle structure (regular or irregular), aggregate structure, and multi-walled structure (Fig. 8.1). A sphere of the active ingredient surrounded by a thick uniform wall or membrane, resembling the shell of a hen's egg, is called a single particle structure. An aggregate structure is formed when several distinct core particles are enclosed within the same capsule wall. When the capsule is a multi-walled structure, different concentric layers have the same, or quite different, composition. In this case, the multiple walls are placed around the core to achieve multiple purposes related to the manufacture of the capsules, their subsequent storage, and controlled release (Shahidi and Han, 1993). In this way, the microcapsule will have the ability to preserve a substance in a finely divided state and to release it as occasion demands.

Microcapsules are involved with two main functions following the formation of a wall around the core material: (a) keeping and protecting the core material inside the shell during storage, and (b) releasing the core material at the right time. Table 8.4 shows the parameters affecting the release of

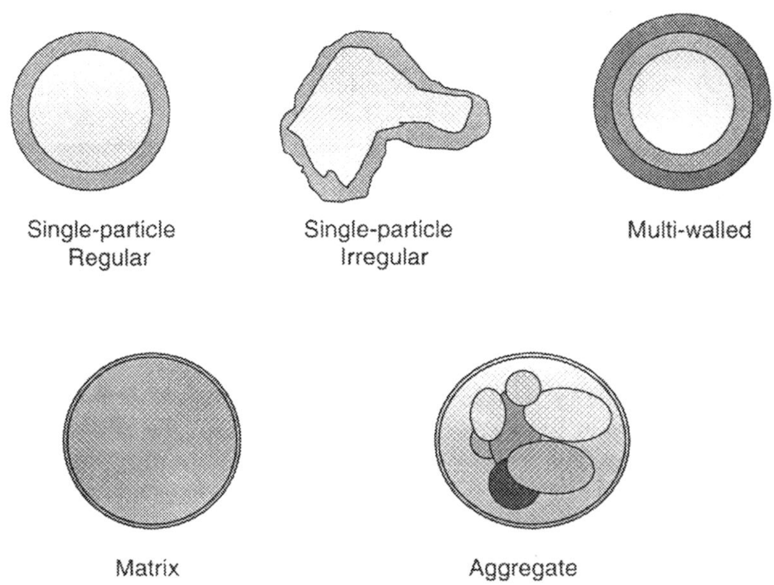

Figure 8.1. Various forms of capsules (adapted from Gibbs et al., 1999).

Table 8.4. Parameters affecting the release rate of core materials (adapted from Shahidi and Han, 1993).

Coating properties	Density, crystallinity, orientation, solubility, plasticizer level, cross-linking, pretreatments
Capsule properties	Size, wall thickness, configuration, conformity, coating layers, post-treatment
Experimental parameters	Temperature, pH, moisture, solvent, mechanical action, partial pressure differential (inside and outside of coating)

core materials. The interior contents of the capsules can be released under four different mechanisms: fracturation, diffusion, dissolution or melting, and biodegradation. These are described as follows:

- *Fracturation*: The coating can be fractured or broken open by external forces, such as pressure, shearing, or extrasonics. Chewing is the most commonly used mechanical release means.
- *Diffusion*: Given that microcapsules are very small, they have a large surface area per unit weight. Microcapsules may function as a semipermeable membrane, releasing the core material by a diffusion-controlled process. The slight heat application or increase of solvent (e.g., increase of moisture content) heightens permeability by changing the crystalline state of the amorphous matrix into a more mobile rubbery state. Thus, flow of the core material through the coating will be facilitated.
- *Dissolution or melting*: The integrity of the coating can be destroyed by dissolution into an appropriate solvent or by thermal means. Water soluble coatings can be easily dissolved from around the core by increasing moisture in the systems. Thermal release is commonly used for fat capsules. In this case, the coating melts away from the core, thus releasing the core ingredients in an environment such as that occurring during baking.
- *Biodegradation*: Release from microcapsules can be accomplished by biodegradation mechanisms. For example, lipid coatings may be degraded by the action of lipases.

There is a wide variety of engineering techniques or processes available for encapsulation purposes; it is extremely difficult to systematically cover all the processes that can be encountered in this field. Among the most remarkable techniques are spray drying, spray chilling and spray cooling, extrusion, coacervation, inclusion in cyclodextrins, air suspension coating, centrifugal extrusion, centrifugal suspension-separation, and freeze-drying. Among all of these methods, spray drying and extrusion are the two major commercial processes usually used in terms of product volume (Reineccius, 1994). Details on some of these processes mentioned above will be discussed in this chapter.

8.3. SPRAY DRYING

In 1932 the English company A. Boake, Roberts & Co., Ltd. produced the first spray-dried flavor powders in which the flavors were encapsulated by a thin film of gum arabic. Since then, spray drying technology has become one of the most important processes used to produce dry flavors from liquids throughout the food and beverage industries (Heath, 1978). The process of spray drying is economical and flexible, uses equipment readily available and produces particles of good quality (Shahidi and Han, 1993). Although it is mostly considered a dehydration process, spray drying can also be used as an encapsulation technique when it entraps core "active" material within a protective matrix formed from a polymer or melt (Dziezak, 1988).

The spray drying process for encapsulation is demonstrated in Fig. 8.2. At the beginning, the selected suitable carrier (shell) material is hydrated and dissolved to form a high solid loading. A

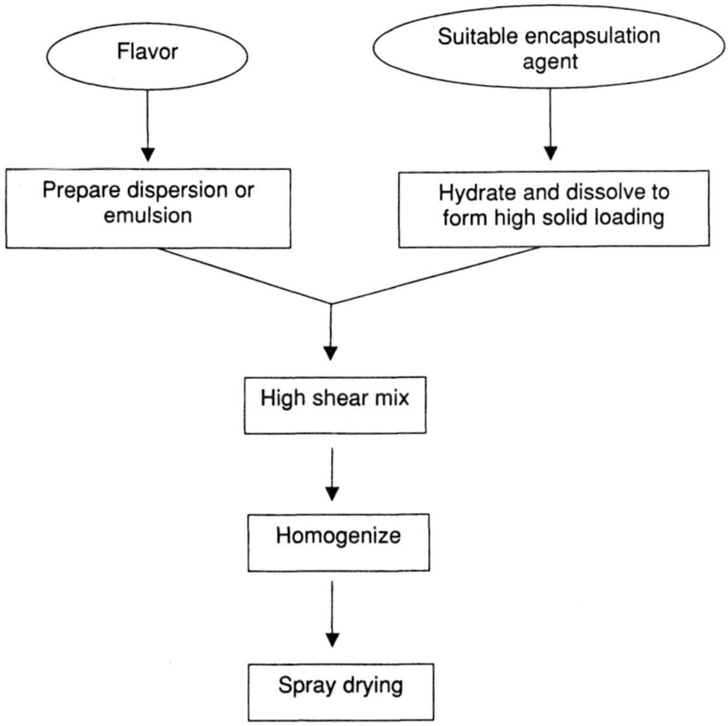

Figure 8.2. Flow diagram of spray drying process for flavor encapsulation (modified from Greenblatt et al., 1993).

dispersion of the core ingredient is added to the carrier solution. All of the ingredients are mixed under high shear, homogenized to form a fine emulsion and then atomized through a heated chamber in a spray drier, which is essentially the same as the one used for the dry milk production. As water evaporates from the particles, the core ingredient is wrapped in the shell and thereby protected (Greenblatt et al., 1993).

The initial step in spray drying of an encapsulated food material is the selection of a suitable carrier (Reineccius, 1994; Shahidi and Han, 1993). Apart from the basic properties mentioned before, the most suitable carrier should have good emulsifying properties, have low viscosity at high solids level (<500 cps at >45% solids level), and exhibit low hygroscopicity. It is worthwhile to note that the proper selection of carriers will yield more than a 1-year shelf life without any antioxidant (Reineccius, 1994).

Gum arabic, hydrolyzed and modified starches, dextrin, gelatin, or non-gelling protein are generally used as carriers. Because of its outstanding natural emulsifying property and higher rate based on the criteria mentioned above, gum arabic is the traditional carrier used in spray drying, even though it is a costly product and has limited supply. Native starches impart no emulsification properties to the spray dried flavors, but they have similar or even superior functional properties compared to gum arabic after certain chemical treatments. Approved chemical treatments include esterifying the partially hydrolyzed starch with substituted cyclic dicarboxylic acid anhydrides or treating the starch with a maximum of 3% octenylsuccinic acid anhydride. Modified starches often have an undesirable off-taste and do not afford good protection for oxidizable flavors. Hydrolyzed starches have a dextrose equivalent range from about 2 to 36.5. Compared to the first two groups of carriers, hydrolyzed starches are inexpensive (about one-third the cost of modified starches), blend in flavor, have low viscosity at high solids content and afford good protection against oxidation. Their major disadvantage is the lack of emulsification properties. Therefore, it is common to use blends of gum arabic/hydrolyzed starches or modified starches/hydrolyzed starches.

Some considerations must also be taken into account when conducting the formulation and preparation of the material being spray dried. First, the infeed solids concentration is the most important determinant of flavor retention during spray drying, and there is a unique optimum infeed solids level in each carrier material. Increasing the solids level up to the point that additional solids are no longer soluble benefits flavor retention by decreasing the required drying time by forming a high solids surface film around the drying droplets. The infeed must first be rehydrated with water. Insoluble solids offer no barrier to the diffusion of flavor molecules and thus do not improve flavor retention during drying. The second consideration is the amount of flavor added into the carrier solution. A 20% flavor load based on the solubilized carrier solids is traditionally used. The infeed temperature is most often kept at or near ambient temperature (Reineccius, 1994). Among industrial atomizers, the single fluid high pressure spray nozzle, and the centrifugal wheel are recognized for industrial coating.

The major advantage of this method is its ability to handle heat-liable materials due to the short heat exposure time (a few seconds at most). In spite of the high air temperature (ranging from 160 to 280°C) in the drying chamber, the rapid water evaporation from the coating material keeps the core temperature below 100°C (Dziezak, 1988; Reineccius, 1994). Spray dried particles usually have a very small particle size (<100 μm), which makes them highly soluble and dispersible in water. They can be readily agglomerated by a separate step in which they are treated with steam to induce their cohesion and form larger particles, and have improved flowability and reduced segregation tendency. In addition, the operation is economical. In modern spray dryers, the viscosity of the solution to be sprayed can be as high as 300 mPa·s, implying that less water needs to be removed from these concentrated solutions (Dziezak, 1988; Graves and Weiss, 1992).

There are some limitations for using the spray drying process. The heat may drive off volatile components which make up the essence of the flavor and also damage the shell by causing micro-cracks that can lead to decomposition of the capsule, poor flavor retention, and undesirable alteration of the flavor. The cost may be high if an expensive carrier is used (Greenblatt et al., 1993).

8.4. EXTRUSION

When the term "extrusion" is used for flavor encapsulation, it refers to the process where a flavor emulsion is forced through a die at pressures less than 700 kPa and temperatures lower than 115°C. It is different from the commonly used high pressure and high temperature extrusion processes for producing cereal-based products (Reineccius, 1994). Extrusion has been used to encapsulate flavors (about 100 different types), vitamin C, and colors (Dziezak, 1988). By using this method, the flavoring ingredients or other core materials are not strictly encapsulated. but locked into a matrix of long-chain molecules having much the same effect as a continuous capsule wall (Heath, 1978). The process consists of dispersing the core material in a molten carbohydrate mass, and then forcing it through a series of dies into a bath of dehydrating liquid. Upon contacting the liquid, the coating material that forms the encapsulating matrix hardens to entrap the core material (Shahidi and Han, 1993). Extruded filaments are then separated from the liquid bath, dried to decrease moisture content, and sized.

A typical extrusion process widely used in today's flavor industry is shown in Fig. 8.3. A sugar–starch hydrolysate mixture is melted by heat, and emulsifiers are added in order to facilitate emulsion formation and to promote stability. Antioxidants are often added to flavoring oils to provide flavor stability during the comparatively high temperature processing. Flavoring oils are incorporated at

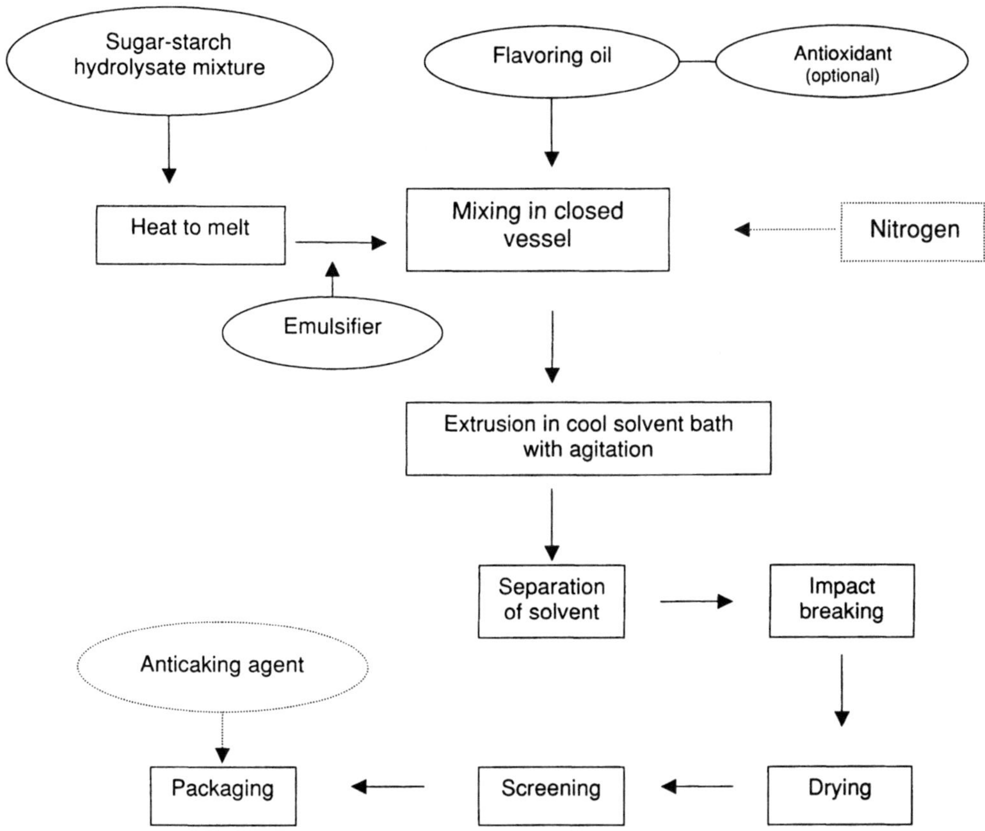

Figure 8.3. Extrusion process for flavor encapsulation (modified from Reineccius, 1994).

about a 10% level based on the hydrolysate mass. The mixture is agitated violently in a closed vessel under nitrogen to form an oxygen-free emulsion, which is later forced through a die into a hot mineral or vegetable oil and then rapidly cooled for solidification and ground to the desired particle size. The ground material is washed with solvent to remove surface oil and then dried under vacuum. An innovation to this process is to extrude the emulsion into a cold solvent bath with sufficient agitation to reduce the particle size. Therefore, it is no longer necessary to grind and wash the product.

High-DE corn syrup and a combination of sucrose and maltodextrin are often used as the encapsulation matrix. Modified food starches with emulsification properties may be used to replace sucrose, producing a "sugar-free" product that has some advantages in marketing a finished food product. Because all of these matrix materials are quite hygroscopic, such anti-caking agents as tricalcium phosphate or pyrogenic silica are recommended to maintain a final product of free-flowing particles (Reineccius, 1994).

The major advantage of extrusion is its outstanding protection of flavors against oxidation. An accelerated shelf-life study conducted by Swisher on extruded orange peel oil showed that its shelf life is more than 4 years. This is much more superior to any other method used for flavor encapsulation (Reineccius, 1994). Because of its extremely long shelf life, extruded flavor products are particularly recommended for use in crystal beverage mixes and table desserts (Heath, 1978). The extruded flavors are soluble in hot water, making them suitable for use in a variety of dry food products such as drink mixes, cake mixes, gelatin dessert mixes, and cocktail mixes (Dziezak, 1988). Other advantages of extrusion over other methods are that it produces visible pieces of flavoring, which might be significant only in terms of marketing (Reineccius, 1994).

The limitations of extrusion include its relatively high cost, low flavor loading, low solubility in cold water, and high process temperature. Its processing costs are estimated to be almost double in comparison to spray drying. Its flavor loading is currently around 8–12%, which is lower than the 20% flavor loading in the standard spray drying process. The extruded product is not readily soluble in cold water and not stable in beverage application because of its large particle size (average of 3.3–6.0 μm). Furthermore, the flavoring to be extruded must be able to tolerate temperatures of 110–120°C for a long period of time (Reineccius, 1994).

8.5. MOLECULAR INCLUSION IN CYCLODEXTRINS

As the name implies, inclusion in cyclodextrins is the only method of encapsulation that takes place on a molecular level (Risch, 1995). Cyclodextrins are enzymatically modified starch molecules (Hedges et al., 1995). The interior of the molecule is formed by hydrogen atoms and glycosidic oxygen bridge atoms, which give the cavity hydrophobic character and interact with various organic molecules or moieties. Guest molecules, with suitable dimensions to fit inside the interior, can be included into the cyclodextrin molecules to form agent–cyclodextrin complexes. This interaction is through the hydrophobic group or groups of the guest molecules with the walls of the cavity of the cyclodextrins.

Cyclodextrins can be produced from starch via fermentation by microorganisms such as *Bacillus macerans* and *Bacillus circulans*. An enzyme in these microorganisms called cyclodextrin-transglycosidase converts the partially hydrolyzed starch into three typical cyclodextrins: alpha-, beta-, and gamma-cyclodextrins containing six, seven, or eight glucose molecules in the ring, respectively (Reineccius, 1994). Depending upon the enzyme used and the conditions under which the reaction is performed, a mixture of different cyclodextrin types or a single type of cyclodextrin can be formed. The β-form is the predominant cyclodextrin produced by most enzymes and has been found to be the most suitable for the inclusion of flavor molecules.

The shape of cyclodextrins is like a hollow truncated cone, as shown schematically in Fig. 8.4. The hydroxyl groups of the glucose monomers are located on the rims of the molecule and directed away from the cavity of the cyclodextrin ring, which interact with water and give the cyclodextrins

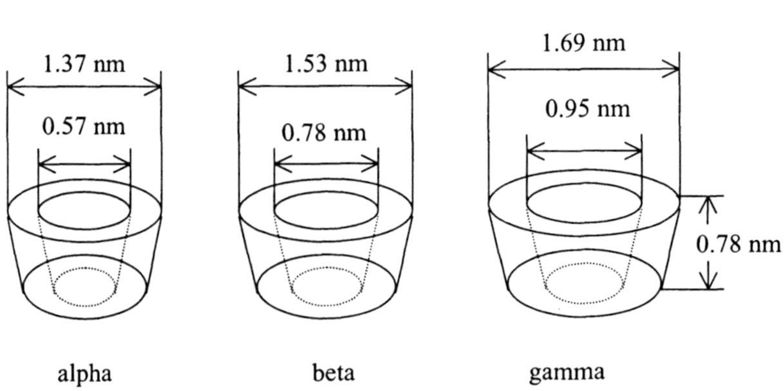

Figure 8.4. Dimensions and schematic structure of three cyclodextrins (adapted from Reineccius, 1994; Hedges et al., 1995).

Table 8.5. Molecular dimensions of some aroma components and the molecular weights of their β-cyclodextrin complexes (modified from Reineccius, 1994).

Aroma component	Molecular size (nm) Length	Diameter	Molecular weight of β-cyclodextrin complex
Allicin	1.2	0.5	162.27
Allylisothiocyanate	0.7	0.4	99.19
Anethole	1.0	0.6	148.20
Benzoic acid	0.9	0.6	122.12
Benzaldehyde	0.9	0.6	106.12
Benzyl alcohol	0.9	0.6	108.13
Borneal	1.1	0.7	154.24
Cinnamaldehyde	0.9	0.7	132.15
Cineol	1.0	0.7	154.24
Citral	1.0	0.6	152.23
Citronellol	1.0	0.6	156.26
Cinnamic acid	0.9	0.7	148.16
Camphene	0.9	0.7	136.23
Carvone	1.0	0.7	150.21
Diallyldisulfide	1.2	0.5	146.26
Eugenol	1.0	0.6	164.21
Fenchone	1.1	0.7	152.23
Geraniol	1.0	0.7	154.24
Linalool	1.0	0.7	154.24
Menthol	0.9	0.6	156.27

their aqueous solubility properties. The binding forces involved in the complex formation include van der Waals forces, hydrophobic interaction, and dipole–dipole interaction (Hedges et al., 1995).

In its aqueous solution, the cyclodextrin interior is occupied by water molecules that can be readily substituted by the less polar guest molecules. Normally, one molecule of cyclodextrin will include one guest molecule; therefore, guest loading depends on the average molecular weight of the guest. Cyclodextrin complexes are relatively stable and their solubility in the aqueous solution is greatly reduced so they will readily precipitate and be separated (Reineccius, 1994). A wide range of organic molecules is capable of forming complexes with cyclodextrins, which includes aliphatic compounds, phenyl derivatives, conjugated rings, and heterocylic compounds (Hedges et al., 1995). The molecular dimensions of some aroma components and the molecular weights of their β-cyclodextrin complexes are listed in Table 8.5.

The following gives a brief description of a process for the molecular inclusion of flavoring by cyclodextrin. First, a 2:1 ethanol:water mixture is prepared and heated to 50–55°C. β-Cyclodextrin is added to the solution at a soluble concentration of more than 10% (by weight). Immediately upon the addition of flavoring material, the β-cyclodextrin complex, enclosing the flavor molecule starts to precipitate. With continuous agitation, the temperature of the solution is allowed to drop to room temperature and finally to 4°C in a refrigerator. For efficient formation and precipitation of the complex, the temperature should be low. The cold, precipitated complex is collected from the solvent by filtering, first dried by air, and then dried at 50°C for 16 h. The final product is a free-flowing cyclodextrin/flavor complex containing 6–15% (w/w) flavoring.

Cyclodextrins provide exceptional protection to enclosed flavors in terms of evaporation loss and oxidation. Flavor compound losses are generally due to lack of inclusion rather than a loss during complex formation and/or further drying steps. Once the complex is formed, it is quite stable in evaporation. It has been reported that only about a 5% loss of included volatiles exists after storage at room temperature for 2 years. The included flavors are also very stable to oxidation. After 10 years of storage, several cyclodextrin/flavor complexes were found to still have good sensory quality in terms of organoleptic quality evaluation.

In addition to their use as an encapsulation matrix for flavors, cyclodextrins have other applications in the food industry due to their variable affinity for different flavor compounds. For example, β-cyclodextrin has been used to remove bitter substances from orange and grapefruit juices. It has also been used to remove off-flavors from old rice, and beany odor and taste from soybean products (Reineccius, 1994). β-Cyclodextrin can also be used to recover flavor oils from natural resources, such as onion and garlic oils. The bitter taste of hops and the astringency of protein hydrolysates can be completely masked by adding β-cyclodextrin (Hedges et al., 1995).

Cyclodextrins are very costly and have low flavor loading (6–15% flavor on dry basis) when compared to other competing processes. At present, they are not approved for food use in the United States and Western Europe as they are in Japan and Eastern Europe (Heath and Reineccius, 1986). They will be used in very special applications where competing processes fail or very high valued flavors must be protected.

8.6. COACERVATION

Coacervation is a colloidal chemical phenomenon which may be defined as "the partial miscibility of two or more optically isotropic liquids, at least one of which is in the colloidal state" and it belongs to one of the oldest techniques for encapsulation. It is considered to be the true microcapsulation process since the coating material completely surrounds the core with a continuous coating (Soper, 1995).

The early commercial use of coacervation was in the production of carbonless duplicating paper. The capsules still produced by this technology contain a colorless dye based in oil and are used to coat the back of one sheet of paper. The front of the other sheet is coated with acid clay that reacts with the dye base and produces color. When a pen is pressed on the surface of the paper, it causes the rupture of the capsules and the release of the dye base (Risch, 1995). This technology was pioneered and developed by the National Cash Register Company in 1957. Since then, it has found many commercial applications (e.g., time-release drugs, scratch-and-sniff perfume samples) in addition to replacing carbon paper. It has been used for the production of microcapsules by the application of thin polymeric coatings to small particles of solids, droplets of liquid, solutions or dispersions (Shay, 1994; Heath, 1978). A limited volume of food flavors is produced by this method.

The basic mechanism involved in this method is the formation of an emulsion and subsequent precipitation of the continuous phase around the droplets of the discontinuous phase. It employs a three-phase system, which includes a manufacturing vehicle (solvent), the material to be encapsulated, and the coating material. As shown in Fig. 8.5, there are three major stages in coacervation processes (Heath, 1978; Heath and Reineccius, 1986; King, 1995; Shay, 1994):

(a) *Formation of the three immiscible phases while mixing under controlled conditions.* In the case of food flavoring, the coating materials are strictly limited by food additive regulations; only gelatin is used in most situations.
(b) *Deposition of the coating material around the core material.* This involves interfacial sorption of the hydrophilic phase on the droplets of the core material. To form the capsules, the pH and temperature must be adjusted to cause the encapsulant to come out of solution so it can coagulate and form a cell wall. At this stage the cell wall is still liquid and needs hardening.

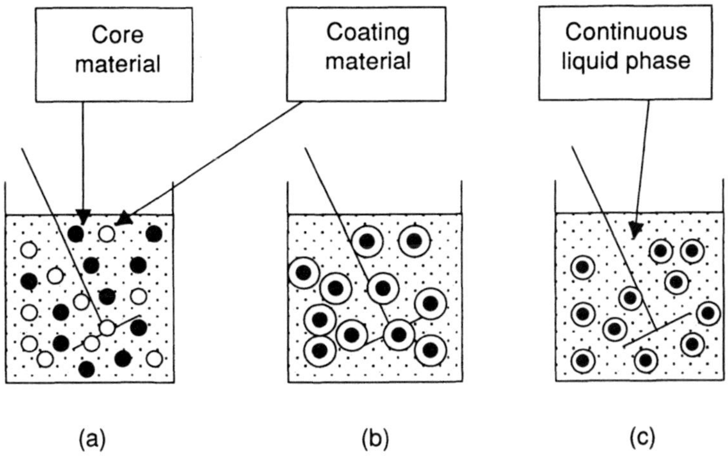

Figure 8.5. Three major stages of microcapsulation by coacervation (adapted from Shay, 1994): (a) formation of the three immiscible phases; (b) deposition of the coating material around the core material; and (c) solid microcapsule formation through shrinkage and solidification.

(c) *Shrinkage and solidification of the liquid coating to form the solid microcapsules.* This can be done through heating, desolvation, or cross-linking techniques.

Coacervation is classified as simple or complex. Simple coacervation deals with systems containing only one colloidal solute (e.g., only gelatin), while complex coacervation deals with systems containing more than one solute (e.g., gelatin and gum acacia). For simple coacervation, a non-solvent or another chemical that competes for solubility with the colloidal solute is added. Then, protein precipitation takes place and a protein-rich coacervation phase is formed. A typical case of simple coacervation is the encapsulation of citrus oil in gelatin. First, the gelatin is dispersed in water. Then, the core material (i.e., hydrophobic citrus oil) is added and the blend is agitated. The solubility of gelatin in water is reduced by lowering the temperature or adding sodium sulfate. Consequently, a two-phase system is created: the colloid-rich phase, appearing as an amorphous cloud, and the colloid-poor aqueous phase. The coalescence of the polymeric colloids occurs around the suspended citrus oil droplets, forming small and still unstable microcapsules. At this stage, the capsule wall is liquid and must be hardened to form the final solid microcapsules. Glutaraldehyde can be used as a cross-linking agent to yield more solid-like capsules. However, excess is not desirable due to toxicity problems. Another optional or complementary approach includes adjusting the temperature or the pH value. The final steps in the process include collecting, washing, and drying the now-stable citrus oil encapsules (Graves and Weiss, 1992; Shay, 1994; Soper, 1995).

Complex coacervation is most commonly used and involves the addition of a second oppositely charged hydrophobic colloid. It is possible only if the pH values are below the isoelectric point of gelatin, because at these pH values, gelatin becomes positively charged and the gum arabic continues to be negatively charged. A typical complex coacervation process begins with the suspension of core material in, for example, either gelatin or gum arabic solution. Then, a gelatin or gum arabic solution (whichever is not used to suspend the core material) is added into the system, the pH is adjusted to 3.8–4.3, and the system is cooled to 5°C. As the gelatin and gum arabic react, viscous liquid microdroplets of polymer coacervate will separate and form a wall on the core particles if the latter are easily wetted by these microdroplets. The still liquid gelled capsule walls can be hardened by glutaraldehyde or other hardening agents. The hardened microcapsules are collected, washed, and dried. This process works well for microencapsulating solids and oily materials (Sparks, 1981; Graves and Weiss, 1992; Versic, 1988a).

Coacervation, also called "phase separation," can be further classified as aqueous- and nonaqueous-phase separation depending on whether water is used as the solvent. Aqueous separation is most common because flavor components (as core material) are hydrophobic in nature. The coating material must be soluble in the solvent and the finished food product is commonly an aqueous system (Heath and Reineccius, 1986). It is possible to microencapsulate hydrophilic core material in oil-soluble coatings, which is called either water-in-oil microencapsulation or organic-phase separation. After a polar core material is dispersed in an organic, nonpolar solvent at an elevated temperature, an oil-soluble coating material is then dissolved in the solvent. By lowering the temperature, the polymeric coating material will emerge as a separate coacervate phase and form a coating around the core particles. The coating will gradually solidify and remain insoluble in cold solvent. This process is typically applied in the pharmaceutical industry, using ethylcellulose as the coating material, and has not been approved for use in the food industry (Graves and Weiss, 1992).

Coacervation is an efficient but expensive encapsulation technique. It is probably the only method that can produce encapsulated particles of sub-micron sizes. The amount of coating can be adjusted; thus, the final flavoring strength of the end product is controllable. Release of the encapsulated flavor may be achieved by physical rupture of the capsule wall, heat or water dissolution, or chemical reaction, depending on its nature and end use (Shay, 1994). With typical high payload (85–90%), it might be expected that it would be economical to have numerous applications in food ingredient encapsulation, but such is not the case: it has been used for only a few specialized flavor encapsulations and evaluated as a potential flavor encapsulation technology. High cost and the problem of finding food approved coating materials are the two main reasons that limit its applications in the food industry (Dziezak, 1988; Graves and Weiss, 1992).

8.7. CENTRIFUGAL EXTRUSION

The Southwest Research Institute (San Antonio, TX) developed this unique encapsulation concept in 1960s and has patented several generations of this method (Sparks, 1981; Schlameus, 1995). Centrifugal extrusion is a liquid co-extrusion process where a rotating extrusion head that contains concentric nozzles is used. Through a concentric feed tube, both coating and core material are pumped separately to the many nozzles mounted on the other surface of the device. Core material flows through the center of the tube, while coating material flows through the other tube. The entire device is attached to a rotating shaft such that the head rotates around its vertical axis. As the head rotates, the core material and coating material are co-extruded through the concentric orifices of the nozzles as the core is enclosed in coating material. As the extrusion head rotates, centrifugal force impels the rod outward, causing it to break into tiny spherical particles. By the action of surface tension, the coating material encircles the core material, forming a continuous coating (Graves and Weiss, 1992). While the droplets are in flight, the molten coating wall is hardened through solvent evaporation from the wall solution. Since the drops are formed by the breakup of a liquid jet, this process is only suitable for materials in liquid or slurry state. The droplets have a narrow particle size range (within $\pm 10\%$ of the mean diameter).

The capsules are collected on a moving bed, which cushions their impact and absorbs unwanted coating moisture. A narrow ring around the spray nozzle is generally used in order to collect the capsules in a ring-shaped bath. If needed, the bath may contain either solids or liquid solvents, which can cushion the impact, protect the particles, and serve additional functions. Starches can absorb

Table 8.6. Typical core and coating material suitable for centrifugal extrusion encapsulation method (adapted from Schlameus, 1995).

Core materials	Vegetable oils, hard fats, flavor oils, vitamins, micronutrients, acids, dyes, seasonings, aqueous systems, air
Coating materials	**Aqueous systems:**
	Gelatin, sodium alginate, cellulose derivatives, gum arabic, starches, carrageenan
	Hot melts:
	Fats, fatty acids, waxes, polyethylene glycol

excess moisture and coatings. Solvents containing suitable cross-linking agents may harden or reduce its solubility if a gelatin coating is used (Graves and Weiss, 1992; Schlameus, 1995; Sparks, 1981).

For good encapsulation results, both core and coating materials must adjust to some special requirements. The liquid core materials must be pumpable, and the following types of liquids can be used: (a) materials that are liquid at room temperature; (b) solids that can be melted at temperatures below 80°C; and (c) a dispersion of finely divided solids in a liquid. Coating materials should be good film formers and capable of rapid hardening in order to collect the capsules, with their viscosity and "stringiness" as controlling factors. For both the core and coating material, their viscosity should be low enough (up to several thousand centipoise) to allow their extruded stream to break into droplets. For coating material, its low viscosity also allows the forming of droplets without excessive tailing or stringing between the individual capsules (Schlameus, 1995). A number of innovative, food approved coating systems have been formulated to encapsulate products such as flavors, seasonings, vitamins, and many others. Typical core and coating materials are listed in Table 8.6.

Centrifugal extrusion is an inexpensive and excellent process for producing particles 400–2,000 μm in diameter. Its production rate is high (up to 22.5 kg of microcapsules per nozzle per hour), and heads containing 16 nozzles are available. It is possible to encapsulate aqueous solutions in waxy wall materials, a difficult problem for some processes. Materials that are stringy and do not permit the clean breakup of the jet are the only wall materials that are not suitable for this process (Sparks, 1981).

8.8. AIR SUSPENSION COATING

Air suspension coating is also called spray coating or fluidized bed processing, and is generally used for solid material encapsulation (Graves and Weiss, 1992). It accounts for the second largest commercial production of encapsulated products for the food industry. Probably because of its flexibility in being suitable for many different core materials, air suspension coating has been successfully applied to encapsulate food ingredients. Wurster, a professor of pharmacy at the University of Wisconsin, first developed this process in the 1950s and used it for coating pharmaceutical tablets. Since then, it has become more popular mainly due to the commercial availability of suitable particle fluidizing equipment. Many modifications have been made to its original form and are mostly focused on the different methods used to disperse coatings to achieve improved encapsulation or special characteristics. However, the basic concept of Wurster has not changed, that is, the spraying of aerosol droplets to impinge on and coat solid particles (Graves and Weiss, 1992).

As shown in Fig. 8.6, solid particles to be encapsulated enter the air stream, heated or cooled, at the bottom of the coating chamber and are lifted by the upward moving air stream to the top of the chamber. At the top of the chamber the particles settle back to the sides of the chamber because of the slower flow of air and wait for recycling. A fine mist of coating material, which is melted

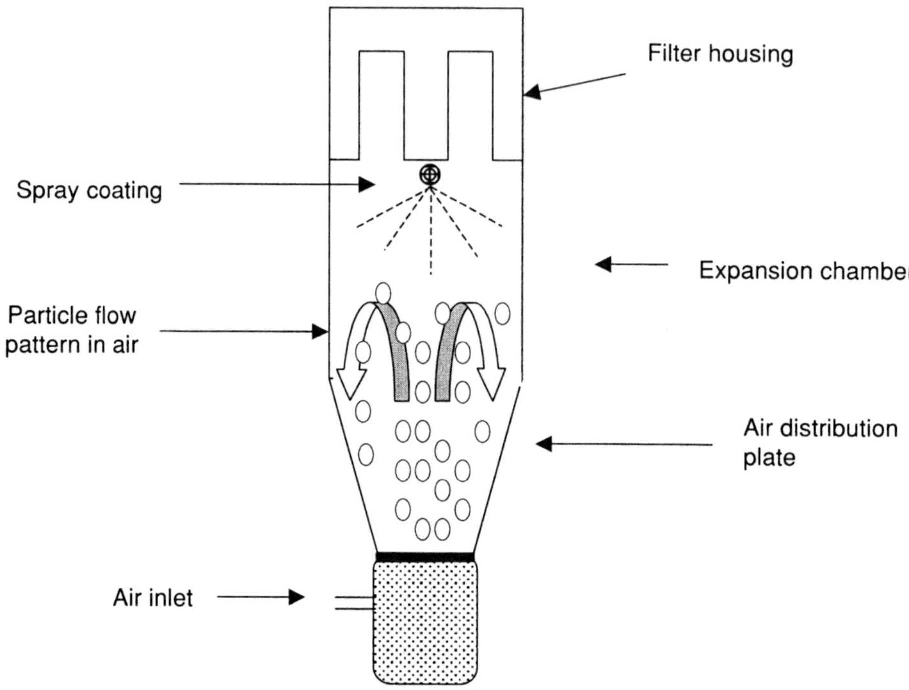

Figure 8.6. Fluid bed coating chamber (adopted from Reineccius, 1994).

or dissolved in an evaporated solvent, is atomized through spray nozzles into the air chamber and deposited on the particles while they are suspended. Each particle is gradually covered with a film of coating material by making numerous passes into the spray region. The movement of the particles from the bottom to the top of the chamber through the aerosol is random, allowing for a rather uniform coating of the particles. The constant flow of air, which has a controlled temperature and humidity setting, allows the melted lipid coating material impinging on the particles to cool and harden or in the case of solubilized coating, the solvent to evaporate. The porous filter on top of the chamber traps the smaller particles entrained in the air and returns them to the process for further coating. The final product is removed from the chamber, cooled, and put through a final drying procedure prior to packaging (Graves and Weiss, 1992; DeZarn, 1995).

A number of important operating parameters should be considered to achieve optimum results. The volume of air used for fluidization is a key processing variable because it controls the height of particles suspended in the coating region and gives proper bed movement for uniform coating. Another critical variable in encapsulation is the air temperature: proper temperature promotes wet-out or surface spreading of the coating material onto the particles, while incorrect temperature results in incomplete coverage by the coating material and subsequently results in products of poor encapsulation quality. Other operating parameters include spray port size, nozzle height, spray rate and temperature, and atomizing air temperature and pressure (DeZarn, 1995).

The physical properties of the particles must also be investigated to determine the feasibility of using fluid bed technology. Particle shape and size are critical to the quality of the final encapsulated product. In general, the more spherical the particle, the better its encapsulation will be. This may be explained in two ways: (a) sharp edges could protrude through the applied coating surface and become vulnerable to release; and (b) the amount of surface area required for coverage is reduced due to the more spherical shape and therefore the coating is thicker. Particles with irregular shapes normally require structure modification to improve their shape prior to using this method (DeZarn, 1995).

It has been found that denser particles with narrow particle size distribution and good flowability are the most suitable for encapsulation in the fluid bed technique. Although it is possible to encapsulate particles of size ranging from 35 to 5,000 μm, the best particle size range is between 50 and 500 μm. The maximum particle size permitted is dependent on the turbine capacity of the fluid bed, while the minimum particle size depends on the porosity of the air exit filter. For fine powders, agglomeration in the beginning stage is unavoidable, but an effective coating can be achieved later. Since the larger particles have less surface area than the smaller ones, they will have a thicker coating material than do the smaller particles on a coating weight basis and are therefore better encapsulated (DeZarn, 1995).

When choosing a suitable coating material, a number of factors must be considered. The viscosity, thermal stability, and film-forming ability of a coating material are critical. It must have an acceptable viscosity to be pumpable and atomizable, must withstand processing temperatures ranging from 15 to 75°C, and be able to spread over the particle surface (DeZarn, 1995). It may be selected from cellulose derivatives, dextrins, emulsifiers, lipids, or protein and starch derivatives, while in a molten state or dissolved in an evaporable solvent (Dziezak, 1988). The ideally suited materials for hot melt coating are hydrogenated vegetable oils or stearines, such as soybean, cottonseed, palm, and canola (low erucic acid rapeseed); fatty acids, various emulsifiers, and bee wax or carnauba wax. The coating levels range from 5 to 50%, depending on the particle size of the core material and the required degree of protection. Commonly used water-soluble coating materials are maltodextrins, starches, gums, and cellulose derivatives. Their coating levels also range from 5 to 50%, but generally do not exceed 30% because of the costly and long processing time included in removing the water from the coating material (DeZarn, 1995).

This method can be used for secondary fat coatings of already encapsulated materials in obtaining flavorings with thermal release properties. It provides excellent additional protection to the encapsulated core material and tailor-made solubility/meltability by the secondary coating process. The primary disadvantage of this method is the additional cost due to the long batch process, which takes 2–12 h to complete. There is also a substantial amount of coating material required to ensure complete coating, which results in the dilution of core materials and additional cost (Reineccius, 1994; Dziezak, 1988).

8.9. SPRAY CHILLING AND SPRAY COOLING

Spray chilling and spray cooling are very similar to the spray drying encapsulation process in that both involve dispersing the core material into a liquefied coating material and spraying it through heated nozzles into a controlled environment (Dziezak, 1988). They differ from spray drying in the temperature of the air (cooled or refrigerated) used in the drying chamber and the type of coating applied. The cool or chilled air in the chamber causes the coating material to solidify around the core. In fact, there is no water evaporation from the coating material in the air drying chamber. In spray chilling, the most commonly used coating materials are molten fractionated and hydrogenated vegetable oils with a melting point of 32–42°C, while vegetable oils or other materials with the melting point of 45–122°C are often used in spray cooling (Risch, 1995).

These two methods differ only in the melting point of the coating material used. They are most often used to encapsulate solid food additives, such as vitamins, minerals, or acidulants. The end products of the spray chilling method are water insoluble but can release their contents at or around the melting point of the coating material. This so-called controlled release property makes this process suitable for secondary coating of spray-dried flavors to retard their volatile component loss during thermal processing (Dziezak, 1988; Graves and Weiss, 1992; Risch, 1995). Spray chilled products have applications in bakery products, dry soup mixes, and food containing high levels of fat (Dziezak, 1988).

The spray chilling method could be used to encapsulate flavors by simply mixing the flavor with fat and then spraying the flavor/fat solution into a cold air spray dryer. The final product is a free-flowing powdered flavor encapsulated by fat. However, there is little or no barrier to the flavor loss by diffusion if the flavor is soluble in fat. Alternatively, a fat or wax that is lipophilic in nature and immiscible with the flavor may be used as the carrier, but it is difficult to find such inexpensive materials in food grade. This method is also useful to encapsulate aqueous flavors and produces a water-in-fat emulsion (Reineccius, 1994).

These two methods are relatively inexpensive and provide the type of encapsulation that can be used for a controlled release of core material. With the possibility of selecting the melting point of the coating material, a desired thermal release property is available, which is important to food ingredients in many food applications.

8.10. CENTRIFUGAL SUSPENSION-SEPARATION

The centrifugal suspension-separation coating (CSS) method, also called rotational suspension-separation, was first developed by Sparks in 1987. It consists of forming a suspension of core particles in a coating liquid and passing this suspension over a rotating disk atomizer. Two different types and sizes of particles, the larger core particles with a layer of coating material and the smaller droplets of pure coating material, are formed at the edge of the rotating disk atomizer, which is mounted at the top of a drying or cooling tower. They can be simply separated by their trajectory movement from the disk and both are solidified when they fall through the tower. The smaller sized coating material is removed for recycle by using a sieve or a cyclone (Sparks et al., 1995). The principle behind this process is illustrated in Fig. 8.7.

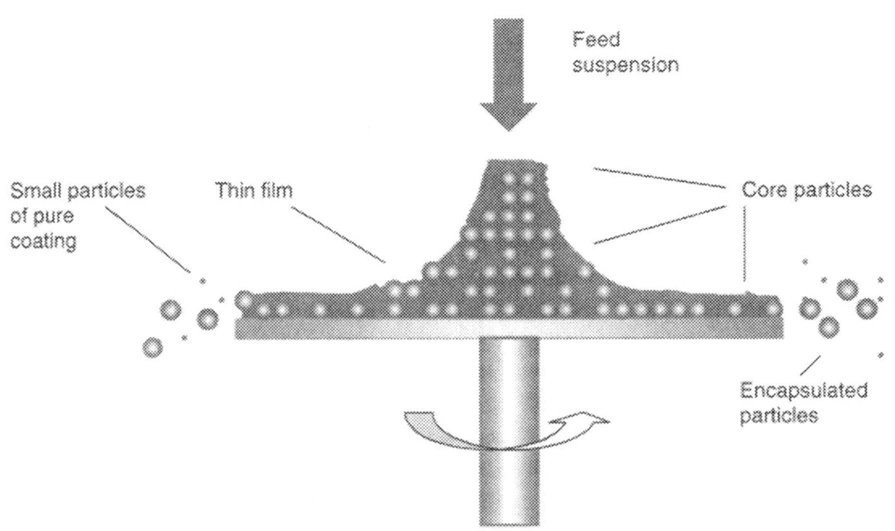

Figure 8.7. Centrifugal suspension-separation sytem (adapted from Shahidi and Han, 1993).

CSS is a continuous, high production rate process that takes seconds to minutes to coat core materials. It has reduced process time and reduced coating material usage (because of recycle) when compared with the air-suspension coating method. It can coat particles of a wide size range, from 30 μm to several millimeters, with coating thicknesses ranging from 30 μm to 2 mm (Dziezak, 1988; Sparks et al., 1995; Reineccius, 1994). The process is very efficient, as it handles each particle only once and produces few uncoated particles. Food-grade solutions, suspensions, or such meltable materials as fats and diglycerides can be used as the coating materials directly without using solvents (Sparks et al., 1995).

CSS is an effective coating method for many food applications. It can be used to coat spray-dried flavors for a good retention of flavor profile and a less volatile compound loss in long-term storage. The coating is efficient if the spray-dried core particles are of a larger particle size (up to 100 μm). However, for smaller sized particles there may be an unacceptable amount of pure coating material droplets in the final product when CSS is used. This is due to the size distribution of the droplets of unused coating material, which has some overlap with that of the coated core material, and thus the recycle cyclones do not give a sharp cut-off in diameter (Sparks et al., 1995).

8.11. FREEZE-DRYING

As will be described in Chapter 11, freeze-drying is a desirable process for the dehydration of almost all heat sensitive materials, including flavors. It has been used for encapsulating water soluble essences and natural aromas (Shahidi and Han, 1993). Because the whole dehydration process is completed under low-temperature and low-pressure conditions, it is considered that the process should have a high retention of volatile compounds. It has been proposed for production of citrus aroma powders as natural flavor ingredients by simply dissolving various blends of corn syrup solids and sugars in the aroma solution and then going through freeze-drying, retaining the aroma in the carrier.

8.12. CO-CRYSTALLIZATION

Co-crystallization is an application process where sucrose is utilized as a matrix for the incorporation of core materials. It involves spontaneous crystallization, which produces aggregates of micro- or fondant-size crystals ranging from 3 to 30 μm, while entrapping non-sucrose materials within or between sucrose crystals (Shahidi and Han, 1993). A saturated sucrose syrup concentrate is mixed with a predetermined amount of core material. The core material itself will provide, through mechanical agitation, the nucleation sites for the sugar crystallization. Encapsulated materials are discharged from the vessel and then dried to a desirable moisture content. The core will remain located primarily in the interstices between crystals. This method is also mentioned in Table 7.4 as an example for size enlargement of cocoa products.

8.13. FINAL REMARKS

The range of applicability using the encapsulation methods described above is extremely wide, and tends to overlap. For instance, one may exclude certain processes and favor others based on the characteristics of the material to be encapsulated, release methods and patterns, health and safety issues, or economic concerns.

In the application of different encapsulation methods for food ingredients, health and safety factors should be considered first. Only the materials approved by the U.S. Food and Drug Administration (FDA) should be used and the "Generally Recognized As Safe" (GRAS) list should be consulted. Any process or material not previously approved must be submitted to the FDA, whose approval requirements extend to methods of solvent removal, storage conditions, handling methods and so on (Sparks, 1981).

Although many techniques are available for food ingredient encapsulation, some needs in the food industry are still unaddressed (Reineccius, 1994). New encapsulation technologies are needed and new coating materials of different functions are desired. It is expected that a newer, improved and more economical encapsulation technique will be developed in the next 10 years. However, if the available techniques stay the same, the market for encapsulates in the food industry will continue to grow at an accelerated rate because of the significant benefits (Graves and Weiss, 1992).

REFERENCES

DeZarn, T.J. (1995). Food ingredients encapsulation. In *Encapsulation and Controlled Release of Food Ingredients*, Risch, S.J. and Reineccius, G.A. (eds.). American Chemical Society, Washington, DC, Chapter 7, pp. 75–86.

Dziezak, J.D. (1988). Microencapsulation and encapsulated ingredients. *Food Tech.* **42**(4): 136–148.

Finch, C.A. (1993). Industrial microencapsulation: polymers for encapsule walls. In *Encapsulation and Controlled Release*, Karsa, D.R. and Stephenson, R.A. (eds.). The Royal Society of Chemistry, Cambridge, pp. 1–17.

Gibbs, B.F., Kermasha, S.K., Alli, I. and Mullingan, C. (1999). Encapsulation in the food industry: a review. *Int. J. Food Sci. Nutr.* **50**: 213–224.

Graves, B. and Weiss, H. (1992). Encapsulation techniques. In *Encyclopedia of Food Science and Technology*, Vol. 2 Hui, Y.H. (ed.). John Wiley & Sons, Inc., New York, pp. 697–703.

Greenblatt, H.C., Dombroski, M., Klishevich, W., Kirkpatrick, J., Garrison, W. and Redding, B.K. (1993). Encapsulation and controlled release of flavors and fragrances. In *Encapsulation and Controlled Release*, Karsa, D.R. and Stephenson, R.A. (eds.). The Royal Society of Chemistry, Cambridge, pp. 148–162.

Heath, H.B. (1978). Flavoring materials. In *Flavor Technology: Profiles, Products, Applications*. AVI Publishing Company, Inc., Westport, CN, pp. 359–366.

Heath, H.B. and Reineccius, G.A. (1986). Flavor production. In *Flavor Chemistry and Technology*. AVI Publishing Company, Inc., Westport, CN, Chapter 11, pp. 354–371.

Hedges, A.R., Shieh, W.J. and Sikorski, C.T. (1995). Use of cyclodextrins for encapsulation in the use and treatment of food products. In *Encapsulation and Controlled Release of Food Ingredients*, Risch, S.J. and Reineccius, G.A. (eds.). American Chemical Society, Washington, DC, Chapter 6, pp. 61–71.

King, A.H. (1995). Encapsulation of food ingredients. In *Encapsulation and Controlled Release of Food Ingredients*, Risch, S.J. and Reineccius, G.A. (eds.). American Chemical Society, Washington, DC, Chapter 3, pp. 26–39.

Reineccius, G.A. (1994). Flavor manufacturing. Part II. Flavor encapsulation.In *Source Book of Flavors*, Heath, H.B. (ed.). Chapman & Hall, New York, Chapter 11, pp. 605–625.

Risch, S.J. (1995). Review of patents for encapsulation and controlled release of food ingredients. In *Encapsulation and Controlled Release of Food Ingredients*, Risch, S.J. and Reineccius, G.A. (eds.). American Chemical Society, Washington, DC, Chapter 18, pp. 197–203.

Schlameus, W. (1995). Centrifugal extrusion encapsulation. In *Encapsulation and Controlled Release of Food Ingredients*, Risch, S.J. and Reineccius, G.A. (eds.)., American Chemical Society, Washington, DC, Chapter 9, pp. 97–103.

Shahidi, F. and Han, X-Q. (1993). Encapsulation of food ingredients. *Crit. Rev. Food Sci. Human Nutr.* **33**(6): 501–547.

Shay, R. (1994). Flavor manufacturing, Part I. In *Source Book of Flavors*, Reineccius, G. (ed.). Chapman & Hall, New York, Chapter 11, pp. 538–605.

Soper, J.C. (1995). Utilization of coacervated flavors. In *Encapsulation and Controlled Release of Food Ingredients*, Risch, S.J. and Reineccius, G.A. (eds.). American Chemical Society, Washington, DC, Chapter 10, pp. 104–112.

Sparks, R.E. (1981). Microencapsulation. In *Encyclopedia of Chemical Technology*, 3rd ed., Grayson, M. (ed.). *Kirk-Othmer* **15**: 470–493. John Wiley & Sons, Inc., New York.

Sparks, R.E., Jacobs, I.C. and Mason, N.S. (1995). Centrifugal suspension-separation for coating food ingredients. In *Encapsulation and Controlled Release of Food Ingredients*, Risch, S.J. and Reineccius, G.A. (eds.). American Chemical Society, Washington, DC, Chapter 8, pp. 87–95.

Vasishtha, N. (2002). *Microencapsulation*. Southwest Research Institute™, San Antonio, TX.

Versic, R.J. (1988a). Coacervation for flavor encapsulation. In *Flavor Encapsulation*, Risch, S.J. and Reineccius, G.A. (eds.). American Chemical Society, Washington, DC, Chapter 14, pp. 126–131.

Versic, R.J. (1988b). Flavor encapsulation. In *Flavor Encapsulation*, Risch, S.J. and Reineccius, G.A. (eds.). American Chemical Society, Washington, DC, Chapter 1, pp. 1–6.

CHAPTER 9

MIXING

9.1. INTRODUCTION

The unit operation in which two or more materials are interspersed in space with one another is one of the oldest and yet one of the least understood unit operations in process engineering. In agriculture and food processing, mixing operations are often used to blend ingredients. Particularly, mixing is used in the food industry with the main objective of reducing non-uniformities and gradients in properties such as concentration, color, texture, or taste between different parts of a system (Uhl and Gray, 1986). The degree of uniformity required may vary somewhat, but most of the time it is important to provide a nutritionally balanced and palatable feed mixture.

Mixing requires the motion of ingredients in order to reach the desired homogeneity or a uniform distribution for a certain food product. It is accomplished by the movement of food particles between various parts of a whole particulate mass (Lindley, 1991a). In some cases, suitable mixing can be simply achieved by placing ingredients together in a common container. However, in other cases a great deal of planning and expenditure of energy is required. The food processing industry relies heavily on mixing to ensure delivery of a product with constant properties. A satisfactory mixing process should produce a uniform mixture in minimum time and with minimum cost for overhead, power, and labor. Currently, almost all food mixing processes are batch, due to their greater flexibility and lower installation costs in comparison to continuous processes. Actual knowledge in mixing is a combination of "know-how" and science, in which "know-how" is predominant and only a few individuals and specialized companies around the world have a thorough knowledge of this matter.

Since the components being mixed can exist in any of the three phases (liquid, gas, or solid), a number of mixing possibilities arise. Mixing cases involving a fluid, e.g., liquid–liquid and solid–liquid, are most frequently encountered, so they have been extensively studied. Despite the predominant importance that mixing of particulate materials has in many processing areas, fundamental contributions of real value either to designers or users of solids mixing equipment is still relatively sparse. Important advances in the understanding of mixing of solids and pastes have been made in very specific fields such as powder technology and multi-phase flow.

Mixing is more difficult to define and evaluate with powders and particulates than it is with fluids. However, advances in quantitative measures of dry solids mixing are helping to control mixer performance. In actual practice, however, a mixer is tested by the properties in the mixed material it produces. A significant proportion of research efforts in the food industry are directed toward the development of new and novel mixing devices for food materials. These devices may be effective for many applications since they deliver a mixed product with the required blending characteristics. Powdered foods are complex systems, and properties can themselves vary during mixing. Developments in mathematical modeling of food mixing processes are scarce and established procedures for process design and scale-up are lacking. As a result, it is virtually impossible to devise relationships between mixing and quality (Niranjan, 1995), especially when blending food powders.

Powder blending is mainly affected by the mixing time, the design of the mixer (including size, shape, paddle geometry, and rotational speed), and type of powders being mixed. Niranjan and de Alwis (1993) mentioned as characteristic features of food mixing the fragile and different-sized nature of food products, as well as the segregating tendency of blended food systems on discharge. These characteristics, along with some others such as cohesiveness and stickiness, make food particulate mixing a complicated operation. A wide range of dry food materials are mixed, including combinations of flour, sugar, salt, flavoring materials, dried milk, and dried vegetables and fruits (Lindley, 1991b).

9.2. MIXING MECHANISMS

As opposed to mixing of miscible liquids, the mixing of particles is often a readily reversible process. A mixture of miscible liquids leaving a mixing unit retains or even improves its mixedness during the transport process, while a well-mixed batch of particles can be separated almost completely at a subsequent process stage. Particles change their relative positions in response to movement and the subsequent rearrangement may be more random. Powder mixing occurs when any particle changes its path of circulation. If two identical particles exchange positions, this is useless mixing; i.e., useful mixing occurs only when particles changing positions (or circulation paths) are different.

Three mechanisms have been recognized in solids mixing: convection, diffusion, and shear. In any particular process one or more of these three basic mechanisms may be responsible for the course of the operation. Other mechanisms such as segregation can also be involved during particle motions. Depending on the equipment used, mixing mechanisms can receive other classifications that will be mentioned further in this section.

9.2.1. Convective, Diffusive, and Shear Mixing

During *convective mixing*, masses or groups of particles transfer together from one location to another, while in *diffusion mixing*, individual particles are randomly distributed over a surface developed within the mixture. In *shear mixing*, groups of particles are mixed through the formation of slipping planes developed within the mass of the mixture. Shear mixing is sometimes considered as part of the convective mechanism.

In *convective mixing*, a circulating flow of powders is usually caused by the rotational motion of a mixer vessel, an agitating impeller (such as a ribbon or a paddle), or gas flow. This circulating flow contributes mainly to a macroscopic mixing of bulk powder mixtures. Large portions of the total mix are moved at relatively high rates, and changes at a microscopic scale are not expected. Therefore, pure convection tends to be less effective, leading to a final mixture, which may still exhibit poor mixing characteristics on a fine scale. Convective mixing is beneficial for batch mode operations but gives unfavorable effects for continuous mode mixing.

Diffusive mixing (or random wall phenomenon) is caused by the random motion of powder particles. The rate of mixing by this mechanism is low compared with convective mixing, but diffusive mixing is essential for microscopic homogenization. It has been concluded that diffusion is the best mechanism for axial mixing, similar to diffusion of particles in gas and liquid phases (Fan et al., 1970). Pure diffusion, when feasible, is highly effective, producing very intimate mixtures at the level of individual particles but at an exceedingly slow rate.

A trough mixer with a ribbon spiral can give almost pure convective mixing, while a simple barrel mixer gives mainly a form of diffusion mixing. These features of diffusion and convective mixing mechanisms suggest that an effective operation may be achieved by combining both, in order to take advantage of the speed of convection and the effectiveness of diffusion.

Shear mixing is induced by the momentum exchange of powder particles having different velocities (differential velocity distribution). Shear mixing is developed by the formation of slipping planes in the bulk material; the originally coherent particle groups are gradually broken along these planes. The velocity distribution develops around the agitating impeller and the vessel walls due to compression and extension of bulk powders. It is also developed in the powder layer in rotary vessel mixers and at blowing ports in gas-flow mixers. Shear mixing can enhance semi-microscopic mixing and be beneficial in both batch and continuous operations. In free-flowing powders, both diffusive mixing and shear mixing give rise to size segregation (or de-mixing), therefore, for such powders, convective mixing is the major mechanism promoting mixing (Rhodes, 1998).

Powder mixing proceeds in a mixer where the three mechanisms described above take place simultaneously. The characteristic curve of mixing is the plot of the degree of mixedness M (on a logarithmic scale) against the mixing time t (on a linear scale). The mixing time is the time measured from the start of mixing in a batch mode operation, whereas it corresponds to the mean residence time (the powder volume in a mixer divided by its volumetric flow rate) in a continuous operation. The characteristic curve of mixing is useful for the performance evaluation of mixers. Figure 9.1 shows a schematic example of the curve, where the standard deviation is plotted on a logarithmic scale. Generally speaking, convective mixing is dominant in the initial stage (I) and the mixing proceeds steadily by both convective and shear mechanisms in the intermediate stage (II). In the final stage (III), the effect of diffusive mixing appears and the dynamic equilibrium between mixing and segregation is reached. The degree of mixedness at this state is called the final degree of mixedness, M_{∞}. Various powder mixers exhibit a variety of patterns in the characteristic curve of mixing. Operating conditions and powder properties significantly influence the value of M_{∞}.

In comparison with fluid mixing, in which diffusion can be normally regarded as spontaneous, particulate systems will only diffuse as a result of mechanical movement provided by gravity, shaking, tumbling, vibration, or any other mechanical mean. Lacey (1954) tried to adjust Fick's equation, the simplest model for molecular diffusion in liquids, to the mixing of solids. Fick's equation

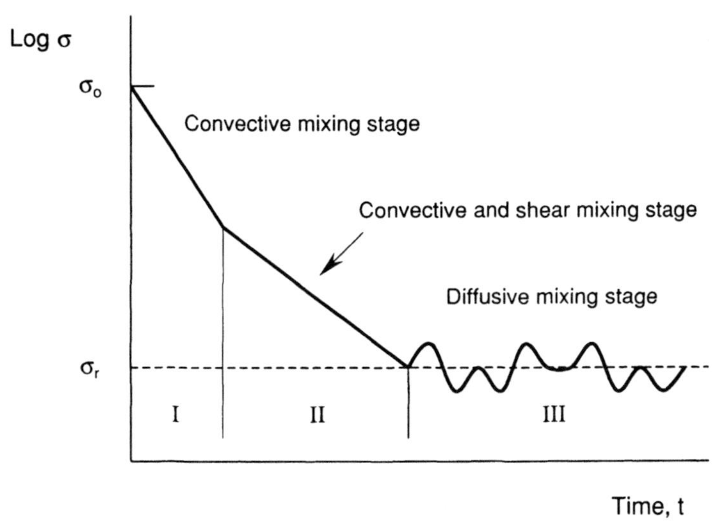

Figure 9.1. Characteristic curve of mixing process (adapted from Miyanami, 1997).

has the following form:

$$\frac{\partial C}{\partial t} = D\frac{\partial^2 C}{\partial x^2} \tag{9.1}$$

where C is the concentration of solids, D is the diffusivity, and x is the distance in the direction of dispersion. It is clear that the diffusivity D in solids does not have the same physical meaning as in liquids, given that D varies with the magnitude and direction of the force impelled to the powder bulk to generate movement. However, the model could describe a binary mixture of particles with the same mean diameter fed into rotating horizontal drum in such a way that a thin layer is perpendicular to the axis of rotation. The equation can be solved as a function of the number of revolutions of the mixer and the distance from one of the sidewalls. The movement of particles during a mixing operation, however, can also result in another mechanism that may retard, or even reverse, the mixing process, known as *segregation*.

9.2.2. Segregation

When particles differing in physical properties, particularly size and/or density, are mixed, mixing is accompanied by a tendency to un-mix. Thus, in any mixing operation, mixing and de-mixing may occur concurrently, and the intimacy of the resulting mix depends on the predominance of the former mechanism over the latter. Apart from the properties already mentioned, surface properties, flow characteristics, friability, moisture content, and tendency to cluster or agglomerate, may also influence the tendency to segregate. The closer the ingredients are in size, shape, and density, the easier the mixing operation and the greater the intimacy of the final mix. Once the mixing and de-mixing mechanisms reach a state of equilibrium, the condition of the final mix is determined and further mixing will not produce a better result.

When non-cohesive particles having significantly different densities are mixed, it can be observed that the denser particles tend to settle to the bottom of the mixture, presumably, lowering the total potential energy of the system. On the other hand, if coarse and fine particles are set in motion, the fines tend to segregate at the bottom via a possible percolation mechanism, in which the fines can pass through the interstices between larger particles. Williams (1968/1969) also showed that a single large particle placed on a vibrating bed of smaller particles will tend to rise towards the bed top, even if its density is greater than that of the finer material. For example, a large steel ball can be made to rise to the top of a beaker of sand by simply shaking the beaker up and down (Rhodes, 1998). One explanation for this is that the large solid particle, which will be generally denser than the loosely packed bed, causes a compaction of the bed immediately beneath it. Consequently, the freedom of movement of the large particle, in response to vibration, will be restricted to the lateral and upward directions, and the net result will be a tendency to rise. These explanations of segregation are only applicable to specific situations.

A general theory of segregation, regardless of the particular circumstances in which the operation takes place, has not yet been offered to explain the segregation phenomena in particulate systems. More information about segregation mechanisms and their influence on conveying, storing, packaging, and even mixing operations are offered in Chapter 12. In any blending operation, the mixing and de-mixing mechanisms will be acting simultaneously. The participation of each of these two sets of mechanisms will be dictated by the environment and the tendency of each component to segregate out of the system. Since these two mentioned sets of mechanisms will be acting against each other, an equilibrium level will be obtained as the final state of the mixture.

The importance of segregation on the degree of homogeneity achieved in solids mixing cannot be over-emphasized. Any tendency for segregation to occur must be recognized when selecting solids mixing equipment. Segregation in a mixture of dry solids is readily detected by use of a heap test. A well-mixed sample of the solids is poured through a funnel to form a conical heap. Samples taken from the central core and from the outside edge of the cone should have essentially the same composition if segregation is not to be a problem. When the two samples have significantly different compositions, it can be assumed that segregation is very likely to occur unless a very careful choice of equipment is made. It is generally accepted that the efficiency of a mixing process must be related to both the flow properties of the components, and to the selection or design of the mixer.

Avoiding segregation is a challenge in the food industry, where materials with a wide range of properties are often mixed. These materials may include spices, liquids, flavors, salt, hydrolyzed vegetable protein, monosodium glutamate, and dehydrated vegetable powders. Flow properties may be modified by the addition of anticaking or flow agents to prevent undesired agglomeration.

9.2.3. Other Classifications for Mixing Mechanisms

Mixing mechanisms can be classified according to the type of motion applied to a bulk as follows: (a) *mixing within bulk material*, by displacing solid particles relative to one another; (b) *centrifugal mixing*, which occurs when some of the solids are detached from the bulk of the material and are rearranged in the air space by gravity or centrifugal forces, and then returned to the surface of the mixture; (c) *mixing in a fluidized bed*, where high speed mixing particles may be thrown off to the air and drawn into the mixture, giving the bed the character of a liquid; (d) *mixing solids in a suspended condition*, when the particles are completely suspended by maintaining a gas flow; (e) *free fall mixing* due to gravity; and (f) *mixing several streams of material*, involving several of the mentioned mechanisms. For each mixing method, a characteristic mechanism determines the rate attainable and the degree of mixing, as well as other important features such as the productivity and energy consumption of the mixer (Gyenis and Ārva, 1989). Mixing mechanisms that powders undergo in industrial common blenders are described in Section 9.4.

9.2.4. Horizontal Drum Blender Mechanism

The mixing mechanism of non-cohesive solids in a horizontal drum mixer has been described by several authors (Lindley, 1991b). Particles are carried around with the mixer until the angle of repose is exceeded, when the particles near the top edge roll down the slope as a thin layer over the rest of the particles (Fig. 9.2). As the rotation speed is increased, the particle velocity may become great enough to project them into the air. Fan and Shin (1979) called this mechanism drift velocity effect, and they proved that it was dependent on the diffusion mechanism. On reaching the end of the slope, the particles are again carried around the mixer walls to complete the circuit. The circuit made by the particles is defined as the path of circulation. This represents the basis for tumbling mixing mechanisms. On the other hand, Hwang and Hogg (1980) used a diffusion–convection model for an inclined drum mixer that includes a term representing the drift velocity V:

$$\frac{\partial C}{\partial t} = D\frac{\partial^2 C}{\partial x^2} - V\frac{\partial C}{\partial x} \tag{9.2}$$

A stochastic diffusion model has also been proposed to examine non-ideal mixing in horizontal drums (Fan and Shin, 1979):

$$\frac{\partial f}{\partial t} = \frac{\partial^2 (Df)}{\partial x^2} - \frac{\partial (Vf)}{\partial x} \tag{9.3}$$

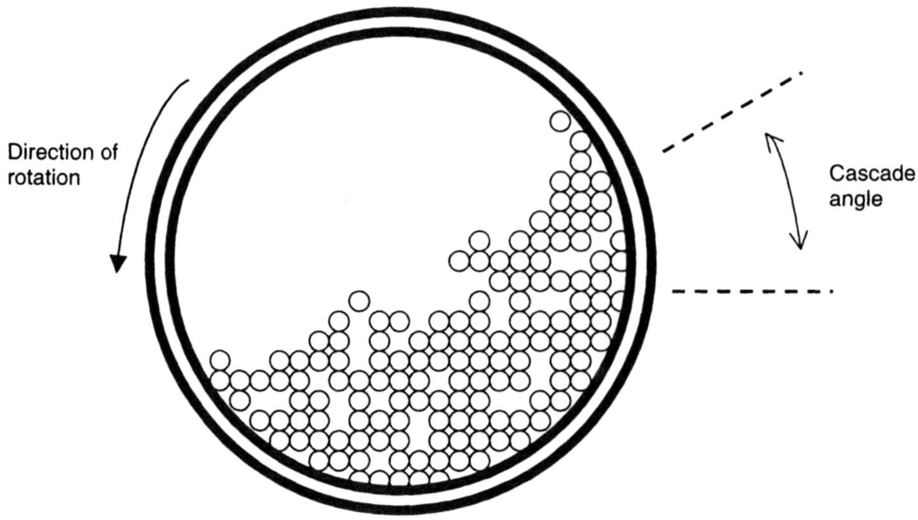

Figure 9.2. Horizontal drum mixer.

where

$$f = f(x_0, t_0, x, t)$$
$$D = D(x, t)$$
$$V = V(x, t)$$

Bridgwater (1976) and Fan et al. (1990) further discussed these models. In all cases, the modeling of powder mixing is still limited to rotating cylindrical drums. In fact, it has been difficult to adapt other models for other industrial applications.

9.3. STATISTICAL APPROACH OF SOLIDS MIXING

In the mixing of particulate solid materials, the probability of getting an orderly arrangement of particles, which would represent perfect mixing, is virtually zero. In practical systems, the best mixture attainable is that in which there is a random distribution of the ingredients. An ideal random distribution of two solid components in equal proportions would resemble a chessboard, i.e., white and black squares in a perfect alternating pattern. In practice, however, a perfectly random mixture is commonly defined as one in which the probability of finding a particle of a constituent of the mixture is the same for all its points, as represented in Fig. 9.3a. Over the years, many workers have attempted to establish criteria for the completeness and degree of mixture. In order to accomplish this, frequent sampling of the mix is usually required and, tending to be statistical in nature, such an exercise is often of more interest to mathematicians than to process engineers. Thus, in practical mixing applications, an ideal mixture may be regarded as the one produced at minimum cost and which satisfies the product specifications at the point of use.

9.3.1. Types of Mixtures

Many food powders are a mixture of two or more particles of different composition (e.g., as in dry beverages, cakes, or ice cream mixes). They may even contain large particulates, e.g., soup mixes

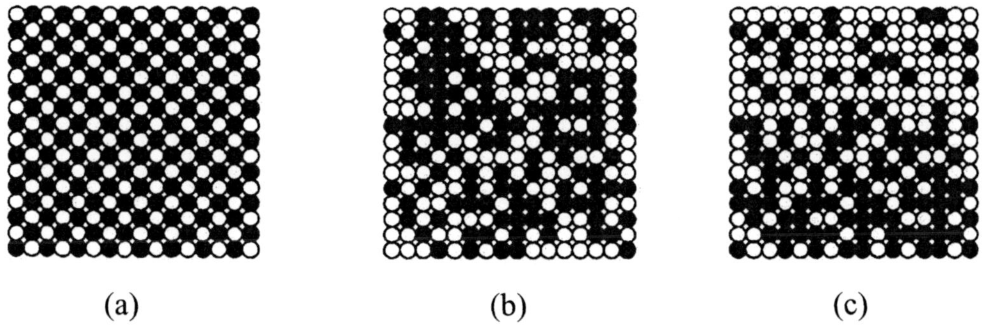

Figure 9.3. Types of mixture: (a) perfect mixture; (b) random mixture; and (c) segregating mixture (reproduced from Rhodes, 1998).

Figure 9.4. An ordered mixture of small particles on carrier articles.

with noodle particles and/or small pieces of dried meat or vegetables. Generally, powder mixtures can be classified into three main types: *random* (non-interactive), *ordered* (interactive), and *segregating*. Figure 9.3 symbolizes these types of mixtures with two components (Rhodes, 1998).

In fine powders (<30 μm), the interparticle forces generated by electrostatic charging, van der Waals forces, and forces due to moisture are large compared with the gravitational and inertial forces on the particles. It is possible to take advantage of this natural tendency for particles to adhere to produce mixtures of better quality than *random mixtures* (Fig. 9.3b). Such mixtures are known as *interactive ordered* mixtures (Fig. 9.4). They are made up of small particles (e.g., <5 μm) adhered to the surface of a carrier particle in a controlled manner (Rhodes, 1998).

When attempting to mix particles that are not subject to segregation, this is generally the best quality of the mixture that can be achieved. In this case, particles of one component have a greater probability of being found in one part of the mixture, and so a *random mixture* can be achieved. As already mentioned, many common food systems have mixed particles with different properties and tend to exhibit segregation (Fig. 9.3c). Although differences in size, density, and shape of the constituent particles of a mixture may give rise to segregation, difference in particle size is by far the most important of these.

Mixing of free-flowing powders differs from mixing of cohesive powders in that it is most difficult to impel movement to cohesive powders and to separate attached particles from one component and mix them with particles from another component. It is particularly difficult to obtain acceptable mixtures in which one component is in very small proportions. Therefore, segregation is more likely

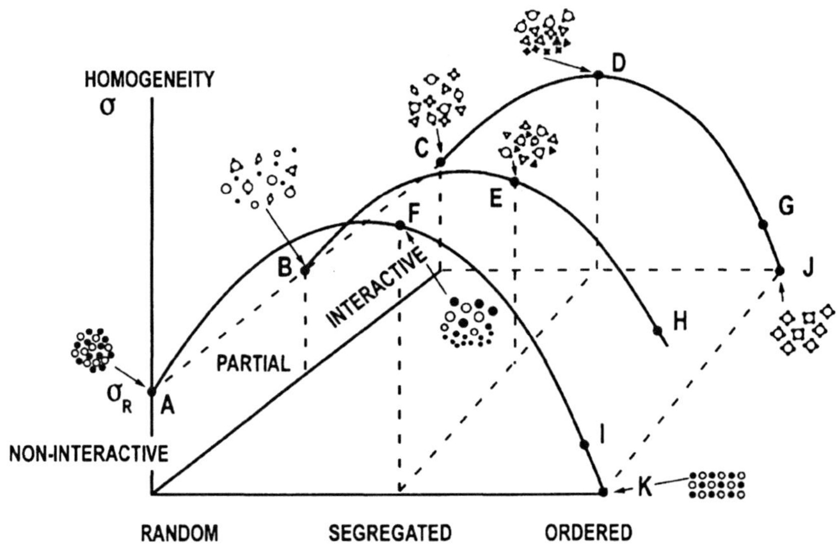

Figure 9.5. Homogeneity (variance) in different types of mixtures (from Thiel, 1984).

to occur in free-flowing mixtures where particles have independent mobility. Figure 9.5 shows the standard deviation (or homogeneity) if component concentration in a mixture is a function of level of interaction and the order of the mixture.

9.3.2. Sampling

Sampling is a crucial step in the mixing process because any form of control of mixing operations involves sampling procedures. The sample must be representative of the mixture and post-sampling handling must not alter it.

As sampling has a statistical aspect, sampling procedures following a pure mathematical approach are not completely practical in industrial situations. The confidence that can be placed in any results obtained from the sampling and analysis of a mixture is greatly influenced by several factors, including the method of sampling, the number of samples, the size of the sample, and the location in the bulk material from which the sample is taken. If sampling is not performed carefully, every mixture determination could be considered meaningless. Harnby (1985) recommended collecting samples from the outflow of a mixer, in the moving stream, instead of taking them from a static mass inside the mixer. In such a way, the possibility of bias in sample retrieval is minimized.

It can be demonstrated by statistical means that the larger the number of samples, the more reliable the results. For example, using statistical theory of sampling, it is stated that a most representative sample would approach an infinite number of samples. In other words, the only way of including every member of a population being sampled is taking this whole population as a sample. Since this is unfeasible and unreasonable, for most practical purposes in mixing of food powders it has been established that at least 50, but not less than 20, samples should be taken to obtain representative results. Fan et al. (1970) suggested taking 5–15 samples, while Weinekötter and Gericke (2000) suggest at least 25.

Determination of the number of samples depends primarily upon the mixer and the mixture, the technique of sampling, the total mass being mixed, and the method and cost of analyzing the sample (Fan et al., 1970). In process quality control, it also depends on the type of analysis (physical, physicochemical, or microbiological) needed for product screening, and its relevance to the final product when defining or comparing to specifications. No systematic methods have yet been developed to determine the optimum number of samples. The random selection of samples likewise places high requirements in terms of the sampling. Random sampling and analysis using variance represents the standard procedure and can be widely applied (Weinekötter and Gericke, 2000).

The size of the sample is also important, given that it represents the resolution by which a mixture can be judged. If a simple particle is drawn from the mixture, no mixing is evident. In contrast, if the whole mixture were to be analyzed, provided the ingredients were present in the correct proportions, complete homogeneity would appear to be achieved. As both of these extremes are impractical and unreliable, the recommended sample volume, often called the *scale of scrutiny* or *characteristic sample size*, falls between them and is defined as the size of sample that may be taken to correspond with the product usage. The smaller the size of a sample, the more closely the condition of the mixture will be scrutinized. The size of the sample can only be meaningfully specified in connection with further application of the mixture (Weinekötter Gericke, 2000). In animal feed manufacture, for instance, feed contains carbohydrates and proteins, balanced with added nutrients. In a particular feed, an animal must receive the correct balance of components. Provided that the required quantities of the necessary ingredients are present in the food consumed at each feed, intimate mixing is not essential. Thus, the volume of sample that would give such a balance would be the useful one, regardless of its perfection in statistical terms.

Specifying the size of the sample is, therefore, an essential step in analyzing the quality of the mixture, since it quantifies the mixing task from the outset. The amount of mixing that has occurred is a function of the scale of scrutiny that determines the sample size: in other words, the maximum sample size that would show non-uniform composition for a satisfactorily mixed mixture. It is intrinsic that good assessment of mixing requires good sampling techniques, which are reviewed in Chapter 1.

9.3.3. Mixture Quality: Mixing Index and Rate

Assessing the extent of mixing is of great interest for both equipment manufacturers and food powder processors. Mixing indices have been proposed to assess the extent of mixing. Mixing indices intend to provide a measure of the performance of a piece of equipment (a blender) or a process as related to ideal desired conditions. Considering some previously discussed aspects of the mixing process, it can be gathered that food mixing is a complicated task not easily described by mathematical modeling. Mixture quality results from several complex mechanisms operating in parallel, which are hard to follow or to fit to a particular model. Dankwertz (1952) defined the scale and intensity of segregation as the quantities necessary to characterize a mixture. The scale of segregation is a description of unmixed components, while the intensity of segregation is a measure of the standard deviation of composition from the mean, taken over all points in the mixture.

In practice, it is difficult to determine these parameters, since they require concentration data from a large number of points within the system. However, they provide a sound theoretical basis for assessing mixture quality. Taking into account the complexity of components and interactions in food solids mixing, it would be difficult to define a unique criterion to assess mixture quality. In fact, there over 30 criteria have been developed to express the degree of mixedness (Fan et al., 1979). A mixing endpoint or optimum mixing time can also be considered as a very relative definition due to the segregating tendency of food powder mixing.

The degree of uniformity of a mixed product may be measured by analysis of a number of spot samples. Food powder mixers act on two or more separate materials to intermingle them. Once a material is randomly distributed through another, mixing may be considered to be complete.

Based on that, the well-known statistical parameters of mean and standard deviation of component concentration can be used to characterize the state of a mixture. If spot samples are taken at random from a mixture and analyzed, the standard deviation of the analyses s about the average value of the fraction of a specific powder $\bar{x}$ is estimated by the following relation:

$$s = \sqrt{\frac{\sum_{i=1}^{N} (x_i - \bar{x})^2}{N - 1}} \tag{9.4}$$

where x_i is every measured value of fraction of one powder and N is the number of samples.

The standard deviation value on its own may be meaningless, unless it can be checked against limiting values of either complete segregation s_0 or complete randomization s_r. The minimum standard deviation attainable with any mixture is s_r, which represents the best possible mixture. Furthermore, if a mixture is stochastically ordered, s_r would equal zero. Based on these limiting values of standard deviations, Lacey (1954) defined a mixing index M_1 as follows:

$$M_1 = \frac{s_0^2 - s^2}{s_0^2 - s_r^2} \tag{9.5}$$

The numerator in Eq. (9.5) would be an indicator of how much mixing has occurred, while the denominator would show how much mixing can occur. A Lancey mixing index M_1 of zero would represent complete segregation, and a value of unity would represent a completely random mixture. Practical values of this mixing index, however, are found to lie in the range 0.75–1. Thus, the Lancey mixing index does not provide sufficient discrimination between mixtures (Rhodes, 1998).

In practice, however, the values of s, even for a very poor mixture, lie much closer to s_r than to s_0. Poole et al. (1964) suggested an alternative mixing index:

$$M_2 = \frac{s}{s_r} \tag{9.6}$$

Equation (9.6) clearly indicates that for efficient mixing or increasing randomization M_2 would approach unity. The values of s_0 and s can be determined theoretically. These values would be dependent on the number of components and their size distributions. This index gives better discrimination for practical mixtures and approaches unity for completely random mixtures (Rhodes, 1998). Equations (9.5) and (9.6) can be used to calculate mixing indices defined by Eq. (9.4).

For two-component systems, the theoretical upper and lower limits of mixture variance are:

$$\text{(a) upper limit (completely segregated)}: \quad s_0^2 = \bar{x}_c(1 - \bar{x}_c) \tag{9.7}$$

$$\text{(b) lower limit (randomly mixed)}: \quad s_r^2 = \frac{\bar{x}_c(1 - \bar{x}_c)}{n} \tag{9.8}$$

where n is the size of sample or the number of particles in each sample, and $\bar{x}_c$ and $(1 - \bar{x}_c)$ are the proportions of the two components determined from the sample. Actual values of mixture variance lie between these two extreme values (Rhodes, 1998). For a binary multi-sized particulate mixture, Poole et al. (1964) demonstrated that:

$$s_r^2 = \frac{pq}{w\Big/\left[q\left(\sum f_a w_a\right)_p + p\left(\sum f_a w_a\right)_q\right]} \tag{9.9}$$

where p and q are the proportions by weight of components within a total sample weight w and f_a is the size fraction of one component of average weight w_a in a particle size range. For a given component in a multi-component and multi-sized particulate system, Stange (1963) presented an expression for s_r, as follows:

$$s_r^2 = \frac{p^2}{w}\left\{\left[\frac{1-p}{p}\right]^2 p\left(\sum f_a w_a\right)_p + q\left(\sum f_a w_a\right)_q + r\left(\sum f_a w_a\right)_r + \cdots\right\} \tag{9.10}$$

For a multicomponent mixture, the degree of mixedness can also be evaluated by a covariance matrix in the same way as in the binary mixture, but its measurement and calculation procedure are complicated. In practice, the multicomponent system is regarded as a mixture of the single most important component (key component) and the others, and it is treated as a binary mixture (Miyanami, 1997).

Another suggestion for the characterization of degree of homogeneity in mixing of powders was reviewed by Boss (1986), with the degree of mixing M_3 defined as:

$$M_3 = 1 - \frac{s}{s_0} \tag{9.11}$$

Some other mixing indices have been reviewed by Fan and Wang (1975). The above-mentioned methods have been recommended to compare powder mixers between samples.

Mixing time can be defined as the time needed for the concentration variance to drop below some specified value. Concentration variance may be expressed as (Lindley, 1991b):

$$\sigma_t^2 = \frac{1}{n-1}\sum_{i=0}^{n}(C_i - C_\infty) \tag{9.12}$$

where C_i is the concentration of the ith sample at time t, C_∞ is the concentration after infinite mixing time, and n is the number of samples in one test.

McCabe et al. (1992) presented the following relationship to evaluate mixing time t for solids blending:

$$t = \frac{1}{k}\ln\frac{1 - 1/\sqrt{n}}{1 - 1/M_2} \tag{9.13}$$

where k is a constant and n is the number of particles in a spot sample. Equation (9.13) can be used to calculate the time required for any required degree of mixing, provided k is known and segregating forces are not active.

Kuakpetoon et al. (2001) utilized an exponential decay model to describe the uniformity of dry flour mixing as a function of mixing time, which is defined by the equation:

$$y = a\,\mathrm{e}^{-bt} + c \tag{9.14}$$

where y is the uniformity parameter (e.g., normalized standard deviation), t is the mixing time, and a, b, c are constants. The constant a is the difference in uniformity between the beginning and the final mixtures, being $(a + c)$ the initial degree of uniformity. Constant b represents the mixing rate, and c is the highest degree of uniformity that a mixture can achieve.

Mixing times should not be very long due to the unavoidable segregation nature of most food solids mixtures. In this case, instead of improving efficiency, long mixing times often result in poor blending characteristics. A graph of degree of mixing versus time is recommended to select the proper mixing time quantitatively. Most cases of mixing of powders will attain maximum degree of homogeneity in less than 15 min, when the proper type of machine and working capacity have been chosen.

9.4. POWDER MIXERS

In general terms, mixers for dry solids have nothing to do with mixers involving a liquid phase, as solid particles are subjected to various interactive forces which are not self-diffusive. Powders cannot be set in motion without an external force such as mechanical agitation (Miyanami, 1997). According to the mixing mechanisms previously discussed, solids mixers can be classified into two groups: segregating mixers and non-segregating mixers. The former operate mainly by a diffusive mechanism, while the latter practically involve a convective mechanism. Segregating mixers are normally non-impeller type units, such as tumbling mixers, whereas non-segregating mixers may include screws, blades, and ploughs in their designs; examples of them include horizontal trough mixers and vertical screw mixers. Furthermore, mixing can operate in batch or continuous modes. The following sections will describe batch type blenders in detail and will mention some aspects of continuous blending.

9.4.1. Tumbler Mixers

Free-flowing non-segregating powders may be readily mixed in batch by use of tumbler mixers. Tumbler mixers operate by tumbling the mass of solids inside a revolving vessel. Blenders are available in various geometries, affecting material movement, mixing efficiency and ease of cleaning between batches. These vessels take various forms, such as those illustrated in Fig. 9.6, and may be fitted with baffles or stays to improve their performance. A tumbling batch blender can be of four types, which are described as follows:

(a) *Horizontal cylinder* (Fig. 9.6a): This cylindrical mixer has a tubular vessel mounted on trunnions. Internal baffles or lifter bars are mounted along the inner walls of the vessel. The inlet is typically located at the top center of the vessel and the outlet at the bottom center. The blender tumbles and the internal baffles gently lift and aerate the material preventing it from sliding along the blender bottom; they also de-lump the material.

(b) *Double cone blender* (Fig. 9.6b): The double cone blender consists of two cone-shaped sections, typically with 45° slopes. The cone sections are welded at their ends to a center band. The blender is mounted between two trunnions that permit the unit to tumble end over end. An opening in one of the ends of the cones serves as inlet and outlet, or the inlet can be in one cone end with the outlet in the other. Cleaning access is through the outlet. The blender tumbles, and the material in the vessel spreads out. The transition area at the band between the cones prevents the material from sliding along the inner wall and instead causes the material to fold over itself. This provides gentle mixing with only very slight shear.

(c) *V-cone blender* (Fig. 9.6c) and *Y-cone blender* (Fig. 9.1d): The V-cone blender is similar to a double cone unit, but consists of two large diameter pipe sections cut at a 45°-angle and welded together to form a V. In the same way, the Y-cone blender has a third section that extends the volume of the blender in a bisectional direction with respect to the other pipe sections. Inlets are typically located at both ends of the V (or of the Y); the outlet is at the V point (or at the bottom of the Y). The unit is also mounted on trunnions to allow it to tumble and can be equipped with a spray line for liquid addition and an agitator for de-lumping. The units tumble end over end as in the double cone blender. The free-falling action combined with increased frictional contact between the material and the long vessel sides result in less gentle mixing than in a double cone blender.

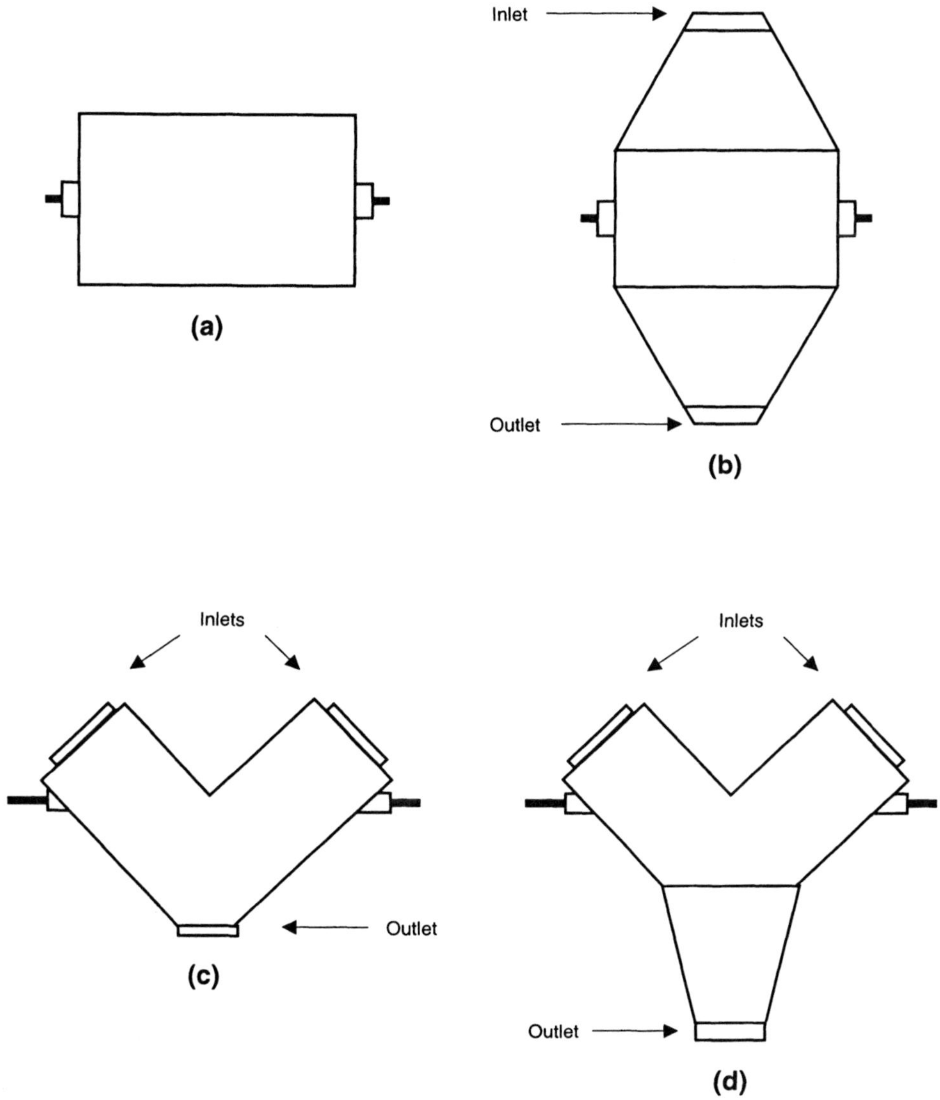

Figure 9.6. Tumbler mixers used in food powder blending: (a) horizontal cylinder; (b) double cone; (c) V-cone; and (d) Y-cone.

From the outlets of the mixers, the batch can be discharged via an optional retractable sleeve into drums or containers for shipping. The mixer's tumbling action distributes the materials along an ever changing angle of repose surface (Dudley, 2001). Tumbler mixers can be used to provide a gentle mixing required to avoid attrition of friable materials. The shells rotate at variable speeds having values up to 100 rev/min with working capacities around 50–60% of the total vessel volume. Rotational speed is set at 50–80% of the critical rotational speed, N_{cr}, given as:

$$N_{cr} = \frac{0.498}{\sqrt{R_{max}}} \ (s^{-1}) \tag{9.15}$$

where R_{max} (m) is the maximum radius of rotation of the mixer. The rate of mixing is rather low, but a good final degree of mixedness can be expected. Tumbler mixers are manufactured using a wide variety of materials, including stainless steel. Batch ingredients can be weighed automatically as they enter the mixing vessel rather than individually prior to mixing. Once the desired weight has been reached, rotation start can be automatically controlled by computerized systems. This type of mixer can create a free falling curtain of material, exposing each particle surface area and allowing uniform dispersion of liquid additives. This mixing is known as free-fall processing, and particles can be coated through spray lines. Liquid addition can include for coating and encapsulating applications (Fuller, 1998). This type of equipment is best suited for gentle blending of powders with similar physical characteristics. Segregation can represent a problem if particles vary, particularly in size and shape. Thus, careful design is required to minimize degree of segregation.

One of the main disadvantages of this type of blender is the demixing possibility due to the funnel flow pattern formation that can occur upon discharging. Very few double cone or V-cone blenders have hopper surfaces steep enough and sufficiently low in friction to provide mass-flow discharge. One possible solution would be to connect the tumble blending unit to a mass flow container (Carson et al., 1996). In fact, a special device called the cone-in-cone bin has tumbling and mass flow discharge possibilities, thereby eliminating segregation upon discharge, while maximizing the capacity of the container.

9.4.2. Horizontal and Vertical Trough Mixers

Horizontal trough mixers consist of a semi-cylindrical horizontal vessel in which one or more rotating devices are located. For simple operations, single or twin screw conveyors are appropriate and one passage through such a system may be good enough. For more demanding duties, a ribbon mixer, like the one shown in Fig. 9.7, may be used. A typical design of a ribbon mixer will consist of two counteracting helical blades (called ribbons or spirals) mounted on the same shaft. Ribbons convey materials inside and outside in opposite directions forcing them to intermix. The mixing tool and vessel are typically made of stainless steel. One moves the solids slowly in one direction, while the other moves it quickly in the opposite direction. There is a resultant movement of solids in one direction, so the equipment can be used as a continuous mixer. Some other types of ribbon mixers operate on a batch basis. The mixer can have more than one discharge so different batches can be discharged to different processing lines. In these designs, troughs may be closed to minimize dust hazard, or may be jacketed to allow temperature control. Due to small clearance between the ribbon and the trough wall, this kind of mixer can cause particle damage and may consume high amounts of power.

Another possibility is that the ribbon mixer is in its vertical position. These mixers operate at relatively low speed to gently handle materials while providing enough intensity to thoroughly blend them in a relatively short time cycle (typically 1/3 to 1/2 the time of conventional horizontal ribbon mixers). The mixer has a filling capacity of up to 90% of the gross volume of the mixer, thus

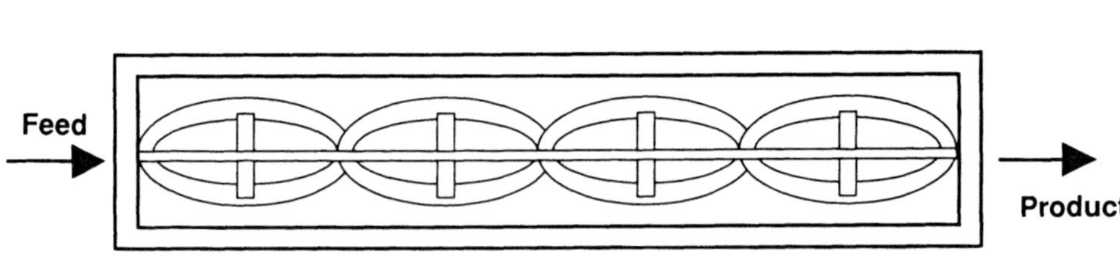

Figure 9.7. Plain view of an open ribbon mixer.

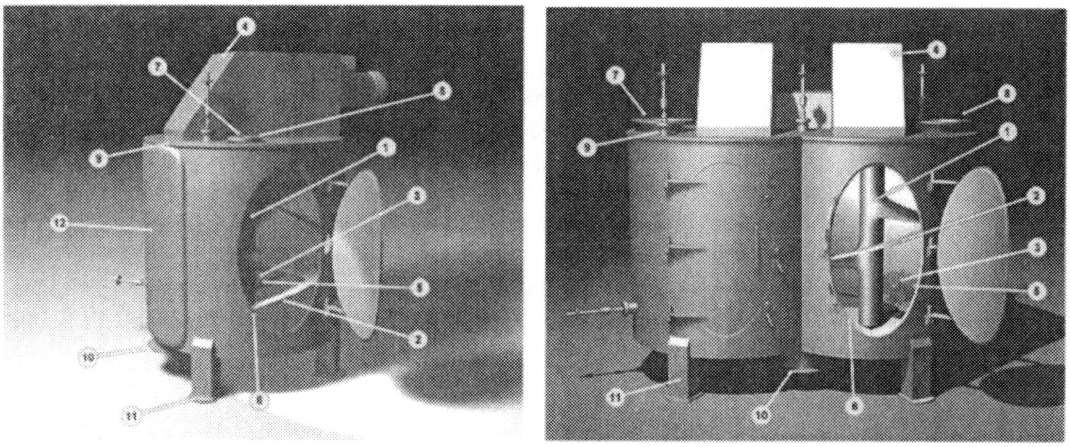

(a) (b)

Figure 9.8. (a) Vertical single-shaft mixer and (b) vertical double-shaft mixer. Description: (1) filling ratios 10–100%; (2) adjustable distance between base and mixing apparatus; (3) cutting rotors; (4) mixing tool motor; (5) liquid spray; (6) oval inspection door; (7) level indicator; (8) chemical aseptic flanges in connection pieces; (9) orbital tank washing heads; (10) discharge flap; (11) optional load cell mounting; (12) double jacket (courtesy of Amixon GmbH).

improving process efficiency (Cavender, 2000). The vertical ribbon mixer is available in two types: single and double shaft. The single-shaft mixer (Fig. 9.8a) has a rotating shaft fitted with horizontal arms that support the ribbons. The shaft, arms, and blades together are the mixing tool. The rotating shaft runs through the center of a cylindrical, vertically oriented air- and watertight vessel. Once the ingredients are in the vessel, the rotation creates an upward screw-like movement of the ingredients along the vessel periphery and a downward movement along the shaft through the vessel center. The double-shaft mixer (Fig. 9.8b) functions similarly to the single shaft unit. However, both shafts rotate in the same direction at somewhat higher speed, creating a synchronized crosswise mixing pattern. Mixing time can be less than 10 min regardless of the bulk properties of the material (the double shaft mixes in less that 4 min).

A common variation of the ribbon—the paddle mixer—has flat paddles rather than ribbons (Fuller, 1998). The mixer operates similarly, but the paddles tend to last longer and are suited to mixing abrasive materials. Large particles are also less likely to wedge between the paddle trough wall, which can prevent material degradation. Another type is a combination of ribbon, paddle, and plow, which increases particle random movement and improves mixing efficiency. Some of the advantages of these types of equipment are the low room they occupy (low headroom in the case of horizontal mixers) and their ability to mix different types of powders. They may also require less power to operate than the tumbler blenders.

One disadvantage of this type of equipment is that the blend may segregate upon discharge depending on the particle size of the mixture. Another disadvantage is that since the outer shell is stationary, transport of the blend to another location requires a conveying system or portable container, resulting in possible mix de-blending (Carson et al., 1996). When the mixer has a small discharge, the mixture can be discharged slowly while the flow is regulated to prevent segregation. This is suitable for applications where the mixture will immediately be packaged into small containers or delivered to a dosing unit.

9.4.3. Vertical Screw Mixers

In vertical screw mixers, a rotating vertical screw is located in a cylindrical or cone shaped vessel. The screw may be mounted centrally in the vessel or may rotate or orbit around the central axis of the vessel near the wall. Materials are lifted from the bottom to the top of the hopper and are then exchanged with materials on the way up (Dudley, 2001). Such mixers are schematically shown in Fig. 9.9. A vertical screw blender (Fig. 9.9a) may be desired for larger batches handled in a small space, while the orbiting screw mixer (Fig. 9.9b) is used for difficult mixes. The latter arrangement is more effective and stagnant layers near the wall are eliminated. Vertical screw mixers are quick, efficient, and particularly useful for mixing small quantities of additives into large masses of material. Specialized atmospheres as well as normal temperatures and pressures are accessible for multipurpose operations.

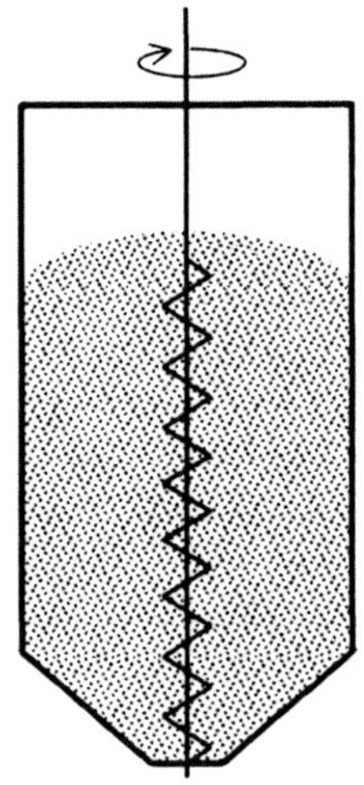

(a)

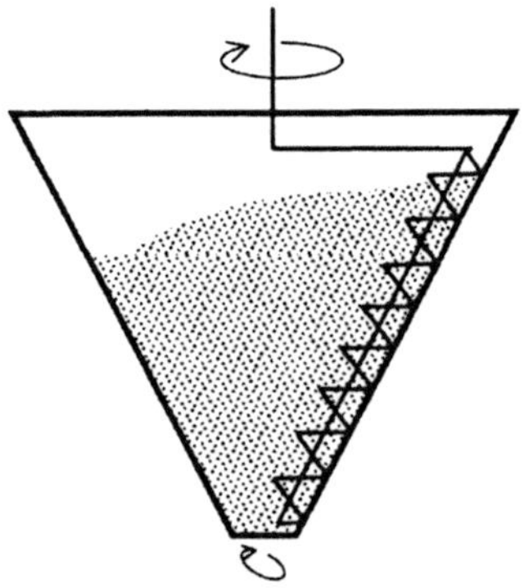

(b)

Figure 9.9. Vertical screw mixers: (a) central screw; and (b) orbiting screw.

9.4.4. Fluidized Bed Mixers

Food powders can also be mixed by aeration using a fluidized bed. The resulting turbulence of passing air through a bed of particulate material causes material to blend. Materials are moved upward by air jets, causing differential movement. Stationary vessels using gas-flow agitation are used primarily for batch mode mixing.

Materials to be mixed have to be relatively fine and fairly narrow in their size distribution, as well as not too cohesive. Powders to be mixed can be charged to more than 70% of the vessel volume. Mixing times required in fluidized beds are significantly lower than those required in conventional powder mixers. The mixing is largely convective with the circulation patterns set up by the bubble motion within the bed. An important feature of the fluidized bed mixer is that several processing steps (mixing reaction, coating, drying, etc.) may be carried out in the same vessel (Rhodes, 1998). Additional equipment can include blowers, dust collectors, and pressure regulators, which will enlarge the system as a whole. Van Deemter (1985) discussed different mixing mechanisms prevailing in fluidized beds.

A particular type of the fluidized mixer is the fluidized paddle mixer (also called fluidized zone mixer). The mixer has twin troughs, each with a center mounted rotating shaft. Flat paddles are welded to spokes on each shaft. The paddles lift the material from the bottom and throw it into a zero gravity, fluidized mixing zone, settling a random displacement pattern for the material.

9.4.5. Hopper Blenders

It has been the practice of some industries where very large quantities of materials need to be blended to use hopper blending. This concept may also be utilized as a premix prior to use of a mechanical mixer. Hopper blending techniques may involve recycling through a single hopper flow, through several hoppers, or metering from multiple hoppers to a combined discharge (Lindley, 1991b). As discussed in Chapter 4, mass-flow type hoppers are the most commonly used in the food industry. Cone shaped and cylindrical inserts in the mixer's hopper and cylinder produce differential flow velocities, promoting mixing at the mixer's discharge (Dudley, 2001). A common type of in-bin blender is the cone-in-cone blender, which works on the principle of pronounced velocity gradients in the bin. Cohesive materials and highly segregating materials can be blended with this unit.

A special feature that can be added in a hopper for blending is the pressure pulse unit blending head, which attaches to the bottom of a hopper by means of a coupling (Fauver, 1996). The unit consists of a cone valve, 16 pneumatically operated injector valves around the perimeter of the hopper head, a high-pressure manifold linked to an air supply, and a discharge flange leading to the hopper's outlet. A control system is linked to electric timers that control the injector valves. During blending, the valves release pulses of low-pressure compressed air into the powder, moving in a slight circular pattern, and the material is fluidized. Depending on the application, the air pulses are filtered as they are vented from the vessel top and exhausted from a dust collector. During discharge, the cone valve, raises opening the outlet to release the blended powder. The cone valve serves as a flow insert that prevents segregation and ratholing.

9.4.6. Continuous Blenders

In general, continuous mixers manage similar principles as the batch mixers, especially in trough systems. Continuous mixers are primarily used in the following situations: (a) if high volume production is needed; (b) when a limited number of streams (two or three) are to be blended and less degree of homogeneity is expected; (c) for particles similar in size, shape, and density that differ slightly in color or some other characteristic from batch to batch; (d) if there is a requirement to provide a gradual, rather than step change in the material being processed.

Two of the most common ways to achieve continuous blending include metering individual streams onto a conveyor or by using an in-bin blender. In the first case, a degree of blending can be achieved simply by depositing several streams of material onto a collecting conveyor such as a belt or screw. Some type of gravimetric (e.g., loss-in-weight or weight belt) feeder can be used. If the collected stream is discharged from the conveyor onto a transfer chute, additional mixing can often be achieved by using a series of offset cones rather than a flat transfer chute. However, segregation can also be exacerbated at transfer points, so all the transfer points handling blended material must be designed carefully. In-blender continuous mixing is commonly performed by using a cone-in-cone blender type (Carson et al., 1996).

9.5. SELECTION AND DESIGN CRITERIA

Design of mixing systems is not based on well-established scientific principles and no general organized argument is available for the methodology of selecting the target mixer. In fact, mixer selection is primarily a trial and error process with due consideration given to prior experience. Design of a mixer system has normally entailed selection of the type of machines that are expected to perform best, followed by laboratory tests on them. The increasing demands imposed on quality, and the trend toward the rationalization and mechanization of mixing processes, have created the need for scientific criteria for comparing performances of mixing. Understanding of the science of power mixing is poor and there is a great need for design equations and techniques for comparison of performance. The main considerations are: (a) rate of mixing; (b) power requirements; (c) efficiency; (d) design methods; and (e) scale-up criteria. Furthermore, the purposes of the mixing operation, powder properties, maintenance problems, unit and running costs, and related matters must also be well defined. The priority of each of these factors depends on individual cases (Miyanami, 1997).

Normally, the mixer performance should be evaluated in terms of the powder properties being handled, acceptable homogeneity level, operating conditions (time and volume capacity), plant headspace and headroom available, and the application purpose (e.g., if liquid addition is needed). Table 9.1 lists characteristics of different batch blenders that have been explained in the previous section. These characteristics include continuous mixing possibilities, predominant mixing mechanism, allowable fill level or batch size range, mixing cycle length, liquid addition configuration, de-lumping agitator configuration, and major advantages that each blender type provides.

A rational approach to design of a mixer system could be to define the input condition and desired output characteristics, and then use a mathematical model that describes the fundamental mixing laws along with mixer performance parameters to select the possible system. Once a system has been designed, it will need to be monitored and controlled, and methods for accomplishing these tasks are not well developed. Some factors that might be important in modeling the performance of mixing are: time, temperature, power input, shear developed, noise emitted, color of the mixture, moisture content, particle size and distribution within the mixture, density, pressure, changes of particle shape, and changes in mixture flowability.

9.5.1. Factors Affecting Equipment Design

Before selecting solids mixing equipment, a careful study ought to be made of several performance characteristics. As previously stated, food solids mixing is a complex operation and mathematical modeling can hardly be used. Many factors affect the operation of solids blending, so

Table 9.1. Some mixers characteristics (adapted from Fuller, 1998; Harnby, 1997; Rielly, 1997; Carson et al., 1996).

Mixer	Type	Batch (B) or continuous (C)	Main mixing mechanism	Percentage of total mixer volume capacity	Mixing cycle length (min)	Liquid addition configuration	De-lumping agitator configuration	Major advantages
Tumbler mixers	Horizontal cylinder	B	Diffusive	<50	5–15	Spray line through trunnion	Agitator through trunnion	Good for friable materials, easy to clean without intensifier
	Double cone	B	Diffusive	30–65	5–10	Spray line through trunnion	Agitator through trunnion	
	V-cone (Y-cone)	B	Diffusive	30–65	5–15	Spray line through trunnion	Agitator through trunnion	
Fluidized bed	Paddle type	B	Convective	40–140	<1	Spray bar above paddles	Pin mills above paddles	No moving mechanical parts
Hopper blenders	Low pressure pulse	B/C	Convective	–	1	None	None	Mixing and storage in one container, eliminates segregation upon discharge, no moving mechanical parts
	Gravity silo	B/C	Depends on application	–	12	None	None	
Trough mixers	Ribbon and paddle	B/C	Convective	40–85	15–20	Spay bar above ribbons	High speed chopper blades at side	Versatile, low headroom

process features such as mixing homogeneity and time, loading and discharging arrangements, power consumption, equipment wear, etc., need to be analyzed and properly weighed in order to make the most convenient decision.

- *Mixing homogeneity*: As has already been mentioned, blending uniformity and mixing rate are best evaluated using design graphs, bearing in mind the tendency to segregate and considering that long mixing times tend to worsen, rather than improve, efficiency. Solids mixing can be a batch or continuous operation. In batch mode, proper mixing design will produce the desired blend in a few minutes.
- *Ingredients and operating conditions*: Ingredients being mixed influence the mixer selection. Particle size ranges, proportions of the materials to be mixed, end product flow properties, hygroscopicity, corrosiveness, temperature sensitiveness, aeration and compaction properties of the powders are to be considered before selection.
- *Residence time*: Determination of residence time in continuous operation is a more difficult task but, considering the main properties of solids blending, such times also tend to be short, in the order of a few minutes or even seconds. The ribbon type mixer is often used for continuous mixing, although it is also employed for batch mixing. Continuous mixing should be considered an option only if a single formulation can be run for an extended period, or when the fluctuations of the outgoing product are within process requirements. When any of these factors are compromised, the batch mode of operation would be preferred to assure the most attainable mixing uniformity.
- *Loading and discharge arrangements*: These are also important aspects, more critical in continuous operation mode. The total handling system must be considered in order to obtain optimum charging and discharging conditions. This includes the efficient use of weight hoppers and surge bins, minor ingredient premixing, location of discharge gates, and any other ancillary device used to aid the continuity of the process.
- *Power requirements*: These are not a major concern when choosing solids mixers, since other considerations usually predominate. Nevertheless, sufficient power must be provided in order to handle the maximum needs as well as to prevent changes during the mixing operation. When materials and operating conditions are subjected to variation, enough power should be made available for the heaviest bulk density materials and for extreme conditions of operation. If the loaded equipment is to be started from rest, there should be sufficient power for this. When speed variations may be desirable, this should be taken into account.
- *Equipment cleaning*: The ease, frequency, and thoroughness of cleaning are crucial when batches of different nature are to be mixed alternatively in the same equipment. Plain tumbling vessels are easy to clean, provided that adequate openings are available. Areas that could be difficult to clean are seals or stuffing boxes, crevices at baffle support, any corners, and discharge arrangements. If cleaning between different batches is time consuming, several small mixers should be considered instead of a large single unit.
- *Dust formation*: This should be avoided for safety reasons and when loss of dust may significantly affect batch composition. Minimization of dust formation can be achieved by using less dusty but equally satisfactory batch ingredients, by employing pelletized forms of extremely dusty materials, by proper venting so as to enable filtering of displaced air rather

than unregulated loss of dust-laden air, or by addition of liquids if tolerable. Water addition in small quantities could be effective in minimizing dust upon discharge from the mixer, and it will also render the batch less dusty in subsequent handling stages. Water or any other liquid should be directed into the batch material instead of onto the bare surface of the mixer, since this could cause buildup. Spraying by using a nozzle is considered the most convenient way of incorporating moisture into the mixing batch. Nozzle spray pressure should be sufficient to penetrate the batch, but not so high as to cause heavy splashing. The liquid should be added to the well-mixed batch, particularly when premature addition of liquid could impair the adequacy of blending. Also, both time of addition and time of application are important and should be carefully considered. More information about dust formation prevention is presented in Chapter 12.

- *Equipment wear*: This should be taken as a crucial issue, especially with abrasive materials such as grinding wheel grains. Abrasion resistant coating such as rubber coating, special alloys, or platings need to be considered in these cases. Any internal agitator device may wear even when operating at slow speed. Particularly when highly abrasive materials are to be mixed, the benefits of an agglomerated breaking device must be weighed against potential contamination and maintenance costs.
- *Equipment capacity*: This is an important factor when the mixed batches vary considerably in size from time to time. There are some features of the mixing operation that are not flexible in terms of capacity. For example, certain agitation devices in tumbler mixers do not function properly unless a given capacity is maintained. In general, the effect of percentage of mixer volume occupied by the batch on the adequacy of mixing should be considered when changes from recommended operating volumes are planned. The mixer volume required per batch is the batch size divided by the average finished batch density (Callaghan, 1996).
- *Equipment selection*: Food powder mixers should be selected or designed for a particular operation, first by analyzing and giving careful examination to the applicable areas discussed above. Mixer selection should also involve consideration of the placement of the mixer within the overall process. Possible consolidation of many steps of food solids processing deserves scrutiny at this time. If there is no machinery available that includes all the necessary requirements, consideration should be given to modifications in order to obtain the most desirable combination of features. It is important to consider how the mixer will integrate with other equipment (e.g., packaging units or agglomeration feeds) and how much floor space and headroom is available for the mixer.
- *Construction materials*: For mixing a food product, mixers are usually made of stainless steel. Strict purity requirements specify a stainless steel interior finish (for instance with welded and ground metal joints). Construction must meet 3-A sanitary standards or USDA (U.S. Department of Agriculture), FDA (Food and Drug Administration), or other approval (Presnell, 1995). According to FDA (2002), "equipment shall be constructed so that surfaces that contact components, shall not be reactive, additive, or absorptive so as to alter the safety, identity, strength, quality, or purity of the food product beyond the official or other established requirements."
- *Auxiliary equipment*: Other equipment might be combined with the blender. For example, sifters, blowers, dust collectors, screw conveyors, and surge bins can be incorporated into the mixing system. Auger, star wheel, or other devices can usually add ingredients volumetrically to a screw conveyor. Automatic weighing machines can be used to provide better control.

Pilot tests are relevant in the final decision of selection or design of a specific mixing process. In general, the larger the pilot unit, the more reliable the prediction of large-scale performance. Published solids mixing scale-up data are very scarce, especially in food applications. With geometrically similar tumblers, if the speeds are adjusted to give comparable motion and the mixer volume fraction occupied by the load is the same, scale-up of results will be straightforward. The presence of internal rotating devices leads to difficulties in scaling up clearances, blade area to mixture volume, and sizes as well as speeds of the rotating devices. The actual materials to be processed in the industrial operation should be used if possible in the scale-up procedure. If substitute materials need to be used, they should have the same mixing characteristics. Differences in materials of construction between the pilot and the production unit should be considered, since these may have a bearing on caking, abrasion, and some other adverse effects.

Certain novel monitoring techniques might prove useful in obtaining information on mixing systems: positron emission tomography, magnetic resonance imaging, isokinetic radio pills, radio frequency reflection, laser Doppler anemometry, light emission via fiber optics, acoustic sensors, surface heat sensors, torque on powder input unit, and strain gauges on the mixer vessel and impeller (Lindley, 1991a).

9.5.2. Mixer Selection Based on Flow Properties

Since mixing involves the flow of materials, mixing mechanisms can be affected by the same properties that affect flow: mechanical interlocking, surface attraction, plastic welding (from high pressures between small contact areas), electrostatic attraction, ambient moisture, and temperature fluctuations. Powder flow properties can simplify blender selection by allowing the prediction of the behavior of materials of specific composition in different types of mixers. Three conditions must exist for a blender to operate efficiently: (a) the blender must have no stagnant regions; (b) the blender must promote different flow velocities in various sections of the blender; and (c) blender operation must not segregate, or demix, mixture ingredients (Dudley, 2001). In many cases, these conditions depend on bulk properties, like cohesiveness and angle of repose, which may change with product formulation. Table 9.2 shows the degree of cohesiveness, or angle of repose difference, necessary to yield a good mix in different types of blenders.

Stagnant regions are areas where materials can sit undisturbed and not enter the mixing process, thus preventing complete mixing from taking place. This generally tends to occur in the area between the powder bed surface and the top of the blender, as well as in the area between the agitator blades and blender walls. The effect of stagnant regions depends on the type of mixture and, therefore, on the flow properties of its individual ingredients.

Tumbler mixers rely on continual pile formation and avalanche flow, in a small region on top of the material pile in the vessel to mix the material. An excessively cohesive material will create thick avalanche layers with little inter-particle motion. As a result, stagnant regions, formed due to stickiness, can reduce mixing effectiveness. However, a completely free-flowing material can have very thin avalanching zones that keep motion all through the mixer. Thus, a tumbler mixer works best with ingredients that have similar angles of repose and only enough cohesiveness to prevent sifting.

Mixer operation can segregate individual ingredients during operation and discharge. For example, since a V-cone blender relies on continual pile formation to mix material, segregation can occur if there is a significant difference in angle of repose among ingredients or if sifting takes place. This produces a non-uniform mixture. Thus, selecting another blender can help decrease segregation.

As indicated in Table 9.2, air currents within the mixer can also segregate ingredients. For example, a ribbon blender typically has several feed ports. Connecting one port to a dust collection

Table 9.2. Common mixer classification in function of flow properties and potential segregation potential dangers (adapted from Dudley, 2001).

Mixer	Materials that mix well in this blender	Potential segregation dangers: Angle of repose[a]	Sifting	Fluidization	Air current
Tumble mixer (double cone, V-cone)	Materials with the same angle of repose within ±2° that are not cohesive or lumpy but have sufficient cohesión to prevent sifting	Very high (>55°)	High	Moderate	Low
Ribbon mixer	Moderately cohesive materials of similar particle densities without fluidizing densities	Moderate (>45°)	High	Moderate to high depending on speed	Low to moderate with a dust collector
Screw mixers	Moderately cohesive materials that do not contain hard lumps	Low (>35°)	High	Low	Low
Gravity mass flow hopper (cone-in-cone)	Low to moderately cohesive materials; a cone-in-cone must also have a fluidizable material to prevent preferential flow patterns	Moderate (>45°)	Moderate	Low	Low
Fluidized bed	Easy-flowing, closely sized materials without fluidizable components	High (>50°)	High	High	High

[a]According to Carr (1976), angles of up to 35° indicate free flowability, 35°–45° some cohesiveness, 45°–55° cohesiveness (loss of free flowability), and 55° and above very high cohesiveness (very limited or no flowability).

system leads to fines accumulation below that port. Since a ribbon blender mixes poorly from one end to the other, operating a dust collection system during blending can cause segregation, reducing blender effectiveness and increasing blending times. Therefore, air currents have to be considered in the design process.

Kuakpetoon et al. (2001) studied the effect of particle size, shape, surface, and mixing ratio on the characteristics of dry flour mixes. For this study they utilized a laboratory drum mixer and a double ribbon mixer. They found that differences in size, shape, and surface of particles affected the mixing characteristics. For example, the flour mixture that contained smaller size particles (5–50 μm) with spherical or oval shapes and smooth surfaces achieved high uniformity (i.e., standard deviation), but required a longer mixing time (determined by using Eq. (9.12)). Contrarily, mixes with larger sizes (50–150 μm) with irregular shapes and very rough surfaces had a low degree of mixing, but required a shorter time to reach uniformity. Furthermore, angle of repose, tensile strength, and true density measurements affected mixing characteristics. Finally, characteristics of dry flour mixing in both the drum mixer and the double ribbon mixer were affected similarly by the studied flow properties. Thus, other aspects have to be taken into consideration to determine mixer selection.

9.5.3. Mixing in Food Powdered Product Development

When developing a powder mix, the narrower the size ranges of the product, the easier it is to mix it with another ingredient. Consider the ingredients in a dried soup mix; consumers like to be able to distinguish familiar, anticipated ingredients, and thus do not prefer a soup that contains carrot powder, which is so small that one cannot see that it is a carrot (Barker and Mehta, 1993).

In machines that dispense soup, the manufacturers have no choice but to make the powdered ingredients all the same size to avoid segregation. This fine pulverization of all ingredients often lends an anemic look to the soup when the powder is mixed with hot water and dispensed by the machine. The segregation of the ingredients in powdered soup is one of the reasons why commercial soup powders are usually individually packaged rather than sold in relatively large canisters from which one can make several servings of soup.

The same problem of differential size of ingredients occurs in soft drink mixes. Vendors sell relatively small packages of the ingredients to avoid segregation, which would occur in a large canister of material. In general, segregation problems from environmental vibration in an assembled mixture are minimized if the ingredient size ratios are not greater than 3 to 1. They become severe, in the absence of electrostatic forces, if the size ratios are greater than 9 to 1 (Kaye, 1997).

Sometimes, modifying the powder before proceeding to create a mixture can improve the product. For example, when making a product with an artificial sweetener with low solubility in water, some previous modifications need to be done in order to obtain the desired mouth feel. Dissolving the sweetener in water and then encapsulating the solution in gelatin can produce a free-flowing powder. Mixing the sweetener in its encapsulated form can provide a better taste than when the fine powdered sweetener is directly mixed with the other ingredients. Sometimes, microencapsulation of minor ingredients not only makes it easier to produce a powder mixture, but also can enhance shelf life and quality of the product. In particular, the widespread use of microcapsules, sometimes called flavor buds, for gelatin dessert production, not only makes it easier to manufacture the product, but the flavor trapped inside the capsule has a much longer shelf life (Kaye, 1992).

9.5.4. Selection Based on Mixing Costs

If more than one mixer can be found that satisfies both the process and the mixture quality requirements, then it is likely that a final mixer selection will be based on the unit cost of mixing. Generally, powder mixing costs represent only a very small percentage of the total product manufacturing costs. Mixing becomes expensive only when production time is lost due to a failure to meet product specification, and it is therefore more important that the mixer meets mixture quality specifications and integrates fully into the process.

The costs of mixing can be split into three components: (a) depreciation on capital cost; (b) power requirements; and (c) labor costs (Harnby, 1997). The most important of these components are the labor costs. Increasing the equipment capacity, thereby diminishing the frequency of each operating cycle, can reduce labor costs. Thus, labor is released for longer periods and can be used for other operations. Furthermore, the implementation of a continuous mixing system can make labor utilization very low. Nevertheless, depreciation costs would increase, and other costs such as ancillary equipment (e.g., for solids flow control) would have to be added.

9.6. APPLICATIONS

Applications of powder mixing in food systems are diverse and varied and include blending of grains prior to milling, blending of flours and incorporation of additives into flours, preparation of custard powders and cake mixes, blending of soup mixes, blending of spice mixes, incorporation of additives into dried products, preparation of baby formula, juice mixes, etc. For example, bakery mixes, chocolate mixes, sweeteners, spice blends, and colorants are commonly mixed in horizontal and vertical ribbon mixers.

REFERENCES

Barker, G.C. and Mehta, A. (1993). Size segregation powders. *Nature*, **361**: 308.

Boss, J. (1986). Evaluation of the homogeneity degree of a mixture. *Bulk Solids Handl.* **6**: 1207–1210.

Bridgwater, J. (1976). Fundamental powder mixing mechanisms. *Powder Technol.* **15**: 215–236.

Callaghan, D.C. Jr. (1996). Achieving gentle mixing with a horizontal rotary drum mixer. *Powder Bulk Eng.* **10**(1): 47–54.

Carr, R.L. (1976). Powder and granule properties and mechanics. In *Gas–Solids Handling in the Processing Industries*, Marchello, J.M. and Gomezplata, A. (eds.). Marcel Dekker, New York.

Carson, J.W., Royal, T.A. and Troxel, T.G. (1996). Mix dry bulk solids properly and maintain blend integrity. *Chem. Eng. Progress* **92**(11): 72–80.

Cavender, J. (2000) Quick, through, and gentle blending with a vertical ribbon mixer. *Powder Bulk Eng.* **14**(1): 46–57.

Dankwertz, P.V. (1952). The definition and measurement of some characteristics of mixing. *Appl. Sci. Res.* **3A**: 279–281.

Dudley, L. (2001). Choosing a blender based on your material's flow properties. *Powder Bulk Eng.* **15**(1) 37–39.

Fan, L.T., Chen, Y.M. and Lai, F.S. (1990). Recent developments in solids mixing. *Powder Technol.* **61**: 255–287.

Fan, L.T., Chen, S.J. and Watson C.A. (1970). Solids mixing. *Ind. Eng. Chem.* **62**(7): 53–66.

Fan, L.T. and Shin, S.H. (1979). Stochastic diffusion model of non-ideal mixing in a horizontal drum mixer. *Chem. Eng. Sci.* **34**: 811–820.

Fan, L.T., Too, J.R., Rubison, R.M. and Lai, F.S. (1979). Studies on multicomponent solids mixing and mixtures. Part III. Mixing indices. *Powder Technol.* **24**: 73–89.

Fan, L.T. and Wang, R.H. (1975). On mixing indices. *Powder Technol.* **11**: 27–32.

Fauver, V.A. (1996). Engineering study: solving a blending problem. *Powder Bulk Eng.* **10**(1): 31–45.

FDA (2002). Equipment. Code of Federal Regulations. Title 21 (4) Part 211 D: 121–122.

Fuller, O.W. (1998). Mixing up a batch: batch mixer types and selection tips. *Powder Bulk Eng.* **12**(1): 48–66.

Gyenis, J. and Ārva, J. (1989). Mixing mechanism of solids in alternately revolving mixers. 1. Change of local concentrations and concentration profiles. *Powder Handl. Process.* **1**(3): 247–254.

Harnby, N. (1985). The selection of powder mixers. In *Mixing in the Process Industries*, Harnby, N., Edwards, M.F. and Nienow, A.W. (eds.). Butterworths, London.

Harnby, H. (1997). *Mixing in the Process Industries*, Harnby, H., Edwars, M.F. and Nienow, A.W. (eds.). Butterworth-Heinemann, Oxford, UK.

Hwang, C.L. and Hogg, R. (1980). Diffusive mixing in flowing powders. *Powder Technol.* **26**: 93–101.

Kaye, B.H. (1992). Microencapsulation: the creation of synthetic fine particles with specified properties. *KONA* **10**: 65–82.

Kaye, B.H. (1997). Mixing of Powders. In *Handbook of Powder Science & Technology*, Fayed, M.E. and Otten, L. (eds.). Chapman & Hall, New York.

Kuakpetoon, D., Flores, R.A. and Milliken, G.A. (2001). Dry mixing of wheat flours: effect of particle properties and blending ratio. *Lebensm.-Wiss. U. Technol.* **34**: 183–193.

Lacey, P.M.C. (1954). Developments on the theory of particle mixing. *J. Appl. Chem.* **4**: 257–268.

Lindley, J.A. (1991a). Mixing processes for agricultural and food materials: 1. Fundamentals of mixing. *J. Agric. Eng. Res.* **48**: 153–170.

Lindley, J.A. (1991b). Mixing processes for agricultural and food materials: 3. Powders and particulates. *J. Agric. Eng. Res.* **49**: 1–19.

McCabe, W.L., Smith, J.C. and Harriot, P. (1992). *Unit Operations in Chemical Engineering*. McGraw-Hill, New York.

Miyanami, Kei. (1997). Mixing. In *Powder Technology Handbook*, Gotoh, K., Masuda, H. and Higashitani, K. (eds.). Marcel Dekker, New York, pp. 609–625.

Niranjan, K. (1995). An appraisal of the characteristics of food mixing. In *Food Process Design and Evaluation*, Singh, R.K. (ed.). Technomics, Lancaster, PA, USA.

Niranjan, K. and de Alwis, A.A. (1993). Agitation and agitator design. In *Encyclopedia of Food Science, Food Technology and Nutrition*, Vol. 1, Macrae, R., Robinson, R.K. and Sadler, M. (eds.). Academic Press, London.

Poole, K.R., Taylor, R.F. and Wall, G.P. (1964). Mixing powders to fine scale homogeneity: studies of batch mixing. *Trans. Instn. Chem. Eng.* **42**: T305–T315.

Presnell, D. (1995). What you should know about selecting and maintaining a single-shaft batch ribbon mixer. *Powder Bulk Eng.* **9**(1): 47–53.

Rhodes, M. (1998). Mixing and segregation. In *Introduction to Particle Technology*. John Wiley & Sons, West Sussex, England, pp. 224–240.

Rielly, C.D. (1997). Mixing in food processing. In *Chemical Engineering for the Food Industry*, Fryer, P.J., Pyle, D.L. and Rielly, C.D. (eds.). Blackie Academic & Professional, London.

Stange, K. (1963). Die mischgute einer Zufallsmischung aus drei und mehr Komponenten. *Chem. Ing. Tech.* **35**: 580–582.

Thiel, W.J. (1984). Comments on 'Ordered mixtures–interactive mixtures'. *Powder Technol.* **39**: 147–149.

Uhl, V.W. and Gray, J.B. (1986). *Mixing*, Vols. 1–3. Academic Press, Orlando, FL.

Van Deemter, J.J. (1985). Mixing. In *Fluidization*, 2nd ed., Davidson, J.F., Clift, R. and Harrison, D. (eds.). Academic Press, London.

Weinekötter, R. and Gericke, H. (2000). *Mixing of Solids*. Kluwer Academic Publishers, Netherlands.

Williams, J.C. (1968/1969). The mixing of dry powders. *Powder Technol.* **2**: 13–20.

CHAPTER 10

SEPARATION AND CLASSIFICATION

10.1. INTRODUCTION TO DRY SEPARATION TECHNIQUES

Separation techniques are involved in a great number of processing industries and represent, in many cases, the everyday problem of a practicing engineer. In spite of this, the topic is normally not covered efficiently nor sufficiently in higher education curricula of some engineering programs, mainly because its theoretical principles deal with a number of subjects ranging from physics principles to applied fluid mechanics. In recent years, separation techniques involving solids have been considered under the general interest of powder and particle technology, as many of these separations involve removal of discrete particles or droplets from a fluid stream.

Separation techniques are defined as those operations that isolate specific ingredients of a mixture without a chemical reaction taking place. Several criteria have been used to classify or categorize separation techniques. One such criterion consists of grouping them according to the phases involved, i.e., solid with liquid, solid with solid, liquid with liquid, etc. A classification based in this criterion is shown in Table 10.1. Dry separation techniques would, therefore, constitute all those cases in which particles are isolated or segregated from a mixture that is not wet. Particular examples include dry separation in solid mixtures and gas–solid mixtures as listed in Table 10.1. The most important dry separation techniques in processing industries have been reviewed by Beddow (1981). In food processing, there are important applications of dry separation techniques, such as the removal of particles from dust laden air in milling operations, the recovery of the dried product in spray dehydration, and the cleaning of grains prior to processing.

The present chapter will cover dry separation techniques used in food processing operations aimed at removing discrete particles suspended within a fluid, or forming part of a mixture of two or more different food solid materials. The objective may be the recovery for further processing or removal in order to avoid pollution, contamination, or safety risks. The most important operations, due to their applications, are screening and use of cyclones or gas filters. Air classification, which has important applications in food processing, will also be included.

10.2. SCREENING

Screening is a separation technique of a mixture of various sizes of solid particles into several fractions, based on size difference. It consists of forcing the mixture through a screen of a specific size aperture. Small capacity plane screens are often called sieves. By vibrating or oscillating a screen, particles smaller than a given aperture pass through, thus being separated from the remaining mixture. Screens are made from metal bars, perforated or slotted metal plates, woven wire cloth, or fabric, such as silk bolting cloth. Metals used include steel, stainless steel, bronze, copper, nickel, and monel. The screen surface may be plane (horizontal or inclined) or it may be cylindrical. The aperture

Table 10.1. Classification of separation techniques according to phases involved.

Type of mixture	Techniques
Liquid–liquid	Distillation Extraction Decantation Dialysis and electro-dialysis Parametric pumping
Solid–solid	Screening Leaching Flotation Air classification
Solid–gas	Cycloning Air filtration Scrubbing Electrostatic precipitation
Solid–liquid	Sedimentation Centrifugation Filtration Membrane separations

size of the screens ranges from about 0.1 to 250 mm, with exceptional cases in which the aperture may be as large as 460 mm. The material passing through a given screen is termed *undersize, fines or minus* (−) *material*, while the material retained in a given size screen is called *oversize, tails or plus* (+) *material*. Either stream may be the desired (*product*) stream or the undesired (*reject*) stream. Screening has two main applications: laboratory technique for particle size analysis, and industrial operation for fractionation and classification of particulate solids. Although screen aperture, defined as the space between the individual wires of a wire mesh screen, is the preferred terminology for screening operations, the former designation of mesh number, defined as the number of wires per lineal inch, is still widely used.

10.2.1. Screening Fundamentals

The objective of a screening operation is to separate a feed stream into two fractions, an underflow that is passed through the screen and an overflow that is rejected by the screen. An ideal screen sharply separates the feed in such a way that the smallest particle in the overflow is just larger than the largest particle in the underflow. Such an ideal separation would define a cut diameter, D_{pc}, representing the point of separation between the fractions. For an ideal operation, a plot of screen opening against cumulative fraction retained would have the shape shown in Fig. 10.1b. As can be seen, the largest particle of the underflow has the same size as the smallest particle of the overflow. In practice, however, the plot would have the shape illustrated in Fig. 10.1c, in which there is an overlap; the underflow has an appreciable amount of particles larger than the desired cut diameter, while the overflow has particles smaller than such desired cut diameter. It has been observed that the overlap is smaller when particles are spherical (or close to a spherical shape) and is larger when particles are needle-like, fibrous or tend to agglomerate. The main problems encountered in screening result from sample stickiness, sieve blockage, and agglomeration. According to Strumpf (1986), all these problems increase exponentially as the screen aperture decreases.

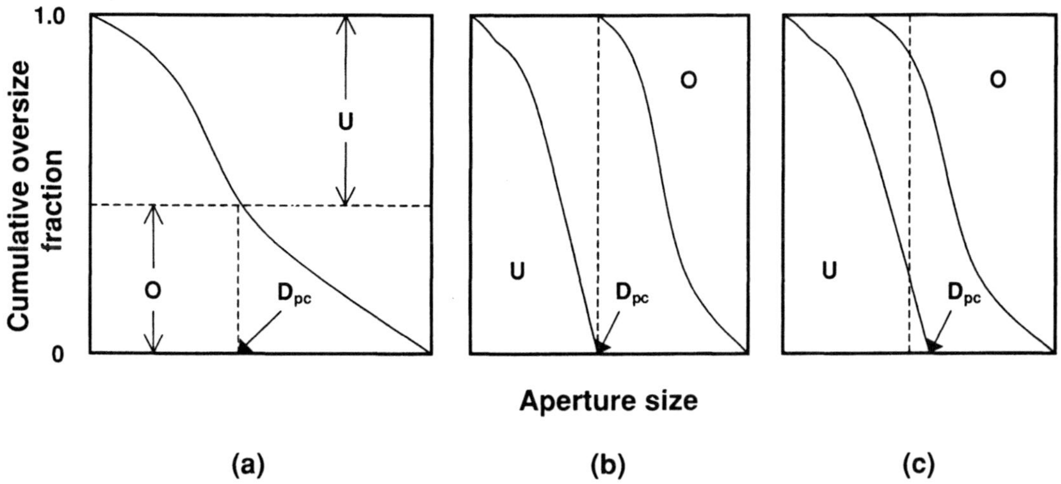

Figure 10.1. Cumulative oversize diagrams describing screening process: (a) feedstock; (b) perfect separation; and (c) actual screening.

10.2.2. Mass Balances in Screening

The efficiency of a screening operation may be evaluated by simple mass balances. Let F be the mass flow rate of feed, O the mass flow rate of tails and U the mass flow rate of fines; also, let X_F be the mass fraction of tails in the feed, X_O the mass fraction of tails in the overflow and X_U the mass fraction of tails in the underflow. Furthermore, the fractions of fines in the feed, overflow and underflow are $1 - X_F$, $1 - X_O$ and $1 - X_U$, respectively. Since the total of the material fed to the screen must leave either as overflow or underflow:

$$F = O + U \tag{10.1}$$

The tails in the feed must also leave in the two streams, so:

$$FX_F = OX_O + UX_U \tag{10.2}$$

Elimination of U from Eqs. (10.1) and (10.2) gives:

$$\frac{O}{F} = \frac{X_F - X_U}{X_O - X_U} \tag{10.3}$$

Similarly, elimination of O gives:

$$\frac{U}{F} = \frac{X_O - X_F}{X_O - X_U} \tag{10.4}$$

The effectiveness of a screen is a measure of how well it performs the separation of tails and fines. If the screen functioned perfectly, all of material O would be in the overflow and all of material U would be in the underflow. A manner of determining screen efficiency is to calculate the ratio of oversize material O that is actually in the overflow to the amount of material O entering with the feed:

$$E_O = \frac{OX_O}{FX_F} \tag{10.5}$$

Similarly, considering the fines material:

$$E_{\mathrm{U}} = \frac{U(1 - X_{\mathrm{U}})}{F(1 - X_{\mathrm{F}})} \tag{10.6}$$

An overall combined efficiency may be defined as the product of Eqs. (10.5) and (10.6), and it may be denoted simply as E:

$$E = \frac{OUX_{\mathrm{O}}(1 - X_{\mathrm{U}})}{F^2 X_{\mathrm{F}}(1 - X_{\mathrm{F}})} \tag{10.7}$$

Substituting Eqs. (10.3) and (10.4) into Eq. (10.7) gives:

$$E = \frac{(X_{\mathrm{F}} - X_{\mathrm{U}})(X_{\mathrm{O}} - X_{\mathrm{F}})X_{\mathrm{O}}(1 - X_{\mathrm{U}})}{(X_{\mathrm{O}} - X_{\mathrm{U}})^2(1 - X_{\mathrm{F}})X_{\mathrm{F}}} \tag{10.8}$$

Equation (10.8) is an alternative expression to evaluate screen efficiency without involving the streams and using only the fractions.

10.2.3. Operating Features

10.2.3.1. Capacity and Efficiency

Efficiency of separation, along with capacity, is the most important variable involved in industrial screening. Capacity and effectiveness are opposing factors as maximum effectiveness is related to small capacity, while large capacity is only attainable at the expense of efficiency. A reasonable balance between capacity and effectiveness is desired in practice. Although accurate relationships are not available for estimating operating characteristics in screen operations, certain fundamentals apply and may be used as guidelines when running and designing a screening process.

The capacity of a screen is measured by the mass of material which can be fed per unit time to a unit area of screen, and can be simply controlled by varying the feed rate to the equipment. The effectiveness obtained for a given capacity is dependant upon the specific nature of the screening operation. The chance of passing through the screen of an undersize particle is a function of the number of times the particle strikes the screen surface, as well as its probability of passage in a single contact. If a screen is overloaded, the number of contacts is small and the chance of passing on contact is reduced by particle interference. The improvement of effectiveness obtained at the expense of reduced capacity is a result of more contacts per particle and better chances for passing through the screen aperture on each contact.

A particle would have an ideal opportunity of passage when striking the surface perpendicularly, which would only be possible if it were oriented with its minimum dimensions parallel to the screen surface. Additional conditions would be no interference by other particles, as well as not sticking to, or wedging into, the screen surface. None of these conditions apply to actual screening, but this ideal situation can be used as a basis for estimating the effect of mesh size and wire dimensions on screen performance. If the width of a screen were negligible in comparison to the size of the openings, the wires would not interfere with particle passage and, practically, the entire screen surface would be active. In such a case, the probability of passage of a striking particle would approach unity. In actual screening, the diameter of the wire, or the fraction of the surface not constituting openings, is significant and the solid meshes strongly affect screen performance, especially by retarding the passage of particles nearly as large as the screen openings.

10.2.3.2. Factors Affecting Efficiency

The probability of passage of a particle through a given screen mainly depends on the fraction of the total surface represented by openings. Other factors are the ratio of the diameter of the particle to the width of an opening in the screen, and the number of contacts between the particle and the screen surface. If all these factors were constant, the average number of particles passing through a single screen opening in a unit time would be constant and independent of the size of the screen opening. The capacity of a screen in mass per unit time divided by the mesh size would therefore be constant for any specified conditions of operation. In practice, however, a number of complicating factors appear and cannot be treated theoretically. Some of these disturbing factors are the interference of the bed of particles with their particular motion, the cohesion of particles to each other, the adhesion of particles to the screen surface, and the oblique direction of approach of the particles to the surface. When large and small particles are present, the large ones tend to segregate in a layer next to the screen, preventing the smaller particles from reaching the screen surface. All these factors tend to reduce capacity and lower efficiency.

Pertaining to moisture, either dry particles or particles moving in a stream of water pass more easily through a screen opening than damp particles, which are prone to stick to the screen surface and to each other. In terms of size aspects, as particle size is reduced, screening becomes progressively more difficult, while capacity and efficiency tend to decrease.

Blinding or clogging of the openings is particularly likely to occur when particles have sizes very close in dimension to the screen aperture. In general terms, there are three defined possibilities of a given particle facing a screen aperture: (a) the particle being too large in relation to the aperture so it would be easily retained; (b) the particle being too small in relation to the aperture so it would go through easily; and (c) the particle having a critical dimension so it would be trapped and promote blinding or clogging of the screen surface. This critical dimension has been defined when the particle has an approximate size of $1.1D$ (see Fig. 10.2), which occurs if the angle θ between the particle and the screen aperture is less than $\tan^{-1}\mu$, with μ being the coefficient of friction between the particle and the screen material. The extreme case of blinding produces complete clogging of the screen with a consequent damage that impairs separation and operation efficiency. Damaged screens should, therefore, be repaired or replaced immediately. It has been demonstrated that clogging is affected by the size of the screen aperture and the particle shape (Beddow, 1980). It has also been reported that particle shape has a significant effect on efficiency for circular and rectangular screen apertures, but only a minor effect for square screen apertures (Nakayima et al., 1978).

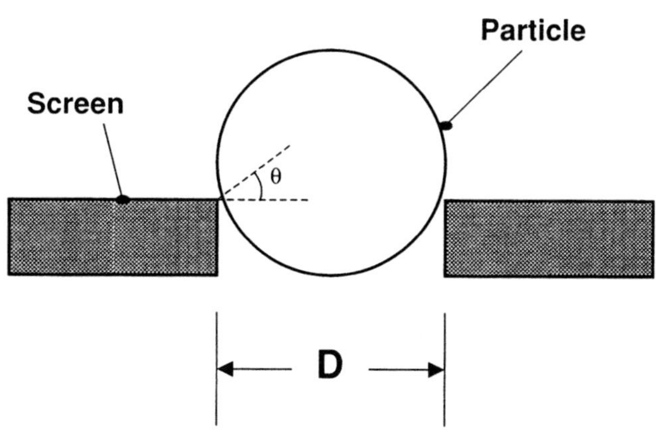

Figure 10.2. Relationship of a particle size to sieve opening to cause blinding in screening.

10.2.4. Equipment Used for Screening

Screening as a unit operation may be carried out in different types of equipment. Three types are more common: grizzlies (bar screens), screens, and trommels. Basic designs of each type of equipment are shown in Fig. 10.3. Grizzlies are used for screening larger particles (pieces greater than 25 mm). They consist of a set of parallel bars, spaced to the desired separation. The bars are often wedge-shaped to minimize clogging. They may be used horizontally or inclined at angles up to 60°. Vibrating grizzlies are available, with the feed material passing over screening surface in a series of jerks.

Screens are of many types: sifter, vibrating, shaking, centrifugal, and revolving, to name only a few. Sifter screens can be conveniently divided into circular-motion, gyratory-motion, and circular-vibrator types. They may be mounted in several decks, and the rate of throughput can be increased by inclining the screen surface. In centrifugal screens, the surface consists of a vertical cylinder rotating at a constant speed with a gyratory motion. Gravity moves the oversize particles down the length of the cylinder as fines are forced through the openings. They are normally inclined to the horizontal and may be multi-deck units, a series of screens being mounted beneath each other, permitting separation of a given feed stock into several size ranges.

Reels or trommels are revolving cylindrical screens mounted almost horizontally. Again, the screening surface may consist of wire mesh or perforated sheets. Hexagonal cross sections are also used since these lead to agitation, which aids the separation of fine material. The capacity of a trommel increases with increasing speed of rotation until a critical speed is achieved. At speeds greater than this, the material does not cascade over the surface but is carried around by centrifugal force, and separation is seriously impaired. The critical speed of a trommel is given by:

$$N = \frac{42.3}{D^{1/2}} \tag{10.9}$$

where N is the number of revolutions of the trommel per minute and D is the diameter of the trommel in meters.

10.2.5. Selection and Design

Based on the information given, a processing engineer should be able to evaluate capacity and efficiency for a particular screening industrial operation. In order to avoid problems and come up with the most suitable choice for a process involving screening, other details can be equally important.

Structural supports will be used under varying conditions, so attention must be given to adequate sizing not only for the loads to be carried out, but for the deflection and vibratory conditions that could prevail in operation. Feed and product chutes also deserve careful consideration. The feed to the screen must be delivered so as to cause a minimal amount of abrasion or disturbance of the bed of material on the screen. The trajectory of material being discharging from feed conveyors or other units of equipment must be considered, and the force of the falling material directed against an abrasive-resistant wear plate, or a dead-bed of the material itself.

The screening operation should be considered within the context of a whole process flowsheet. A description of the unit operations immediately preceding and following is quite relevant. Description of the equipment adjacent to the screening step must be carefully considered. For example, there is little point in dry screening a dusty material if the succeeding stages are wet. The screening operation involved should also describe the methods used for controlling feedrate, product collection, required screen efficiency, number and size of products, etc.

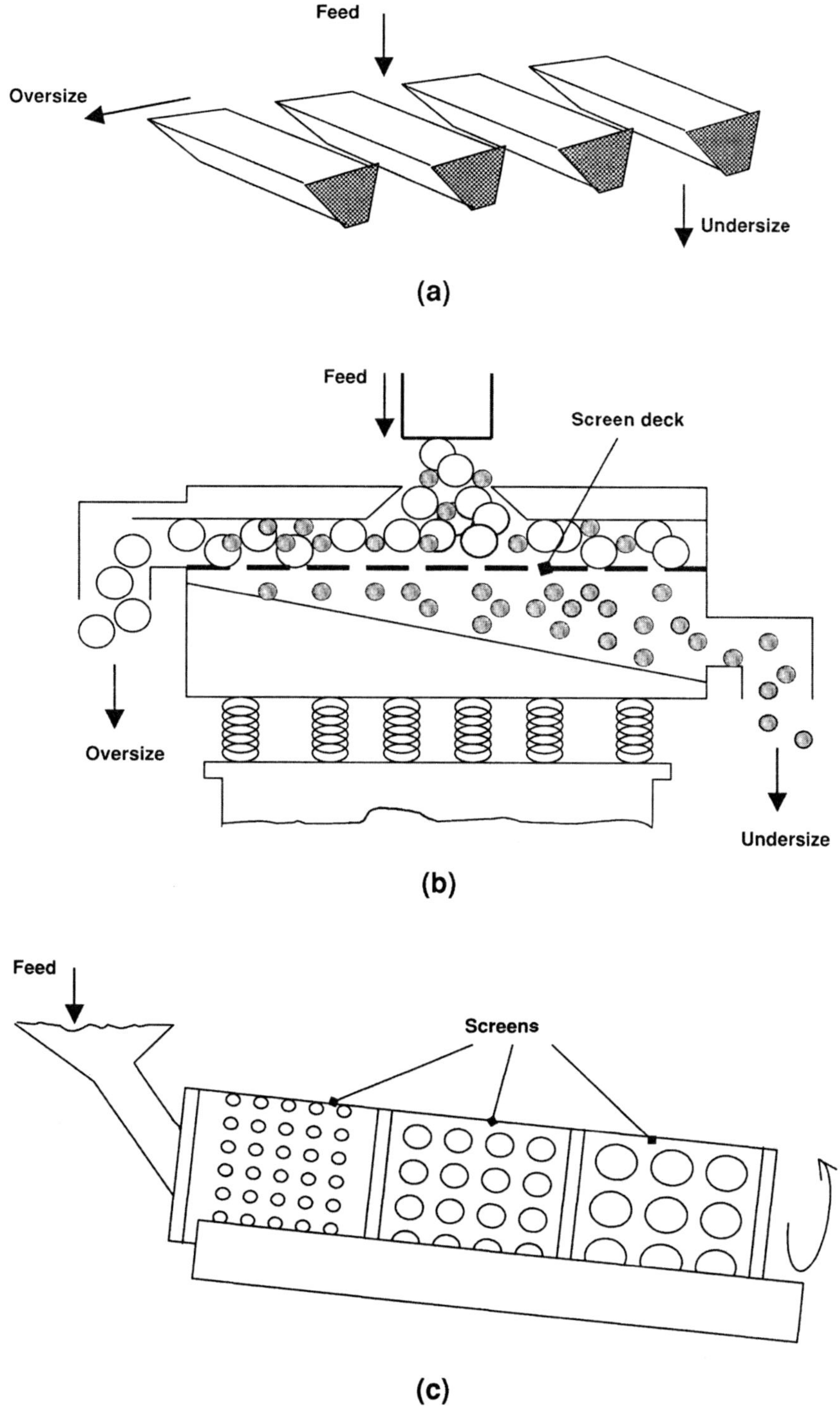

Figure 10.3. Different types of industrial screens: (a) grizzlies (parallel bars); (b) high-capacity sifter with steep inclined plane; and (c) revolving trommel.

Several relevant properties of the material being separated must be known or determined to properly select or design adequate screening equipment. Some of these include particle size distribution, particle shape, bulk density, moisture content, abrasiveness, and corrosiveness. Particle size distribution is essential to correctly size the screening unit, as well as to specify the type of screen to be installed. As previously mentioned, particle shape is determinant in promoting or avoiding blinding or clogging of screen units. Long or splinter-like, round or oval, or cubic particles will have slightly different screening characteristics, and may have a great influence on choice of cloth opening. Bulk density permits the determination of the volume of flow, and a measure of the load to be carried by the screen. Moisture content, as has been stated earlier, may cause difficulties due to stickiness. Information on moisture content, along with data on the process following screening, will allow selection of dry, damp, or wet screening. In some dry screening applications, when moisture content is low, the choice of a proper screen medium will eliminate problems. Pre-drying of materials using heated screen cloths, or going wet by adding water sprays, are all equally effective in damp screening. The abrasive characteristics of the material have great influence on choice of materials of construction, as well as in selection of methods of loading, collecting, and transporting products. Corrosive features will also influence materials of construction used in screen frames, media, chutes, feedboxes, and other elements of the system. This property is relevant for choosing between dry and wet screening.

10.2.6. Applications

One of the main applications of screening is in the flour industry, to separate the different fractions of flour. Specific terms relate to the fractions being removed as a function of the equipment used, as presented in Table 10.2. In this important application, the term "scalping" is often simply used to refer to removal of large particles, while "dedusting" is employed when referring to separating small particles (Brennan et al., 1990). As will be discussed in the following section, dedusting is also used to describe solids–gas separations in general.

In the food industry, screening is also widely used for cleaning and sorting of diverse commodities. Cleaning may be carried out in trommels or flat-bed screens, the latter being, in its simplest form, a pitched stationary frame clad with a screen bed. The operation may be arranged so as to retain oversize material such as string, bag-hairs, etc., from flour, salt or sugar, while discharging a cleaned product. Alternatively, the screen may be used to retain the cleaned material as oversize, while discharging undesired material, e.g., in the removal of weed seeds, grit and small stones from cereals.

Sorting by size of fruits and vegetables is extensively also performed in flat-bed screens, as well as in trommels or drum screens. Simple deck flat-bed screens are used for preliminary sorting

Table 10.2. Screening operations in flour processing related to equipment.

Operation name and details	Type of screen
Scalping: removal of small amount of oversize from a feed predominantly with fines	Grizzly
Coarse separation: separation of fractions larger than 4 mesh (4.76 mm)	Vibrating screen
Fine separation: separation of fractions between 4 and 48 mesh (4.76 and 0.297 mm)	Vibrating screen, high speed, low amplitude
Ultrafine separation: separation of fractions smaller than 48 mesh (0.297 mm)	Same as previous plus sifter screens, static sieves, or centrifugal screens

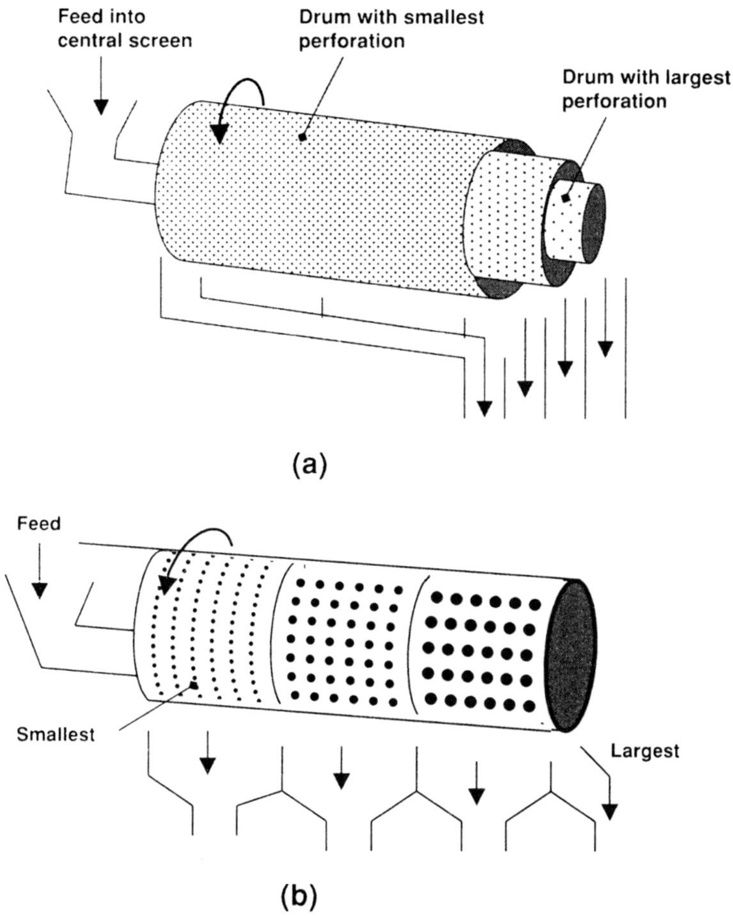

Figure 10.4. Drum screens for sorting of foods: (a) concentric drum screen and (b) series-type consecutive drum screen.

of potatoes, carrots and turnips, while multi-deck screens of this type find extensive use in the size sorting of raw materials such as cereals and nuts, as well as in part-processes and finished foods like flour, sugar, salt, herbs and ground spices. Drum screens are used extensively as size sorters for peas, beans and similar foods that can withstand the tumbling action produced by the drum rotation. Drum sorters are usually required to separate the feedstock into more than two streams and, thus, two or more screening stages are needed. In order to attain this, the screens may be arranged to operate concentrically or consecutively.

The concentric drum screen illustrated in Fig. 10.4a has the advantage of compactness but, because it is fed at the center, the highest product loading goes through the smallest screen area. The series-consecutive drum screen shown in Fig. 10.4b has the disadvantage of requiring a large floor area. Also, since the feed enters at the end that has the smallest aperture screen, the whole screen tends to become overloaded at the inlet end, resulting in inefficient sorting. There is a parallel-consecutive drum screen arrangement that overcomes the disadvantages of the previously described designs by first contacting the inlet material with the large-aperture screen, leaving the following smaller-aperture screens to deal with a reduced quantity of undersized material. Another type of drum screen that is reported to reduce damage during pea sorting uses spaced, circumferential, wedge-section rods instead of perforated-screen drums. The spacing of these rods increases in steps from inlet to outlet, giving a series-consecutive system. Built in flights ensure smooth transfer of peas through the sorter.

10.3. DEDUSTING TECHNOLOGY

In many processes of the food and related industries, separating solids from a gas stream is very important. The typical example is the risk of dust explosion in the dry milling industry. It has been found that not only in this industry, but also in many others that handle fine powders, the atmosphere may become dust laden with particles from different sources, representing a health risk. In other cases, the suspension of particles in a gas stream has been promoted, as in pneumatic conveying or spray drying, but at the end of the process there is a need to separate the phases. Separation of solids from a gas is accomplished using many different devices, most commonly cyclones and bag or gas filters.

10.3.1. Cyclone Separation

Cyclones are by far the most common type of gas–solids separation device used in diverse industrial processes. They have no moving parts, are inexpensive compared to other separation devices, can be used at high temperatures, produce a dry product, have low energy consumption, and are extremely reliable. Their primary disadvantage is that they have a relatively low collection efficiency for particles below about 15 μm. As illustrated in Fig. 10.5, a cyclone consists of a vertical cylinder with a conical bottom, a tangential inlet near the top, and outlets at the top and the bottom, respectively. The top outlet pipe protrudes into the conical part of the cyclone in order to produce a vortex when a dust laden gas (normally air) is pumped tangentially into the cyclone body. Such a vortex develops centrifugal force and, because the particles are much denser than the gas, they are projected outward to the wall, flowing downward along the wall in a thin layer in a helical path. They are eventually collected at the bottom of the cyclone and separated. The inlet gas stream flows downward in an annular vortex, reverses itself as it finds a reduction in the rotation space due to the conical shape, creates an upward inner vortex in the center of the cyclone, and then exits through the top of the cyclone. In an ideal operation in the upward flow there is only gas, while in the downward flow there are all the particles fed with the stream. Cyclone diameters range in size from less than 0.05 to 10 m, feed concentrations cover values from 0.1 to about 50 kg/m^3, while gas inlet velocities may be on the order of 15–35 m/s.

10.3.1.1. Theoretical Aspects

A cyclone is in fact a settling device in which a strong centrifugal force, acting radially, operates instead of the relatively weak gravity force, acting vertically. Due to the small range of particles involved in cyclone separation (the smallest particle that can be separated is about 5 μm), it is considered that Stokes law primarily governs the settling process. The common form of Stokes law is:

$$u_t = \frac{x^2(\rho_s - \rho_g)g}{18\mu_g} \tag{10.10}$$

where u_t is the terminal settling velocity, x is the particle diameter, ρ_s is the solids density, ρ_g is the gas density, and μ_g is the gas viscosity.

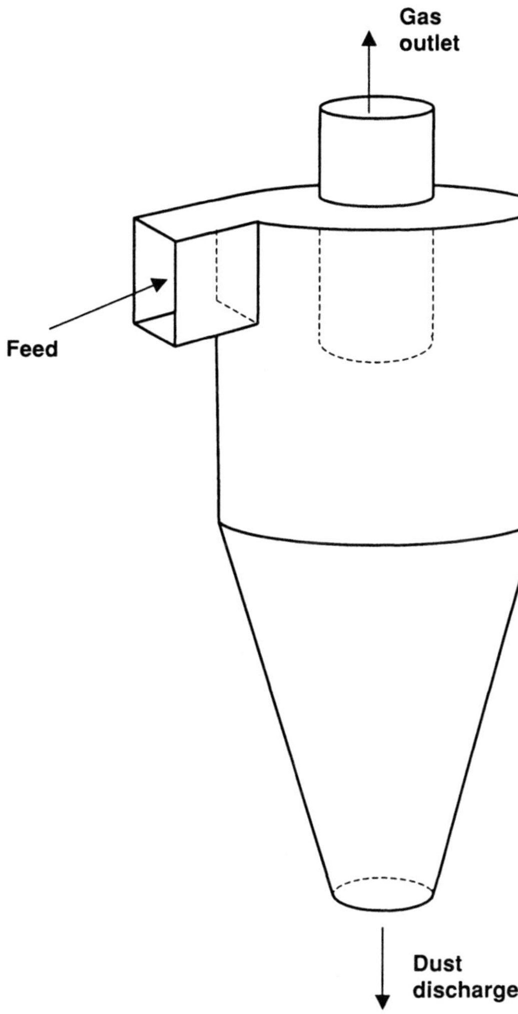

Figure 10.5. Schematic diagram of a cyclone.

Cyclones can generate centrifugal forces between 5 and 2,500 times the force of gravity, depending on the diameter of the unit. When particles enter into the cyclone body, they quickly reach their terminal velocities corresponding to their sizes and radial position in the cyclone. The radial acceleration g in a cyclone depends on the radius of the path being followed by the gas and is given by the equation:

$$g = \omega^2 r \tag{10.11}$$

where ω is the angular velocity and r is the radius.

Substituting Eq. (10.11) into Eq. (10.10) gives:

$$v_t = \frac{x^2(\rho_s - \rho_g)\omega^2 r}{18\mu_g} \tag{10.12}$$

where v_t is the terminal velocity of the particle.

Also, the centrifugal acceleration is a function of the tangential component of the velocity $v_{\text{tan}} = \omega r$, and thus, considering this, Eq. (10.12) becomes:

$$v_t = \frac{x^2(\rho_s - \rho_g)v_{\text{tan}}^2}{18\mu_g r} \quad (10.13)$$

Multiplying Eq. (10.13) by g/g, the resultant equation gives:

$$v_t = \left[\frac{x^2(\rho_s - \rho_g)g}{18\mu_g}\right]\frac{v_{\text{tan}}^2}{gr} = (u_t)\frac{v_{\text{tan}}^2}{gr} \quad (10.14)$$

where u_t is the terminal settling velocity defined by Eq. (10.10). As can be inferred, according to Eq. (10.14), the higher the terminal velocity the easier to "settle" a particle within a cyclone.

For a given particle size, the terminal velocity is a maximum in the inner vortex, where r is small, so the finest particles separated from the gas are eliminated in the inner vortex. These migrate through the outer vortex to the wall of the cyclone and drop, passing the bottom outlet. Smaller particles, which do not have time to reach the wall, are retained by the air and carried to the top outlet. Although the chance of a particle for separation decreases with the square of the particle diameter, the fate of a particle also depends on its position in the cross section of the entering stream and on its trajectory in the cyclone. Thus, the separation according to size is not sharp. A specific diameter, called the *cut diameter* or *cut size*, can be defined as that diameter for which half of the inlet particles, by mass, are separated, while the other half are retained by the gas. The cut size is a very useful variable to determine separation efficiency of a cyclone. Since a given powder to be separated in a cyclone would have an extremely fine half of its distribution, such half may not be easily separated using conventional pressure drops. Therefore, it is advisable to make the cut size coincide with the mean size of a powder particle size distribution to guarantee separation of the coarse part of such distribution, as the fine part may be unattainable due to the small range involved.

10.3.1.2. Dimensionless Approach

Experience and theory have shown that there are certain relationships among cyclone dimensions that should be observed for efficient cyclone performance (Geldart, 1986), and which are generally related to the cyclone diameter. There are several different standard cyclone "designs." A very common design is called the Stairmand, whose dimensions are shown in Fig. 10.6. Using standard geometries of cyclones makes it much easier to predict effects on variable changes, and scale-up calculations are greatly reduced. Such calculations may be carried out by means of dimensionless relationships. Selection and operation of cyclones can be described by the relationship between the pressure drop and the flow rate, and the relationship between separation efficiency and flow rate (Svarovsky, 1981). The pressure drop versus volumetric flow rate relationship is usually expressed as Eu $= f(\text{Re})$, where Eu is the Euler number, and Re is the Reynolds number.

The Euler number is in fact a pressure loss factor, easily defined as the limit on the maximum characteristic velocity v obtained by a certain pressure drop ΔP across the cyclone. It can be expressed as:

$$\text{Eu} = \frac{2\Delta P}{\rho_g v^2} \quad (10.15)$$

where, as previously defined, ρ_g is the gas density.

The well-known Reynolds number defines flow characteristics of the system and, in the case of cyclones, the characteristic dimension may be taken as the cyclone body diameter D_c. The Reynolds

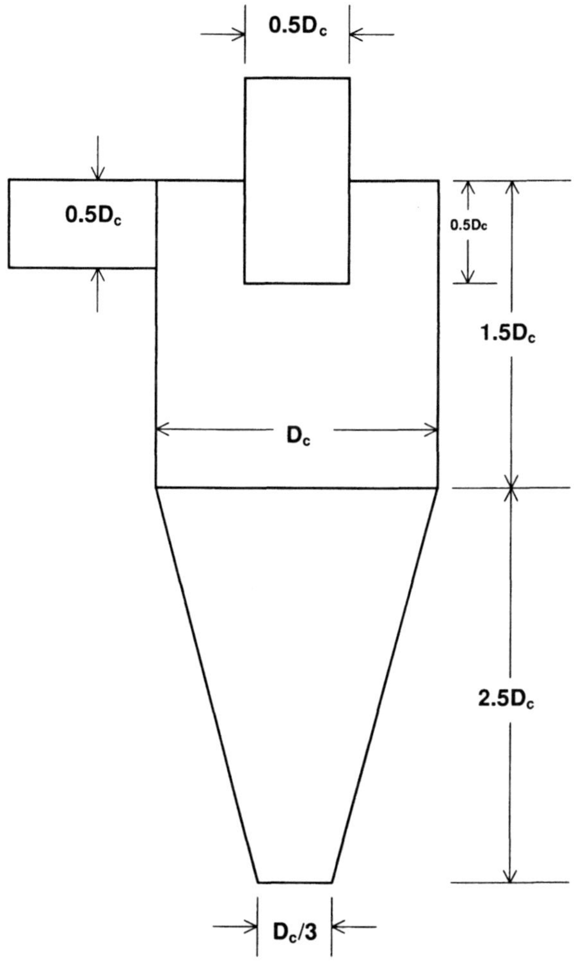

Figure 10.6. Dimensions of a Stairmand standard cyclone.

number for this case is, therefore, represented by:

$$\mathrm{Re} = \frac{D_c v \rho_g}{\mu_g} \tag{10.16}$$

where, as already defined, μ_g is the gas viscosity.

The relationship between separation efficiency and flow rate is not significantly influenced by operational variables, so it is commonly expressed in terms of cut size x_{50}. The use of cut size to define efficiency of cyclones is of the utmost importance, since their performance is highly dependent on particle size. Considering that cut size implies size of particles to be separated, it follows that such particles must be influenced by forces exercised on the suspension. The forces developed in a cyclone can be analyzed by sedimentation theory, and a dimensionless group thus derived, the Stokes number, Stk, will include the cut size. The Stokes number is a very useful theoretical tool and, for the case of cyclones, its derivation may be carried out as follows.

The radial settling velocity in a cyclone is due to the centrifugal acceleration, which is proportional to the square of the tangential velocity of the particle and indirectly proportional to the radius of the particle position. As the tangential motion of the particle is unopposed, the tangential particle velocity can be taken as equal to the tangential component of the fluid velocity at the same point. For the same flow regime, the velocities anywhere in the flow in a cyclone are proportional to a characteristic velocity v, a function of the cyclone cylindrical geometry, also called the body velocity. The position radii are proportional to the cyclone diameter D_c. Under such assumptions, Eq. (10.13) can be approximated to:

$$v = \frac{x^2(\rho_s - \rho_g)v^2}{18\mu_g D_c} \tag{10.17}$$

Re-expressing Eq. (10.17) in dimensionless form, the Stokes number, Stk, is obtained as:

$$\text{Stk} = \frac{x^2(\rho_s - \rho_g)v}{18\mu_g D_c} \tag{10.18}$$

Since the value of the gas density, usually air, is negligible in comparison with the solids density, Eq. (10.18) can also take the following form:

$$\text{Stk} = \frac{x^2 \rho_s v}{18\mu_g D_c} \tag{10.19}$$

Furthermore, if the dimension x is replaced by the specific cut size x_{50}:

$$\text{Stk}_{50} = \frac{x_{50}^2 \rho_s v}{18\mu_g D_c} \tag{10.20}$$

Equations (10.15), (10.16) and (10.20), defining Euler Eu, Reynolds Re, and Stokes Stk_{50} numbers, respectively, are related by specific functions, which can be plotted as shown in Figs. 10.7 and 10.8, for a given cyclone geometry. The cyclone inside diameter D_c is shown in Fig. 10.6 and, as previously mentioned, all geometrical proportions are related to it. In the case of scale-up procedures, proportions must be maintained. The cyclone body velocity v is the characteristic velocity which can be defined in various ways, but the simplest one is based on the cross section of the cylindrical body so that:

$$v = \frac{4Q}{\pi D_c^2} \tag{10.21}$$

where Q is the gas flow rate.

10.3.1.3. Operating Features and Selection Criteria

When designing cyclone systems, it is essential for optimum results to have full process data available, as there are a number of variable factors that will determine final performance. Cyclone inlet velocity not only affects efficiency of separation, but also reflects pressure loss and possible erosion. Gas viscosity has an important effect on particle efficiency, so it is advisable to check its dependency with temperature and to consider those cases in which a gas different from air is involved in the process. Smaller cyclone diameters increase overall efficiency, but will promote erosion. In addition to this, it is sometimes necessary to consider possible attrition of solids in the cyclone, which will result in production of fines and considerable losses. Erosion occurs primarily where the particles first impact the cyclone wall, but will also occur at the bottom of cyclones too short to accommodate the length of the naturally occurring vortex.

In order to remove as much of the dust from a gas stream as possible, cyclones are often placed in series with each other to increase the overall collection efficiency relative to a single unit. Although

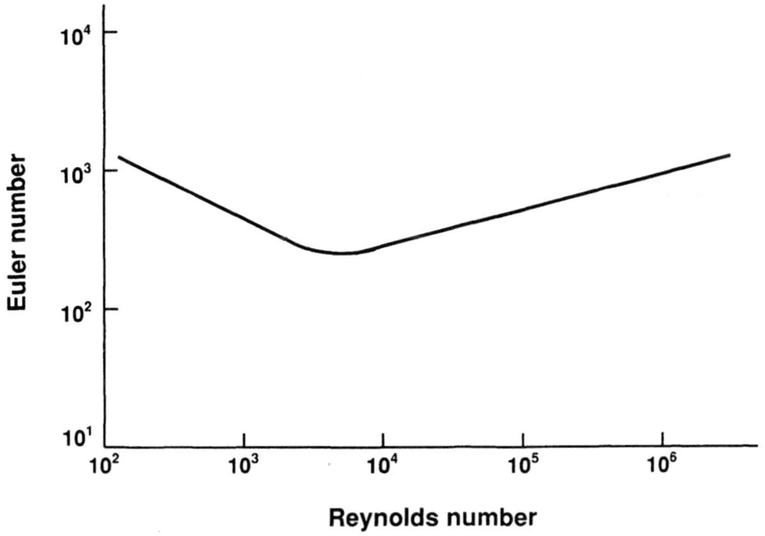

Figure 10.7. A typical plot of Eu versus Re for cyclones.

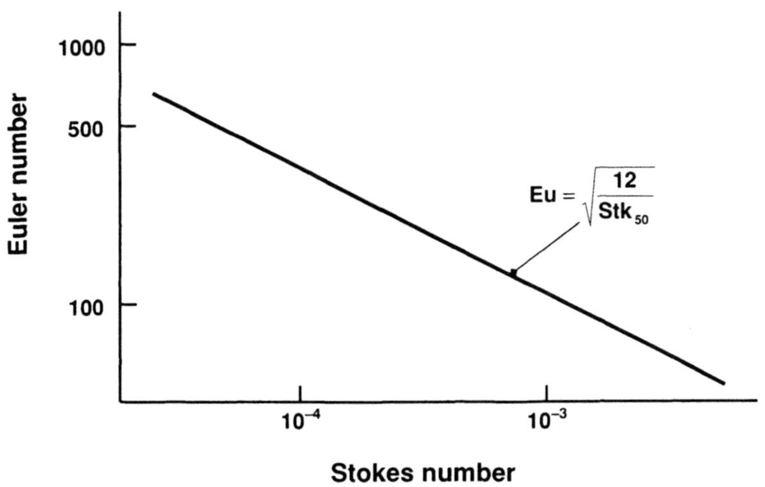

Figure 10.8. A typical plot of Eu versus Stk_{50} for cyclones.

this is theoretically true, efficiency is increased at the expense of pressure drop, so extreme care should be taken to consider whether great efficiency will render a costly process. There is also a great complexity when trying to improve collection efficiency using many small cyclones in parallel to replace a large cyclone. A disadvantage of parallel cyclones is that equal gas distribution to each cyclone can be difficult to achieve. If this occurs, collection efficiency is reduced, with the advantages of parallel cyclones being greatly negated.

There is an increasing trend towards using cyclones for final cleaning for environmental purposes in processes where it is becoming economically impractical to use other methods. There are many applications where this can be done subject to certain limitations, by the use of one, two, or even three stages of separation. In many cases, a first stage cyclone can be used to deal with high inlet loadings and will produce excellent results at the expense of reasonable pressure loss. The addition of a second stage to deal with the first stage losses can then often achieve the required results, but where further cleaning is necessary a third stage cyclone may provide the answer. The use of a third stage cyclone invariably means that a high degree of clean-up is necessary and, therefore, the third stage should give the best possible efficiency and be capable of maintaining this efficiency for a long time. The use of several stages to try to improve efficiency or remove very fine particles may become impractical by employing only cyclones. This difficulty can be alleviated by combining methods so, for environmental purposes, when certain particles need to be removed, regardless of their fineness, bag filters may be coupled with cyclones.

10.3.1.4. Applications

As mentioned before, cyclones are extensively used in the food industry to reduce particle load to safe levels in dry milling, as well as in classification of particles in closed circuit grinding operations. They are also employed in recovering fines from spray drying and fluidized bed drying processes. Another important application is in pneumatic conveying of diverse food products, such as grains and flours.

10.3.2. Gas Filtration

Gas–solids separations can be performed using filtration means. Gas filters are used for final particulate removal in many processes of the food industry. These filters can capture particles much smaller in diameter than a cyclone can, so they are commonly placed downstream of a cyclone in diverse applications. A gas filter generally consists of a porous fabric that can be woven to conform to the shape of a cylinder, or may be supported in a frame. The former is called a bag filter, while the latter is known as an envelope filter. The main difference in these designs is in the way solids are accumulated. In bag filters, dust may accumulate inside, whereas in envelope filters, it forms a cake outside. The filtering arrangements are placed in a matrix so that their total area will result in a low gas velocity through the bags and, therefore, a low pressure drop through the filter. Gas velocities through the filtering media are of the order of 0.005–0.02 m/s. Particulate loadings to these filters generally lie in the range of 0.2–250 g/cm^3. A diagram of the two main types of gas filters is given in Fig. 10.9. Filters used in gas–solids separations may be woven or felted fabrics of natural or synthetic fibers. There are tables listing properties of filtering media to determine whether they are suitable for applications under diverse conditions, such as high temperatures, corrosive and chemical resistance, etc. (Green and Maloney, 1999). Filtrations may also be made with granular solids in the form of stationary or moving beds. Many other types of materials that are porous in nature, or are capable of providing a screening effect after weaving or fabrication, may be found suitable for certain filtration applications.

10.3.2.1. Filtering Fundamentals

In gas filtration, the collected particles build up on the surface of the filtering medium and cause a gradual increase in the pressure drop through the filter. After a certain limiting pressure drop is reached, the bags are cleaned by pulsing gas back through the filter to remove the solids buildup on the bag. As described above, standard commercial types of filters use a bag or an envelope, the latter actually being a retainer for the filter cloth. The bag filters are, in fact, elongated cylinders that may be open at both ends to allow a cleaning cycle by using air jets blowing into the top. On the other hand, envelope filters, with a frame for support and with large flat surfaces exposed, do not present the cleaning capabilities of bag filters, which can be given more shaking or can be collapsed

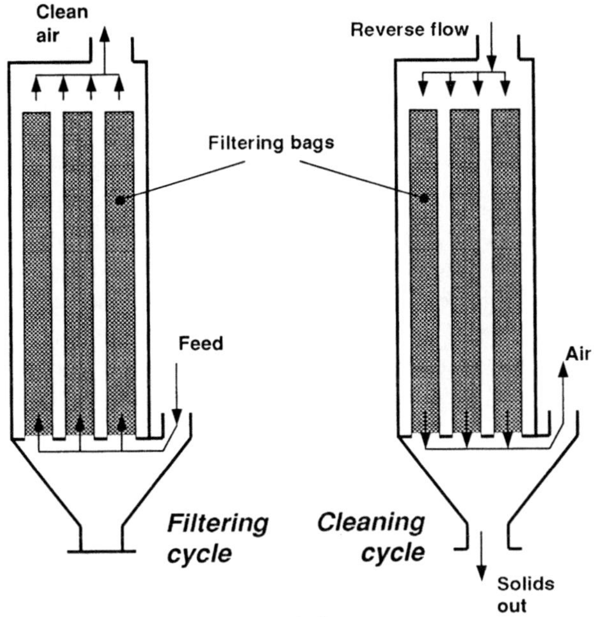

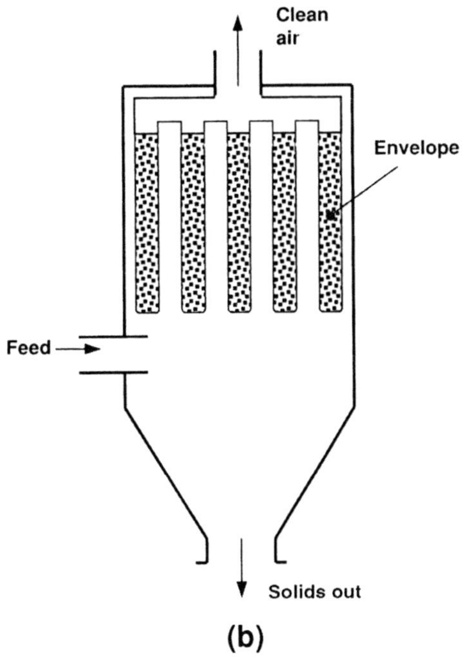

Figure 10.9. Schematic diagram of gas filters: (a) bag filter and (b) envelope filter.

without risking any harm. When air is blown in reverse direction to filtration flow for cleaning purposes, such cleaning can be made more effective. Envelope filters are, therefore, best used on dust that is easily shaken or removed from the cloth surface. The dust is removed in the envelope type by beating the screen supporting the filter, and by shaking or rocking the frame by mechanical means.

The use of a blow ring in some bag- or cylinder-type mechanical filters allows for continuous operation while performing the cleaning duty. Dust is collected on the inside while the blow ring travels up and down along the outer surface of the bag. The blow ring has an inside slot used to blow gas against the bag wall. The ring is tight enough to partially collapse the bag in order to break the dust cake and to provide a tight seal so the gas blown is fully delivered through the filter when blowing back. Dust may be collected on the outside of filter tubes or bags if a support is provided inside to prevent the filter from collapsing. Sometimes, it is possible to remove the cake from the outer wall by periodically using a jet of compressed air from inside the filter to produce a shock to break the cake from the outer wall whence it can be settled into the bin. The envelope- or frame-type filter collects the dust on the outer wall of the filter, as expected, since the outer wall is easier to get to in this arrangement.

10.3.2.2. Operation Characteristics

The operating variables of gas filtration are resistance to flow, permeability of air to the filtering medium, and resistance due to particle accumulation. With regard to resistance to flow, the pressure drop across the filtering medium ΔP_{f} can be represented by:

$$\Delta P_{\mathrm{f}} = K_{\mathrm{c}} \mu_{\mathrm{g}} V_{\mathrm{f}} \tag{10.22}$$

where K_{c} is a constant depending on the filtering medium and V_{f} is the superficial velocity of the gas through the filtering medium.

The resistance to the layer of particles accumulated during the filtration cycle can be calculated by determining a variable known as cake resistance factor K_1:

$$K_1 = \frac{\Delta P_{\mathrm{c}}}{V_{\mathrm{f}} w} \tag{10.23}$$

where ΔP_{c} is the pressure drop through the powder layer and w is the powder mass flow rate approaching the filter.

10.3.2.3. Applications

As previously mentioned, bag filters have practically the same applications of cyclones, being normally coupled with these in order to remove the finest tails of particle size distributions in diverse food powders (e.g., when using ground sugar as an ingredient). A promising application of gas filtration in food processes is the use of ceramic candle filters, composed of multiple porous ceramic cylinders, because they can withstand high temperatures.

10.3.3. Other Gas–Solids Separation Techniques

10.3.3.1. Scrubbers

Solid particles are often scrubbed from a gas stream by spraying a jet of liquid, usually water, into the stream. The particles are intercepted by the droplets of the water spray and are removed by the scrubber in the form of a slurry. Although scrubbers are normally more efficient than cyclones, they have the disadvantage of collecting the solids wet instead of dry. If this feature is unacceptable or impractical for processing reasons, the solids must be separated from the liquid. Thus, the gas–solids collection problem is replaced by a solid–liquid separation difficulty. Scrubbers are standard accessories for spray dryers, as illustrated in Section 11.1.1.

10.3.3.2. Electrostatic Precipitators

This type of equipment separates solids or liquids from a gas stream by passing it through a strong, high voltage field produced between two electrodes of opposite polarity. The field imposes a charge on the particles so that they migrate toward the collecting electrode. The particles are usually removed from the electrode by periodic rapping. The advantage of an electrostatic precipitator is that it can collect solids of very small size in a dry form. Electrostatic precipitators are generally large units of equipment, because collection efficiency is proportional to the area of the collecting electrodes. For this reason, capital costs are very high, although operating costs are low and may justify the use of a precipitator instead of a filter for some particular applications (Svarovsky, 1981).

10.4. AIR CLASSIFICATION

10.4.1. Introduction

Air classification is a method of separating powdery, granular, or fibrous materials in accordance with the settling velocity, combined with the influence of particle size, particle density, and particle shape. The procedure of winnowing or aspiration is a traditional way to separate chaff from grain after threshing and is one of the simplest forms of air classification. Ideally, the separation effect of an air classifier should be such that all particles which exceed the cut point are transported into the coarse fraction, while the smaller particles are transported to the fines fraction. In this sense, air classification basically consists of dividing particle size distributions of given powders and, as such, is a technique commonly used in combination with size reduction equipment, normally to eliminate fines that may affect properties like wettability and dispersibility. The major interest in air classification is that it provides a means for separating small particles, in a dry manner, which cannot be readily achieved by sieving, i.e., below 50 μm. Ortega-Rivas and Svarovsky (2000) reported a successful sharp split of the particle size distribution of calcium carbonate into a fine fraction with mean particle size as fine as 6 μm.

10.4.2. Operating Principles

The mode of operation of a typical air classifier is as follows. The inlet air is mixed with the material being separated. The feed particles are subjected to a centrifugal force operating from a revolving rotor and a drag force produced by the air current, which moves in a spiral direction towards the central shaft. As previously mentioned, separation is based on differential mass, density and shape. The larger and more dense particles are influenced by the mass-dependent centrifugal forces and move towards the outside of the chamber, where they are removed by a screw conveyor or some means. The smaller, lighter particles are more subjected to the frictional forces of the air current, therefore, moving with the air stream and leaving from the center of the classifying chamber to be separated by a cyclone from the air stream. The relative magnitude of these two forces can be

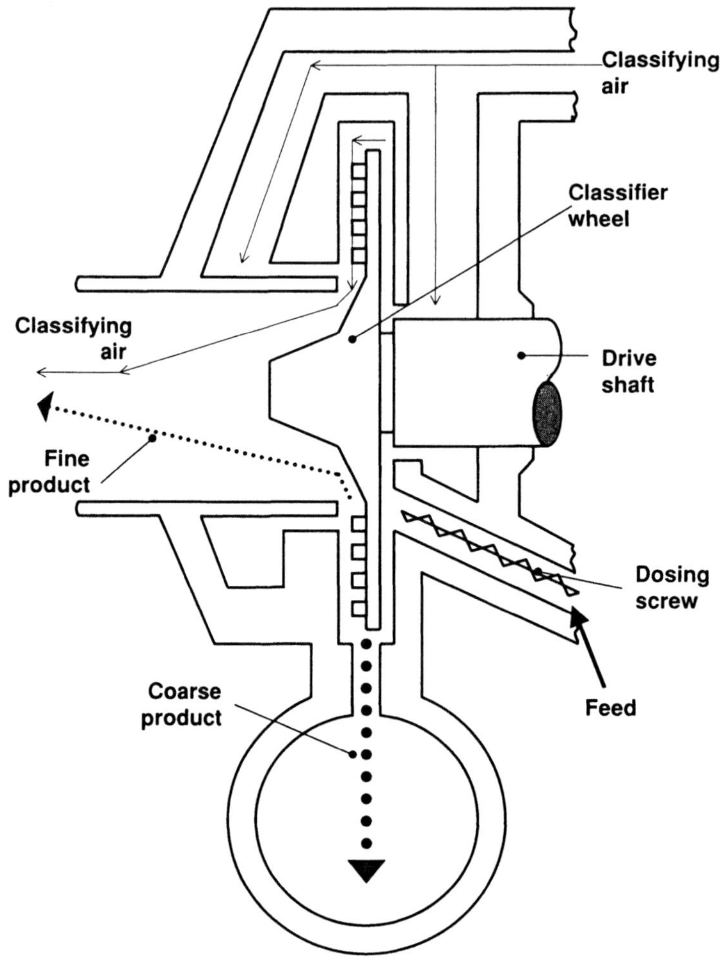

Figure 10.10. Schematic diagram of rotating zigzag wheel air classifier.

changed by altering the rotational speed of the disc and the air velocity. By varying either of these, there will be an effect on the cut size. A common design of air classifier consists of a rotating wheel with zigzag channels over its surface, each of them comprising six components. A diagram of this type of classifier is shown in Fig. 10.10.

Classifiers with the facility to change rotational speed and air velocity independently will be very flexible in operating terms. Although the separation mainly takes place within the classifying chamber, some preliminary removal of the coarsest particles may be achieved outside the chamber. A disc or turbine can be mounted on a horizontal or vertical axis. The latter produces a centrifugal force in the horizontal plane, favoring high throughputs, but low precision in cut size (Fedoc, 1993). Air classifiers are categorized by reference to a number of factors, such as the presence or absence of a rotor, drag force on the air, relative velocity and direction of the air and particles, the use of directional devices such as vanes or cones, and location of the fan and fines collection devices. Other important features include the capacity of the classifier and energy utilization. A comprehensive treatise on classifier types has been presented by Klumpar et al. (1986).

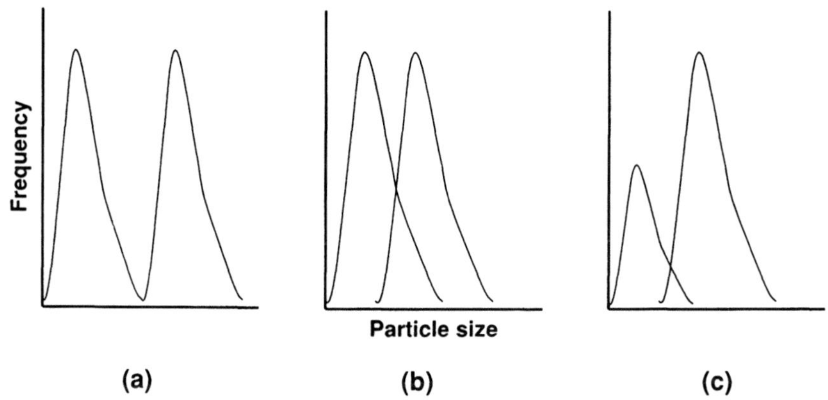

Figure 10.11. Graphical determination of cut size in air classifying: (a) ideal separation; (b) real separation; (c) weight frequency distribution corrected for yield.

10.4.3. Efficiency and Cut Size

The obvious method of characterization of separation capability of air classifiers is by using the cut size concept. Ideally, all particles below the cut size would end up in the fines stream while all particles above the cut size would follow the coarse stream. However, there will be always misplaced material, i.e., a small fraction of particles smaller than the cut size would be in the coarse stream and an equally small proportion of particles larger than the cut size would appear in the fines stream. The extent of overlap due to misplaced material, as well as the cut size, can be determined by measuring the particle size distributions of both streams and presenting their data as a weight frequency distribution. The yields of fines Y_f and coarse Y_c streams need to be identified. When they are equal, the point of overlap (Fig. 10.11b) gives the cut size. When they are not equal, which is most likely, the frequency distribution for the fines stream must be multiplied by the yield for the fines stream, while the yield for the coarse stream must be multiplied by the yield for the coarse stream (Fig. 10.11c). The cut size is thus given by the point of intersection of these curves.

In air classifiers a solid-state element in a rotor is radially affected by two forces: a centrifugal force toward the outside and the frictional force of the air toward the inside. Since quite fine particles are of interest, assuming Stokes law applicability, the resulting equilibrium between the two mentioned forces for a certain grain size would be the so-called cut size. Particles larger than the cut size are centrifugally extracted as oversize toward the outside, while smaller particles are carried inside by the air stream. Assuming a constant air throughput, the fineness of the separation in the classifier depends on the peripheral speed of the particle, which in turn is in conformity with the peripheral speed of the rotating wheel. With the speed remaining constant, an increasing air throughput changes the cut point within the coarser range. In principle, any cut point can be attained by combining two matching values of speed and air throughput.

By equating the above-mentioned forces when they are in equilibrium, a relation for the cut size x_{50} based on the Stokes law can be derived as follows:

$$x_{50}^2 = \left[\frac{18\mu_g v_a r}{\rho_s v_p^2}\right] \tag{10.24}$$

where μ_g is the air viscosity, v_a is the radial speed of air, r is the clearance of the classifier wheel, ρ_s is the particle density, and v_p is the peripheral speed of particle, which is equivalent to rotational speed.

Equation (10.24) is useful in predicting how air flow and rotational speed may influence the cut size. Basically, it predicts that increasing the air flow rate would increase the cut size, while increasing the rotational speed would decrease this cut size. Due to the complicated flow conditions in air classifiers, the classifying results with the given values of speed and air throughput cannot be predetermined. For this reason, the assignment of cut point and operating data of the classifier are determined by experiments with a calibration curve. The various particle sizes of a given material of known distribution are used to plot the calibration plot. These particle sizes are separated under certain conditions of speed and air throughput. The coarse fraction is weighed at the same time. The particle size corresponding with the coarse grain proportion and legible from the fineness characteristics is the cut point according to the operating conditions of the individual classifier, also known as the analytical cut point. For inert materials with densities of the order of 2,700 kg/m^3, cut points between 1 and 100 μm are obtained with a "normal" air throughput and by varying the speed between 2,500 and 20,000 rpm. A calibration curve for this type of material is shown in Fig. 10.12.

Cut size alone does not provide information on how sharp a separation is obtained. An alternative method of evaluation is to determine the grade efficiency, which also has the advantage of indicating the sharpness of the separation. Particle frequency distribution may be determined by weight for the coarse stream $q_c(x)$ and the feed material $q_f(x)$; the yield is determined for the coarse stream Y_c and the grade efficiency $G(x)$ indicates for any particle size the mass fraction of feed material appearing

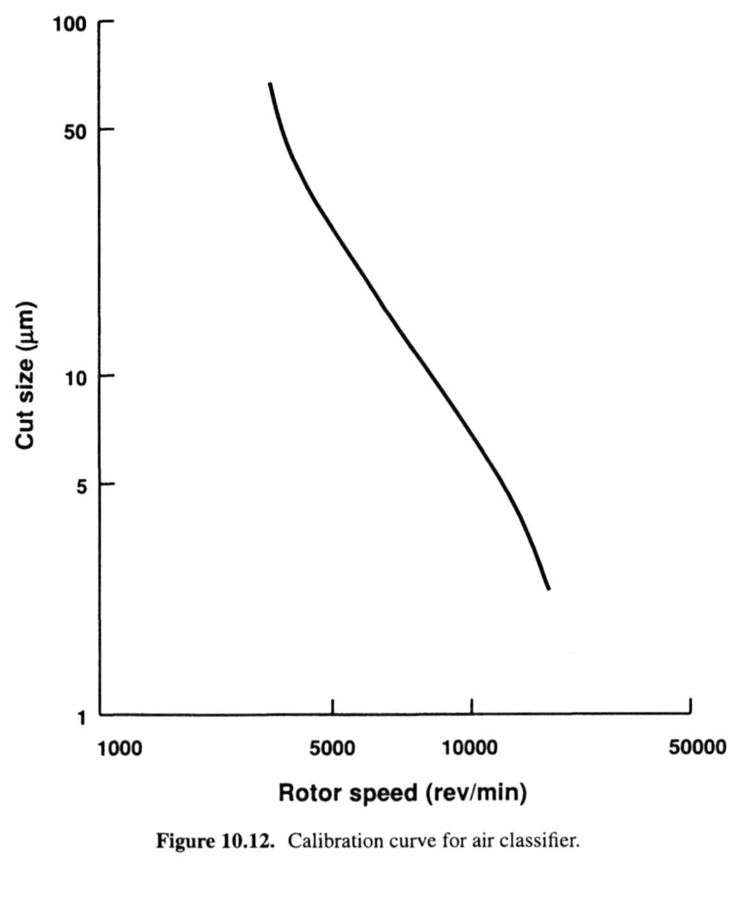

Figure 10.12. Calibration curve for air classifier.

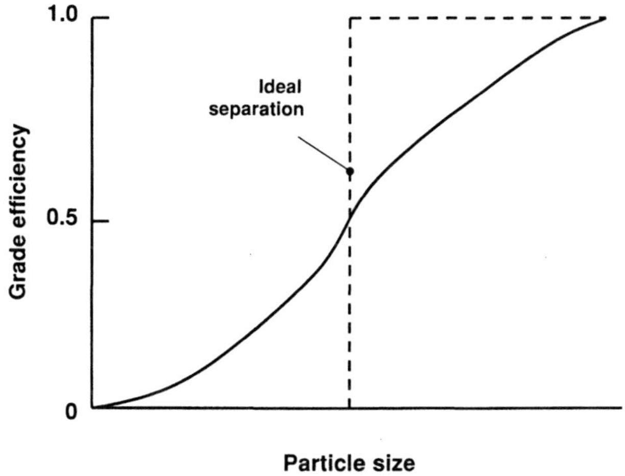

Figure 10.13. Grade efficiency curve.

in the coarse fraction:

$$G(x) = \frac{Y_c q_c(x)}{q_f(x)} \tag{10.25}$$

Grade efficiency can be plotted against particle size, giving a graph like the one shown in Fig. 10.13. The cut size is located at $G(x) = 0.5$, indicating the size of the particles, half of which appear in the coarse stream and half of which appear in the fines stream. The cut size thus obtained is called the graphical cut size, in contrast to the analytical cut size described above. The sharpness of the separation is measured by the ratio:

$$k = \frac{x_{25}}{x_{75}} \tag{10.26}$$

where x_{25} and x_{75} indicate grade efficiencies of 0.25 and 0.75, respectively, as read from the graph represented in Fig. 10.13. Ideally, $k = 1.0$, but the best air classifiers would achieve $k = 0.7$. Typically, commercial air classifiers only show k values between 0.3 and 0.6 (Schubert, 1987).

10.4.4. Applications

Air classification is used in the food industry in important applications such as wheat flour fractionation to separate the coarse, low protein fraction from the fine, high protein fraction. Other applications include classification of confectionery products, soy flour, potato granules, rice flour, lactose and oleaginous fruits, removal of shells or hulls from disintegrated peanuts, cottonseed, rapeseed or cocoa beans, preparation of oat-bran, and separation of gossypol from cottonseed protein.

REFERENCES

Beddow, J.K. (1980). *Particulate Science and Technology*. Chemical Publishing, New York.

Beddow, J.K. (1981). Dry separation techniques. *Chem. Eng.* **88**: 70–84.

Brennan, J.G., Butters, J.R., Cowell, N.D. and Lilly, A.E.V. (1990). *Food Engineering Operations*. Elsevier, London.

Fedoc, P. (1993). *Air Classification, Encyclopedia of Food Science and Technology*. Academic Press, London.

Geldart, D. (1986). *Gas Fluidization Technology*. John Wiley and Sons, London.

Green, D.W. and Maloney, J.O. (1999). *Perry's Chemical Engineers' Handbook*. McGraw-Hill, New York.

Klumpar, I.V., Currier, F.N. and Ring, T.A. (1986). Air classifiers. *Chem. Eng.* **93**: 77–92.

Nakayima, Y.N., Whiten, W.J. and White, M.R. (1978). Method for measurement of particle shape distribution by sieves. *Trans. Inst. Min. Met.* **87**: C194–C203.

Ortega-Rivas, E. and Svarovsky, L. (2000). Centrifugal air classification as a tool for narrowing the spread particle size distributions of powders. In *Processing Part I*, Wöhlbier, R.H. (ed). Trans Tech Publications, Clausthal-Zellerfeld, Germany.

Schubert, H. (1987). Food particle technology part I: Properties of particles and particulate food systems. *J. Food Eng.* **6**: 1–32.

Strumpf, D.M. (1986). Selected particle size determination techniques. *Manuf. Confect.* **66**: 111–114.

Svarovsky, L. (1981). *Solid–Gas Separation*. Elsevier, Amsterdam, Netherlands.

CHAPTER 11

DRYING

Drying is one of the oldest methods of food preservation used by human beings and also the most widely used method for food preservation purposes. The exact date when our ancestors began to dry food for preservation is unknown, but the first recorded dried vegetables appeared in the 1700s. During drying processes, water in food products is reduced to a level where the growth of spoilage microorganisms, as well as the occurrence of chemical reactions, are halted or slowed down. In addition to preservation, the reduced weight and bulk volume of dried products and their longer shelf stability reduce the costs and/or difficulties of product packaging, handling, storage, and distribution (Barbosa-Cánovas and Vega-Mercado, 1996; Toledo, 1991).

Solar drying, cabinet dryers, spray-drying, freeze-drying, osmotic dehydration, drum-drying, microwave, extrusion, fluidized-bed drying, and pneumatic drying, just to name a few, are the drying processes commonly used for food products. At present, the major dried foods include vegetables, fruits, milk, whey, and spices. Among the drying processes mentioned above, spray-, freeze-, and drum-drying processes, which are closely related to the production of foods as powders, are extensively covered in this chapter.

11.1. SPRAY DRYING

Spray drying is defined as the transformation of liquid state feed into a dried particulate form. The feed is atomized (i.e., sprayed) into a hot drying medium, resulting in moisture evaporation. The feed can be either a solution, suspension, or paste, while the final product is in the form of powders, agglomerates, or granules. Samuel Percy first patented the concept of spray drying in 1872, and its industrial application in milk and detergent production first occurred in the 1920s. Spray drying blossomed during World War II and is still undergoing continuous development. Spray drying has found extensive applications in all major industries, including the production of chemicals, pharmaceuticals, fragrances, or pesticides (Masters, 1985).

The spray drying process can produce free flowing particles of a spherical shape with a well-defined particle size distribution. In addition, its relatively short drying time, when compared with other drying processes, makes it suitable for drying heat sensitive materials. Spray drying is widely used in the food industry because many food products are heat sensitive, and powdered products are attractive to consumers (Oakley, 1997). Among dried food products produced by spray drying processes, the most popular are dairy products such as whole milk, skim milk, whey, and fat-enriched milk. Other food products dried by this process include coffee and tea extracts, baby foods, egg, cheese, enzymes, microorganisms, yeast, and ice cream mixes (Barbosa-Cánovas and Vega-Mercado, 1996).

Figure 11.1 shows a typical spray drying process. The drying process begins by pumping a liquid feed to an atomizer, which breaks up the feed into a spray of fine droplets and ejects it into the drying chamber. The spray is then contacted with and suspended by a heated drying medium (usually air), allowing the moisture to evaporate and the droplets to be transformed into dry particles

Table 11.1. General features of spray drying (adapted from Masters, 2002).

Process stage	Features	
1. Concentrate feed	• Concentration through evaporation	
2. Atomization		
	• Rotary	√ Wheel, disc
	• Nozzle	√ Pressure √ Pneumatic √ Sonic
	• Combined	√ Rotary—pneumatic
3. Spray-air contact	• Co-current flow chamber	√ Vertical with conical/flat base √ Horizontal (box) with V-trough base
	• Counter-current flow chamber	√ Vertical with conical base
	• Mixed flow chamber	√ Vertical with conical base
4. Drying of spray	• Moisture evaporation • Particle formation	
5. Dried product separation	• Primary discharge from drying chamber • Secondary discharge from particulates collector (cyclone, bag filter, electrostatic precipitator) • Total discharge from particulates collector • Final exhaust air cleaning in wet scrubber, dry police filter	
6. Dried product handling	• After (post) drying • Product cooling • Conveying • Packing, storage • Agglomeration • Dedusting • Coating • Fines recycling	

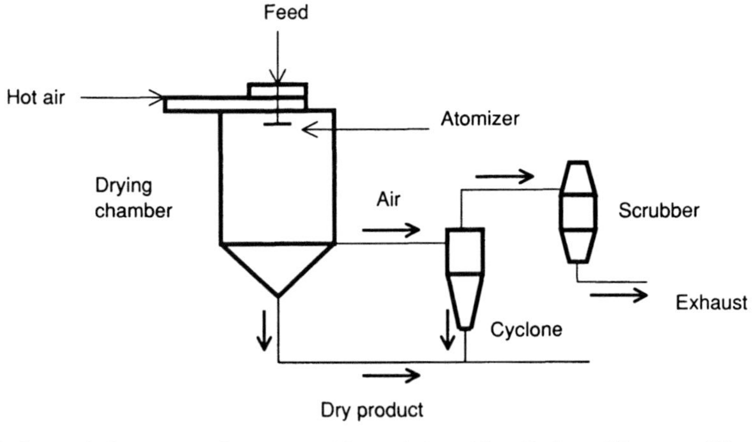

Figure 11.1. Typical spray drying (open cycle, co-current) layout (adapted from Barbosa-Cánovas and Vega-Mercado, 1996).

of almost the same shape and size. Finally, the dried particles are separated from the drying medium and collected as final products. The drying medium is cleaned by means of a cyclone or a scrubber and released to the environment or, in some cases, recirculated to the drying chamber. Table 11.1 shows the main features involved in spray drying, which will be described in this section.

The advantages of spray drying are: (a) it is possible to maintain constant powder specifications throughout the dryer when drying conditions are held constant; (b) it is an easy and continuous drying operation; (c) it is adaptable to full automatic control; and (d) a wide range of dryer designs are available to treat these difficult solutions and slurries that are heat sensitive, heat resistant, corrosive, and/or abrasive. High installation costs, low thermal efficiency, energy waste, and the handling of powder laden air at saturated or near saturated conditions are the main disadvantages of spray dryers (Masters, 1985).

11.1.1. Drying Process Layouts

Spray drying is a unique process in which particles are formed at the same time as they are dried. Many spray dryer designs are available to meet a wide range of product specifications. Although drying designs are diversified, each contains standard equipment that can be classified into four categories:

(a) *Air heaters* provide heat to the drying air with accompanying fans, air filters, dampers and ducts;
(b) *Atomizer*, which connects the feed supply system of pumps, tanks, and feed pre-treatment equipment and transforms the feed into a spray;
(c) *Drying chamber*, in which air contacts the sprayed feed and dries it. The drying chamber has air dispersers and outlets for the dried product and the exhausted air. Pictures of chambers set in different spray drying systems are shown in Fig. 11.2;
(d) *Equipment for product discharge, transport, packaging and air exhausting* provide full product recovery of dried products and final air cleaning. A complete air exhausting system contains fans, wet scrubbers, dampers and ducts.

(a) (b)

Figure 11.2. Examples of spray drying chambers: (a) industrial scale and (b) pilot scale (courtesy of *Niro Group* www.niro.dk).

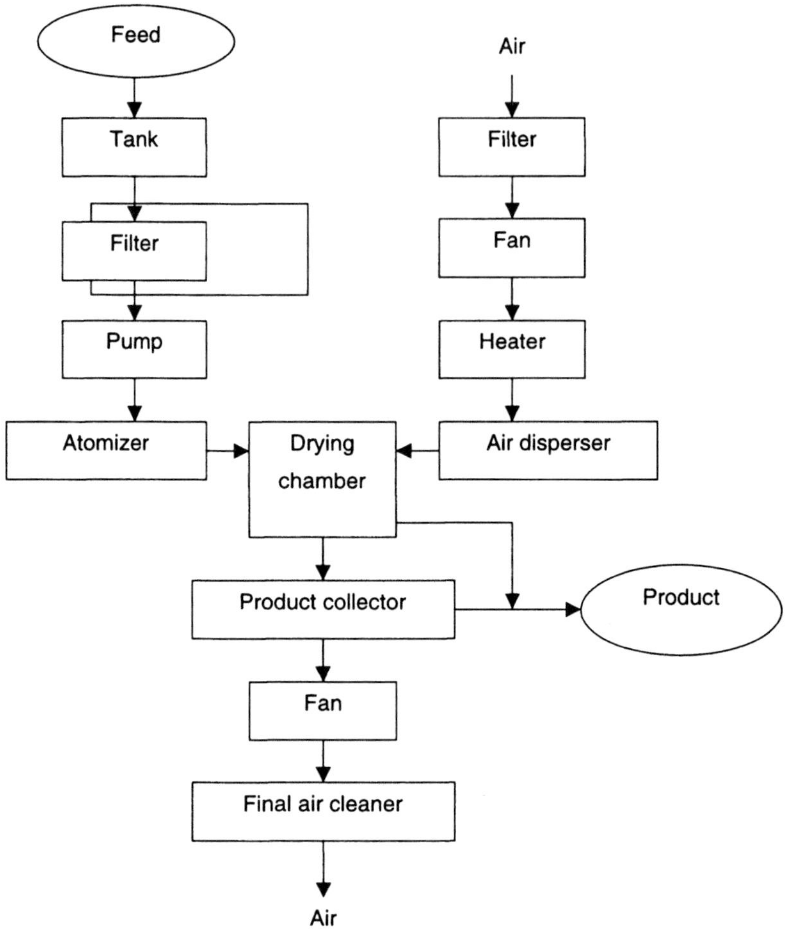

Figure 11.3. Flow chart of product and air in open-cycle layouts (adapted from Masters, 1985).

Different drying designs can be grouped based on their process layouts. Among these, open cycle, closed cycle, semi-closed cycle, multiple stage, or other special layouts such as aseptic layouts can be cited. The *open cycle* layout is the standard design and is most widely used. As shown in Fig. 11.3, the air is drawn from the atmosphere, heated, used once as a drying medium mixed in the drying chamber, cleaned by means of cyclones and scrubber, then released again to the atmosphere. This type of layout wastes energy contained in the heated air because the air is exhausted without recycling. The whole unit usually operates under a slight vacuum. Cyclones and scrubbers are explained in detail in Chapter 10.

The *closed cycle* layout features the recycling and reuse of the gaseous medium, which can be air or an inert gas such as nitrogen. This design has gas tight dryer component fabrication and is operated at a slight pressure to prevent any inward leakage of air. The flow chart of product and air in a closed cycle layout is shown in Fig. 11.4. This type of layout makes it possible to produce new products and has helped to solve problems associated with traditional production methods. A closed cycle design is usually chosen under the following conditions: (a) feed contains flammable organic solvents; (b) complete solvent recovery is required; (c) air pollution caused by odor, solvent vapor, and/or emission of particles is not permitted; (d) powder/air mixture is explosive; (e) explosive/fire

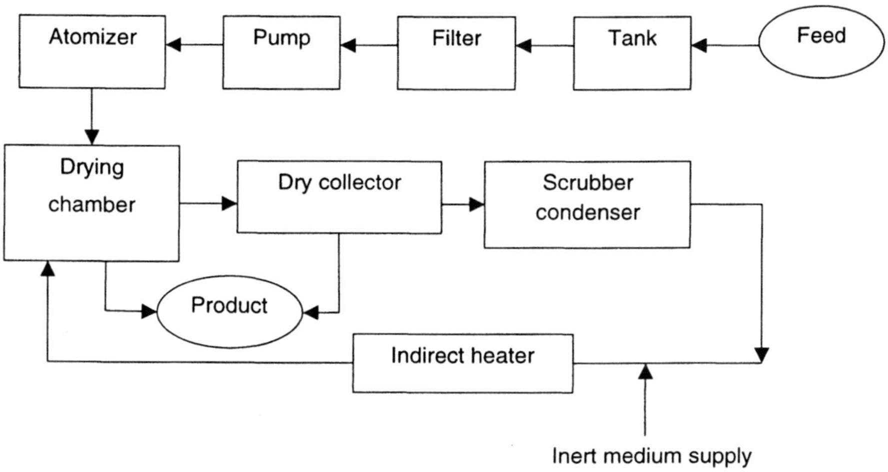

Figure 11.4. Flow chart of product and air in closed-cycle layouts (adapted from Masters, 1985).

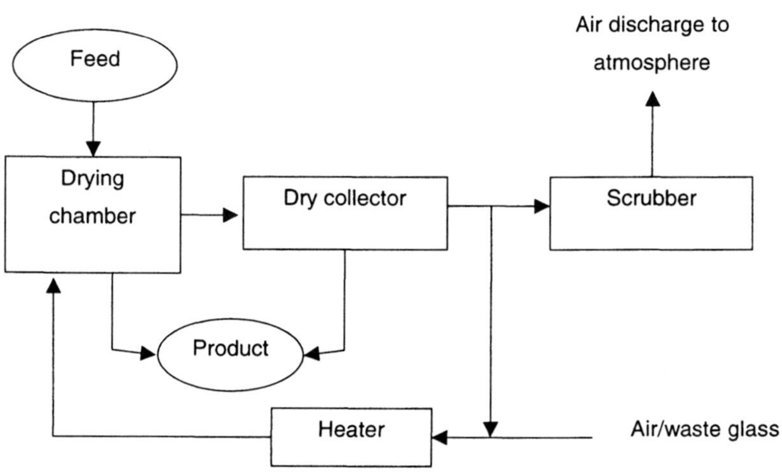

Figure 11.5. Flow chart of product and air in semi-closed-cycle layouts (adapted from Masters, 1985).

risk must be prevented; and (f) inert gas is used to prevent product oxidation (Masters, 1985). The energy efficiency in this type of layout is higher than that in the open cycle system. Furthermore, it is very environmentally friendly because the output is only the dried products.

Semi-closed cycle layouts were developed to improve thermal efficiency by utilizing waste heat in exhaust air leaving the dry collectors, enabling to handle materials of active, odorous, or explosive characteristics. They can be further regrouped based on whether the drying medium is partially or mostly recycled. The passage of product and air through semi-closed cycle spray dryers is shown diagrammatically in Fig. 11.5. The feed system is similar to that in Fig. 11.3. The recycled air can be heated again directly or indirectly before it re-enters the drying chamber. When the drying medium is heated directly, oil or gas may be used. However, when the dryers are used to process food products such as milk and coffee, direct heating is not used because nitrosamines, as well as other nitrogen-containing compounds, can be formed, resulting in undesirable food contamination (Land, 1991).

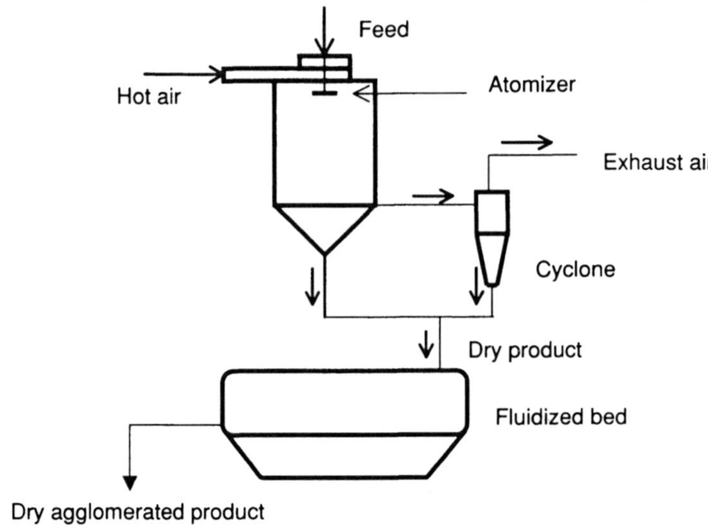

Figure 11.6. Two-stage layout (adapted from Barbosa-Cánovas and Vega-Mercado, 1996).

All layouts mentioned above are only one-stage drying layouts and represent the majority of spray dryers in operation. Although they are capable of producing dried products that meet quality specifications, the constant need for improved product quality and higher thermal efficiency has brought out the idea of developing a *two-stage* drying system. In two-stage layouts, as shown in Fig. 11.6, spray dryers like the one-stage layouts mentioned above are combined with a fluidized air bed acting as an agglomerator, after-dryer, or cooler. This type of layout is often used if it is necessary to (a) achieve very low moisture content in the final product; (b) improve the thermal efficiency; (b) lower the temperature of powder leaving the dryer; or (d) improve the physical properties of powder by classification or agglomeration.

The two-stage layout is widely used in the dairy industry to manufacture products with instant properties. The agglomeration of fines is carried out by operating the spray dryer in such a way that the powder is still slightly wet when leaving the drying chamber; agglomeration is promoted by the self-adhesion forces of particles. The moist powder is discharged from the conical base of the chamber and completely dried and cooled in an attached fluid bed. The final product has coarse particle size, few fines, and a high degree of flowability, wettability and dispersibility. With a few additional attachments, this type of layout can also be used to produce other instant food products, such as protein powder, instant beverage whiteners, powder with high fat content, and baby food formulas (Masters, 1985).

11.1.2. Atomization Classification

The formation of small droplets or spray, i.e., atomization, is the major step in spray drying. Actually, the atomizer is considered to be the heart of the spray dryers. Selection and operation of the atomizer is of supreme importance in producing a spray for optimum evaporation conditions and, subsequently, a final product that meets specifications. During atomization, 1 m^3 of liquid bulk can be broken up into approximately 2×10^{12} individual droplets of uniform 100 μm diameter and the total surface area over 60,000 m^2 (Masters, 1985). The high specific surface area of the spray allows high drying rates and low residence times.

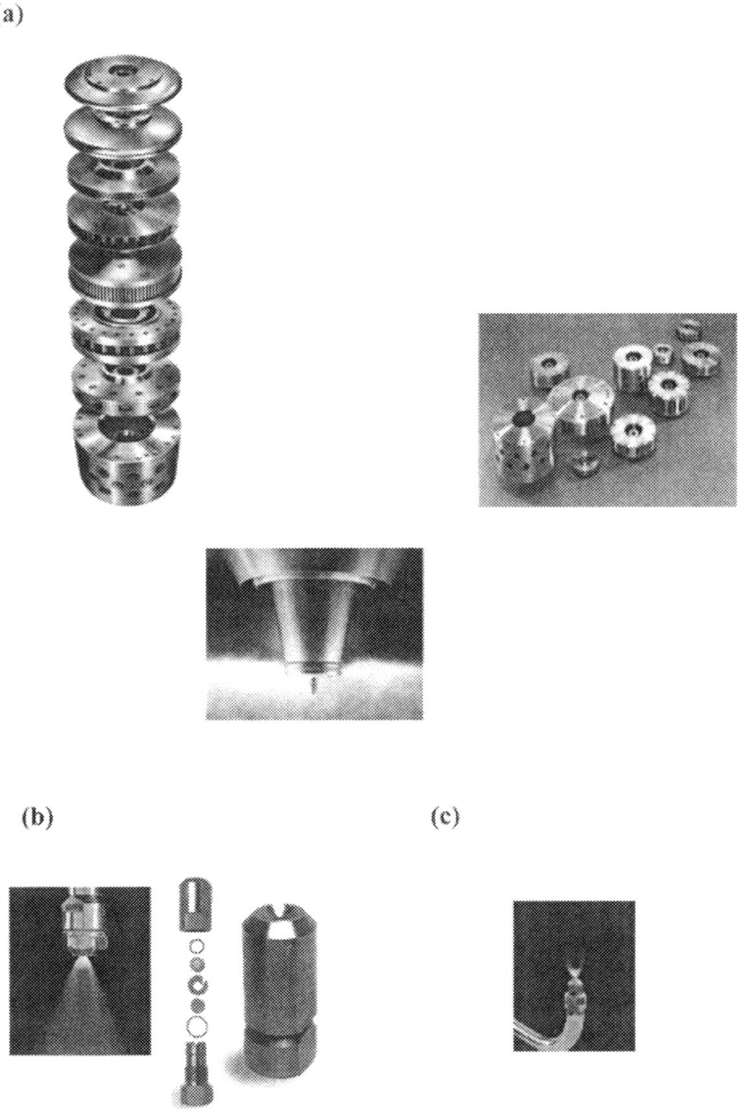

Figure 11.7. Common commercially available atomizers utilized in industrial drying: (a) centrifugal atomizers and atomizer wheels; (b) pressure atomizers; and (c) kinetic atomizers or two-fluid nozzle (courtesy of the *Niro Group:* www.niro.com, *KILBURN:* www.kilburnengg.com, *JICON:* www.jicon.com, and *Drytecheng Engineering:* www.drytecheng.com).

The type and efficiency of atomization affects not only the parameters of dryer design, such as chamber size, air temperature, exposure time, evaporation time, and drying efficiency, but also the final product properties, such as air content, moisture, bulk density, particle size and distribution, and reconstitutability (Hall and Hedrick, 1971). In any type of atomization, energy is needed to break up liquid bulk to create individual spray droplets. Even though there are different atomization techniques available, atomizers can be classified into four main categories according to the type of energy used to produce the spray: centrifugal, pressure, kinetic, and sonic. Figure 11.7 shows the atomizers that are currently available for processing.

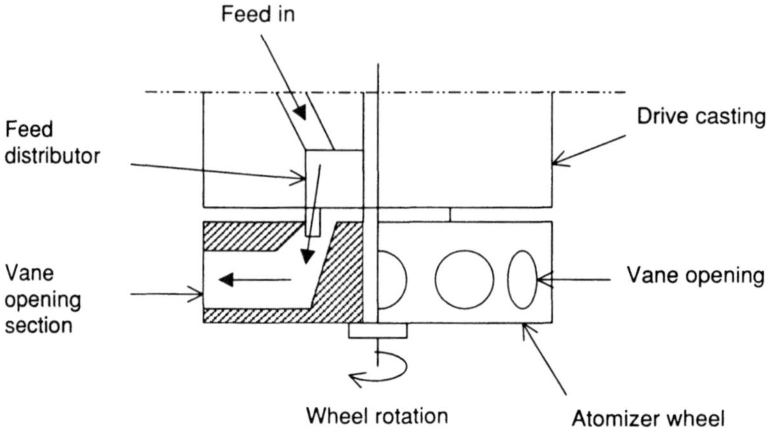

Figure 11.8. Schematic view of a wheel centrifugal atomizer (adapted from Masters, 1985).

11.1.2.1. Atomizers Using Centrifugal Energy

This type of atomizer is called rotary or centrifugal nozzles. A rotating wheel atomizer is shown schematically in Fig. 11.8. The energy used for atomization, which is supplied by an atomizer motor, rotates a wheel-, cup-, or disk-shaped nozzle at high speed. The feed is introduced onto the rapidly spinning nozzle, accelerated outwards over the surface, and ejected as a thin sheet of liquid that is subsequently broken into droplets. The mean particle size ranges from 30 to 120 μm, depending on the amount of energy transmitted to the liquid. The mass flow rate, viscosity, solid content, and surface tension influence particle size directly, but not to the degree that the wheel speed and its diameter do. Rotary nozzles normally operate at 5,000–25,000 rpm with wheel diameters of 5–50 cm. These devices are available in many sizes, ranging from laboratory units dealing with 1–10 liter liquid feed per hour to the largest commercial units dealing with 200,000 kg/h (Masters, 1985; Shaw, 1994).

Rotary atomizers operate under low feed pressure; hence, they are easy to operate and maintain. They are reliable and practically non-clogging, and can be run for long periods of time without operator interface. They can handle abrasive feeds and are resistant to wear. The feed rates may be very high without atomizer duplication, and particle size can be controlled by merely changing the rotation speed of the atomizer. The rotary atomizer is usually positioned at the center of the ceiling of the drying chamber and used only in co-current spray dryers (Land, 1991; Shaw, 1994).

Since the liquid is thrown horizontally, wide chambers are required to prevent droplets from hitting the walls before dried. In addition, rotary atomizers generate droplets of a broad particle size distribution, forcing the dryer to be relatively larger in diameter. The cylindrical height of dryers with rotary atomization is about equal to the chamber diameter (Land, 1991). The rotary atomizer is sophisticated and expensive when compared with other types of atomizers. Nevertheless, its ability to deal with difficult solutions and slurries encountered in spray drying makes it the first choice in many spray drying applications (Oakley, 1997).

11.1.2.2. Atomizers Using Pressure Energy

In this type of atomizer, the feed is forced through an orifice under pressure as a high speed film that readily disintegrates into a spray in the shape of a flat sheet to a 120° cone. A schematic view of a typical pressure nozzle is shown in Fig. 11.9. The orifice sizes are usually between 0.5 and 3.0 mm. Pressures from 1,500 to 5,000 psi (8.49–28.3 MPa) are commonly supplied by piston-type

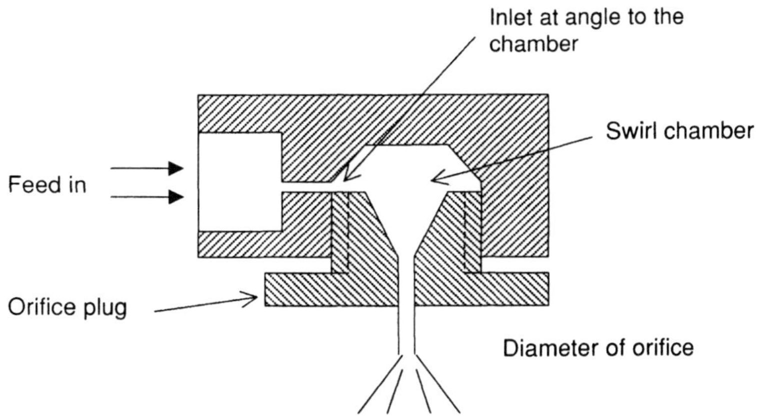

Figure 11.9. Pressure nozzle with a swirl chamber (adapted from Hall and Hedrick, 1971).

feed pumps. As a result, a single pressure nozzle can handle 750 kg feed per hour as an upper limit, depending on the pressure, orifice size, viscosity, and solids content. Pressure nozzles are generally used to form coarse particles of mean size 120–250 μm with the narrowest particle size distribution (Hall and Hedrick, 1971; Masters, 1985; Shaw, 1994).

The principle of pressure atomization is to rapidly supply enough energy to overcome the surface tension of the feed, because at high pressure, surface tension is the controlling factor in atomization. As the atomized feed moves through the air, atomization is further assisted by the friction force of the air (Hall and Hedrick, 1971). Feed rate and spray characteristics are controlled by changing the nozzle pressure. Mean size of droplets is proportional to the feed rate and viscosity and inversely proportional to nozzle pressure. The smaller the desired size of droplets, the smaller the orifice size used, and the higher the pump pressure is required to achieve the same mass flow rate through the nozzle (Masters, 1985; Shaw, 1994).

Pressure nozzles are applicable for feeds that do not contain large particles and have low viscosity. They are not as suitable for viscous liquids as rotary atomizers, due to their small orifices. They are small (i.e., a few centimeters in diameter), easily replaceable, simple to maintain, and low in cost. High pressure pumps are required. Furthermore, feed filtration must be effective in order to avoid orifice clogging. There are specific restrictions on the type of feed material to be used so that nozzle abrasion can be prevented. Nozzle durability has been prolonged considerably utilizing specially abrasion- and/or corrosion-resistant materials such as aluminum, tungsten, silicon, and chromium carbides to design the flow surface. Conversely, there are certain parts of the nozzle that might still be affected by abrasion and need regular replacement (Masters, 1985; Oakley, 1997).

Multiple nozzle dryers are used if the precise particle size control is not required and the desired mean particle size is quite large. In addition, it is practical to have dryers fitted with multiple nozzle arrangement because of their low space occupancy (Oakley, 1997). As many as 40 nozzles may be used in a very large spray drying system. Problems with one nozzle will not cause the whole operation to shut down in such multiple nozzle dryers (Shaw, 1994). When more than one nozzle is used, they must be arranged in a way such that the spray patterns are not overlapped to avoid droplet combination and uneven or difficult drying results (Hall and Hedrick, 1971).

Pressure nozzles can be used in co-current, counter-current, and mixed-flow systems. Usually, the cylindrical height is three to four times larger than the chamber diameter for a spray dryer equipped with pressure atomization (Land, 1991).

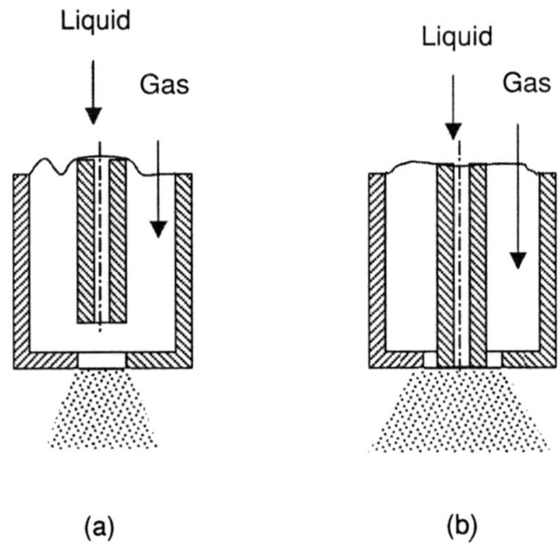

Figure 11.10. Pneumatic nozzles: (a) internal mixing type and (b) external mixing type (adapted from Filkova and Cedik, 1984).

11.1.2.3. Atomizers Using Kinetic Energy

Kinetic energy atomizers consist of a two-fluid nozzle atomizer with two phases being fed into the nozzle: the liquid feed and the gaseous atomizing medium; the atomization process is called pneumatic or two-fluid atomization. When the feed and a gaseous atomizing medium (usually air) are passed separately to the nozzle head, the liquid feed is broken down into small droplets due to high friction forces over the liquid surface induced by the high velocity gas. The air stream is rotated within the nozzle, contacting the feed either inside the nozzle (internal mixing) or outside the nozzle body (external mixing), as shown in Fig. 11.10. In internal mixing nozzles, the liquid phase is disintegrated inside the nozzle body and the spray has already formed at the orifice, while in external mixing nozzles, the liquid and gaseous phase meet outside the nozzle body and form the spray. The latter type is more commonly used (Filkova and Cedik, 1984).

Pneumatic nozzles can be classified into two groups according to the operating pressures: high pressure and low pressure nozzles. The former's operating pressure is between 0.3 and 1.0 MPa, with the specific air consumption ratio ranging from 0.3 to 1.0 kg/kg liquid feed, while the latter's operating pressure is under 0.01 MPa and specific air consumption ratio from 4 to 10 kg/kg liquid feed. Factors affecting the atomizing process include properties of the feed (surface tension, density, and viscosity), properties of the gas (velocity and density), direction of the gas, liquid turbulence degree, and ratio of liquid to gas. This type of atomizer is particularly useful for materials of high viscosity, which produce particles of medium coarseness (Filkova and Cedik, 1984; Masters, 1985).

The main advantage of this type of atomizer is that the liquid leaves the atomizer at a relatively low velocity and therefore has a shorter flight path, requiring a smaller drying chamber. This makes them ideal to be used in plot- or laboratory-scale drying processes. This atomization system is inexpensive in initial cost, but expensive to operate. Especially for medium or large production scales, the operating cost associated with providing a high velocity air stream makes these two-fluid nozzles prohibitively expensive. The energy they require is generally two to three times that required for pressure nozzles, another major disadvantage of the pneumatic nozzles (Hall and Hedrick, 1971; Oakley, 1997; Shaw, 1994).

11.1.2.4. Atomizers Using Acoustic/Pulsation Energy

This device belongs to the ultimate generation of atomizers, and it is still not applied in industrial applications (Masters, 2002). A sonic generator is part of the nozzle head; when the feed passes through the head it makes the liquid break up into droplets. Due to capacity restrictions, unpredictable continuous operation, low rate feeds and acoustic environmental problems, sonic energy can be combined with pneumatic nozzle principles into a gas dynamic atomization and pulse combustion system. During atomization, the combustion system within the atomizer heats the drying air prior to contacting the feed in a high resonance environment. This form of atomizer imposes no shear on the product during the liquid disintegration process into droplets, and immediate contact with hot air followed by quenching with a cooler air stream ensures high drying rates without heat degradation (Masters, 2002). Droplet size characteristics are similar to pneumatic nozzles, and through control of the pulsating stage, both fine and coarse spray dried particles can be produced.

Choosing the best atomization method suitable for a particular application is a difficult task. Marshall and Seltzer suggested comparing these atomization methods on the following basis: (a) capacity in terms of square feet of surface area per pound of liquid per minute; (b) power on the basis of square feet of surface area produced; (c) particle size distribution; and (d) weight flow distribution curve. By analyzing each atomization method and considering all the factors related to the drying process, one of these methods may be superior for the application, but it is impossible to say that one is superior for the majority of applications (Hall and Hedrick, 1971).

No matter what kind of atomization method is used, the atomizer must be positioned inside the drying chamber and operated so that the droplets have intimate contact with the hot drying medium and can be dried sufficiently before they touch the chamber wall. Different atomizer positions are shown in Fig. 11.11.

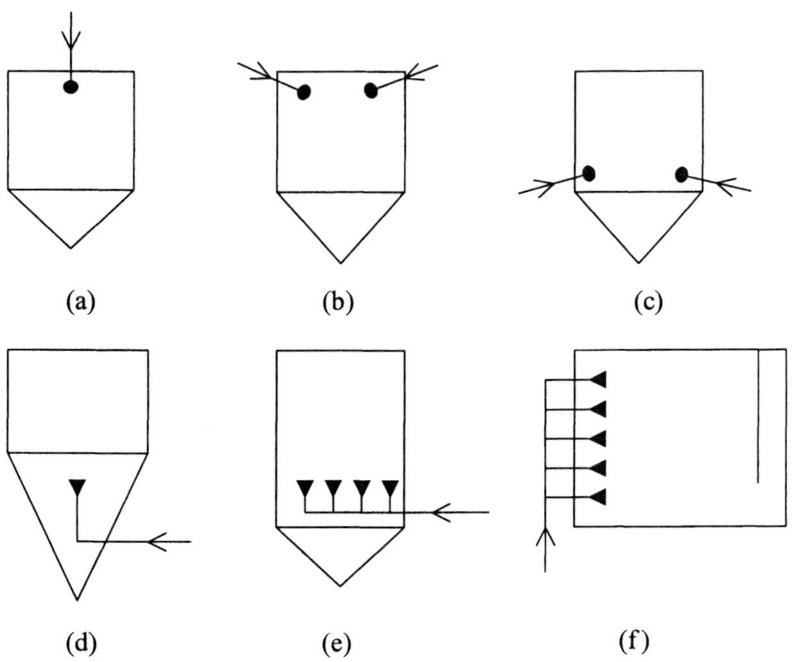

Figure 11.11. Different atomizer positions in the drying chamber: (a) top; (b) side top; (c) side base; (d) middle; (e) base; and (f) end (adapted from Masters, 1985).

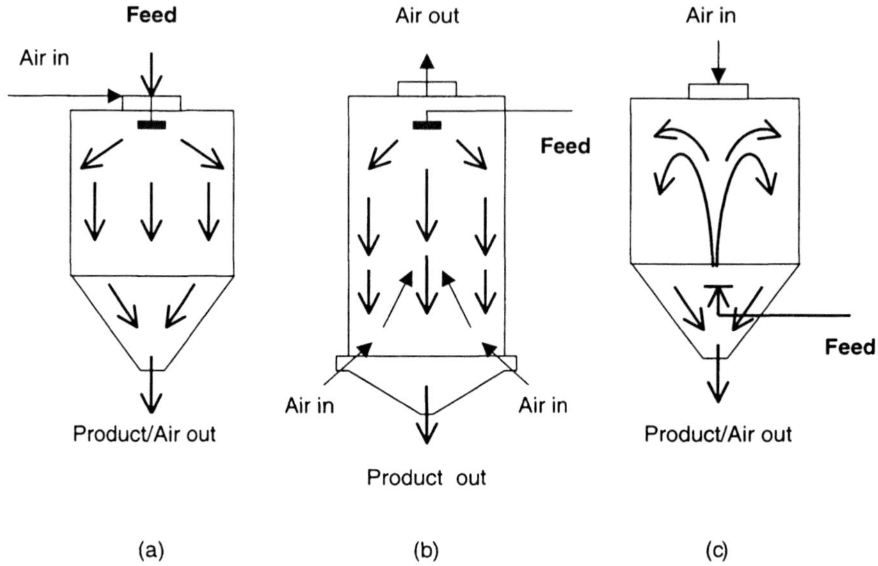

Figure 11.12. Typical product–air flow patterns in spray dryers: (a) co-current; (b) counter-current; and (c) mixed flow (adapted from Masters, 1985).

11.1.3. Spray-Air Movement

Spray-air contact pattern is determined by the relative position between the atomizer and the drying air inlet. The ideal resulting spray-air movement provides droplets with sufficient residence time in the drying chamber to obtain the desired moisture level without heat degradation and unwanted wall deposits. There is no general relationship to express the effect of operating and design variables on spray-air movement because so many variables are involved in the drying processes. Spray-air movements are usually classified as co-current, counter-current, and mixed flow even though these designations are not a true representation of actual conditions (Masters, 1985). They are schematically shown in Fig. 11.12.

In co-current flow drying chambers, the droplets and air pass through the dryer in the same direction. The atomizer and the inlet air disperser are located at the same position in the chamber, at the bottom, top, or end. The spray of feed encounters the drying medium of the highest temperature, which causes rapid surface evaporation, while it is still wet. Thus, the co-current flow arrangement is particularly suitable for heat sensitive materials, since they will be protected from the high inlet gas temperature by evaporative cooling (Oakley, 1997). By the time evaporation slows down and is limited by diffusion of liquid from the center of the droplets to the surface, the particles have reached a cooler region of the drying chamber. Therefore, heat sensitive materials can be spray dried in a drying medium of elevated temperatures, even though such temperatures would cause quality damage on the products in other drying processes, such as using an oven or drum dryer.

In the counter-current flow pattern, spray and drying media enter at opposite ends of the drying chamber, offering a most efficient utilization of heat available in the drying chamber. This type of flow pattern is mainly restricted to use with pressure nozzle atomization in tall, narrow diameter towers, since the upward streamline of the drying medium reduces the downward velocity of the larger droplets in the spray, resulting in longer residence time for complete evaporation in the drying chamber. The driest particles are exposed to the hottest drying medium; thus, this design can only be used for non-heat sensitive materials and is less commonly used than the co-current flow arrangement (Masters, 1985; Oakley, 1997).

Mixed-flow is a combination of co-current and counter-current flow patterns. A nozzle is positioned in the bottom half of the chamber, forcing the spray travel to upward until overcome by gravity and the downward flow of drying medium. The spray then flows in the reverse direction and finally falls to the bottom cone of the chamber. The advantage of this arrangement is that the spray has a long path length through the chamber, making it a good method to dry relatively coarse droplets in a small chamber at small production rates (Oakley, 1997). In mixed flow designs, the drying begins in a cooler part of the dryer chamber and continues in the hottest zone; thus, it is not suitable for heat sensitive materials because the evaporative cooling effect is lessened for partially dried particles and the chance of thermal degradation becomes greater (Shaw, 1994).

11.1.4. Mass and Heat Balances

For evaluating the performance of spray dryers, air-product flow and temperature data are available from heat and mass balance calculations. Based on an assumption of no product accumulation for a continuous operation, the mass input (from both the air and the product) per unit time equals the mass output, and the heat input equals the heat output plus the heat loss in the drying chamber. Thus, the moisture balances around an open cycle, co-current flow drying system as shown in Fig. 11.13, can be expressed as:

$$M_{\mathrm{P}} W_{\mathrm{P1}} + G_{\mathrm{A}} H_{\mathrm{A1}} = M_{\mathrm{P}} W_{\mathrm{P2}} + G_{\mathrm{A}} H_{\mathrm{A2}} \tag{11.1}$$

or

$$M_{\mathrm{P}}(W_{\mathrm{P1}} - W_{\mathrm{P2}}) = G_{\mathrm{A}}(H_{\mathrm{A2}} - H_{\mathrm{A1}}) \tag{11.2}$$

where M_{P} is the flow rate of dry solid in a feed with a moisture content of W_{P1} (dry basis) entering the dryer, G_{A} is the flow rate of dry air of absolute humidity H_{A1} at the inlet, W_{P2} is the moisture content of the dried solid leaving the dryer, and H_{A2} is the absolute humidity of the air leaving the dryer.

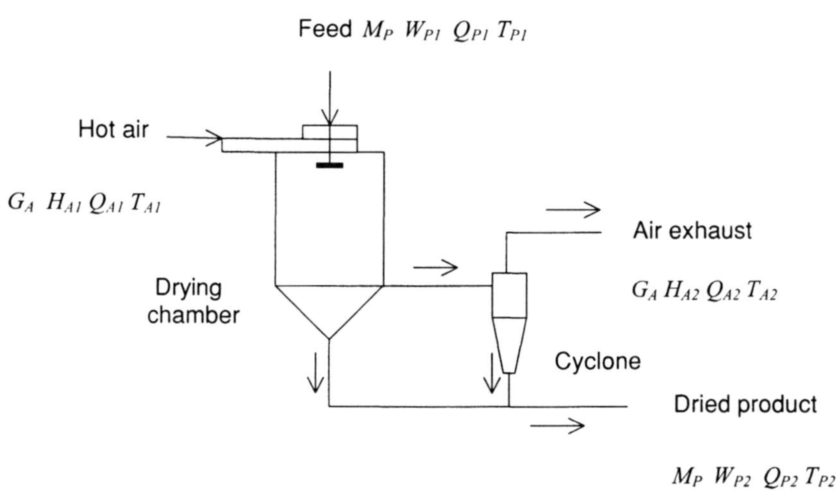

Figure 11.13. Flow chart of an open cycle, co-current spray drying layout for calculation of mass and heat balance.

The heat or enthalpy balance around the system leads to:

$$M_P Q_{P1} + G_A Q_{A1} = M_P Q_{P2} + G_A Q_{A2} + Q_L \tag{11.3}$$

where Q_{P1} and Q_{P2} are the enthalpies of the solid entering and leaving the system, Q_{A1} and Q_{A2} are the enthalpies of air entering and leaving the system, and Q_L is the heat loss in the drying chamber.

Thermal efficiency is related to the heat input required to produce a unit weight of dried product of desired specifications. Achieving the dried product of desired specifications at the highest possible thermal efficiency is the main purpose of dryer design. Two types of thermal efficiencies, overall thermal efficiency and evaporative efficiency, are useful to evaluate spray drying performance. The overall thermal efficiency η_{overall}, defined as the ratio of heat used in evaporation to the total heat supplied to the dryer, can be approximately calculated as:

$$\eta_{\text{overall}} = \frac{T_H - T_C}{T_H - T_0} \tag{11.4}$$

where T_H is the temperature of the hot air entering the dryer, T_C is the temperature of the cooled air exiting the dryer, and T_0 is the atmospheric air temperature.

The following data show that overall efficiency is greatly affected by the inlet hot air temperature when the other two temperatures are fixed. When T_C is 85°C and T_0 is 20°C, the overall efficiency is 45.8% for T_H at 140°C, 75% at 280°C, and 89.8% at 660°C. Although higher inlet air temperatures mean higher overall efficiencies, there are limitations for the highest inlet temperature based on some considerations: (a) high inlet temperature can cause severe heat effects on the atomized spray, resulting in loss of product quality; and (b) from an economic point of view, the maximum inlet temperature should be controlled below a certain level to avoid extra cost for air heating (Masters, 1985).

Evaporative efficiency is defined as the ratio of the actual evaporative capacity to the capacity obtained in the ideal case of exiting air at saturation:

$$\eta_{\text{overall}} = \frac{T_H - T_C}{T_H - T_{\text{sat}}} \tag{11.5}$$

where T_{sat} is the adiabatic saturation temperature corresponding to the inlet air temperature T_H.

Despite its popularity in many drying processes, the spray drying process remains one of the most difficult to design because of the complexity of atomization and spray-air interaction. High capital costs make it impractical to test the spray dryer on full-scale and scale-up dryer designs; in the conventional sense it is not possible because it is difficult to independently control the initial spray momentum, drop size, and drying rate. Traditional spray dryer design uses empirical design rules based on pilot plant work and the past experience of dryer manufacturers. This approach works well for established designs, but makes innovation extremely risky (Oakley, 1993).

11.2. FREEZE-DRYING

Freeze-drying, or liophilization, was first used by Altmann in 1890 to prepare organ and tissue samples for histological examination under microscope. In the 1940s, it was initially introduced for the production of dry plasma and blood products on a large scale. Today, freeze-dried plasma is well accepted, not only as equivalent to, but also superior, when compared with stored liquid plasma. In the food industry, freeze-drying, followed by ordinary vacuum drying of fish fillets, was carried out in 1954. Freeze-dried coffee was first introduced into the market in 1964 with great success. Freeze-drying has made a considerable impact on the food preservation field in recent years, and has been shown to be an attractive method for extending the shelf life of foods (Mellor, 1978; Nair and Sivetz, 1973; King, 1975).

Freeze-dried products are dry, light, and porous, almost retaining their original shape and texture. Properly packaged freeze-dried products can be stored for more than a year and retain most of the physical, chemical, biological, and organoleptic properties of their fresh state. Pharmaceuticals, biological materials, and foodstuffs are the three broad categories to which freeze-drying methods are often applicable, especially when some of these materials may not be heated to moderate temperatures in conventional drying processes (Mellor, 1978).

The freeze-drying process mainly consists of two steps: (a) the product is frozen; and (b) the product is dried under sublimation of ice under reduced pressure. Being a cold process, freeze-drying is especially useful for drying heat sensitive foods, such as coffee and tea extracts, in which the volatile compounds responsible for aroma and flavor are easily lost during ordinary drying operations. People can tell the taste differences between the freeze- and spray-dried instant coffees available on the market. The low drying temperatures used in the process minimize the degradative reactions that usually occur in ordinary drying processes (Barbosa-Cánovas and Vega-Mercado, 1996; Geankoplis, 1993; Mellor, 1978; Land, 1991).

The obvious advantage of freeze-drying is that the ice structure in the product during the drying process minimizes shrinkage of the product and thus promotes a rapid and nearly complete rehydration (Bruin and Luyben, 1980). Therefore, it is also very useful for drying some food materials (e.g., vegetables, fruits, meat) to be used for instant foods, such as soup mixes, whose good reconstitution properties are highly desired. Their good reconstitution properties allow them regain their original shape and structure rapidly by the addition of liquid, exhibiting the characteristics similar to those of fresh products. With the exception of instant coffee and soup mixes, many commercially freeze-dried food products are mainly used for military and camper's rations (Mellor, 1978).

Traditional dehydration processes usually cause physical and structural changes in the dried products due to heat application. Among these changes, shrinkage of cells, loss of rehydration ability and wettability, migration of solids, case hardening, and loss of volatile aroma components, are some of the most common examples. In fact, caramelization, discoloration, loss in texture and physical form, loss of volatile flavoring characteristics, and poor rehydration ability of many dried foods have left an imprint on the mind of consumers (Desrosier, 1977).

The freeze-drying method is included in this chapter mainly because of its popularity in preparation of dehydrated food powders from their liquid state. The product's porous structure resulting from ice sublimation provides excellent instant properties to freeze-dried food powders. Convenience factors, changing life style and improved product quality have given increased opportunities for utilization of freeze-drying for the preparation of instant food powders of high quality. Coffee and tea are the most important products when freeze-dried liquid extracts are considered. Other freeze-dried food powders include fruit juices (such as citrus, apple, grape, and pineapple), whole egg, flavor concentrates, and even beer (Flink, 1975).

Freeze-drying is recognized as the best method of producing dried food products of the highest quality. However, it is very expensive because of its slow drying rate and the use of vacuum. Energy costs are high because the material must be frozen first and heat supplied to sublime the ice and bound water. The long drying time is caused by resistance to heat and mass transfer, and other factors. In addition, it is difficult to set a heat-removal system that forms a homogeneous ice crystal distribution throughout the frozen-food; a fact that adds to preventing its extensive use (Geankoplis, 1993; Land, 1991; Liapis and Marchello, 1984).

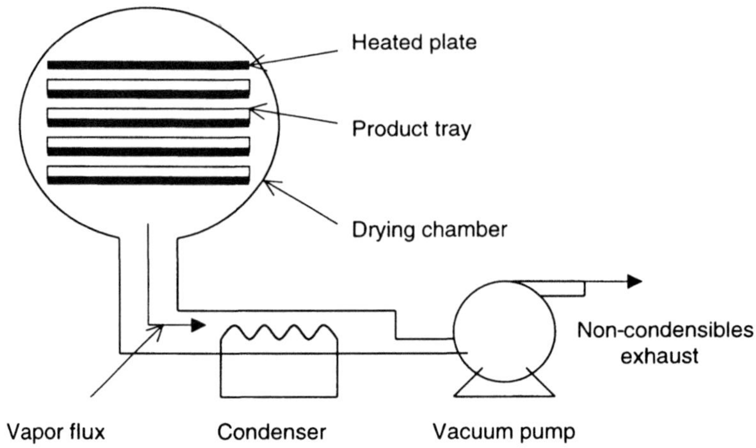

Figure 11.14. A common freeze-drying system (adapted from Liapis and Marchello, 1984).

11.2.1. Freeze-Drying Process

In freeze-drying, water is first removed from the frozen material as a vapor by sublimation in a vacuum chamber and then recovered by mechanical vacuum pumps or steam jet ejectors (Geankoplis, 1993). Freezing of the product, sublimation of ice, and removal of water vapor are the three main steps in a freeze-drying operation. A schematic diagram of a common freeze-drying system is shown in Fig. 11.14. The frozen material is loaded on the top of heated plates that provide the energy for ice sublimation and "bound" water desorption. Heat transfer occurs by conduction from the heated plates, convection from the air inside the drying chamber to the exposed surfaces, and radiation. The vacuum pump is used to evacuate the drying chamber at the start of the operation and to remove the non-condensing gases and any air that leaked into the system. The absolute pressure inside the chamber is determined by the temperature at which the vapor trap is maintained. When the ice starts to sublime, the water vapor is transported through the chamber to the refrigerated condenser that prevents the return of the water vapor to the product and also reduces the volume of gases to be removed from the system by the vacuum pump (Liapis and Marchello, 1984; Toledo, 1991).

The liquid to be evaporated is usually water, but sometimes solvent. Figure 11.15 shows the phase diagram of water at different temperatures and pressures. Since the vapor pressure of ice is very small, very low pressure or high vacuum is required for freeze-drying. As can be seen from the diagram, if water is in a pure state, the absolute pressure in the drying chamber must be lower than 627 kPa and the temperature less than or equal to 0°C in order to carry out the drying process successfully. In the case of food, the water usually exists in a solution or combined state; the frozen temperature of the food material is lower than that of the pure water, and the material must be cooled below 0°C to keep the water in a solid state. Most freeze-drying is carried out at a temperature of −10°C and absolute pressure of 2 mm Hg or less (Barbosa-Cánovas and Vega-Mercado, 1996; Land, 1991; Geankoplis, 1993; Okos et al., 1992).

A systematic freeze-drying process for food products includes five successive steps (Barbosa-Cánovas and Vega-Mercado, 1996; Land, 1991):

(1) *Preparation and pre-treatment*: Food materials are often cut, ground, blanched, precooked, or preconcentrated at the beginning of the drying process. Food extracts, which may be obtained by pressing, pulping, mixing, or contacting the raw material with a liquid phase

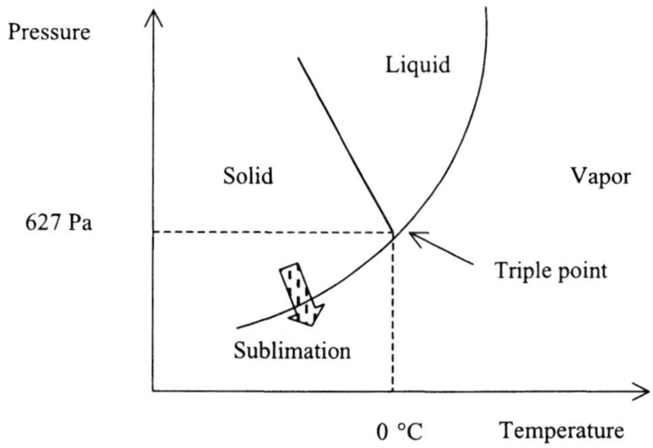

Figure 11.15. Phase diagram of water (adapted from Barbosa-Cánovas and Vega-Mercado, 1996).

(classical definition of extraction), have total solids content of 8–15%. Thus, it is desirable to increase the solids content to 30–50% by implementing a preconcentration step, thus improving the economy of the process (Flink, 1975).

(2) *Pre-freezing*: Minimum temperature and freeze rate are the two important variables in this process. Food materials have lower freezing temperatures than that of pure water. The structure, consistency, color, and aroma retention of the final product are greatly affected by the freezing rate. The freezing rate is usually in the range of 0.5–3 cm/h. The entire mass should become rigid, consisting of ice crystals and food components, at the end of this step.

(3) *Primary drying*: This refers to the ice sublimation under vacuum that happens when the energy for the latent heat is supplied. The driving force for sublimation is essentially the pressure difference between the water vapor pressure at the ice front and the partial pressure of water vapor in the drying chamber. The initial drying rate is high because of the low resistance to both heat and mass transfer at the beginning. As drying proceeds, the dry layer around the frozen material serves as an insulation material, blocking the heat transfer to the ice front and the mass transfer from the ice front.

(4) *Secondary drying*: This begins when there is no more ice in the product and the bound water in the drying material needs to be removed. The temperature and pressure in the drying system should be the same as those for the primary drying. Usually, the drying process kept going until the residue moisture is in the range of 2–3%. For food products, 2–10% moisture content has been recommended (Mellor, 1978).

(5) *Packaging*: Dried food products should be packed properly to prevent moisture pick-up and lipid oxidation reactions. They are commonly packaged in foil or cans (King, 1975).

There are two major characteristics of freeze-drying processes. First, the drying process is carried out under vacuum conditions, so there is a virtual absence of air. The absence of air prevents deterioration due to oxidation or chemical modification of the products during drying. Secondly, drying temperature is lower than ambient temperature. Therefore, products that decompose or undergo changes in structure, texture, appearance, and flavor as a consequence of high temperature can be dried under vacuum with minimal physical and chemical damage (Barbosa-Cánovas and Vega-Mercado, 1996).

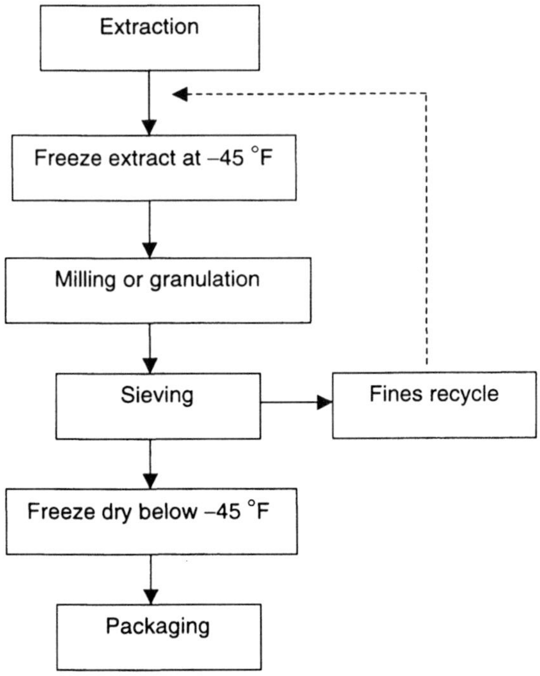

Figure 11.16. Flow chart for freeze-dried coffee production (modified from Nair and Sivetz, 1973).

A typical process for freeze-dried coffee production is illustrated in Fig. 11.16. The coffee extract is prepared by a standard method and concentrated by low temperature evaporation to contain at least 30% of soluble solid. The concentrated extract is then frozen to about −45°F at a rate that gives a desired color; the slower the freezing rate, the darker the color. Four different methods are available to freeze the coffee extract: use of a moving stainless steel belt, brine-cooled on the upper side by a cold air blast from above; a moving Teflon-coated rubber belt in a −45°F room; Teflon-coated aluminum trays placed in a −45°F room, or a vertical rotating refrigerated cylinder. The frozen extract is then passed through a milling and sieving system to give the final product the physical appearance of roasted and ground coffee. The frozen and dried coffee extract will have almost the same particle size, except that there is a small amount of attrition during drying, handling and packaging processes. After milling and sieving, the coarser particles are sent for freeze-drying, while the fines are recycled (Dalgleish, 1990; Nair and Sivetz, 1973).

11.2.2. Fundamentals of Freeze-Drying

11.2.2.1. Freezing

Freezing is the initial step of freeze-drying. At this stage, the removal of sensible heat in chilling occurs first, followed by the removal of the latent heat of freezing (Toledo, 1991). Once the product is frozen, its freezing conditions are maintained until the drying process is started. Freezing processes are classified into two groups: (a) direct contact cooling and (b) indirect contact cooling. In direct contact cooling, every effort is made to bring the refrigerant into contact with as many product surfaces as possible, while in indirect contact cooling, the product is separated from the refrigerant by some type of barrier, such as a plate or product package surfaces (Heldman, 1992).

Table 11.2. Values of characteristic size D and shape constants P and R for different shaped solids in Plank's model for freezing time calculation (modified from Heldman, 1992; Toledo, 1991).

	D	P	R
Infinite slab	Thickness	0.5	0.125
Sphere	Diameter	0.167	0.04167
Infinite cylinder	Diameter	0.24	0.0625
Rectangular brick	Smallest dimension	—	—

Freezing time, the basic design criteria for freeze-drying, is defined as the time required to reduce the initial product temperature to some fixed final temperature at the slowest cooling location. A full analytical solution of the freeze drying rate is difficult to obtain because of the variation of physical properties with temperature, the amount of frozen and unfrozen zones varying with temperature, and many other factors, (Geankoplis, 1993). Numerous equations and approaches have been proposed, and an approximate model developed by Plank in 1913 is the best known and most used, even though it is not an accurate way to predict freezing time. However, the Plank's equation is simple and accurate enough for engineering purposes. Its general form may be expressed as:

$$t_{\mathrm{f}} = \frac{\lambda\rho}{T_{\mathrm{f}} - T_{\mathrm{a}}}\left(\frac{PD}{h_{\mathrm{C}}} + \frac{RD^2}{k}\right) \tag{11.6}$$

where t_{f} is the freezing time for a solid (s), T_{f} is the freezing temperature (K), T_{a} is the environment temperature (K), D is the characteristic size of the solid (m), λ is latent heat of fusion (kJ/kg), ρ is density of the unfrozen material (kg/m^3), h_{C} is convective heat transfer coefficient (J/(s m^2 K)), k is thermal conductivity (J/(s m K)), and P and R are shape constants. Values of D, P, and R for solids of different shapes are listed in Table 11.2 (Toledo, 1991; Geankoplis, 1993).

The most obvious difficulties related to the application of Plank's model come from the selection of a latent heat magnitude and a value for the thermal conductivity. In addition, this model does not account for the time required for sensible heat removal (Heldman, 1992)

Freezing has a great influence on the final quality of the product. The two important independent variables are the freezing rate and the minimum temperature. The structure, consistency, color, and aroma retention of the final product are greatly affected by the freezing rate (Land, 1991). Slow freezing results in large ice crystals, causing cellular damage, while fast freezing results in smaller ice crystals, causing less cellular damage for such structured foods as vegetables and fruits. It is worthwhile to use slow freezing to obtain large crystals, resulting in shorter drying times, but a lower quality in the rehydrated product (Persson, 1975). The selection of fast or slow freezing is determined based in the characteristics of the product. In general, the faster the freezing rate, the longer the freeze-drying time, and the better the quality of the final product (Mellor, 1978).

Minimum temperature can be determined by the eutectic point, at which time the whole mass should become rigid. Ideally, a frozen mass in a rigid eutectic phase is necessary to ensure that the primary drying occurs by sublimation only, and is not combined with partial evaporation from a liquid state. However, the last traces of water in the mass do not have to be frozen (Land, 1991; Mellor, 1978).

11.2.2.2. Ice Sublimation

At the beginning of freeze-drying, the original material is composed of a frozen core of materials. The plane of sublimation starts at the outside surface and recedes to the inner layer, leaving a porous shell of dried material through which the water vapor diffuses and goes to the atmosphere in the

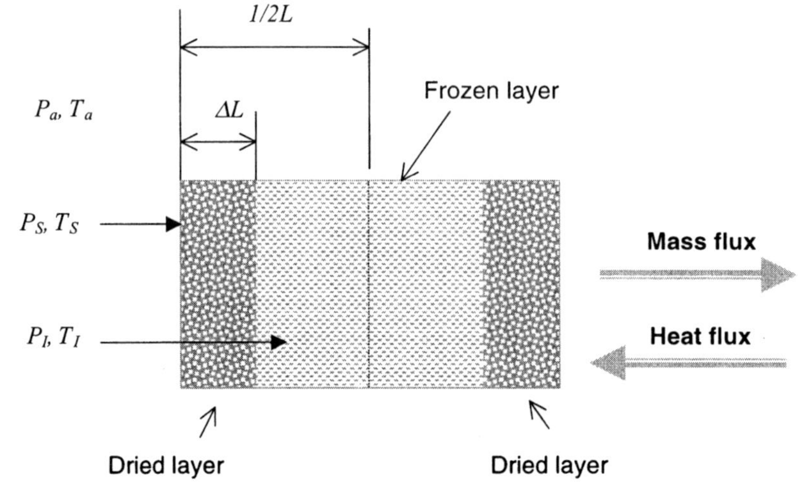

Figure 11.17. Schematic representation of freeze-drying of a simplified model for heat and mass transfer calculations (adapted from Geankoplis, 1993).

drying chamber. The heat used for sublimation is first transmitted to the dried surface of the product by conduction, convection, and/or radiation from the gas phase, and then reaches the icy layer by conduction. In some cases, ice sublimation may also be promoted by heat conducted through the frozen material. Thus, heat and mass transfer are occurring simultaneously (Geankoplis, 1993).

Assuming a pseudo-steady state, the heat flux Q to the sublimation surface shown in Fig. 11.17, which is equal to that conducted through the dry solid layer, is given by:

$$Q = E(T_a - T_S) = k_d \frac{T_S - T_I}{\Delta L} \tag{11.7}$$

where E is external heat-transfer coefficient (J/(s m^2 K)), T_a, T_S, and T_I are the temperature of the gaseous environment, the surface of the dried solid, and the ice front or sublimation front (K), respectively, k_d is the thermal conductivity of the dried layer (J/(s m K)), ΔL is the dried layer thickness (m), and L is the thickness of slab (m). The heat flux has the unit of J/(s m^2).

The mass flux of water vapor M_A from the front ice layer is given by:

$$M_A = \frac{D}{RT\Delta L}(P_I - P_S) = F(P_S - P_a) \tag{11.8}$$

where D is the average effective diffusivity of water vapor in the dry layer (m^2/s), R is the gas law constant (82.057×10^{-3} m^3 atm/(kg mol K)), T is the average temperature in the dry layer (K), and P represents partial pressure of water vapor, in the air outside the solid (P_a), at the dry layer surface (P_S), and at the ice surface in equilibrium with the ice sublimation (P_I, atm), respectively, and F is the external mass-transfer coefficient (kg mol/(s m^2 atm)). M_A has the unit of kg mol/(s m^2).

The two equations mentioned above may be rearranged to express the heat flux and mass flux in terms of the external operating conditions as follows:

$$Q = \frac{T_a - T_I}{1/E + \Delta L/k_d} \tag{11.9}$$

$$M_A = \frac{1}{1/F + RT\Delta L/D}(P_I - P_a) \tag{11.10}$$

The constants E and F are determined by the gas velocities and characteristics of the dryer, while k_d and D are determined by the nature of the dried material. T_a and P_a are set by the external operating conditions. Equations (11.9) and (11.10) can be related through the latent heat of ice sublimation ΔH_S (J/(kg mol)):

$$Q = \Delta H_S M_A \tag{11.11}$$

Substituting Eqs. (11.9) and (11.10) into (11.11) gives:

$$\frac{T_a - T_I}{1/E + \Delta L/k_d} = \frac{\Delta H_S}{1/F + RT\Delta L/D}(P_I - P_a) \tag{11.12}$$

From these equations, it should be noted that temperature and pressure are the two major control variables. The drying rate is increased by raising T_a and hence T_S, but there are some important constraints for temperature increase. First, the surface temperature T_S must not be too high to cause thermal degradation. Second, the ice front temperature T_I must be kept well below the melting and/or collapsing point (Geankoplis, 1993).

The dried layer is highly insulating to heat flux, but usually less resistive to mass transfer at low pressures. To further increase the drying rate, k value, i.e., the thermal conductivity of the dried solid, must be raised. Hence, the freeze-drying process is considered to be heat-transfer controlled. Because the dried layer's resistance to heat and mass flux is a function of pressure, it is possible to raise the pressure from the minimum value to enhance heat transfer at the expense of increasing resistance to mass transfer. However, the overall drying rate may be increased because the water vapor driving force is increased (Geankoplis, 1993; Liapis and Marchello, 1984).

Finally, the freeze-drying rate is related to the mass flux of water vapor M_A by:

$$M_A = \frac{L}{2m_w V_S}\left(-\frac{dZ}{dt}\right) \tag{11.13}$$

where m_w is the molecular weight of water, L is the total thickness of the solid (m), V_S is the volume of the solid occupied by a unit kg of water ($V_S = 1/X_0\rho_S$, X_0 is the initial moisture content in the unit of kg water/kg dry solid, ρ_S is the bulk density of the dry solid in kg/m^3), and Z is the material fraction containing the initial moisture at time t. Fraction Z is the ratio of the length of the remaining frozen layer with respect to the total length L, i.e., $Z = (L - 2\Delta L)/L$. When ΔL equals 0, then Z is 1.

By combining Eqs. (11.9), (11.11), and (11.13), substituting ΔL in terms of Z[being $\Delta L = (1 - Z)(L/2)$], separating variables, and integrating time t and fraction Z within the intervals $[0, t_f]$ and $[1, Z']$, respectively, the necessary drying time in order to dry a fraction Z' can be calculated from:

$$t_f = \frac{\Delta H_S}{2V_S m_w(T_a - T_I)}\left[-\frac{Z' - 1}{E} + \frac{L}{2k_d}\left(\frac{Z'^2}{2} - Z' + \frac{1}{2}\right)\right] \tag{11.14}$$

As concluded by Sandall and King, this so-called uniformly retreating ice-front model satisfactorily predicted the drying times for removal of 65–90% of the total initial water. During the removal of the last 10–35% of the water, the actual drying time was considerably greater than that predicted by the model. The sublimation interface temperature did remain constant, as that assumed in the model derivation (Geankoplis, 1993; King, 1973; Okos et al., 1992).

11.2.2.3. Water Vapor Condensation

Because of the low chamber pressure, the water contained in freeze-dried products will expand by a factor of 10^7 in volume upon sublimation; consequently, a very large volume of vapor must be handled. The sublimated water vapor must be removed for ultimate drying purposes. The removal of water vapor from the drying chamber is the most expensive and critical step in the entire freeze-drying process (King, 1973). Usually, the sublimed vapor is removed from the system either (a) by condensation on a cold refrigerated surface at a temperature of about −40°C; or (b) by direct pumping to the atmosphere by a vacuum pump that maintains a constant pressure below 133.3 Pa (Mellor, 1978).

In conventional plants, a refrigerated, metal-surfaced condenser is often used. Surface condensers are simple heat exchangers that allow uniform contact between the water vapor with the cooling medium through some heat transfer surfaces. Water vapor condenses on these surfaces as ice and builds up as a thick ice layer that must eventually be removed or defrosted to maintain the desired heat transfer rate. The defrosting may be done by passing hot air, water, or steam through the refrigerant side of the condenser. Different types of surface condensers are available: shell, tube, spiral plate, spiral tube, etc. By using surface condensers, the cooling medium is not contaminated with the process streams and it is easy to recover the condensate (Hartel, 1992; King, 1973; Mellor, 1978).

The efficiency of the vapor condenser is dependent on the difference between the pressure in the freeze-drying chamber and the pressure in the vapor condenser, the thickness of the ice buildup on the condenser, its temperature, and the temperature difference between the condenser surface and the evaporating refrigerant (Okos et al., 1992). The water vapor migration rate V_g between the product and the condenser can be expressed in terms of Stefan's model (Mellor, 1978):

$$V_g = \frac{Dm_w P}{HRT} \ln\left(\frac{P - P_C}{P - P'}\right) \tag{11.15}$$

where D is the diffusion coefficient of water vapor in air, m_w is the molecular weight of water, H is the distance between the product and the condenser, R is the universal gas constant, T is the absolute temperature, P is the total pressure, and P_C, P' are the partial pressure of water vapor near the outer surfaces of the condenser and the product, respectively.

Freeze dryers may be classified into three types based on the position in which the condenser is placed relative to the main drying chamber: distal, medial, or proximal, as shown in Fig. 11.18. In the distal type of dryer, the condenser can be isolated to enable another adjunct process to be carried out in the drying chamber, while the proximal type provides a close proximity of the evaporator to the condenser, thus helping to accelerate the ice sublimation process. In the last two types of dryers, condensers are placed within the chamber; a suitable radiation shield should be designed to minimize transfer of radiant heat from the heating elements to the condenser surfaces (King, 1973; Mellor, 1978; Barbosa-Cánovas and Vega-Mercado, 1996).

Generally, the condensers are placed between the drying chamber and the vacuum pump or steam jet ejector so that the vapor flowing along the surfaces sweeps the non-condensable gases into the vapor stream recovered by the vacuum pump. The vacuum pump is used to maintain sub-atmospheric pressures in the drying chamber, and also to remove the non-condensable gases to reduce the resistance of the water vapor migrating to the condenser (Okos et al., 1992).

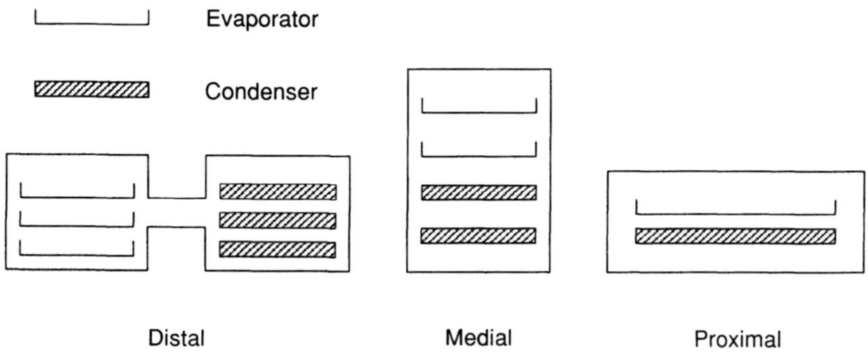

Figure 11.18. Basic types of freeze dryers based on the relative positions of condenser to the main drying chamber (adapted from Mellor, 1978).

11.2.3. Drying Equipment

11.2.3.1. Batch Freeze Dryer

The conventional freeze dryer is operated on a batch basis. The most commonly used batch freeze dryer consists of a cabinet with trays connected to a condenser and a vacuum pump, as shown in Fig. 11.14. Batch dryers deal with materials of fixed quantity at a stated time. The product is arranged to be static when the whole drying process is performed. Environmental variables are altered around them to suit accordingly.

The drying process is carried out in a chamber that has a door at one end, which can be removed for loading and unloading. After the pre-frozen food to be dried is loaded onto flat trays, the cabinet is closed and evacuated (vacuum pressure is less than 27 Pa), and the drying temperature is set for the specific duration of the cycle. The heating medium is supplied to a movable platen truck on which the trays are placed, providing heat to the material both by radiation from platens located above each tray and by conduction from the tray itself. Propylene glycol and steam may be used as the circulating heating medium. At the end of the drying cycle, the chamber is filled with an inert gas and opened. The dried product is removed and packaged in order to prevent moisture and oxygen pick-up.

11.2.3.2. Continuous Freeze Dryer

Compared to the batch drying processes, the continuous freeze-drying concept is more attractive, based on two main reasons. First, any continuous process is economic, in terms of less labor intensity and larger capacity. Secondly, the condenser system in a continuous dryer handles a relatively constant water vapor load over the entire drying process, while it must be sized large enough to handle the initial water vapor output at the beginning of a batch cycle. The disadvantages of continuous freeze-drying processes are related to the difficulty of operating a continuous vacuum system where feed must be received and the product discharged at atmospheric conditions (King, 1973).

There are two types of continuous freeze dryers: static or tray dryer, in which the freeze-dried food product remains stationary on trays and moves along the dryer in a continuous manner; and dynamic or trayless dryer, in which the product is moved through the dryer by means of belt conveyors, circular plates, vibrating plates, a fluidized bed, and sprays. Figure 11.19 shows a diagram of a continuous tray dryer. It consists of a tunnel with a vacuum lock at each end, one for loading pre-frozen lumps of food into the tunnel, and the other for discharging the dried product automatically by

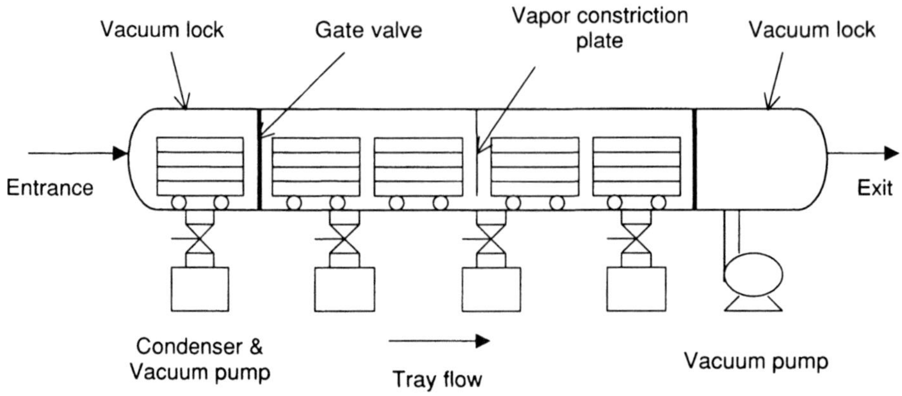

Figure 11.19. Schematic diagram of a continuous tray freeze dryer (adapted from Mellor, 1978).

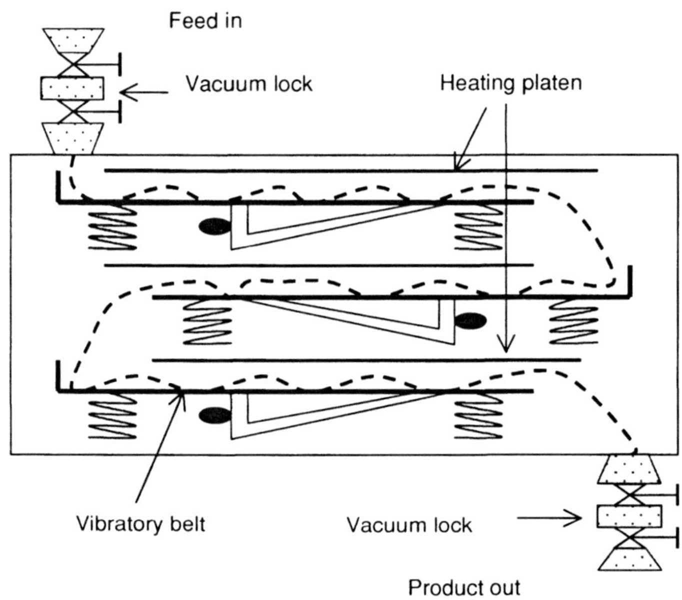

Figure 11.20. Schematic diagram of a trayless continuous freeze dryer (adapted from Barbosa-Cánovas and Vega-Mercado, 1996).

machinery into an air conditioned room before packaging. The heating energy is transported to the trays by an overheated rail connected to them. The trays are washed and recycled for use. In addition to lumps of food (such as coffee or tea extract), food slices or granules can also be loaded into the tray. The chamber pressure is in the range of 0.1–1 mbar. The dryer capacity is about 50–80 kg of fresh product per square meter per 24 h (Land, 1991).

A trayless continuous dryer is shown in Fig. 11.20. The pre-frozen material, which must be free flowing and granular, is loaded and dries as it passes along vibrating beds where vibration is used to shuffle and gently throw the material onward. The vibrating action results in particle mixing that can aid heat transfer through the particle layer and provide a more even exposure to the heat sources. However, vibration may cause particle attrition in some cases, and recycling of fines happens far more than that in static drying. Continuous product flow is obtained by using entrance and exit vacuum locks. In addition to the heating plates in the dryer, heating is also provided by the vibratory conveyors (Barbosa-Cánovas and Vega-Mercado, 1996; King, 1973; Land, 1991).

The pre-frozen material is usually fractured, cut, or broken to get a classified particle size before entering the feeding hopper to make the operation of the vacuum lock easier. Because of the short drying time, the particle size of the granular feed should be small (around 1.5–2.5 mm) and uniform. Powdered beef, banana, and vegetables, cottage cheese, and coffee are reported to be dried by this type of dryer (Dalgleish, 1990).

11.2.3.3. Microwave-Heating Freeze Dryer

The attempt to use microwave energy to provide internal heating in the freeze-drying chamber is aimed at overcoming the limitations on heat transfer rates by conventional heating methods, i.e., thermal contacting heating of the frozen layer, radiation or plate heating of the dry layer, and convection heating in atmospheric freeze-drying (Mellor, 1978). Microwave heating offers a method of heating the product selectively by taking advantage of the difference in dielectric properties between the ice and the dry portion in the product. The heating process is less dependent on thermal gradient and conductivity (Barbosa-Cánovas and Vega-Mercado, 1996). The heat transfer depends on the degree of excitability of the molecules in the absorbing medium and the microwave frequency (Toledo, 1991).

Microwave heating refers to the use of electromagnetic waves of very high frequencies (usually 2,450 and 915 MHz) to transmit energy through space. The heating effect by microwaves is due to the presence of molecular dipoles (primarily water in the case of foods). When placed in a rapidly changing electric field, these molecular dipoles undergo orientation changes that result in friction and, consequently, heat generation. The energy absorbed by the food products P (J/(s cm^3)) in a microwave field is given by (Barbosa-Cánovas and Vega-Mercado, 1996; Toledo, 1991):

$$P = 0.556 \times 10^{-12} E^2 f \varepsilon \tan \delta \qquad (11.16)$$

where E is the electric field strength (V/cm^2), f is microwave frequency (Hz), ε is dielectric constant (dimensionless), i.e., an index of the rate at which energy penetrates a solid, and $\tan \delta$ is called loss tangent (dimensionless), i.e., an index referring to how much energy entering the solid is converted to heat. E and f are dependent upon the frequency of the microwave generator used. Both ε and $\tan \delta$ are intrinsic properties of the materials and depend strongly on temperature and composition. The dielectric constant and the loss tangent of water, some foods, and packaging materials are listed in Table 11.3.

A general scheme of a microwave heating freeze dryer is shown in Fig. 11.21. Although this heating method has been frequently proposed, it has only occasionally been tried. Its application has not been successful because of the following reasons (King, 1973; Mellor, 1978):

(a) Concentrated microwave energy can cause a glow discharge within the dried layer, causing deleterious changes in the product (discoloration and flavor damage) and loss of useful power.
(b) Energy supplied in the form of microwaves is expensive and no profitable microwave equipment is available for freeze-drying of foods on a large continuous scale.
(c) The drying process is very difficult to control. Any localized melting can cause a rapid chain reaction that results in "runaway" overheating. Furthermore, the continuous impedance change in the chamber during drying requires matching-up changes in the microwave output.

Table 11.3. Dielectric properties of some food and other materials (modified from Toledo, 1991; Mellor, 1978).

Material	Temperature (°C)	Dielectric constant ε	Loss tangent $\tan \delta$
Ice	−12	3.2	0.0009
Raw beef	−15	5.0	0.15
Boiled pea	−15	2.5	0.20
Raw pork	−15	6.8	0.12
Boiled potato	−15	4.5	0.20
Boiled squash	−15	5.0	0.30
Porridge	−15	5.0	0.74
Pear	−15	4.17	0.27
Suet	25	2.5	0.07
Steak	25	40	0.30
Water	1.5	80.5	0.31
Water	25	76.7	0.15
Pyrex	25	4.8	0.0054

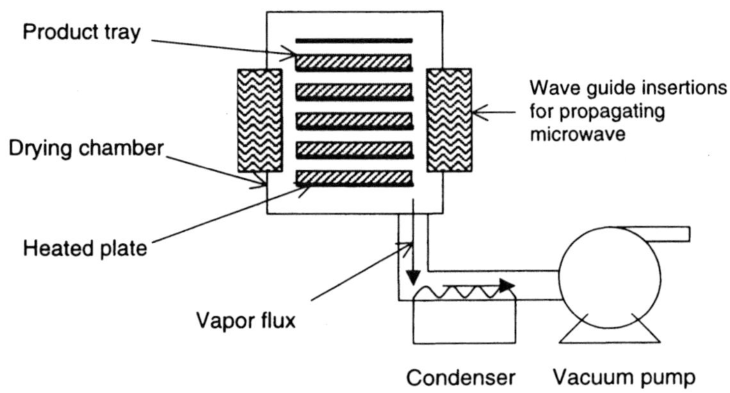

Figure 11.21. Schematic diagram of a microwave-heating freeze dryer (adapted from Barbosa-Cánovas and Vega-Mercado, 1996).

As suggested by King (1973), microwave heating may become attractive economically when conventional freeze-drying is used to remove the first 60–80% of the water and microwave-heating freeze-drying used to remove the remaining water in the product. The operating costs of microwave heating are higher than using steam or other heat sources, but its capital costs can be lower because of its shorter drying cycle.

11.2.3.4. Modified Freeze Dryers

To overcome the disadvantages of traditional vacuum freeze-drying, research has been carried out to redesign the drying process. A new process for freeze-drying based on a fluidized bed type of dry adsorbent, working at atmospheric pressure, was studied by Wolff and Gibert (1990). The experimental atmospheric pressure freeze-drying pilot plant used for research is shown in Fig. 11.22. A comparison has been carried out with vacuum freeze-drying. Processing time is longer than under vacuum drying, but energy savings of 35% can be expected from a continuous duty facility.

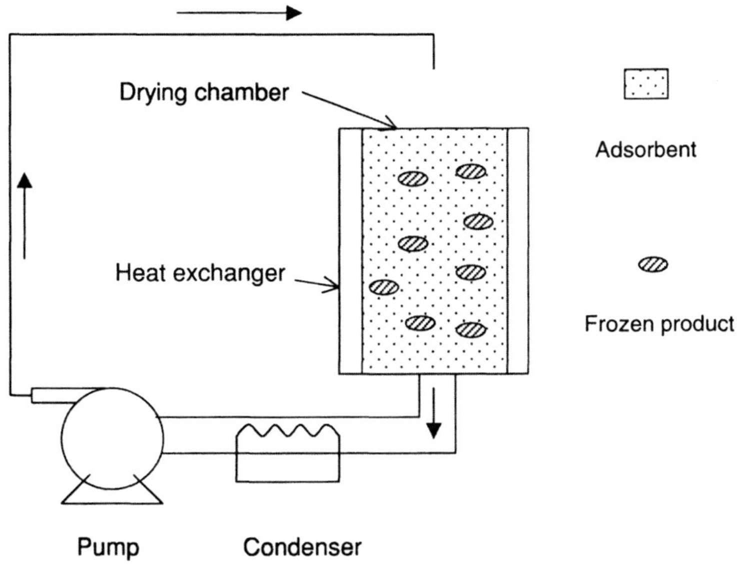

Figure 11.22. Schematic diagram of an atmospheric-pressure freeze dryer (adapted from Barbosa-Cánovas and Vega-Mercado, 1996).

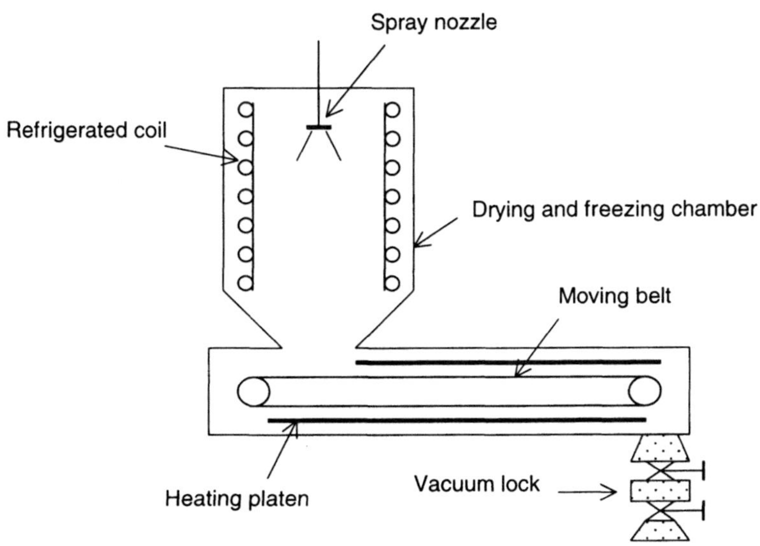

Figure 11.23. Schematic diagram of a vacuum-spray freeze dryer (adapted from Mellor, 1978).

A prototype of a vacuum-spray freeze dryer, as shown in Fig. 11.23, has also been developed in the United States for coffee extract, tea infusion, and milk. The product, which has been pre-concentrated up to 30% solids, is sprayed from a single jet upward or downward in a cylindrical tower in which a refrigerated helical condenser is coiled between the inside wall and a central hopper. The liquid droplets are solidified into small particles by evaporative freezing caused by the condenser, collected at the tower bottom, and then transported to a stainless steel belt. The radiant heaters above and under the traveling belt provide the energy to dry the partially dried particles completely.

Table 11.4. Comparison of drying temperature and time for different drying processes (modified from Adams, 1991).

Process	Typical operating temperature (°C)	Drying time (min)
Spray drying	+80–100	<10
Fluid bed drying	+80	<100
Freeze-drying (1 cm depth)		
Milk	−5	600
Blood serum	−25	1,200
Cytomegalovirus vaccine	−40	3,600

Finally, the product passes into a hopper and is removed from a vacuum lock intermittently for packing. The whole plant is operated under a vacuum of 67 Pa. Frozen particle sizes are about 150 μm.

The freeze-drying process is very time consuming. Table 11.4 illustrates the comparative drying time for typical products of spray, fluid bed or freeze-drying. The considerable increased processing time for freeze-drying should be noted (Adams, 1991).

In addition to the normal running costs involved with an industrial plant, the freeze-drying process depends on the extraction of energy to freeze, sublime and condense water vapor. Incidental factors that increase production costs include the requirement for high dosage containers and the additional costs of sealing the dried products. A cost comparison for processing foodstuffs would indicate a 5–to 8-fold increase in drying cost between freeze-drying and conventional drying. However, when the entire process is compared, this differential may be significantly reduced to 1:1.33 (Adams, 1991). On purely economic grounds, freeze-drying is likely to compare unfavorably with alternative drying techniques, and the process can only be justified on the basis of superiority of product quality.

Freeze-drying is recognized as the best method of producing drier material of high quality. The freeze-drying process has several advantages that make it desirable for food processing. By maintaining the material in the frozen state until it dries, shrinkage and migration of dissolved constituents are eliminated. Physical and chemical changes are inhibited, thereby minimizing loss of volatile components. Freeze-dried products have a porous texture, and are readily rehydrated to their original size and shape (Gutcho, 1977). Freeze-dried coffee and tea are also popular for their excellent instant properties.

The dried products obtained from freeze-drying processes have good flavor and appearance, and a high preservation of nutrition. In addition, freeze-dried products, as opposed to agglomerated powders, are adaptable to simple packaging, storage and shipment, and can be kept for a long time. Temperature consideration for storage of dried foods is fairly flexible (Dalgleish, 1990). Moisture level as low as 2% can be reached with freeze-drying. This makes the products much lighter than those dried by other drying methods and they do not require refrigeration. However, the oxidation of lipids, which is induced by the low moisture level, is a major concern for freeze-dried products. By packaging in oxygen-impermeable containers, lipid oxidation reactions may be controlled.

Energy cost and long drying time are the major disadvantages of the freeze-drying process, but it still has been shown to be a good alternative way to preserve foods with high quality (Liapis and Marchello, 1984). Using microwaves for heating and freeze-drying at atmospheric conditions are additional ways to improve the process efficiency and reduce operation costs (Barbosa-Cánovas and Vega-Mercado, 1996).

11.3. DRUM DRYING

Drum dryers were introduced into industries about 100 years ago. Starting with the double-drum dryer which featured the feeding by nipping between two drums. J. A. Just was one of the first inventors to receive patent rights on a drum dryer with two rolls in 1902. The top feeding single-drum dryer, which is more suitable for viscous feed materials, was introduced in 1945. This drying process is usually called drum drying, film drying or, in the dairy industry, more often called roller drying.

Drum dryers are simple in operating principle, but become physically complex when they are modified to treat particular products (Brown et al., 1973; Hall and Hedrick, 1971; Land, 1991). Typical drum-dried food products include milk, milk products, soup mixes, ingredients for baby foods, navy bean meal and pea meal, potato slurries, and instant cereals (Barbosa-Cánovas and Vega-Mercado, 1996; Nair and Warren, 1973). Many of these products can be quickly rehydrated from their final flakes or powders.

Drum drying is one of the cheapest drying methods available for industries and has many applications for small and medium production runs (Brown et al., 1973). It saves space and is more economical than spray drying for small volumes. These drum dryers have high drying rates and thermal efficiency. Their main limitations include that the raw material has to be in the form of liquid or slurry and it must withstand a relatively high temperature for a short period of time. More specifically, materials that do not adhere to a metal surface cannot be processed by drum dryers (Barbosa-Cánovas and Vega-Mercado, 1996). For milk or milk products, the major disadvantages of drum drying include a scorched flavor and much lower product solubility due to protein denaturization (Hall and Hedrick, 1971).

11.3.1. Drum Drying Process

Drum dryers consist of one or more hollow metal cylindrical rolls or drums that are mounted to rotate on horizontal axes at a variable speed. A typical single-drum dryer is shown in Fig. 11.24. Circled through the drums, the heating medium is usually steam, but occasionally hot water or special high temperature heat transfer liquid may be used, although direct gas or oil heating has been developed (Brown et al., 1973; APV, 1989). The feed material is spread onto the drum surface as a thin film and dried by the heat transferred through the metal thickness from the heating medium inside the drum. A knife is fitted to the drum at an appropriate position to scrape off the thin layer of

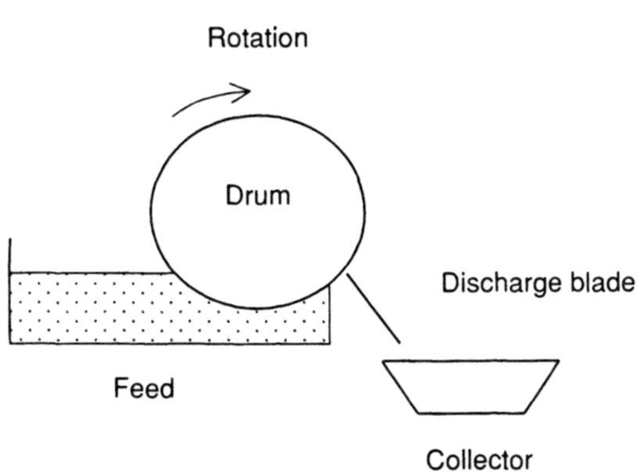

Figure 11.24. A typical single-drum dryer with dip feeding.

dried material from the drum surface. The feeding materials can be slurries, paste, or solutions and final dried products are in the form of powders, flakes, or chips (Land, 1991).

The diameter of a drum may be up to 2 m and a length of 5 m. Their capacities are in the range of 5–30 kg of product per square meter and per hour, i.e., maximum capacity is 1 ton/h for a single-drum dryer. The rotation speed is in the range of 1–30 rpm and the resident time can be fixed from 2 s to 1 min by changing the rotation speed. The maximum steam pressure is normally 12 bar (Land, 1991).

11.3.2. Mass and Heat Transfer

Three steps of heat transfer occur in the thin material layer on the drum: (a) heating the thin layer to its boiling point; (b) water vaporizing and material changing from liquid state to solid state; and (c) product temperature approaching that of the drum. The drying rate for the thin film is determined by the rate of heat transmission from the drum to the product and the thin layer of material presents no restriction to the vaporizing water (Okos et al., 1992).

The overall drying rate of the material film over the drum surface can be expressed as (Barbosa-Cánovas and Vega-Mercado, 1996):

$$\frac{\mathrm{d}X}{\mathrm{d}t} = \frac{M_{\mathrm{S}}(X_0 - X_{\mathrm{f}})}{t} = \frac{KA(T_{\mathrm{W}} - T_{\mathrm{E}})}{\lambda} \tag{11.17}$$

where M_{S} is the mass of solid in the feed, X_0 is the initial moisture content of the feed material, X_{f} is the final moisture content of the product, t is the time needed to reach X_{f}, A is the drying surface area, T_{W} is the temperature of the heated drum surface, λ is the latent heat at the temperature of the evaporating surface T_{E}, and K is the overall heat transfer coefficient.

During drying, heat flows to the product simultaneously by conduction, convection, and radiation. However, the heat transfer rate is affected by several factors: water condensed inside the drum, the drum wall, the food material adhered to drum, and the outside surface. Thus, the total heat transfer coefficient K (J/(s m^2 K)) is given by (Okos et al., 1992):

$$\frac{1}{K} = \frac{1}{h_{\mathrm{W}}} + \frac{1}{h_{\mathrm{M}}} + \frac{1}{h_{\mathrm{P}}} + \frac{1}{h_{\mathrm{C}} + h_{\mathrm{R}} + h_{\mathrm{S}}} \tag{11.18}$$

where h_{W} is the heat transfer coefficient of the condensate, h_{M} is the heat transfer coefficient of the metal material, h_{P} is the heat transfer coefficient of the feed material, h_{C} is the heat transfer coefficient of convection, h_{R} is the heat transfer coefficient of radiation, and h_{S} is the heat transfer coefficient due to vaporization.

The operation of drum dryers is controlled by steam pressure, rotational speed, film thickness, and the characteristics of the feed material. Steam pressure regulates the drum temperature, while rotation speed determines the drying time. The capacity of a dryer is the function of the drying rate of the material in the form of a thin layer and its total amount adhering on the drum surface. The drying rate depends on the feeding type, steam pressure, and the drum rotation speed, while the material adherence is affected by the viscosity, surface tension, and wetting power of the material (Okos et al., 1992).

11.3.3. Types of Drum Dryers

Drum dryers can be classified according to the number of drums (single or double), the pressure surrounding the product (atmospheric or vacuum), the drum turning direction of the two rolls (double or twin), methods of placing material on the drum surface (nipping, dipping, splashing, or spraying) and even the material of construction (steel, alloy steel, or cast iron) (Hall and Hedrick, 1971). In this chapter, they are broadly classified into three types: single drum, double drum, and twin drums.

11.3.3.1. Single-Drum Dryer

Single-drum dryers comprise one roll or drum for drying purposes. A variety of feeding methods can be applied to feed the material to the drum. In the case of simple dip feed in a large tank, good liquid circulation in the tank is needed in order to avoid the feed condensing by evaporation. Rotating devices, such as a spray feeding system, are used for very mobile feeding materials. Dipping feeding from a very small pan can supply the drum with fresh material that cannot tolerate prolonged heat exposure. The top feeding is done by putting the material in a trough or reservoir on the top of the drum, as shown in Fig. 11.25a (APV, 1989; Brown et al., 1973).

When feeding pasty materials, such as starches, top applicator rollers are usually used to coat the material on the drum surface uniformly (Fig. 11.25b). Those rollers are small and unheated. Belonging to part of the feeding system, they are located around the periphery of the main roll and their number is determined by the characteristics of the material being handled. Up to five spread rollers may be equipped on a single-drum dryer to increase the thickness uniformity of the film on the drum (Brown et al., 1973; Hall and Hedrick, 1971).

In the past years, skim milk was conventionally dried on single-drum dryers and its production quantity was enormous. In recent years, new uses for single-drum dryers have extended to drying starches, instant potatoes, dyestuffs, and gelatin (Brown et al., 1973; Land, 1991).

11.3.3.2. Double-Drum Dryer

In double-drum dryers, the material is fed into the nip between two cylindrical rolls. The rolls may rotate toward each other or run away from one another at the top, depending on the different discharge positions of the dried material (Fig. 11.25c and d). The product dries on the drum surface as the drum rotates and is then scraped off by knives. The layer thickness of the product can be controlled by carefully adjusting the space between the two rolls, ranging from 100 to 400 μm (Land, 1991).

Double-drum dryers are most widely used because of their wide applicable range, high production rate, and low labor requirement. Materials ranging from heavy pastes to dilute solutions, especially if they are heat sensitive, can be effectively dried by these dryers. However, they are less suitable for drying viscous materials because of the considerable pressure on the rolls. They are most commonly used in the dairy industry. Other typically dried food products by the double-drum dryers include baby food, glues, applesauce, fruit purees, precooked breakfast cereals, mashed potatoes, dry soup mixes, and devitalized yeast (Brown et al., 1973; Land, 1991; Okos et al., 1992).

11.3.3.3. Twin-Drum Dryer

Twin-drum dryers have a similar appearance to double-drum dryers, but are quite different in their operation and in the products they can handle. Their two rolls are not spaced closely together and rotate away from each other at the top. They are fed by a dipping or splashing system underneath the drums, as shown in Fig. 11.25e, or by a flow system located above the drums (Brown et al., 1973). The discharge blades are located 270° away from the rotary feeding system. They are suitable

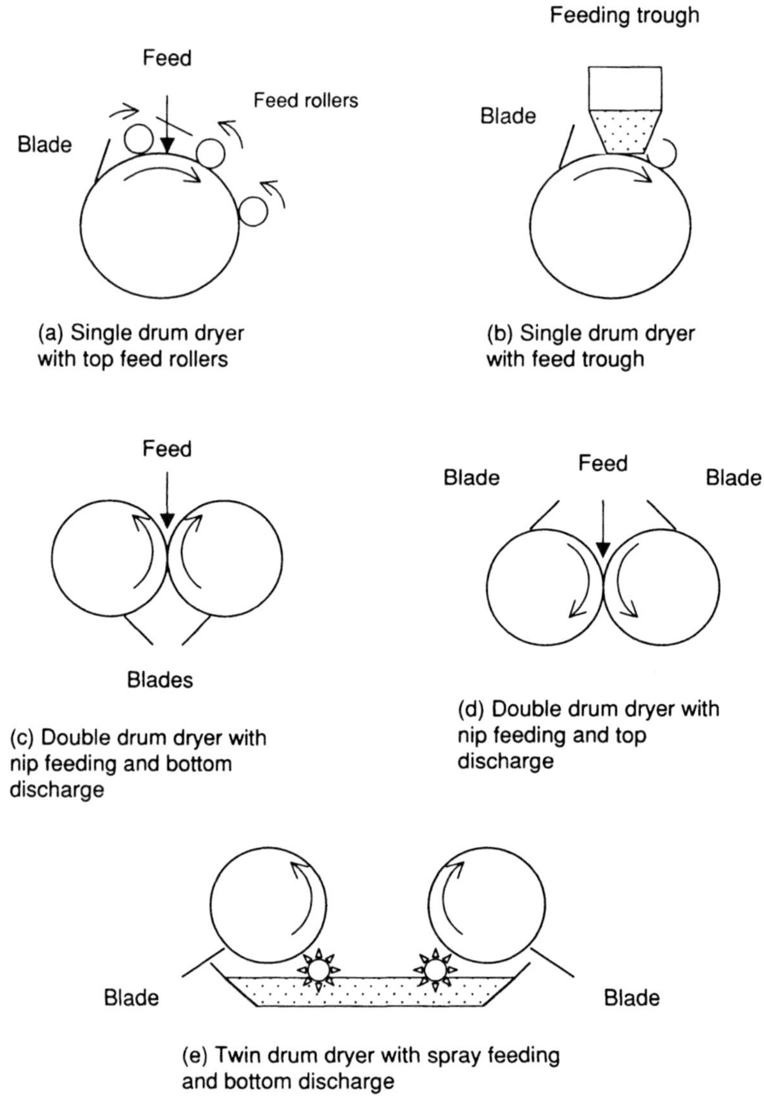

Figure 11.25. Different types of drum dryers with feeding systems (adapted from APV, 1989).

for materials that are dusty when dry, such as salt solutions or clay slips. They can also be used as a pre-dryer when top fed (Okos et al., 1992).

During the drying process, drum dryers are usually fitted with hoods to collect and exhaust water or solvent vapor. When necessary, drum dryers can be totally enclosed to protect workers against toxic products or vapors, to prevent cross contamination when different products are dried on different dryers, or to allow the recovery of valuable materials in the vapor. Furthermore, the entire dryers can be enclosed in a vacuum-tight chamber to operate under a vacuum condition to evaporate solvents or keep the temperature down; thus, some heat sensitive materials can be dried without adversely affecting enzymes, vitamins, and proteins (Brown et al., 1973; Okos et al., 1992).

11.4. FINAL REMARKS

Selection of the appropriate drying system is not a straightforward process. It needs a thorough knowledge of the properties of the feed as well as the desired specifications of the final product. Even after a certain method is selected, there are still many decisions to be made for a detailed process design. It must be kept in mind that the selected drying process is to preserve product quality while controlling moisture content and avoiding thermal or mechanical degradation.

REFERENCES

Adams, G.D.J. (1991). Freeze-drying of biological materials. *Drying Technol.* **9**(4): 891–925.

APV (1989). *Dryer Handbook (DRH-889)*. APV, Rosemont, IL.

Barbosa-Cánovas, G.V., Vega-Mercado, H. (1996). *Dehydration of Foods*. Chapman & Hall, New York.

Brown, A.H., Arsdel, W.B.V., Lowe, E. and Morgan Jr., A.I. (1973). Air drying and drum drying. In *Food Dehydration*, Arsdel, W.B.V., Copley, M.J. and Morgan Jr., A.I. (eds.). The AVI Publishing Co., Inc., Westport, CT, Chapter 5, pp. 82–160.

Bruin, S. and Luyben, K.Ch.A.M. (1980). Drying of food materials: a review of recent developments. In *Advances in Drying*, Mujumdar, A.S. (ed.). Hemisphere Publishing Co., Washington, Chapter 6, pp. 155–215.

Dalgleish, J.M. (1990). *Freeze-Drying for the Food Industry*. Elsevier Applied Science, London.

Desrosier, N.W. (1977). *The Technology of Food Preservation*. The AVI Publishing Company, Inc., Westport, CT.

Filkova, I., Cedik, P. (1984). Nozzle atomization in spray drying. In *Advances in Drying*, Mujumdar, A.S. (ed.). Hemisphere Publishing Co., New York, Chapter 5.

Flink, J.M. (1975). Application of freeze-drying for preparation of dehydrated powders from liquid food extracts. In *Freeze-Drying and Advanced Food technology*, Goldblith, S.A., Rey, L. and Rothmary, W.W. (eds.). Academic Press Inc., London, Chapter 20, pp. 309–329.

Geankoplis, C.J. (1993). Drying of process materials. In *Transport Processes and Unit Operations*, 3rd ed. Prentice-Hall, Inc., Englewood Cliffs, NJ, Chapter 9, pp. 520–583.

Gutcho, M.H. (1977). *Freeze-Drying Process for the Food Industry*. Noyes Data Corporation, NJ.

Hall, C.R. and Hedrick, T.I. (1971). *Drying of Milk and Milk Products*. The AVI Publishing Co., Inc., Westport, CT.

Hartel, R.W. (1992). Evaporation and freeze concentration. In *Handbook of Food Engineering*, Heldman, D.R. and Lund, D.B. (eds.). Marcel Dekker, Inc., New York, pp. 341–392.

Heldman, D.R. (1992). Food freezing. In *Handbook of Food Engineering*, Heldman, D.R. and Lund, D.B. (eds.). Marcel Dekker, Inc., New York, Chapter 6, pp. 277–315.

King, C.J. (1973). Freeze-drying. In *Food Dehydration*, Arsdel, W.B.V., Copley, M.J., Morgan, A.I. (eds.). The AVI Publishing Company, Inc., Westport, CT, Chapter 6, pp. 161–200.

King, C.J. (1975). Applications of freeze-drying to food products. In *Freeze-Drying and Advanced Food Technology*, Goldblith, S.A., Rey, L. and Rothmary, W.W. (eds.). Academic Press Inc., London, pp. 333–349, Chapter 21.

Land, C.M. (1991). *Industrial Drying Equipment: Selection and Application*. Marcel Dekker, Inc., New York.

Liapis, A.I. and Marchello, J.M. (1984). Advances in the modeling and control of freeze-drying. In *Advances in Drying*, Vol. 3, Mujumdar, A.S. (ed.). Hemisphere Publishing Co., Washington, DC, Chapter 6, pp. 217–244.

Masters, K. (1985). *Spray Drying Handbook*, 4th ed. George Godwin, London.

Masters, K. (2002). Aps. In *Process Stages and Spray Dry Systems*. SprayDryConsult. Intl., pp. 39–62 (extracted from Niro Electronic Library: www.niro.com)

Mellor, J.D. (1978). *Fundamentals of Freeze-Drying*. Academic Press, London.

Nair, J.H., Sivetz, M. (1973). Coffee and tea. In *Food Dehydration*, Vol. 2, Arsdel, W.B.V., Copley, M.J. and Morgan, A.I. (eds.). The AVI Publishing Company, Inc., Westport, CT, pp. 384–436.

Nair, J.H. and Warren, D.H. (1973). Dry soups and other dry mixes. In *Food Dehydration*, Vol. 2, Arsdel, W.B.V., Copley, M.J. and Morgan Jr., A.I. (eds.). The AVI Publishing Co., Inc., Westport, CT, pp. 437–464.

Nath, S. and Satpathy, G.R. (1998). A systematic approach for investigation of spray drying processes. *Drying Technol.* **16**(6): 1173–1193.

Oakley, D. (1993). Sprays reveal their secrets. *Chem. Eng.* **539**: 18–21.

Oakley, D. (1997). Produce uniform particles by spray drying. *Chem. Eng. Progress* **93**(10): 48–54.

Okos, M.R., Narsimhan, G., Singh, R.K. and Weitnauer, A.C. (1992). Food dehydration. In *Handbook of Food Engineering*, Heldman, D.R. and Lund, D.B. (eds.). Marcel Dekker, Inc., New York, Chapter 10, pp. 437–562.

Persson, P. (1975). The freezing of vegetables. In *Freeze-Drying and Advanced Food Technology*, Goldblith, S.A., Rey, L. and Rothmary, W.W. (eds.). Academic Press Inc., London, Chapter 10, pp. 109–120.

Shaw, F. (1994). Fresh options in drying. *Chem. Eng.* (July): 76–84.

Toledo, R.T. (1991). *Fundamentals of Food Process Engineering*, 2nd ed. Chapman & Hall, New York.

Wolff, E. and Gibert, H. (1990). Atmospheric freeze-drying, Part I: Design, experimental investigation and energy-saving advantages. *Drying Technol.* **8**(2): 385–404.

CHAPTER 12

UNDESIRABLE PHENOMENA AND THEIR RELATION TO PROCESSING

Different chapters in this book have covered basic unit operations utilized for food powders production and handling. Comprehensive descriptions of several processing features ranging from devices for bulk storage, conveying systems, and size reduction equipment to a series of methods for size enlargement, powder encapsulation, mixing, or classification have been discussed. Each of these food powder production or processing related activities can be directly or indirectly connected to physically and chemically undesirable phenomena that can be deleterious to the final products, to the ingredients utilized during the process, and/or to the equipment in the factory. This chapter will cover four different concepts with negative connotations in the production and logistics fields, known as: attrition, segregation, caking and dust explosion.

Powder attrition is product related and involves the breakdown of particles. In food powders, it is more frequent in agglomerates, mainly because of their multi-particulate structure. Food agglomerates possess brittle characteristics that make the product susceptible to vibrational, compressive, shear, or even convective forces applied to the particles during processing. Descriptions of how these forces intervene in food powders production, handling, and storage and how the final product will be affected are given. Measurement methods as well as known and recently developed mechanisms of accidental particle breakage will be explained. Segregation, although related to attrition, is a more general phenomenon that affects both the products and production. The unwanted separation of fine particles from coarse particles during different handling and storage activities such as conveying, mixing, discharging, charging, bulk storing, and packaging will be explained from different aspects. Various circumstances where segregation can affect production, along with its physical mechanisms and kinetics, will be shown in this chapter. Furthermore, different evaluation methods and minimization techniques will illustrate powder segregation prevention.

Caking in dry products during high shelf temperature and moisture can also lower the quality of different food powders. This phenomenon is very complex and occurs during operations where both temperature and humidity have a significant role. Product formulation, spray drying and size enlargement systems, as well as packaging and storage activities, are mainly affected by caking. Mechanisms of action will be explained through different applications. Laboratory techniques used during research will also be described. Last but not least, a hazardous phenomenon caused by dry fine particulates under specific pressure conditions is dust explosion. This type of explosion may affect close storage installations, bringing high economical losses and even death. The subject of dust explosion is too complicated to be covered in depth in one section of this chapter, but its mechanism of action, related factors, its classification, and its prevention are discussed.

12.1. ATTRITION

Attrition is unwanted particle breakdown by fragmentation or abrasion, which increases the number of particles and reduces particle size. Except for particle size reduction during comminution or grinding processes, attrition is undesirable in most processes. As a matter of fact, it is one of the most ubiquitous problems for a wide range of processing industries that deal with particulate solids. It is not rare that a product leaves the manufacturing site with its specified particle size distribution but exhibits dustiness and difficulty in discharge from storage silos when it reaches the hands of users. Attrition can also result in the production of dust or lead to the loss of material in the system. As a result, costly remedial operations, such as plant shut-down, can be required in order to filter hazardous dusty material formed through this process (Shipway and Hutchings, 1993). Some attrition is inevitable during the manufacture, consumption, and transportation of particulate materials. In fact, attrition occurs whenever there is relative movement between particles in contact with each other or with the container wall (Gundepudi et al., 1997). This section will give more attention to the attrition effect of agglomerated food powders rather than other groups of particulate materials.

As seen in Chapter 7, the powder size enlargement technique by agglomeration is widely used in many industries to improve important physical properties of powders, such as flowability, instant properties, bulk density, appearance, and dust behavior (Hoornaert et al., 1996). Many food agglomerates (such as instant coffee and milk) are generally less hard and have lower mechanical strength than inorganic or polymeric materials, which makes them more susceptible to attrition (Bemrose and Bridgwater, 1987). The most remarkable adverse consequences of attrition on agglomerated food powders include: (a) *changes in bulk properties* (bulk density, flowability, particle size distribution, surface area, angle of repose, and angle of internal friction); (b) *reduction of instant properties* (dispersibility, wettability); and (c) *product segregation*. As will be further described in Section 12.2, particle segregation can bring problems in the intended functionality or nutrition of a certain powder mix due to lack of homogeneity in the final product, and it can affect product appearance. As fine particles are created due to attrition, product can be lost during processing. Consequently, the risk of dust explosion will be created, as well as contamination of the processed particles by debris from wear of the containment system (Barletta et al., 1993a; Barbosa-Cánovas et al., 1985).

12.1.1. Attrition Mechanisms

Although the attrition phenomenon is widespread in different engineering fields, it is not properly understood due to its complexity. However, it is certain that it is influenced by variables involving both the particles and the environment system. Factors such as particle size and shape, moisture content and porosity, particle surface condition and hardness, composition, fluid or gas velocity during processing, environmental temperature and pressure, and intensity of mechanical impact and/or friction determine the severity of the attrition effect (Bemrose and Bridgwater, 1987; Gundepudi et al., 1997; Peleg and Normand, 1987; Shipway and Hutchings, 1993).

As described by Chraibi and Flamant (1989), the stresses causing attrition may be systematically summarized as thermal, chemical, static mechanical, or kinetic. Thermal stresses come from rapid temperature changes in particles, which cause uneven particle expansion or liquid evaporation. Chemical stresses are caused by particle phase changes and/or chemical reactions between the particle components. External loads on particle surfaces will cause static mechanical stresses, while impacts inside particles and/or with the container walls are associated with kinetic stresses.

Three main mechanisms governing the particle attrition process have been distinguished (Fig. 12.1): fragmentation or shattering, surface erosion or abrasion, and a combination of the former,

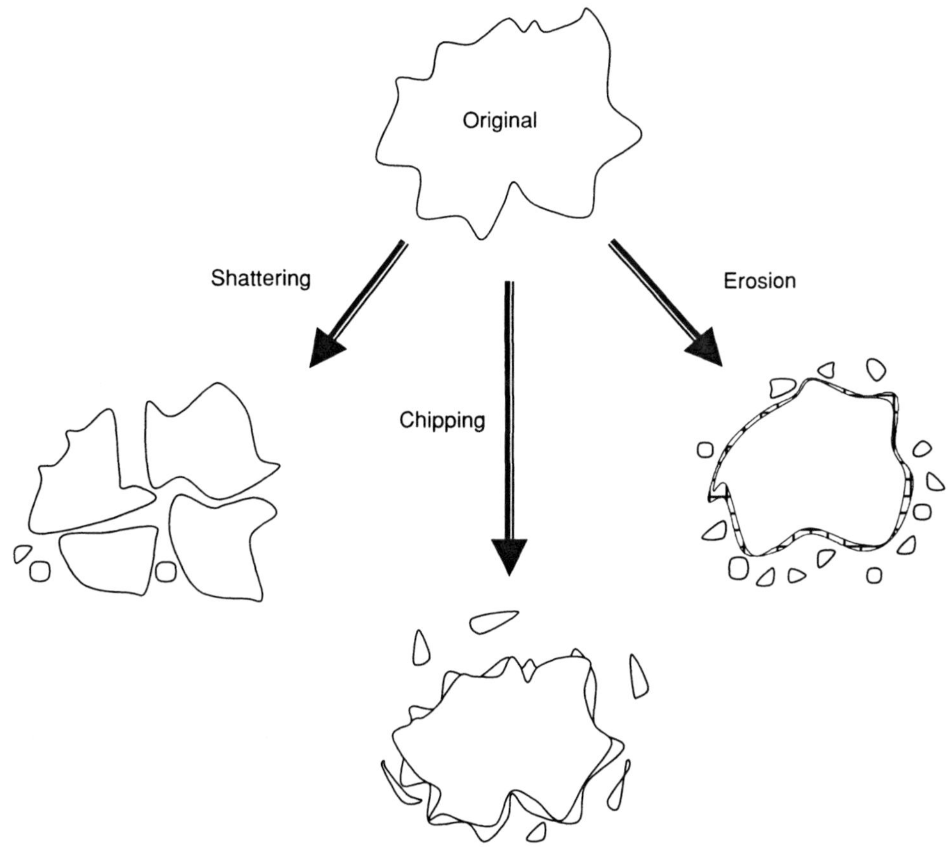

Figure 12.1. Schematic illustration of three main attrition mechanisms (adapted from Popplewell et al., 1989; Menacho, 1986).

i.e., chipping (Menacho, 1986). Particle shattering produces several mid-sized particles as compared to their parent particles. In erosion, a great number of very fine particles are removed from the surface layer and edges or corners of the parent particles. As a result, parent particles remain slightly smaller, and very fine particles are produced. The third attrition mechanism is characterized by partial fracture, which loosens small products of fine particles, plus a "chipped" product near the parent size. In this sense, chipping resembles erosion rather than a shattering process (Menacho, 1986; Neil and Bridgwater, 1994; Biscans et al., 1996).

The three attrition mechanisms are in turn governed by different failure modes: brittle, semibrittle, and ductile (Ghadiri, 1997). The *brittle* failure mode is due to the presence of pre-existing internal or surface cracks. Particle breakdown due to internal cracks is dominant when the stress (or compliance) during elastic deformation at the contact surface is low, while surface cracks play a major role in damage when the elastic compliance of the contact surface is high (Shipway and Hutchings, 1993). *Semibrittle* failure can be identified when limited plastic flow occurs. It is responsible for flow initiation and occurs when the impact stress exceeds the onset yield. As characterized by crack morphology and extension, the formation of median and radial cracks causes particle fragmentation, and the formation of lateral cracks causes chipping. Soft materials are usually damaged under *ductile* failure mode, but the study for particulate solids under this mode has not been investigated so far (Ghadiri, 1997).

12.1.2. Attrition Tests

Many different types of experimental tests have been developed to study the attrition tendency of particulate systems, and they can be divided into two main groups: those dealing with single particles and those with multiple particles. Single-particle tests use either compression or impact methods in which, particle breakage or fragmentation is mainly due to crack propagation. In such tests, surface wear or erosion occur in very small amounts. Multi-particle tests include these methods using shear cells, confined compression cell, rotating drum, vibration, fluidized bed, drop shatter, paddle wheel, and enhanced sieving, in which particles undergo both fragmentation and surface erosion. On the other hand, single-particle tests have the advantage of applying a well-characterized load on the particle and enhancing the understanding of particle breakage. However, variations among particles in a certain bulk make it difficult to compare results with attrition observed in practice.

Multi-particle tests are more closely related to real attrition situations and they are primarily empirical in nature. Consequently, it is difficult in many cases to correlate results obtained in different systems (Shipway and Hutchings, 1993; Mullier et al., 1991). Most multi-particle test methods rely on some mechanically induced motions between particles, or particles and large stationary or moving objects, to assess attrition tendencies for experimental or product quality control purposes. Other theoretically possible but not experimentally available methods have been suggested by using chemical reactions, pressure change, heating, and fluid transport tests (Bemrose and Bridgwater, 1987).

12.1.2.1. Single-Particle Tests

The first type of single-particle test is the *particle compression or crushing* test. The conventional test is the so-called "Brazilian test," as shown in Fig. 12.2, in which a single particle is crushed between two platens and the load required causing the fracture is recorded. The single-particle compression test was found to be useful in a study of the agglomerate strength by Adams and McKeown (1996); it has been established that a particle's resistance to attrition is closely related to agglomerate strength. When a single agglomerate is compressed between two platens at a constant speed, there is a distinct first maximum force in the force–distance data that corresponds approximately to the point at which the agglomerate is observed to fracture. The nominal agglomerate strength τ_{AS} can be calculated from the maximum force F_{max} by using the following expression (Adams and McKeown, 1996):

$$\tau_{AS} = \frac{4F_{max}}{\pi d^2} \tag{12.1}$$

where d is the diameter of the agglomerate.

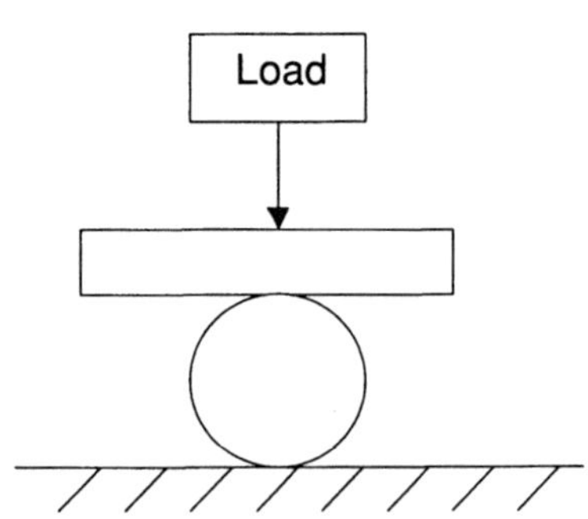

Figure 12.2. Schematic view of Brazilian test for a single particle.

This method is time-consuming since a large number of agglomerates have to be measured, and there are always wide variations in the fracture loads measurements. In addition, the cross-area of the agglomerate is hard to determine accurately if the agglomerates to be tested have irregular shapes. An alternative and more convenient method is the confined uniaxial compression test that uses a confined bed of similar agglomerates instead of single ones, which will be described in the following section.

Another single-particle test is called the *impact test*, which introduces impact between two particles or between a particle and a fixed target. The impact velocity is recorded and fragments collected for further stress studies (Bemrose and Bridgwater, 1987). The impact can be introduced in different ways. One alternative is to use a flat-faced cylindrical aluminum projectile to impact a flat rock specimen in their attrition experiments. The projectile is propelled by a gas gun to give an impact velocity ranging from 10 to 50 m/s on the specimen. The fractured specimen is collected for size distribution analysis by a set of sieves (Bemrose and Bridgwater, 1987). Shipway and Hutchings (1993) developed another method consisting of loading brittle spheres in a sabot. The sabot is accelerated by compressed gas and stopped at the muzzle. Spheres start flying and impact a target that is a short distance away. The fragments and spheres are recovered and the number of unbroken spheres are noted. Scanning electron microscopy is used to examine the broken spheres to elucidate the mechanisms of fracture.

These methods have been used for many years to characterize particles and to assess grindability and associated phenomena. In particular, Shipway and Hutchings (1993) found that the fracture results of single brittle spheres, by both compression and impact, were relevant to the attrition processes. For instance, brittle spheres were in shot and bead-peening when subjected to impaction or uniaxial compression; in packed bed when exposed to multi-axial compression; and/or in bulk flow when impacted with other particles, stirrers, bluff bodies or vessel walls.

12.1.2.2. Multiple Particle Tests

Multi-particle tests comprise a different group of techniques that expose a bed of particles to shear, impact, compression, vibrational, or air convection motions. Some of these tests tend to reproduce the conditions of handling and storage operations during processing in order to give a realistic approach to attrition. All of the tests presented below will allude to particle strength.

Annular shear cell

Among the many different types of shearing cells, such as Jenike's direct single shear cell and rotational shear cell (as reviewed in Chapter 3), the annular shear ring tester developed by Paramanathan and Bridgwater is often used for attrition testing (Bemrose and Bridgwater, 1987). Its schematic diagram is illustrated in Fig. 12.3. The two main parts of this annular shear cell are the top-loading ring and base unit with inner and outer surface diameters of 120 and 160 mm, respectively. Both ring and base form the testing annular zone with a width of 20 mm. The top loading ring is restrained from rotating but can move freely in the vertical direction so that controlled normal force can be applied to the test material. An electric motor is used to rotate the base unit at a required rotation speed (up to 15 rpm). The upper and lower faces of the annulus contain radial saw-tooth grooves to prevent slippage, and "O" ring seals are used to prevent fine loss between the annular faces and the stationary horizontal pressure face.

A routine experimental procedure of the annular shear cell can be described as follows: (a) a reasonable amount of material sample is placed in the cell so that the distance between the two rings is between 2 and 20 particle diameters; (b) the base ring is rotated several times in the absence of normal

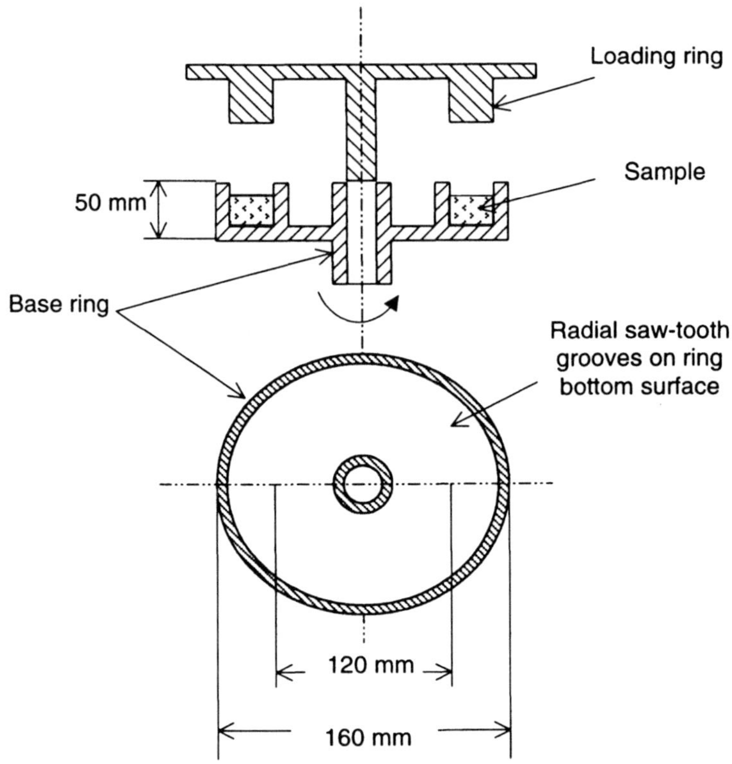

Figure 12.3. Schematic diagram of the annular shear ring for attrition test (adapted from Neil and Bridgwater, 1994).

force to distribute the material evenly in the ring annulus; (c) a normal force of known magnitude is applied to the material through the top ring, and the base ring is slowly revolved a controlled number of revolutions to apply a determined strain to the sample; (d) the sample is removed from the cell and prepared for particle size analysis (Potapov and Campbell, 1997).

By comparing the particle size distribution before and after the experiment, particle attrition can be evaluated in different ways, for example, by measuring the change of unbroken fraction mass with attrition time, or the formation of the fine fraction as a function of the applied stresses (Neil and Bridgwater, 1994). As described in Chapter 3 for rotational shear cells, the annular shear ring tester has a number of major advantages. Its design surpasses Jenike's shear cell in that it can be applied to a wide range of materials. Furthermore, shear surface area remains unchanged during the test, with no angular displacement limit. Apart from the fact that it requires a small sample size (<100 g for 2 mm in particle size), the simple design of the cell retains all of the attrition products (Suzuki, 1991; Neil and Bridgwater, 1994).

Confined impact and crushing tests

Two types of confined compression tests are often used in the coal industry to measure powder strength: those with high rates of force application and those with low rates. The high rate compression test is defined as the impaction on a bed of particles by an object moving under gravity. A cylindrical cell containing the sample receives a specified weight that is dropped into the powder from a certain height and for a certain number of times. The percentage of remaining powder within the initial size range is the attrition measurement, denominated as the *impact strength index*. In a slow compression test, a close fitting piston is applied continuously over a fixed period of time. The percentage of the remaining powder greater than a specified particle size is taken as the *crushing strength index* (Bemrose and Bridgwater, 1987).

Uniaxial confined compression test

As mentioned by Nuebel and Peleg (1994), the mechanical evaluation of some instant food powders (such as instant coffee) by single-particle compression tests is difficult and can be impractical for small particles. Clearly, it is much more convenient to assess their mechanical properties by testing particles in bulk. In this respect, the uniaxial confined compression test is most often used. A bed of powders is compressed in a rigid cylinder by a piston. The piston is in contact with the top surface of the powder bed and moves at a constant speed. Experimental data are recorded in terms of force–deformation or pressure–density relationships. More details about the compression test can be found in Chapter 3.

In certain cases, it can be assumed that the static pressure of agglomerates subjected during storage is similar to the compression pressure in a confined uniaxial compression cell, especially when the compression speed is very low. Thus, it is possible to study the attrition caused by static pressure by using compression tests (Gerhards et al., 1998). Compression tests may also be used to study the padding effect when some kind of cushioning materials, such as plastic foam, are added in the powder bed to reduce attrition caused by mechanical impact or vibration in real practice (Yan and Barbosa-Cánovas, 2002).

Vibration tests

Attrition can also be caused from the relative movement between particles induced by vibration in a bed of particles. It is necessary to determine the effect of handling processes, such as jarring, jostling and vibrating on the attrition of agglomerates in a controlled and reproductive manner. Particle movement may be introduced by using either a form of resonance or a simple mechanical motion transmitted from a container to the particles in it. Bemrose and Bridgwater (1987) described a 40 mm cylindrical drum mounted to a "vibro-saw," which can give a vertical vibration of 6 mm in amplitude at a frequency of 50 Hz. In these tests, the vibration intensity of the particles depends on particle size and density, vibration frequency and amplitude, and the depth of the bed.

A Tapping Density Tester (Vankel Industries, Inc., Edison, NJ) that provides vertical vibration is commercially available and often used for vibration tests (Fig. 3.15). A sample with known quantity (weight or volume) is freely poured into a graduated cylinder that rotates and taps simultaneously at controlled speed and amplitude, as shown in Fig. 12.4. Each tap is the result of the rotation of a cam that lifts the cylinder 3 mm high and drops it the same distance vertically. Therefore, the sample in the cylinder undergoes volume reduction and attrition because it is exposed to compaction compression and/or mechanical vibration. After a number of taps (i.e., vertical motions), the level of the sample in the cylinder is recorded and/or the particle size distribution is analyzed. This method is often used to study the compaction characteristics of powders and the attrition tendency of agglomerates. Research has been conducted for instant coffee, milk powders and other agglomerated food powders (Malave-López et al., 1985; Barletta et al., 1993b; Barletta and Barbosa-Cánovas, 1993a; Yan and Barbosa-Cánovas, 2001).

Fluidized bed tests

Fluidized bed tests are widely used in attrition research because they are directly related to the attrition behaviors of agglomerates or catalyst pellets under fluidized conditions. It is still argued

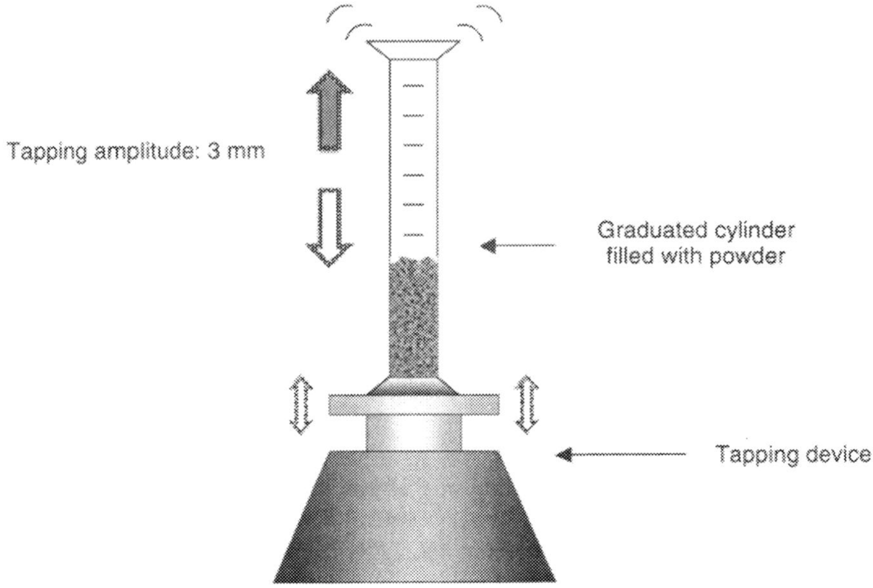

Figure 12.4. Graphical illustration of a tap density tester used for attrition test.

whether this test can simulate the attrition process to a reasonable degree for particles in hydraulic or pneumatic transportation pipes.

Designed by Forsy and Hertwig in 1949, the simple high-velocity air jet becomes a basic part for many subsequent procedures in fluidized bed tests. By using different fluidized bed designs (conventional fluidized bed, spouted bed, combination of spouted and fluidized bed), the attrition behavior of particles may be studied under a wide range of temperature, pressure, and operational and geometric parameters in order to simulate the conditions particles will face in a particular type of handling operation (Bemrose and Bridgwater, 1987).

A schematic diagram of a fluidized bed attrition apparatus used by Mullier et al. (1991) to study the effect of agglomerate strength on attrition during processing is shown in Fig. 12.5. It consisted of a fluidized bed, cyclone, and gas distributor. A cylinder of 0.145 m diameter and 0.6 m height was used as the fluidized bed. A flat-drilled plate with 139 holes (each hole = 1.5 mm diameter) was set on a 12 mm triangular pitch. The compressed air was used to fluidize the particles in the bed. The 0.15 m diameter cyclone was operated at a sufficiently high inlet velocity to achieve complete capture of the particles elutriated from the fluidized bed after undergoing attrition.

Many other methods are available for assessing the attrition tendency of particulate materials. Some of these include drop shatter tests, enhanced sieving, grindability tests, drum tests, and paddle wheel tests (Bemrose and Bridgwater, 1987). In addition, because of the popularization of computer use in many fields and the great availability of related software, it is possible to study attrition by computer simulation. For example, the attrition patterns of food fine particles and agglomerates were studied via computer simulation by Peleg and Normand (1987), while the computer simulation of shear-induced particle attrition was described by Potapov and Campbell (1997). Interested readers can get more information about computer simulation methods from the related references.

Knight and Bridgwater (1985) noticed that different methods for assessing attritability of particles might rank powders differently. They compared three tests for assessing attrition of some spray-dried detergent powders: single-particle compression, annular shear ring, and one test in which particles were blown around by air inside a spiral air classifier. They found that the single-particle

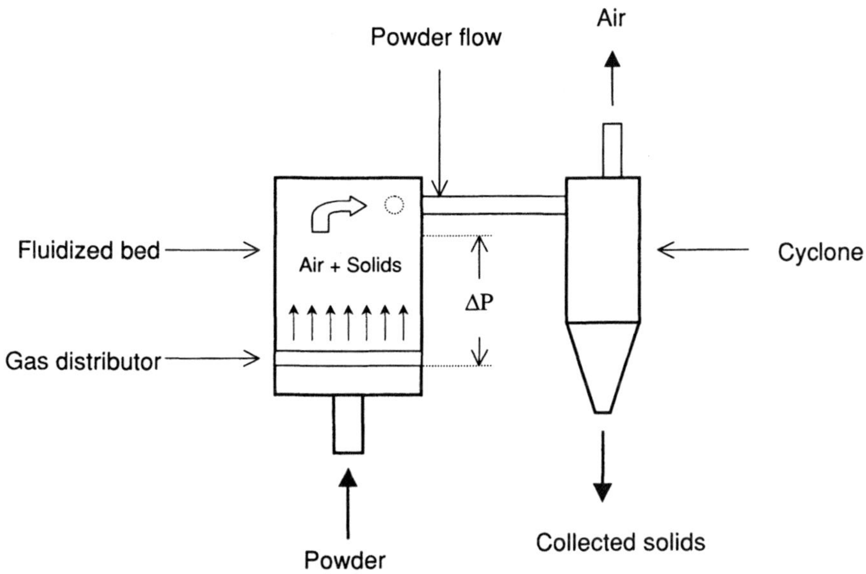

Figure 12.5. Schematic view of fluidized bed for attrition study (adapted from Mullier et al., 1991).

compression test did not provide a reliable attrition assessment in comparison with the other two methods, which agreed with their results. Even though the main common mode of breakage in these three methods was brittle fracture, quantitative differences were found. The total amount of tested particles were broken in the single-particle compression test, while particle breakage is presumably stochastic with a higher probability for only weaker particles in both shear ring and spiral classifier.

12.1.3. Attrition Theory and Modeling

The previous section mainly dealt with test methods to assess or simulate the attrition process. This section introduces a theoretical and systematic approach of attrition behavior from widely used attrition methods utilized for agglomerated food powders. Five perspectives will be analyzed: particle behavior under confined uniaxial compression, particle size distribution variations, kinetic study of the attrition process, compaction characteristics evaluation, and analysis of morphology changes of the particles by fractal analysis method. Different approaches to agglomerate attrition study in physical, mathematical and morphological aspects will be reviewed qualitatively and quantitatively.

12.1.3.1. Particle Behavior Under Confined Uniaxial Compression

The compression process usually takes place in three distinct stages: (a) rearrangement of agglomerates to fill voids equal to or bigger than the size of the agglomerates; (b) deformation or breakdown of agglomerates to fill voids smaller than the size of the agglomerates; and (c) rearrangement of primary particles by elastic, plastic, and fracture (Mort et al., 1994; Nuebel and Peleg, 1994). Those stages may be correlated to the different regions based on the force–deformation relationships of the agglomerates. A general compressive force–deformation relationship of instant skim milk of different particle size ranges is schematically demonstrated in Fig. 12.6. It has a characteristic sigmoid shape, with an almost linear force–deformation relationship at a small deformation region, followed by an upward concave force that increases over the continued deformation, and a very

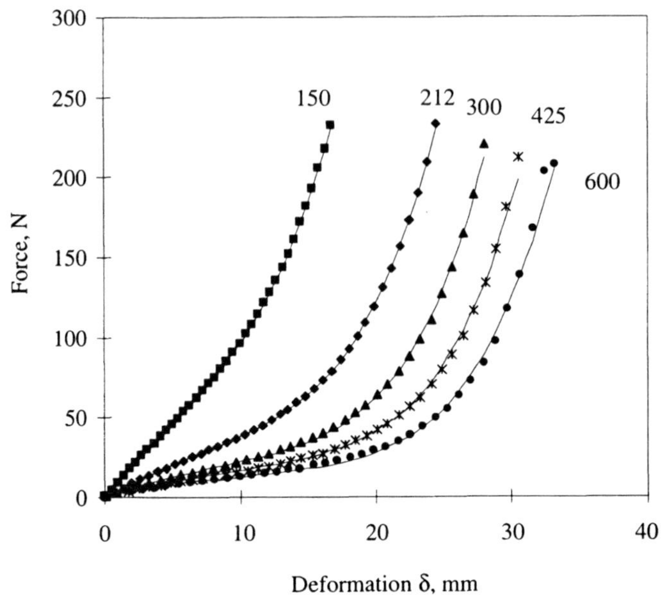

Figure 12.6. Typical force–deformation relationships for instant milk of different particle size ranges during uniaxial confined compression tests. Experimental data are represented by dots while curve fitting by Eq. (12.3) is represented by solid lines. Numerical values at the end of curve represent the agglomerate size range expressed as the sieve opening size in microns (adapted from Yan and Barbosa-Cánovas, 1997).

rapid force increment at higher deformation values. The first region represents the deformation of the still intact agglomerates, the second the breakdown or fracture of the agglomerates, while the last represents the further compression needed to increase the density of the agglomerates. It was noticed that larger agglomerates have a longer second region than the smaller agglomerates (Yan and Barbosa-Cánovas, 1997).

Many models have been developed to describe the force–deformation or pressure–volume relationships for agglomerated powders under compression. The model proposed by Swyngedau et al. (1991) was used by Yan and Barbosa-Cánovas (1997) to describe the force–deformation relationships of selected agglomerated food powders and to evaluate the effect of agglomerate size and water activity on the compression characteristics. It is expressed as (Nuebel and Peleg, 1994):

$$\sigma = A\varepsilon^C + B\varepsilon^D \tag{12.2}$$

where σ is the compression stress, A, B, C, and D are constants, and ε is the engineering strain (ratio of the absolute deformation with respect to the initial height). For the example shown in Fig. 12.6, Eq. (12.2) was used in its modified version, where force and deformation were applied instead of engineering stress and strain. This equation, shown below, was found suitable to fit experimental data of selected agglomerated food powders under compression (Yan and Barbosa-Cánovas, 1997):

$$F = A_1\delta^{B_1} + A_2\delta^{B_2} \tag{12.3}$$

where F is the compressive force, δ is the deformation length unit, A and B are constants, $B_1 < 1$ and $B_2 > 1$. A has the unit of the force over deformation and B is dimensionless.

Other models often used to describe the exponentially shaped compressive stress–strain relationships in many sponges, plastic forms and other cellular solids may also be used to describe similar relationships in agglomerated food powders under compression. Three models are included

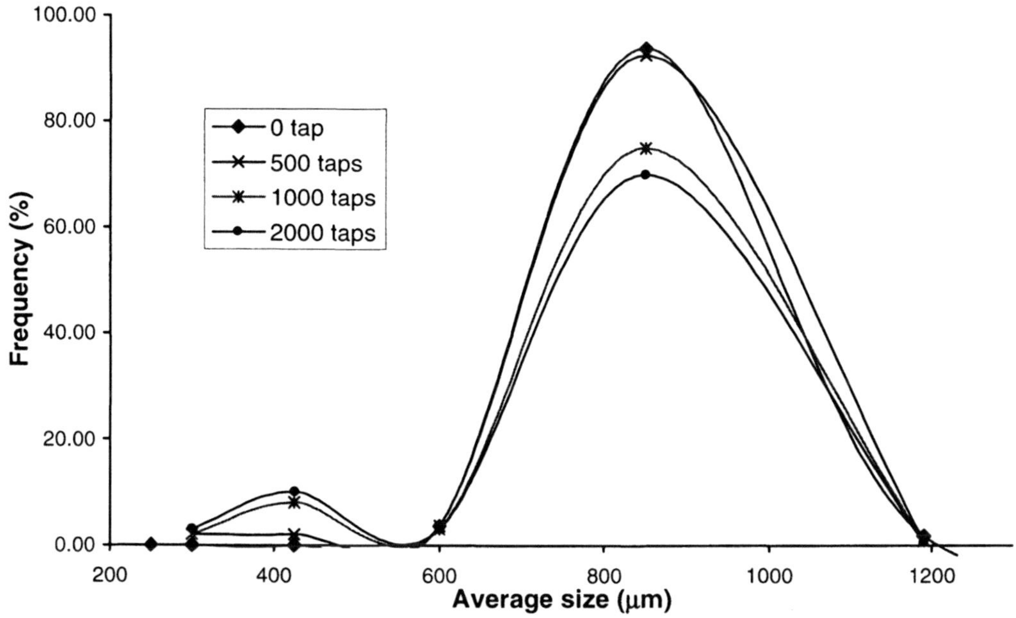

Figure 12.7. Bimodal particle size distribution of agglomerated milk exposed to tapping.

(Swyngedau et al., 1991):

$$\sigma = \frac{A\varepsilon}{(1 + B\varepsilon)(C - \varepsilon)} \tag{12.4}$$

$$\sigma = A\left(\frac{\varepsilon}{B - \varepsilon}\right)^{C} \tag{12.5}$$

$$\sigma = \frac{A[1 - \exp(-B\varepsilon)]}{C - \varepsilon} \tag{12.6}$$

where σ is the stress, A, B, and C are constants, and ε is the natural strain or Hencky's strain. The Hencky's strain is expressed as:

$$\varepsilon = \ln\left(\frac{H_0}{H_0 - \Delta H}\right) \tag{12.7}$$

where H_0 is the initial specimen height, and ΔH is the absolute deformation.

It has been suggested that uniaxial compression behavior could be related to agglomerate strength. Adams et al. (1994) used a simple first-order parameter to analyze the compression process by treating the process as purely dissipative. Strain results were studied as individual agglomerate failure, which was analyzed through the Mohr–Coulomb macroscopic failure criterion. They found that there was a linear correlation between the agglomerate strength measured directly by single agglomerate compression tests and those calculated from the pressure–volume data of uniaxial compression tests by using the model they proposed. They also suggested that any strength expression from the force–deformation relation in compression tests must include some measure of strength when tested individually. However, as there are many variables (such as initial filling density, compression rate, agglomerate size distribution, binder effect, and so on) involved in the compression process, this type of test often yields little comparative information as to the strength of different agglomerates (Park et al., 1993).

Usually, brittle and fragile particles (especially agglomerated ones) have jagged compressive stress–strain relationships when they are tested individually or in bulk. The compressive stress–strain relationships of instant coffee were studied by Nuebel and Peleg (1994), who found that the jaggedness signature of single particles could not be directly correlated to that of the stress–strain relationship when they are compressed in bulk. This is mainly because of the effects of averaging and bed crushing. In other words, the measurement of stress–strain is influenced simultaneously by both the breakage properties of each individual particle and the deformation ability of their assembly as a whole in the compression cells.

12.1.3.2. Particle Size Distribution Variation

Particle size analysis is a useful tool for assessing attrition because particle fragmentation and fine formation are involved in this process. Some of the conventional size reduction theories take the mean particle size as the principal variable. Other theories only account for the disappearance rate of the original fraction or the formation rate of a selected fraction. As described before, production of mid-sized particles by shattering will lower the particle population's mean size and increase its size spread. On the other hand, formation of fines by surface erosion will make the overall size distribution bi-modal or multi-modal. Furthermore, progressive changes in the size, spread, and weight ratio of the fines and coarse particle populations depend on the mechanical energy input and the relative contribution of the two attrition mechanisms (Popplewell et al., 1988). As seen in Fig 12.7, milk powder shows increasing bi-modal distribution with increasing tapping applications. The attrition curves show the relationship between the percentage of agglomerates that retain their original size as a function of a certain attrition time or number of taps (Barletta et al., 1993a). Therefore, the application of bimodal size distribution functions appears to be a very suitable and sensible way to evaluate attrition in particles.

A commonly used mathematical form $f(x)$ for bimodal distribution functions is expressed as follows (Popplewell et al., 1988, 1989):

$$f(x) = wf_{\rm f}(x) + (1 - w)f_{\rm r}(x) \tag{12.8}$$

where $f_{\rm f}(x)$ represents the size distribution of fines, $f_{\rm r}(x)$ is the size distribution of size-reduced particles, and w is the fines weight fraction. Both the distribution functions $f_{\rm f}(x)$ and $f_{\rm r}(x)$ can then be expressed in the form of modified β distribution $f_{am}(x)$:

$$f_{am}(x) = \frac{x^{am}(1 - x)^m}{\int_0^1 x^{am}(1 - x)^m \, dx} \tag{12.9}$$

where a and m are constants, and x is the normalized length given by:

$$x = \frac{X - X_{\min}}{X_{\max} - X_{\min}} \tag{12.10}$$

where $X_{\min}$ and $X_{\max}$ are the smallest and largest particle sizes in the fine or size-reduced particle group, respectively, and therefore $X_{\min} < X < X_{\max}$, $0 < x < 1$.

A hybrid model of the β and normal distribution functions can also be used in Eq. (12.1.7) to describe the uni-modal size distribution either in the produced fines or size-reduced particles (Peleg and Normand, 1987). It is defined in the form:

$$f(x) = \frac{\dfrac{x(1-x)}{x(1-x)+C}\exp\left[-\left(\dfrac{x-M}{S}\right)\right]^2}{\int_0^1 \dfrac{x(1-x)}{x(1-x)+C}\exp\left[-\left(\dfrac{x-M}{S}\right)\right]^2 \mathrm{d}x} \tag{12.11}$$

where C is an arbitrary constant which is $\ll 1$, M is the approximate mode, S is the spread measure of the particle size distribution, and x is the same normalized particle size as described in Eq. (12.11).

The application of bi-modal size distribution functions has proven to be a successful way to provide quantitative and comprehensive information for the attrition process. The magnitude and change of distribution parameters can provide quantitative measure of the effects of the two principal attrition mechanisms and they also provide a mathematical tool to monitor the attrition kinetics (Popplewell et al., 1988). Furthermore, by introducing some time factors to represent the weight increase of fines in Eq. (12.8) or the decrease of coarse fraction mode with time in Eq. (12.11), the modified mathematical model can be applied for computer simulation of the attrition process (Peleg and Normand, 1987; Popplewell et al., 1989).

A more convenient way to characterize bi-modal size distributions while avoiding detailed description is by using the simple erosion index E proposed by Popplewell and Peleg (1989). E is defined as the ratio of fines accumulation A_f to the size reduction of the coarse fraction R_c. They are expressed as:

$$E = \frac{A_\mathrm{f}}{R_\mathrm{c}} \tag{12.12}$$

$$A_\mathrm{f} = \frac{F(t) - F(0)}{1 - F(0)} \tag{12.13}$$

$$R_\mathrm{c} = \frac{M_\mathrm{c}(0) - M_\mathrm{c}(t)}{M_\mathrm{c}(0) - M_\mathrm{f}} \tag{12.14}$$

where $F(t)$ is the weight fraction of fines and $M_\mathrm{c}(t)$ is the coarse fraction mode after time t, $F(0)$ is the initial fine weight fraction, $M_\mathrm{c}(0)$ is the initial coarse fraction mode, and M_f the constant mode of fines. If $F(0) = 0$ (no fines at the beginning) and M_f is much less than $M_\mathrm{c}(0)$, Eq. (12.11) may be simplified as:

$$E = \frac{F(t) \times M_\mathrm{c}(0)}{M_\mathrm{c}(0) - M_\mathrm{c}(t)} \tag{12.15}$$

From Eqs. (12.12) and (12.15), it can be deduced that E is a dimensionless value ranging from zero ($F(t) = 0$ or $F(t) = F(0)$), i.e., there is no erosion, to infinity, ($M_\mathrm{c}(0) = M_\mathrm{c}(t)$), which means the coarse fraction mode has no change after a certain attrition processing time.

12.1.3.3. Attrition Kinetics

In attrition kinetics study, using attrition curves is one of the most convenient ways to present attrition data. A one-term exponential model is expressed as (Barletta and Barbosa-Cánovas, 1993a):

$$A_\mathrm{I} = CT^\alpha \tag{12.16}$$

where A_I is the attrition index, C and α are constants, and T is the number of taps. A_I is defined as:

$$A_\mathrm{I} = \frac{F}{R} \tag{12.17}$$

where F is the weight fraction of fines generated in the attrition test and R is the size stability related to agglomerates retaining original particle size. R can be calculated by the following formula:

$$R = \frac{\sum_{i=1}^{i=L} W_i S_i \mid_{T=n}}{\sum_{i=1}^{i=L} W_i S_i \mid_{T=0}} \tag{12.18}$$

where W_i is the amount of material retained on each sieve, S_i is the normalized sieve opening size, and T is the number of taps. The one-term exponential model (12.16) was found suitable and effective to study the effects of agglomerate size and water activity on the attrition kinetics of agglomerated coffee and nonfat milk (Yan and Barbosa-Cánovas, 2001).

A single exponential decay model is expressed as (Austin et al., 1981):

$$W_R = W_0 \exp(-AT) \tag{12.19}$$

where W_R is the weight of particles retaining their original particle size, W_0 is the initial sample weight, T is time, and A is rate constant. It has been found that this model only holds for the initial part (<3 min) of the attrition curve for agglomerated coffee; after that, it predicts higher attrition values than the agglomerates actually have (Barletta et al., 1993a).

A two-term exponential model has the following expression (Malave-López et al., 1986):

$$W_{R/O} = F \exp(-A_1 n) + (1 - F) \exp(-A_2 n) \tag{12.20}$$

where $W_{R/O}$ is the weight ratio between agglomerates retaining their original particle size and the total sample, n the number of taps, and F and $(1 - F)$ are the fraction of materials that have undergone attrition at the rates of A_1 and A_2, respectively. This model was found to be better in describing the attrition kinetics of agglomerated coffee with original particle size bigger than 16 mesh or 1,180 μm. The reason for such a good fit may be explained as that found for large agglomerates; the attrition is governed by two different mechanisms (i.e., shattering and surface erosion) and each one has its own characteristic rate.

A non-exponential model is suggested by Malave-López and Peleg (1986):

$$W_R = 1 - \frac{n}{B_1 + B_2 n} \tag{12.21}$$

where W_R is the weight fraction of particles retaining their original size after n taps, and B_1 and B_2 are constants; and when $n \to 0$, $1/B_1$ can be considered as the initial attrition rate and $n \to \infty$, $1/B_2$ is the asymptotic weight fraction of material under attrition. This non-exponential model has some advantages over both exponential models. First, it has a residual (asymptotic) fraction of surviving agglomerates, which exist in real attrition tests, while exponential models imply that all agglomerates with the original size will disappear. Second, it generally fits the experimental data very well, and its two constants can be easily calculated by regression models (Barletta et al., 1993a).

12.1.3.4. Compaction Characteristics

Powders are known to be compressible, and their considerable density increase can occur as a result of the application of vibration, impact, or mechanical compression. When agglomerated powders are under such compaction load, their compaction behaviors will be much different from those of non-agglomerated particulate materials. Agglomerates will undergo a relatively higher degree of attrition. In addition, particle position rearrangement will cause the common compaction effect (Barletta et al., 1993a). Thus, compaction models that deal with density changes should be able to hold the combined effects of both attrition and compaction mechanisms for agglomerated powders.

In the case of compaction by tapping, the following relationship between bulk density changes and number of taps was proposed by Sone (Barletta et al., 1993b):

$$\rho_\infty - \rho_n = C \exp\left(-\frac{n}{K}\right) \tag{12.22}$$

where ρ_∞ is the asymptotic bulk density, ρ_n is the density after n taps, and C and K are constants.

Another model, also proposed by Sone, is widely used and commonly referred to as Sone's model. It relates the volume (or density) reduction fraction $Y(n)$ and the number of taps n as follows (Peleg, 1983):

$$Y(n) = \frac{V_0 - V_n}{V_0} = \frac{\rho_n - \rho_0}{\rho_n} = \frac{n}{A + Bn} \tag{12.23}$$

where V_0 is the initial volume, V_n is the volume after n taps, ρ_0 and ρ_n are the respective densities ($\rho_0 = m/V_0$), and A and B are constants.

A single exponential model proposed by Malavé-López and Peleg (1986) is expressed as:

$$Y(n) = C\left[1 - \exp\left(-\frac{n}{N}\right)\right] \tag{12.24}$$

where C is a constant, n is the number of taps, and N is a constant characteristic of the system.

Some models containing three or four parameters have also been proposed. A three-parameter model, known as Chapman–Richard's model, is defined as:

$$Y(n) = A\left[1 - \exp\left(-\frac{n}{N}\right)\right]^C \tag{12.25}$$

where A, C, and N are constants.

A double exponential model proposed by Barletta et al. (1993b) is expressed as:

$$Y(n) = C_1\left[1 - \exp\left(-\frac{n}{N_1}\right)\right] + C_2\left[1 - \exp\left(-\frac{n}{N_2}\right)\right] \tag{12.26}$$

where C and N are constants.

A four-parameter model, known as the modified Sone's model, is expressed as (Barletta et al., 1993b):

$$Y(n) = \frac{n}{A + Bn} + \frac{n}{C + Dn} \tag{12.27}$$

where A, B, C, and D are constants.

The compaction characteristics of agglomerated coffee during tapping were studied by Barletta et al. (1993b) using some of the models mentioned above. They found that the two-parameter models (i.e., the single exponential model in Eq. (12.18) and Sone's model in Eq. (12.22)) did not well represent the density change data, even though those models were found suitable to describe the compaction kinetics in non-agglomerated powders. The three- and four-parameter models have shown considerable improvement in experimental data fitting. These results suggest that a model of three or four parameters is more suitable to describe the compaction phenomenon of agglomerated powders because there is a combined effect of simultaneous compaction and attrition, while the two controlling attrition mechanisms (i.e., particle shattering and erosion) have different characteristic times (Barletta et al., 1993a).

In addition to describing compaction behavior by using these models, the simple and commonly used Hausner ratio, which deals directly with density changes after tapping, may be introduced by correlating it to the parameter B in Eq. (12.23). The Hausner ratio, defined as the ratio of the asymptotic bulk density ρ_∞ to the initial density ρ_0, can be related to a particular B which represents the asymptotic value of $V(n)$ (i.e., when $n \to \infty$, $V(n) \to 1/B$). Their relationship is expressed by:

$$H = \frac{\rho_\infty}{\rho_0} = \frac{B-1}{B} \tag{12.28}$$

The Hausner ratio has been employed as a useful quality criterion to evaluate powder flowability because it is well correlated with the presence of attractive forces and friction in the powder bed (Hayes, 1987). It is also an important practical parameter to qualify the maximum powder compressibility under vibration, which has obvious implications in handling and filling operations of food powders (Malave-López et al., 1985). In the attrition study of some selected agglomerated food powders carried out by Yan and Barbosa-Cánovas (2001), the Hausner ratio was calculated after each tapping test and compared to the attrition index under different test conditions. It was found that the Hausner ratio was closely related to the attrition index at high numbers of taps (i.e., more than 5,000 taps) for both agglomerated coffee and milk. At low numbers of taps, the Hausner ratio (around 1.4) was mainly a measure of volume reduction caused by agglomerate rearrangement or reorientation in the test cylinder. But at high numbers of taps, the Hausner ratio was greater than 1.4 as a result of agglomerate attrition that caused the agglomerate breakdown. Consequently, fines passed through the powder bed, promoting further powder volume reduction. Thus, the Hausner ratio might be used as a simple index to evaluate the attrition severity of agglomerates at extreme test conditions, i.e., very high tap numbers.

12.1.3.5. Fractal Approach

As attrition causes both surface erosion and particle shattering, evaluation of particle shape changes may be a useful tool to describe the eroding process in particulate materials, in addition to traditional particle size analysis methods (Barletta et al., 1993a). Like other particle shape factors, fractal dimension can also provide useful information about particle shape in the form of single numerical numbers (Simons, 1996). The "fractal dimension" concept was first proposed by Mandelbrot (Kaye, 1989) and has been successfully applied in many disciplines such as sedimentology, geology, and morphology. Over the years, the use of fractal dimension to characterize the agglomerate structures has received much attention because of its ability to provide quantitative information on the openness and ruggedness of agglomerates (Simons, 1996). Fractal dimension can be used to characterize the attrition process in agglomerated food powders. It is a sensitive attrition index, based on the fact that attrition causes changes in particle shape and surface, on the scale that the fractal approach is applicable (Peleg and Normand, 1985a; Olivares-Francisco and Barbosa-Cánovas, 1990, Gerhards et al., 1998).

When the fractal approach is applied to analyze two-dimensional images of particles with a ragged shape, it relies on the fact that the perimeter of a self-similar silhouette is dependent on the "step length" with which it was measured. Here, the true self-similar profile is the one that exhibits (either approximately or statistically) the same degree of roughness at any level of scrutiny in the examination of particle boundary lines (Barletta et al., 1993a). Since there are no fractal particles with such a true self-similar profile in nature, the "natural" or "apparent" fractal concept was developed to overcome the theoretical difficulties of self-similarity. By definition, the objects of natural fractals have some form of self-similarity in a given range of magnification under which the profile of the objects is scrutinized. Different ranges of magnification in agglomerated particles have been suggested in order to find self-similarity. A range from 0.001 to 0.5 times the maximum Feret's diameter is commonly recommended (Barletta and Barbosa-Cánovas, 1993b).

In early days, a manual technique known as the "structural walk" or Richardson's "hand and dividers method" was used to estimate the boundary length of the profile of a particle to give an indication of the ruggedness of the particle. In this method, an arbitrary start point is chosen on the edge of the particle silhouette. The next point is determined with the dividers set at a certain step length, using either a clockwise or counter-clockwise protocol. The measuring procedure is processed until the start point is again encountered to form a polygon. An estimation of the silhouette perimeter is calculated by adding the length of all the steps plus the exact length of the last side of the polygon (in case the last side of the polygon is not of the step length). The measurement of the perimeter of the silhouette is repeated by using several different step lengths and it gives different perimeters in each case. The perimeter L_P of the corresponding polygon is expressed as (Peleg and Normand, 1985a):

$$L_P = n\lambda \tag{12.29}$$

where n is the number of the polygon's sides for a particular selected step length λ.

A logarithmic plot of the perimeter L_P against different step length λ is called a Richardson plot. The plot produces a curve with a negative slope, demonstrating that the perimeter increases as step length decreases. Thus, the curve steepness is an indication of the ruggedness of that particle silhouette, and the fractal dimension D_F of the profile is expressed as:

$$D_F = 1 + \tan|\alpha| \tag{12.30}$$

where α is the slope of the logarithmic plotted curve.

Determination of the fractal dimension of objects by the manual method mentioned above is simple in principle, but extremely laborious and time consuming in practice. Furthermore, its requirement for concentration and repetitive work makes it unsuitable for processing a large number of images (Peleg and Normand, 1985a). Therefore, computer image analysis techniques are much more desirable and various computer algorithms have been suggested by a number of researchers to undertake this task (Allen et al., 1995). Thus, fractal analysis can be accomplished by writing computer algorithms when a set of edge points describing the particle outline is available by using equipment (such as scanning electron microscopy or optical microscope) to take a digitized particle image. There are three computer algorithms usually used in calculating fractals of a two-dimension profile: *exact*, *fast*, and *hybrid* algorithms. Figure 12.8 shows how each algorithm traces a given contour expressed by a set of edge points.

The *exact* algorithm, as shown in Fig. 12.8a, starts at an arbitrarily selected point at the contour with a fixed yardstick. It is a computerization process similar to Richardson's hand and dividers method. The perimeter of the contour is calculated by adding the total side length of the polygon constructed with equal side length, plus the last step length. A fixed yardstick in each iteration makes the final perimeter calculation simple and fast but it may produce a polygon whose points are not necessarily the exact ones on the contour. Some details of the contour are missed when a relatively large stride length is used. This algorithm can produce an accurate fractal analysis but it is still cumbersome and time consuming when compared to the other two algorithms mentioned below.

The *fast* method chooses a starting point arbitrarily and constructs an irregular polygon with each point along the outline. The step length is obtained as the average distance between two consecutive points in the particle contours. The step length is then increased to a distance equal to an average of every second consecutive point; therefore, the second point of the polygon will be the third point of the boundary, the third point in the polygon will be the fifth point in the boundary, and so on. In the same way, the step length may be increased to the average distance of every three, or four consecutive points, until a distance of 30 consecutive points is reached. The sum of all the polygon sides is the perimeter, and the average value of each step distance is the step length.

Figure 12.8b shows how the algorithm follows the contour when the step length is equal to three consecutive points. This method is simple to implement and substantially faster than the exact

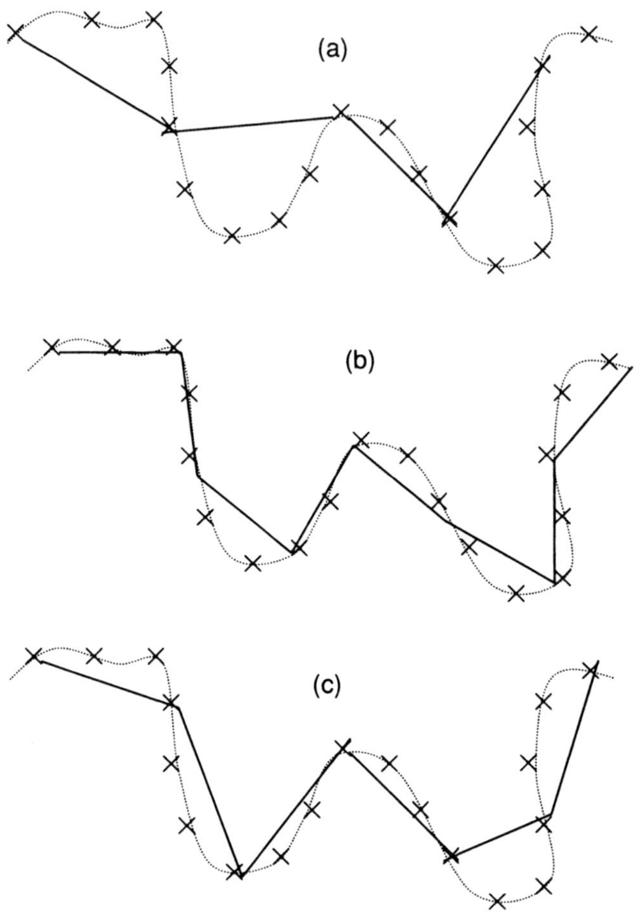

Figure 12.8. Three algorithms often used to trace a set of edge points on a two-dimension profile to calculate the fractal dimension: (a) exact; (b) fast; and (c) hybrid (adapted from Barletta and Barbosa-Cánovas, 1993b).

method but can be considerably inaccurate when the profile has a varied degree of ruggedness. Another disadvantage is that the distance between consecutive points is not always the same, so an average step length must be calculated, therefore introducing substantial scatter in subsequent Richardson plots, and hence hindering the accuracy of fractal dimension evaluation (Allen et al., 1995).

The *hybrid* algorithm was developed to overcome the major disadvantages that both the *exact* and *fast* algorithms have and to compromise between the speed of the *fast* and the accuracy of the *exact* (Allen et al., 1995). As shown in Fig. 12.8c, the step length, rather than a number of profile points, and directional protocol are given at a starting point. The algorithm then finds the next discrete point in the sequence of edge points that is the closest to the step length from the starting point, and that discrete point becomes the second vertex of the polygon. Because only discrete edge points are considered as vertex points, no interpolation occurs. The exact distance between the vertexes is stored as the actual step length, and the process is repeated until the starting point is encountered.

Theoretically, the value range for D_F is from 1 to 2 when a two-dimension contour is examined. When a curve with a D_F of 1 is representative of a straight line, a curve with D_F of 2 is too rugged and re-entrant as to fill some area of a plane entirely (Clark, 1986). But for real particles with different origin and shape, such as carbon aggregates, mineral, protein, catalysts, and instant coffee, their fractal dimensions are limited to the range of 1.05–1.36. Even though there have been reports about producing individual particles with a contour fractal dimension of about 1.5, no evidence has shown that they are stable in bulk. Therefore, it appears that particles with D_F close to 2.0 are not likely to exist because of inherent mechanical instability (Peleg and Normand, 1985b; Normand and Peleg, 1986).

By using commercially available instant coffee and instant skim milk as model agglomerated food powders, Barletta and Barbosa-Cánovas (1993b) used the fractal analysis method to characterize the ruggedness changes in powders after an intended attrition process in a tap density tester. They found that the fractal dimension was sensitive enough to detect the ruggedness changes in the model powders even when small tap numbers were used in the attrition test. It has also been found that the fractal dimension of the agglomerate silhouette decreases as the number of taps increases. It is believed that this method can provide a useful tool for quality assessment of instant food powders and other particulate materials where attrition is unwanted.

12.1.4. Attrition Prevention

Unlike other unwanted processes mentioned in the following sections, attrition prevention has not been studied systematically, even though great effort has been made by manufacturers to minimize attrition during processing and handling of agglomerated powders. Many variables have been involved in attrition mechanisms throughout product-handling history. Therefore, more attention has been paid to solving the problems caused by attrition rather than preventing it in real practice.

The effect of agglomerate size and water activity on attrition of selected agglomerated food powders has been studied based on an attrition index after conducting tapping tests (Yan and Barbosa-Cánovas, 2001). It was found that the larger agglomerates with higher water activity suffered more severe attrition under the same number of taps. Thus, it is reasonable to say that the attrition of agglomerated food powders can be minimized by properly controlling the product particle size, the water activity of the product, and the environment relative humidity.

Alternatively, a widely used technology to reduce impact damage in fruits and vegetables during harvesting and processing might provide a promising and practical solution to minimize attrition. It is reasonable to assume that some kind of cushioning material in the powder bed or on the container wall could absorb or reduce the mechanical or static impact on the agglomerates and, consequently, reduce their degree of attrition. The padding effect on agglomerated food powders can be assessed by using the commonly used confined uniaxial compression test or tapping test (Yan and Barbosa-Cánovas, 2002).

12.2. SEGREGATION

Segregation usually refers to the separation of the material from coarse material during the motion of a powder system or mechanical vibrations (either horizontal or vertical) of a powder bed. It also refers to the phenomenon that particles of different properties are distributed preferentially in different parts of the powder bed, which has been simply described as demixing (Brown and Richards, 1970; Peleg, 1983). In a closed system, segregation is often featured by the fine particles migrating to the bottom and larger particles floating and concentrating at the top of a powder bed. On many occasions in plants and equipment dealing with particulate materials, segregation occurs in free-flowing powders if there are differences in particle size, density, or shape. Segregation also occurs when there is electrostatic charging on the equipment even when the powder is in motion. An internal source of continuous production of fines and a simultaneous continuous change in particle size distribution in the powder system can also cause segregation (Barbosa-Cánovas et al., 1985).

Segregation can lead to changes in the material's appearance, texture, and flowability resulting in production difficulties, plant shutdown, and safety hazards (McGlinchey, 1998). The most frequent problem is demixing, i.e., particle resistance to mixer actions by segregation. Even after a satisfactory mixture is achieved in, the uniformity of the mixture will not remain unchanged during subsequent handling processes. In packaging machines in which equal volumes of material are discharged, segregation causes fluctuations in particle size distribution; this in turn results in variations in bulk density and content weight in packages (Williams, 1976). The completeness and speed of chemical reaction in a powder bed can be adversely affected if particles of different size are non-uniformly distributed in the bed as a result of segregation (Brown and Richards, 1970). A more common problem caused by segregation is related to sampling, i.e., it is impossible to get an unbiased sample when there is segregation in the target material, raising the question of reproducibility of particle size distribution measurement. In fact, segregation can occur in industrial situations. When a powder is shipped from the supplier to the customer, the average particle size value of the sample can be shifted (Parsons, 1976). In this case, it is hard to tell whether the shifting is caused by wrong sampling, segregation, or attrition.

Many food powders are mixtures of basic ingredients, nutrients, flavoring agents, colorants, and other materials of specific functionalities. Thus, segregation in food powders may have severe consequences. Segregation can cause functional or nutritional problems when the segregating ingredients are at low concentration due to the production of uneven quality products. For example, segregation control is crucial in baking powder addition in cake mixes or in yielding the correct proportion of vitamins in baby formulas. The segregation of fines to the container's bottom is one of the most common causes for removal of instant coffee from supermarket shelves due to its adverse effect on product appearance (Barletta et al., 1993). Segregation is usually undesirable when the homogeneity of mixtures is required. Avoiding segregation is a challenge in the food industry where materials with a wide range of physical and chemical properties are often mixed and a completely homogenous final mixed product is expected.

12.2.1. Segregation Applications

Segregation is encountered in various operations accompanied by formation or flow of solid mixtures. It usually occurs when free or easily flowing powders, having a significant range of particle size, are exposed to gravitational, rotational, vibratory, or aeration operations, or other types of mechanical motion. In *gravitational* operations, such as food powder mixtures in a hopper during discharge or moving powder beds during handling, segregation influences the piling up of solids during storage. Air current segregation during bin or hopper filling is a common occurrence in high volume handling. During free fall into a bin, air drag forces create a downward current, which proceeds along the surface of the powder in the bin towards the bin walls and the top of the bin. In this way, the falling powder induces fairly strong air currents, and the flow of air will carry the small particles towards the silo walls, where they settle out (Enstad, 2001). The opposite will tend to occur when emptying the silo, when percolation takes place, as will be further explained below.

As seen in Chapter 9, different dynamics are observed during *rotational* operations or mixing, where finer or less coarse particles congregate in the center of the rotating device if angular speed is low. When tumbling or stirring, if particles are coarse enough, rolling and sieving mechanisms will occur in the sliding layers on sloping surfaces continuously created in the equipment, thus decreasing mixing efficiency (Enstad, 2001). Particularly, in a V-shaped mixer, the denser component is apt to settle in any stagnant region (Shinohara, 1997).

In a container under vertical *vibrations* (e.g., a vibrating trough) smaller particles tend to descend below coarser ones, while larger particles lift up through the particle bed. Regarding *aeration* convective operations, if the bed is fluidized at low air velocity, the larger or denser particles become the segregating components, as in a liquid. In this case, segregation is caused by a difference in density rather than in particle size, and can be reduced at higher air velocity. On the other hand, in pneumatic conveying, larger or denser particles congregate near the center of the pipe cross-section. At higher gas velocity, the situation in the radial direction can become reversed, depending on the food particle composition, density, and shape.

12.2.2. Segregation Mechanisms

It has been noted that if a given system contains particles that differ in their properties, it will tend to show segregation. Among the properties that result in segregation, including differences in particle size, density, shape, and resilience (or elastic recovery), particle size difference is the most important; density difference is not as important, even though most people may expect it to be (Williams, 1976). The segregation process usually occurs in free-flowing powder systems of particles with a wide particle size distribution. Under motion or vibrating conditions, smaller particles will migrate to the bottom of the powder bed, while the coarser particles will remain on the top. The smaller particles could be the original fine ingredients in the powders or they may come from the mechanical attrition of the original coarse product itself. This kind of segregation depends on the mobility of the smaller particles and their ability to sift through the coarse particle array. Segregation is less likely to occur in cohesive powders because the fines in such powders usually adhere to the surface of coarse particles. Thus, inter-particle bonding mechanisms permit particles to move only with an associated cluster of particles, thereby impeding segregation (Lindley, 1991; Peleg, 1983).

The segregation process is governed by the physical properties of particles, the geometrical features of conveying, mixing, and storage systems, and the intensity of mechanical motions (Peleg, 1983). Three main segregation mechanisms were identified by Williams (1976): (a) trajectory segregation; (b) inter-particle percolation or sifting; and (c) the rise of coarse particles on vibration or upthrusting. These mechanisms are represented in Fig. 12.9.

Trajectory segregation takes place when particles of different sizes are projected horizontally with velocity V_0 into a fluid of viscosity η. Along the horizontal distance L, the particles will travel in laminar flow, called "stopping distance," and is expressed as (Williams, 1976):

$$L = \frac{V_0 \rho_S D^2}{18\eta} \tag{12.31}$$

where ρ_S is the particle density, and D is the particle chamber diameter. From this equation, it is clear that a larger or denser particle will travel a longer distance and cause segregation based on particle size and/or density difference. This kind of segregation will also occur when particles have different initial velocities or when their velocity changes as a function of their physical properties. This mechanism can explain why some segregation occurs in a mixer where particles are lifted out of a mass and thrown with the intention of scattering them across the surface. There are some circumstances in which a cloud of particles is flying or feeding, and fine particles may travel further and have a wider spread than coarse particles. This is because the fluid drag effect on fine particles in the cloud is not the same as on individual coarse particles (Williams, 1976). From Eq. (12.31), it can be seen that particles that double in diameter will travel four times as far; some segregation will occur, and this will limit the quality of mixing achieved.

Inter-particle percolation is the most prevalent mechanism when a mixture of particle sizes is being handled. When there is a big particle size difference, smaller particles will be allowed to drain

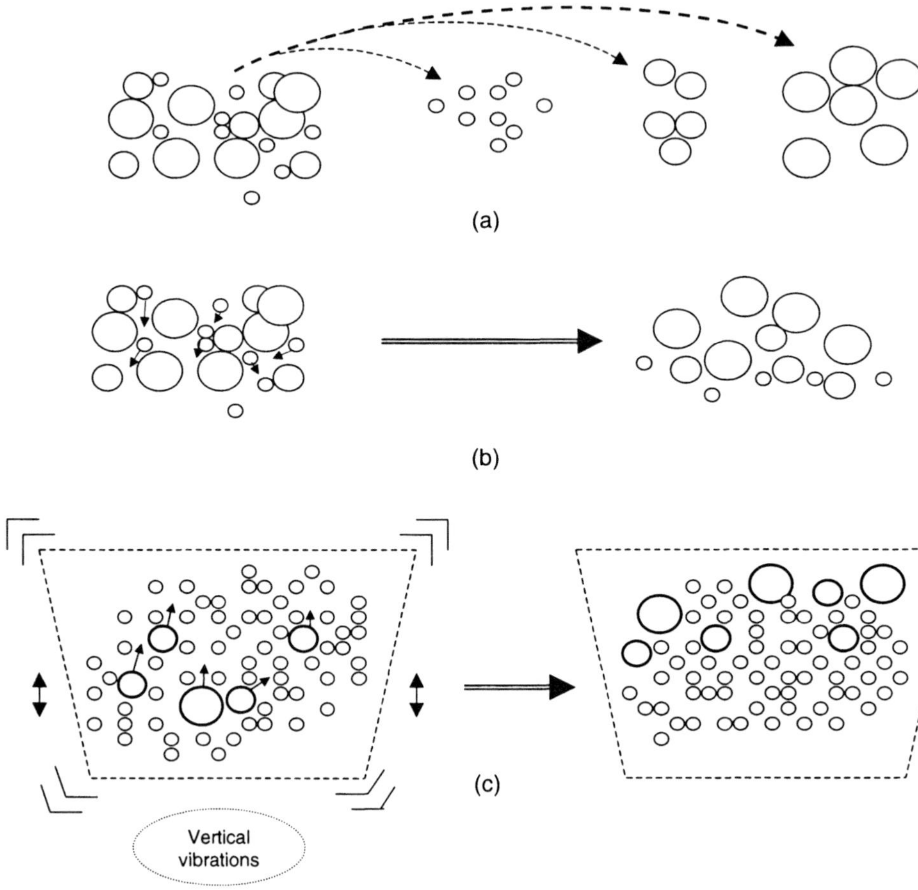

Figure 12.9. Three segregation mechanisms: (a) trajectory segregation; (b) percolation segregation; and (c) upthrusting or rise of particles upon vibration.

through the lattice of larger particles, simply due to gravity or motion. For this effect to occur, it is not necessary for the fines to be smaller than the voids that larger particles form when they are at rest. A very small difference in size is enough for measurable segregation to take place, even when the void fraction is high and/or the powder is in motion (Lindley, 1991). Percolation can also take place when a particle bed is vibrated. When particle shearing, shaking, or vibration occurs, the gaps between particles will change to allow particles from above to fall down and from other places to move up.

Segregation during the formation of a heap of powders is a good example to illustrate particle percolation, and it deserves particular attention. In general, there is remarkable segregation in a heap of free-flowing powders containing particles of different sizes because the fine particles are concentrated in a region near the axis of the heap, with the coarse ones in the outer part. The percolation mechanism (i.e., sifting) describing particle rearrangement in a hopper or bin was reviewed in Chapter 4. During powder storage in bins and hoppers, sifting can be significant when there is a horizontal shear movement of particles (e.g., rolling or sliding) on a free pile surface. In particular, Fig. 4.3 shows particle segregation followed by remixing in a mass flow bin, when flow occurs along the bin and hopper walls. The powder heap has a predominance of fines towards the center of the bin due to

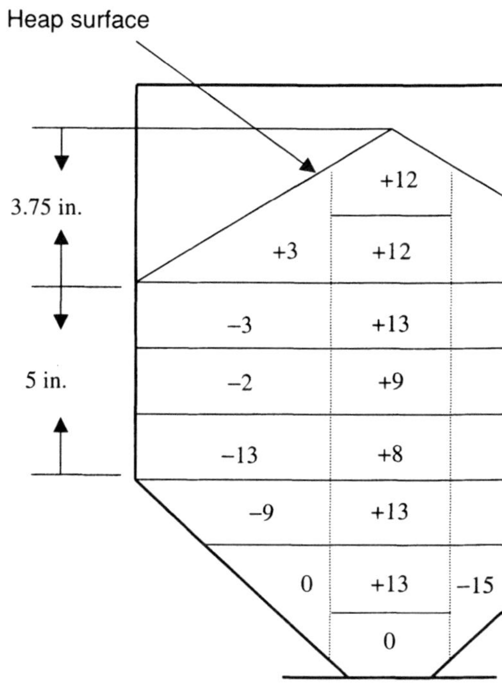

Figure 12.10. Segregation in a heap formed after central loading in a hopper is demonstrated as the excess fines in the center and its deficiency near the walls (expressed as percentage of total), when compared to the standard mix (adapted from Brown and Richards, 1970).

sifting during bin discharge. The excess of fines in the center and its deficiency near the walls of a hopper after central loading is illustrated schematically in Fig. 12.10.

Segregation occurring during angle of repose measurement in a heap (see Chapter 3) is especially relevant, given the importance of this test, which is widely used in the food powder industry for flow characterization. In this case, percolation mechanisms also apply. As mentioned, fine easy flowing particles (i.e., with low angle of repose) tend to displace to the center of the heap. When uniformly sized particles of different materials are mixed, each having different angles of repose, the material having the steeper angle will tend to concentrate at the center, while that having the flatter angle concentrates on the outside.

Percolation velocity is remarkably affected by particle size ratio and shape (Shinohara, 1997). This situation is in agreement with a diffusion mechanism; under study conditions, the dispersion is independent of the diameter of the smaller particles (Lindley, 1991). A mathematical model for this kind of radial dispersion was proposed by Bridgwater et al. in 1969 (Shinohara, 1997) as follows:

$$\ln\left(\frac{N_0}{N_0 - N}\right) = \frac{R}{4E_R t} \tag{12.32}$$

where N_0 is the total number of percolated small particles through a packed bed and N is the number of the small particles having centers within radius R at time t.

When a large particle is placed at the bottom of a bed of fine particles, it will move up to the surface if the intensity of vibration is suitable, which represents another segregation mechanism called *upthrusting*. The phenomenon also occurs if the large particle is denser than the fine particles. During upthrusting, particle rearrangement occurs due to large particles, which cause an increase in pressure in the region below them, compacting the material and stopping particles from moving downwards (Williams, 1976). Any upward movement allows fines to run under the coarse particle, and these in turn are locked in position.

Some other mechanisms are also suggested in the literature, including elutriation, lateral segregation, transverse segregation, stumbling, pushing-away, rolling, sliding, bouncing, and capture (McGlinchey, 1998; Shinohara, 1997). For example, smaller particles can be more easily captured in the surface voids of a powder bed than larger ones (McGlinchey, 1998). Inter-particle percolation or sifting mechanism also plays an important role in keeping fines near the center and allowing the coarse particles on the outer part of a container (e.g., during hopper storage). With regard to particle shape, the rolling effect can explain why larger, more rounded particles go farther down a heap of material than smaller or more angular ones. Particles of different sizes, shapes, or elastic properties will bounce at various distances when they are impacted against each other during pouring. When solids deposit on a pile with significant impact velocity, a flat spot or crater forms on the top of the pile. This flat spot catches particles that are not resilient enough to bounce out. If resilience varies considerably among particles, the more resilient particles will tend to concentrate on the outside of the pile. Since fines are usually less resilient than coarse particles, dynamics segregation can accentuate a central fines concentration.

Shinohara (1997) defined the rolling effect obtained when a single particle is moved down a plate at a given inclination. The travel distance on a horizontal plane described both rolling and sliding motions. It was suggested that the larger or less frictional particles roll down farther, and that only friction segregation takes place during sliding motion, irrespective of particle density.

12.2.3. Segregation Evaluation

Because of the random nature of the segregation process, most of the meaningful attempts to study this phenomenon numerically have a statistical foundation where the primary concern is some form of variance in the spot samples from the population mean (McGlinchey, 1998). Segregation evaluation is usually based on two testing groups: (a) particle size or composition distribution analysis of the powder after it is subjected to a known duration of vibration or rotation, and (b) sampling the size or composition distribution at various locations in a flowing powder stream or in a container after it is filled with the mixture (Barbosa-Cánovas et al., 1985). Many segregation indices have been proposed to quantify the degree of segregation in a powder system after undergoing one of these segregation tests. Some of them are discussed in detail in this section.

A simple and noteworthy example is the Williams segregation coefficient. The Williams segregation test cell can be separated into two halves from the middle, with one called the upper cell and the other the bottom cell. The segregation tendency for a binary powder mixture can be evaluated by monitoring the segregation intensity in the Williams cell when it is subjected to tapping. Based on the coarse fractions in the upper and lower cells, the segregation coefficient can be evaluated as the S_{index} (Williams, 1976):

$$S_{\text{index}} = \frac{X_{\text{CT}} - X_{\text{CB}}}{X_{\text{CT}} + X_{\text{CB}}} \tag{12.33}$$

where X_{CT} and X_{CB} are the weight fraction of the coarse material at the top half and the bottom half of the cell, respectively. S_{index} is 1 for total segregation and 0 for no segregation.

Another segregation index model was proposed by Popplewell et al. (1989) to study the segregation in the Williams test cell. The theoretical comparison of those two models showed that the latter is more segregation sensitive, especially when almost all of the fines are confined to the bottom

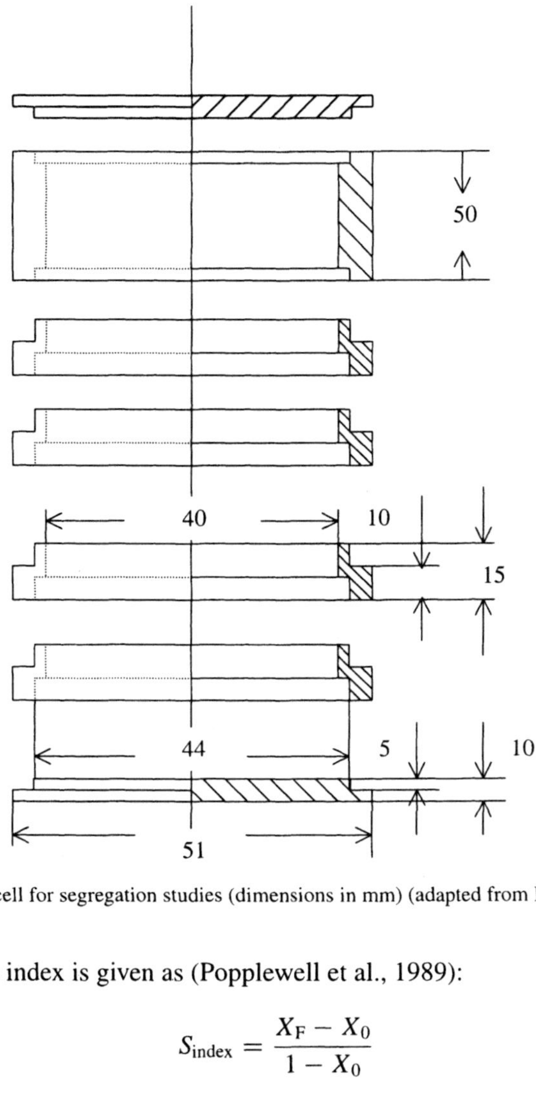

Figure 12.11. A split-ring cell for segregation studies (dimensions in mm) (adapted from Barbosa-Cánovas et al., 1985).

half of the test cell. The index is given as (Popplewell et al., 1989):

$$S_{\text{index}} = \frac{X_{\text{F}} - X_0}{1 - X_0} \tag{12.34}$$

where X_{F} is the weight fraction of the fines at the very bottom of the cell and X_0 is the initial or overall fines concentration. This index has a value range from 0 to 1, with 0 representing no segregation, while 1 represents complete segregation.

Instead of the cell-only split at the middle, a multi-ring split cell, as shown in Fig. 12.11, is also used for segregation studies. The multi-ring split cell makes it possible to study the overall distribution of the fine and coarse fractions along the cell height after the mixture is subjected to vibration or tapping, i.e., undergone segregation. Since the contents of each ring can be weighed separately, density and content changes along the cell height can be easily detected. The segregation index S_{index} for the multi-split cell is calculated by (Olson and Rippie, 1964; Barbosa-Cánovas et al., 1985):

$$S_{\text{index}} = \left[\frac{\sum_{i=1}^{n} W_i (X_i - X)^2}{\sum_{i=1}^{N} W_i}\right]^{1/2} \tag{12.35}$$

Table 12.1. Maximum theoretical segregation index S_{index} values for a binary mixture with same bulk densities in a multi-split cell of different ring number (modified from Barbosa-Cánovas et al., 1985).

	Number of rings				
A:B (weight ratio)	4	5	6	7	20
5:95	0.087	0.100	0.112	0.122	0.218
10:90	0.173	0.200	0.224	0.245	0.300
25:75	0.433	0.387	0.382	0.401	0.433
40:60	0.424	0.490	0.447	0.407	0.490
50:50	0.500	0.447	0.500	0.463	0.500

where W_i is the weight of powder in the ith ring, X_i is the concentration of a given component, and X is the mean concentration of the component in the mixture.

It can be shown that S_{index} can theoretically vary between 0 (total mixing) and 0.5 (total segregation) if the mixture of two components with a 1:1 density ratio is uniform along the cell vertical axis after vibration and the split cell number is even. The maximum theoretical S_{index} values for binary mixture components with the same bulk density but different weight ratios in a multi-split cell with different rings are listed in Table 12.1. If the densities of the binary components are different, the height ratio at complete separation H is given by:

$$H = \frac{L_A}{L_B} = \frac{X_A \rho_B}{X_B \rho_A} \tag{12.36}$$

where L_A and L_B are the heights, X_A and X_B are the weight fractions, and ρ_A and ρ_B are the densities for components A and B, respectively (Barbosa-Cánovas et al., 1985).

An apparatus developed by Harris and Hildon (1970), as shown in Fig. 12.12, is used to determine the segregation extent in a mixture after filling or discharging. First, the whole mixed sample is poured from vessel A to vessel B when gate 1 is removed. After the whole sample has been transferred into vessel B, gate 2 is removed and the sample is poured into vessel C. When the entire sample has run into vessel C, four samples are taken from the sampling slots positioned differently at the base of the heap, and the proportion of one mixture component in each sample is analyzed. The degree of segregation S is defined as:

$$S = \sum_{i=1}^{i=4} |A_i - 25| \tag{12.37}$$

where A_i is the proportion of one mixture component in the ith sample, and $\sum A_i = 100$. The value of S varies from 0 (all the samples have the same composition) to 150 (only one sample contains the entire target component).

Another unsophisticated and inexpensive laboratory test to assess segregation after flow was proposed by Shinohara et al. (1972). As shown in Fig. 12.13, this test has a simple construction structure and thus a simple sampling procedure. The samplers are positioned at different angles from the vertical axis of the bin and samples are taken for concentration or size distribution analysis purposes. The most convenient way to present the results is in the form of concentration (or size distribution characteristics) versus the sampler position angle relative to the vertical axis.

The effect of particle-permeation on the segregation of solid mixture in a rotating cylinder was studied by Sugimoto and Yamamoto (1976), and it was found that the radial and axial segregation

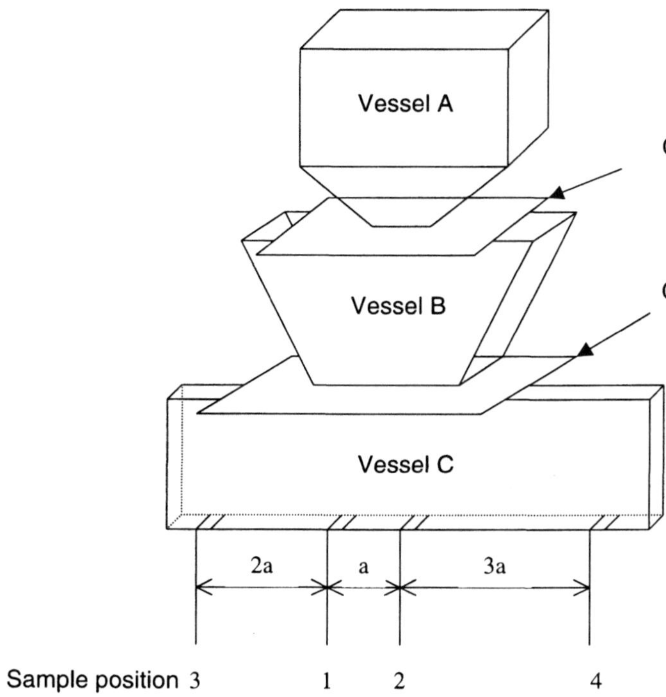

Figure 12.12. A test device to measure the segregation caused by central pouring (adapted form Harris and Hildon, 1970).

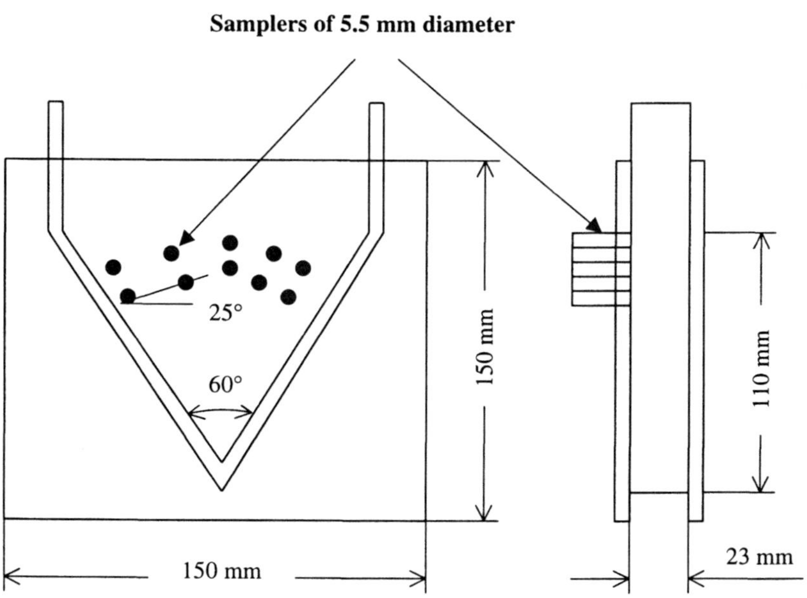

Figure 12.13. A laboratory tester to assess segregation in powders after flow (adapted from Shinohara et al., 1972).

of binary solid mixtures in a rotating vessel were closely related to the permeation effect measured in a moving bed. The radial and axial segregation indices of a spherical binary solid mixture are expressed as follows:

$$S_r = \frac{C - C_{min}}{C_{max} - C_{min}} \tag{12.38}$$

$$S_a = \left[\frac{\sum (C_i - C)^2 V_i}{\sum V_i}\right]^{1/2} \tag{12.39}$$

where S_r is the radial segregation index, C is the average volume concentration of component A sampled in the radial or axial direction of the cylinder, C_{min} and C_{max} are the minimum and maximum concentration of component A sampled in the radial direction, respectively, S_a is the axial segregation index, and V_i is the bulk volume of mixture sampled in each section.

A recent method for identifying segregation, introduced by Rollins et al. (1995), is based on the use of the analysis of variance (ANOVA) statistical test, which has been claimed to be a superior approach to evaluate segregation. The method is based on controlling error rates for false conclusions and specifying confidence intervals for true values (Rollins et al., 1995). In order to calculate the F-value, which is related to the magnitude of segregation, special steps must be taken to prepare the sample. First, the entire volume of the target material is divided into a number of regions, and then a number of samples are taken from each region. The composition of some particular mix is measured, and the variance of composition calculated for each region is compared to the variance of all samples by calculating the F-value:

$$F = \frac{\left[\sum_{j=1}^{M} (X_j - X)^2\right] \Big/ (M - 1)}{\left[\sum_{j=1}^{M} \sum_{i=1}^{N} (X_{i,j} - X_j)^2\right] \Big/ (MN - M)} \tag{12.40}$$

where X_j is the average measured composition in the jth region, X is the average composition in all regions, M is the number of regions, N is the number of samples taken from each region, and $X_{i,j}$ is the ith measured composition taken from the jth region.

From the F-value equation, it should be noted that its variation is only due to the variance caused by segregation. Based on the null hypothesis (i.e., H_0: there is no segregation) and the alternative hypothesis (i.e., H_A: there is segregation) for this ANOVA statistic test, the null hypothesis is rejected in favor of the alternative hypothesis if $F > F_{(\alpha,\ M-1,\ M(N-1))}$. The $F_{(\alpha,\ M-1,\ M(N-1))}$ is the upper 100 αth percentile of the $F_{(M-1,\ M(N-1))}$ distribution, and its values can be found in many statistics textbooks (Ott, 1992).

Segregation tendency in some food mixtures caused by vertical and horizontal vibrations was studied by Barbosa-Cánovas et al. (1985). The results showed that segregation occurs not only in free-flowing powder mixtures, but also in some cohesive powder mixtures. Segregation intensity depends not only on the mixture composition, particle size, and mechanical vibration history, but also on whether the mixture is "ordered" or not. For example, the segregation tendency in a starch–sugar mixture was inhibited by the attractive inter-particle forces at certain mixing weight ratios, while sugar–instant coffee mixture segregated almost completely under similar conditions. It was also noticed that increasing the vibration frequency intensified the mixture's segregation, and large fluctuations in segregation index were observed in the segregation curves when the vibration frequency was higher than a certain value. This was mainly because the segregation testing system acted partly as a non-efficient mixer, resulting in segregation in the mixture and creation of a new kind of mixture occurred at the same time (Barbosa-Cánovas et al., 1985).

12.2.4. Segregation Kinetics

Although there are many proposed equations modeled for segregation study, few attempts have been made to mathematically represent the rate mechanisms for the powder segregation process. This is probably because more attention is paid to predicting the segregation effect after it occurs, rather than describing its progress (Barbosa-Cánovas, 1985). One noticeable segregation rate equation was proposed by Olsen and Rippie (1964), which was characterized by the following first-order kinetic model (Olsen and Rippie, 1964; Rippie et al., 1964):

$$\log\left(\frac{S_\infty - S(t)}{S_\infty}\right) = -kt \tag{12.41}$$

where S_∞ is the standard deviation of the sample composition from the mean composition of the system at equilibrium, $S(t)$ is the standard deviation after time t, and k is the rate constant.

In their study, the rate constant k was defined as a function of different experimental parameters such as particle size, particle size distribution, particle density, and wall effects. They explored the demixing effect, the kinetics in a two- and three-particle size system, the segregation between particles of different weight, and the effect of the wall and particle size on the segregation rate constant k. From their conclusions, the following theoretical inferences can be highlighted:

(a) *Demixing effect*: There is an equilibrium state between mixing and demixing; therefore, segregation does not proceed to completion;
(b) *Two-particle size system*: The segregation rate k in a binary system is directly proportional to the ratio of the particulate volumes;
(c) *Three-particle size system*: The presence of the third component of intermediate size reduces the segregation rate k in the system;
(d) *Particle density*: Segregation rate is markedly affected by the weight ratio of the components;
(e) *Wall effects and particle size*: Wall effects resulting from particle–cylinder contact cause no apparent change in the relationship between segregation rate k and particle size.

Based on a diffusive mixing mechanism and employing a modified Fick's diffusion equation, Rose (Fan et al., 1970) proposed the following mixing rate equation to describe the process in which both mixing and demixing (i.e., segregation) occur simultaneously in a mixer:

$$\frac{\mathrm{d}M}{\mathrm{d}t} = A(1-M) - B\lambda \tag{12.42}$$

where A and B are the mixing and demixing rate constants, respectively, M is the degree of mixedness defined as:

$$M = 1 - \frac{\sigma}{\sigma_0} \tag{12.43}$$

where σ is the standard deviation, σ_0 is the standard deviation at $t = 0$, and λ is:

$$\lambda = \pm\sqrt{1-M} \tag{12.44}$$

12.2.5. Segregation Minimization

The segregation process is undesirable and inevitable in many processes. However, there are some ways to minimize it to certain practical and acceptable levels, based on segregation mechanisms. As suggested by Williams (1976), the most urgent and important step to minimize segregation should be taken at the beginning of the plant design. The engineers responsible for the design of solids-handling plants and equipment should understand the causes of segregation and minimize their effects by careful planning.

The selection of filling technology, the geometric features of storage bins, and discharge patterns and aids can play a substantial role in either introducing segregation or mitigating it. In pneumatic conveying, equipment can be designed so that particles are introduced tangential to the line, rather than in a radial direction, so that trajectory segregation can be reduced. When filling a bin or hopper, the filling point can be moved in order to prevent the formation of a sloping surface, thereby reducing segregation. Furthermore, design of tall bins or hoppers with a small apex angle can shorten the flow length during filling and promote mass flow during discharge (Brown and Richards, 1970; Shinohara, 1997).

Changing particle size by either size enlargement or size reduction processes can modify segregation. Dry materials with a particle size greater than 75 μm tend to be free-flowing, therefore having high individual mobility and increasing tendency to segregate. However, segregation is unlikely for particles of less than 10 μm (Lindley, 1991). Therefore, reducing particle size by grinding is another method for minimizing segregation. Even so, particle size reduction may result in a powder too fine to rehydrate and highly susceptible to caking (Barbosa-Cánovas et al., 1985). Size enlargement methods such as coating, as seen in Chapters 7 and 8, can be used to increase particle stickiness. A very fine mixture component can coat coarser particles, losing their freedom of movement and forming the so-called "ordered" or "adhesive" mixture, thus forming a non-segregation mixture.

Besides dealing with particle size, particle surface properties can be adjusted beforehand to minimize segregation through surface treatment and humidification. In fact, the final mixture can be surprisingly free of segregation if its coarse particles have a rough surface or fibrous shape (Shinohara, 1997). As explained before, wet mixing techniques can be facilitated into the process in order to help increase particle–particle adhesion. Related to this, spray drying of mixed ingredients is also a useful method to minimize segregation. The last option is a commonly employed process to overcome segregation of food mixtures, which is usually an agglomeration process. It is the most effective method in eliminating segregation, but it requires considerable technical complication and adds extra drying costs.

During food processing, when dealing with storage bins, mixers, and mechanical and pneumatic conveyers, different techniques can be applied in order to prevent segregation of food ingredients and final powdered products. Among these, the reduction of vibration during mechanical discharge operations and storage can be a feasible measure that is actually implemented in hopper design. Furthermore, the handling of lower bulk volumes is another option that will reduce segregated fractions. For example, closed packing management of powdered product, rather than storage tower discharge, reduces particle mobility and the probability of segregation. This method is of common use in the food industry for ingredient handling as well as for hygienic prevention practices.

12.3. CAKING

Caking is a deleterious phenomenon by which amorphous food powders are transformed into a sticky undesirable material, resulting in loss of functionality and lowered quality (Aguilera et al., 1995). A powdered mass is defined as caked "when two or more macroparticles, each capable of independent translational modes, contact and interact to form an assemblage in which the particles are incapable of independent translations" (Griffith, 1991). This assemblage is mainly formed as a result of compositional and physical state changes. Caking can result in different composites, ranging from small, soft aggregates that can be broken easily to rock-hard lumps that need a sledgehammer to disperse. Only rarely is caking desirable, such as during processes like agglomeration, sintering (i.e., particle welding due to heat application), or tableting, where particle cohesion plays a positive role. Caking not only affects food powdered products, but also products like detergents, pigments, fertilizers, or chemicals. When packaging food powders, one of the main quality aspects, is whether the powders will cake under normal storage conditions. In fact, caking can be typically observed in everyday life, for example, when salt does not flow out from a salt shaker due to lump formation, or when a box of brown sugar turns into a piece of hard-to-break solid.

Physicochemical changes like sticking, collapse, caking, agglomeration, loss of volatiles, browning and oxidation are of concern for the food powders industry due to economical losses during the production and storage of these products. Important losses in money and time are related to downtimes in production lines due to subsequent disposal of the caked products as waste and/or for their remanufacture (e.g., caked coffee product sent for re-dissolution during spray drying operations). As a matter of fact, records in a powdered gravy plant showed that 25% of the downtime of the packaging line was attributed to caking problems in the product (Griffith, 1991). It is estimated that the total cost of unproductive caked products is more than one billion dollars each year in the United States alone (Griffith, 1991).

Flowing problems of powders in silos or bins caused by caking are very relevant in agricultural and food operations. Many food powders, especially those containing soluble or amorphous components such as sugar, salt, lactose, or lipids have a very strong tendency to cake when they are exposed to an atmosphere of a high humidity or elevated temperature during storage. Products such as powdered milk, egg, avocado, flour, oatmeal, tapioca, and starches also possess caking capabilities (Griffith, 1991). The crystallization of lactose in fresh dried milk powders is one of the most commonly observed reactions that decrease the quality of the dried milk powders when they are exposed to relatively high humidity and/or increased temperature (Lai and Schmidt, 1990).

Some of these ingredients are utilized in the food industry every day, and can be under the effect of caking, affecting varied types of manufacturing processes, from raw material utilization to packaging dosage operations. With the aim of understanding caking influence in food processing, this section will cover inside aspects of the phenomenon. It will describe factors affecting it and the physical changes they provoke in powders, the role of glass transition temperature, existing agents for caking prevention, as well as analytical determination methods such as tools for caking characterization, among other applications.

12.3.1. Caking Mechanisms

Caking is caused by various inter-particle forces developed under moisture absorption, elevated temperature, or static pressure during processing, transportation and storage (Peleg, 1983). Although caking development may vary among powders, material composition, particle size and shape, moisture content, pressure, and variations in temperature and humidity are governing factors of general incidence (Chen and Chou, 1993). These factors can be combined to form bonded powders through more than one mechanism, depending on the environmental conditions involved in a given powder system. Therefore, a strict definition of caking is difficult to formulate due to all the changes occurring in a system, which can involve many overlapping stages. Most caking phenomena, similar to interactions in agglomeration processes, can be classified as the following four major mechanisms (Schubert, 1981; Griffith, 1991) reviewed in Chapter 7:

- *Chemical caking*: This is the most common type of caking mechanism. It may be caused by chemical reactions in which a compound has been generated or modified, such as decomposition, hydration, dehydration, recrystallization, or sublimation. During recrystallization, for example, any traces of water in the saturated solution on intercrystalline surfaces can be evaporated by a small temperature and/or humidity fluctuation in the environment, leading to the formation of recrystallized solid bridges between particles.

- *Plastic-flow caking*: This occurs when the particles' yield value is exceeded and they stick together or merge into a single particulate form. It usually occurs with amorphous materials such as tars, gels, lipids, or waxes. Furthermore, some soft crystalline substances can also stick together when they are subjected to either pressure or higher temperature.
- *Electrical caking*: As the name implies, electrical caking is caused by electrical charges on powders. Except for static electrical charges, most electrical charges are the result of the non-symmetrical properties (either physical or chemical) in the particle's crystal structure. This type of caking can be demonstrated by a simple experiment, in which a crystal is suspended between two plates attached to a high voltage DC power supply, then contacting the crystal with one of the plates and then with the other. Since the crystal has been charged, it will oscillate between the plates as long as power is supplied.
- *Mechanical caking*: This is particle-shape related and caused by particle interlocking, or "bird nesting." It usually occurs with fibrous or plate-shaped particles, but not with spheres. Cotton wad production is a good example of a desirable caking practice of this type.

For food powders, the most commonly used term to explain the caking mechanism is "humidity caking," which may also be classified as chemical caking or plastic-flow caking, as mentioned above, according to its particular characteristics. In these cases, the caking is initiated by the formation of liquid bridges extending through the neighboring particles that later solidify, by cooling and drying, forming aggregates (Peleg and Hollenbach, 1984).

Aguilera et al. (1995) related the "humidity caking" mechanism in food powders to four different stages: bridging, agglomeration, compaction and liquefaction. *Bridging* is the initial stage in humidity caking, and it occurs as a result of surface deformation and stickiness at contact points between particles, without a measurable decrease in system porosity. Adhesion without material bridges is primarily due to van der Waals' and electrostatic forces, especially in very fine particles of 10 μm or less (Adhikari et al., 2001). *Agglomeration* is a later stage that involves an irreversible consolidation of bridges while the high porosity of the particulate system is maintained, resulting in particle clumps with structural integrity and larger size (Schubert, 1981). During *compaction* reduction of inter-particle spaces occurs and deformation at particle clumps under pressure leads to a loss in system integrity as a result of the thickening of the already formed inter-particle bridges. *Liquefaction* is the final stage of humidity caking, where inter-particle bridges disappear because of the high moisture content. This stage usually involves solubilization of low molecular weight fractions provoking hygroscopic behavior. At any given stage, lumps may be few or numerous, of different sizes and of varying degrees of hardness.

The common sources that cause liquid bridge formations in food powders are: (a) water addition due to moisture adsorption, moisture condensation, or accidental wetting of the powder or equipment; (b) melting of lipids or amorphous sugars at the particle surface due to elevated temperature; (c) release of absorbed water from the crystallization of amorphous sugars; (d) chemical reactions that produce liquids; and (e) water immigration from excessive liquid ingredients in the powder (Peleg, 1983; Roos, 1995). The schematic presentation of this caking mechanism is demonstrated in Fig. 12.14.

In practice, a quantitative measure of caking is useful for following changes in the condition of a powder over time. A caking index (Aguilera et al., 1995) can be defined as the state of the system at any time in relation to an initial state. Two morphological indicators define the state of the system: the

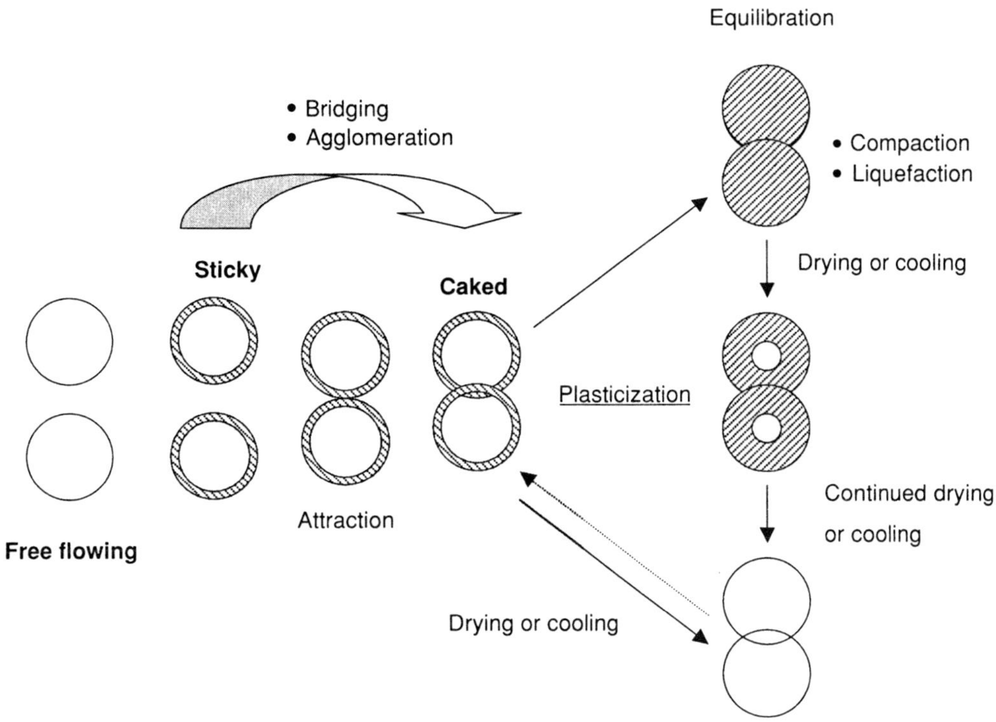

Figure 12.14. Schematic demonstration of the most common "humidity caking" mechanisms in food powders (adapted from Peleg, 1983).

ratio of instant system porosity to initial system porosity ($p(t)/p_o$), and the ratio of the inter-particle bridge diameter to particle diameter ($D_{bridge}/D_{particle}$). A decrease in porosity is observed as soon as bridging occurs, because voids between particles become smaller. Particle bridges start increasing in diameter as soon as caking advances from bridging to compaction stages. Particle size of a sample, in a system with constant properties, also influences caking in which the smaller the particle the more likely it is to cake. Therefore, the surface-to-surface ratio becomes greater as the particles become smaller. Once the particles interact physically, bridging occurs as a result of surface deformation (Griffith, 1991).

12.3.2. Caking Evaluation by the Glass Transition Temperature

The concept of glass transition process in polymer science was introduced by Levine and Slade to the study of collapse phenomena in amorphous foods and other biological materials about 15 years ago (Barbosa-Cánovas and Vega-Mercado, 1996). Known as food polymer science, these collapse phenomena, which include sticking, caking and shrinkage, are considered to involve time-, temperature-, and moisture-dependent viscous flow resulting in structural transformations; these transformations can occur above their glass transition temperature T_g (Aguilera et al., 1993; Roos, 1995). With the introduction of the T_g concept, caking of amorphous food materials comes from the change from the glassy state to a less viscous liquid-like state, which allows the flow of liquid and formation of liquid bridges between particles.

Based on the T_g concept, food powders are treated as amorphous materials with their own glass transition temperature. When a food powder is stored at a temperature below its T_g, it is said that the

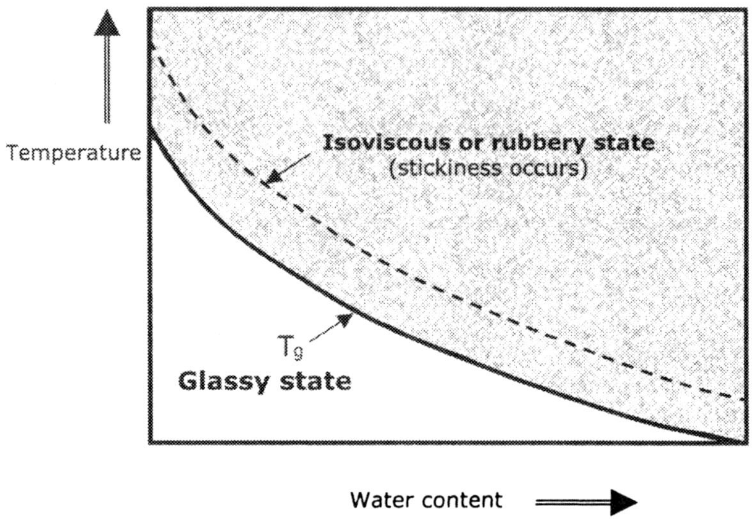

Figure 12.15. Schematic diagram of relationship between glass transition temperature T_g, viscosity, water content, and occurrence of stickiness in food powders (adapted from Roos, 1995).

powder is in its glassy state with an extremely high viscosity and low mobility. Under this crystalline state, it will take a very long time for plasticizers (usually water) to acquire the necessary mobility to facilitate polymers or for low molecular weight compounds to diffuse through the particle to form a sticky surface. As a result, particle-binding times may be short, and therefore caking may not happen. On the other hand, when the powder is stored at a temperature higher than its T_g, the dramatically reduced viscosity (or mechanical modulus) and increased mobility will reduce the contact time and promote stickiness between particles and, finally, caking in the powder (Chung et al., 2000). As compared to traditional or empirical methods dealing with the caking phenomenon, the T_g method is more useful to predict the onset of stickiness or caking of powders (Chuy and Labuza, 1994).

The relationship between the glass transition temperature, water content, and occurrence of stickiness followed by caking in food powders are schematically shown in Fig. 12.15. The T_g decreases as the water content increases. Thus, T_g is lowered by the addition of plasticizers such as water, the universal plasticizer of biomaterials (Bhandari and Howes, 1999). Water decreases T_g of amorphous foods below the T_g for the food in the anhydrous state at a rate of $10°C/0.01$ g water/g material. Amorphous food powders maintain their free-flowing properties when their temperature is kept below the T_g curve (crystalline state). The sticking is observed to occur at an isoviscous (or rubbery) state, which is defined by a constant difference between T and T_g (i.e., a critical temperature difference above T_g must be exceeded for the initiation of structure collapse within a time period of practical importance). Therefore, humidity caking is considered to be a consequence of an increase of water content, or plasticization (either by thermal or water adsorption), and depression of T_g below ambient temperature (Roos, 1995). The rubbery state, due to increased mobility, allows the flow of liquid and formation of liquid bridges between particles.

Furthermore, above T_g, the product's glassy state is changed to a less viscous liquid-like state that leads to an increased rate of physicochemical changes related to caking in dried products, such as crystallization, loss of volatiles, browning and oxidation (Bhandary and Howes, 1999). Different research supports the fact that the rate of crystallization and loss of volatiles is high at collapse temperatures during storage, from its relation with the $T - T_g$ difference (Aguilera et al., 1995; Bhandari and Howes, 1999). Many other physical properties have been observed to suddenly change as temperature increases above T_g. The most important are increase in the free molecular volume, increase in the heat capacity (C_p), increase in the thermal expansion coefficient (α) and dielectric coefficient (ε), and changes in viscoelastic properties (Genin and Rene, 1995).

Viscosity (η) of amorphous powders may decrease from $\eta_g = 10^{12}$–10^{13} Pa s at T_g to an approximate value of $\eta = 10^6$–10^8 Pa s, known as the sticky point in its rubbery state ($T > T_g$) (Wallack and King, 1988; Bhandari and Howes, 1999). A mechanistic relationship gives an idea of the stability of a dried product during the drying process and storage by estimating the critical viscosity (η_c, Pa s) (Wallack and King, 1988; Bhandari and Howes, 1999):

$$\eta_c = \frac{k\gamma t}{KD} \tag{12.45}$$

where k is the dimensionless proportionality constant of order unity, t is the contact time (s), γ is the surface tension, for interstitial concentrate 0.07 N/m, K is the fraction particle diameter required as a bridge width for a sufficient strong inter-particle bond, usually 0.01–0.001, and D is the particle diameter (m). This model relationship is particularly important in relation to the stability of the dried product during the drying process and storage.

Collapse, stickiness and caking appear to be related phenomena. A solid matrix (usually freeze dried) reaches its collapse temperature (T_c) when an initial occurrence of inter-particle bridging (or surface caking) is manifested as a loss of structure and change in sample volume (Aguilera et al., 1995). In the freeze-drying process, during sublimation, reduction of pore size and volume of the food material occurs, resulting in the loss of desirable appearance, texture and volatile substances. If during any freeze-drying operation the temperature or water content of the porous layer is increased (the product being above T_c), the viscosity is not high enough to support the structure and the solid material collapses or shrinks (Levi and Karel, 1995; Bhandari and Howes, 1999). The collapse temperature, as well as the caking temperature, can be raised by the addition of high molecular weight materials (Tsourouflis et al., 1976).

12.3.3. Caking Kinetics

The caking of an amorphous powder is a time-dependent phenomenon. Modeling the kinetics of caking is a difficult task because many factors are known to affect it, both internal (such as particle size, charge, hygroscopic behavior, and the physical state of particles) and external (temperature and relative humidity of the environment, applied mechanical stress and the presence of moisture barriers, among others).

The rate of caking due to viscosity of the material is a function of $T - T_g$. The viscosity of amorphous food materials between T_g and (T_g+ 100) K may be related to sample temperature using the Williams–Landel–Ferry (WFL) model (Roos and Karel, 1991a; Wallack and King, 1988; Aguilera et al., 1993):

$$\log_{10} \frac{\eta}{\eta_g} = \frac{-C_1 \Delta T}{C_2 + \Delta T} \tag{12.46}$$

where η is the viscosity, η_g is the viscosity at glass transition temperature T_g, $\Delta T = T - T_g$, T is the temperature, and C_1 and C_2 are universal constants. This model is applicable in food systems with the constant $C_1 = 17.44$°C and $C_2 = 51.6$°C (Aguilera et al., 1993). WLF equation has received especial attention in mono and polysaccharide research (Peleg, 1994).

Roos and Karel (1991d) found this relationship using universal constant values to be valid for sucrose and lactose. Soesanto and Williams (1981) also validated the WLF equation for an amorphous mixture of sucrose and fructose. Aguilera et al. (1993) also successfully used the WLF equation to predict the viscosity of fish protein hydrolyzate matrix at the onset of collapse during freeze-drying. Nevertheless, Peleg (1992) debated the validity of these fixed constants after experimenting with some polymers and amorphous sugars. He suggested that the WLF equation could not be used as a generally valid model. It was demonstrated that for certain polymers and sugars this model leads to considerable error in the magnitude of viscosity, particularly 20–30 K above the T_g. Peleg (1996) further reported that the upward concavity of changes in a translational region, which cannot be predicted by WLF or by Arrhenius equation, can be described by a model with other known functions. Although the WLF model has been validated for various food systems, a general application of this model in a food system needs to be carefully examined.

Rates of change in mechanical properties are defined by relaxation times. As caking is attributed to viscosity, the relaxation time may be modeled by WLF kinetics; the relaxation time for caking can be related to the difference between the environmental temperature (T) and T_g ($\Delta T = T - T_g$) by a WLF-type relationship keeping the same constants by replacing the ratio η/η_g for the relevant property (e.g., crystallization time, a_w) (Aguilera et al., 1995; Bhandari and Howes, 1999). At the moment of establishing actual caking kinetics, water vapor spreading rate should be controlled and particles should have a uniform exposure to the vapor phase.

An alternative model tested by Peleg (1994) has been used successfully to characterize and compare the transition patterns of different materials (casein and sodium caseinate, coffee creamer, bakers, yeast, etc.):

$$Y(T) = \frac{Y_s}{1 + \exp[(T - T_c)/a]} \tag{12.47}$$

where $Y(T)$ is the magnitude of mechanical integrity or strength parameter (caking) at the corresponding temperature T, Y_s is the magnitude of this parameter in the unplasticized (glassy) state, T_c is the characteristic temperature, and a is an empirical constant. The model is useful in describing rheological behavior of material just before, during and/or soon after transition. This is of extreme importance, since powders cake well before the particles are fully plasticized (Peleg, 1994).

12.3.4. Food Powders Affected by Caking

Many food products have caking problems. Powders like instant coffee, powdered tea, powdered cream and even powdered soup must be free-flowing, especially when vending machines are used. Food powders produced by spray drying are likely to include components in a rubbery state. Starch, amylose, amylopectin, starch hydrolizates and some non-starch polysaccharides, low molecular weight sugars, food proteins and their hydrolyzates are prone to form glasses (Aguilera et al., 1995). Given that powder composition plays a major role in caking behavior, much research work has been focused on different powders. As already mentioned, glass transition temperature theory has contributed significantly to the understanding of caking mechanisms and caking prevention.

12.3.4.1. Carbohydrates

Carbohydrate constituents have the largest effect on influencing the glass transition temperature of an amorphous dried food material. Low molecular weight sugars such as fructose, glucose and sucrose have a relatively low glass transition temperature, so their influence on depression of T_g is very notable in sugar-rich foods. During the drying of sugar-rich products, they may either remain as syrup or stick on the dryer chamber wall. There is also the problem of unwanted agglomeration in the dryer chamber and conveying system. The most common approach to drying such products has been to add high molecular weight additives (e.g., maltodextrin), which has the effect of raising T_g.

The solids in dehydrated fruit juices comprise mostly fructose ($T_g = 5°C$), glucose ($T_g = 31°C$) and sucrose ($T_g = 62°C$), and the T_g of fructose/glucose/sucrose blends (similar composition to dehydrated fruit juices) can be estimated to be below typical room temperature values (Roos and Karel, 1991a, b, c). This also explains why it is recommended that walls of spray dryers used to hydrate this type of products be cooled to reduce stickiness problems.

Cohesion of food powders (Moreyra and Peleg, 1981), and the mechanism of caking in onion powder (Peleg and Manheim, 1977) and amorphous carbohydrates (Wallack and King, 1988) were studied to evaluate the steps leading to collapse and stickiness. Brown sugar has been a worst-case example of a difficult product to use, because of its great caking tendency. Even badly caked brown sugar can be rendered free-flowing by exposing it to microwave radiation for only a few seconds (Griffith, 1991). Radiation is tuned to the frequency of water molecule absorption and the microwaves destructively transform the syrup bridging between the particles of sugar, allowing the sugar to be free-flowing again.

12.3.4.2. Milk Powders

Troy and Sharp (1930) distinguished the following stages in the process of the caking of milk powder particles: (a) absorption of moisture by lactose; (b) sticking together of the powder particles; and (c) solidification and crystallization of lactose. The crystallization of lactose, which is in the amorphous state in fresh dried milk powders, is one of the most commonly observed reactions that decreases the quality of the dried powders when they are exposed to relatively high humidity and/or increased temperature (Lai and Schmidt, 1990). The T_g of amorphous lactose is 101°C and it has a T_g that is below room temperature at 0.05 g water/g material, which is usually the maximum moisture content of dried milk products showing adequate chemical stability. As expected, the stability against caking of several dairy-based infant formula powders increased with increasing amounts of high molecular weight carbohydrates (maltodextrines), which raised the T_g of the sample (Joupilla and Roos, 1994).

Chuy and Labuza (1994) related the caking and stickiness of some typical dairy-based food powders to their glass transition temperature. They made a plot for two infant formula powders at different water activity conditions where caking did not occur during storage and predicted the water activity at which caking should begin to occur at room temperature by linear extrapolation. They found that the predicted water activities for the two powders were close to those observed experimentally during 5 weeks of storage at 20°C, concluding that the T_g is a useful index for assessing the physical stability of food powders during processing and storage.

12.3.4.3. Protein-Based Powders

The viscosity of a lyophilized fish hydrolizate powder dried to 33% relative humidity (RH) was estimated to be 107 times smaller after equilibration to RH = 52% for a change in moisture from 0.075 to 0.150 g water/g material (Aguilera et al., 1993). Hydrolyzed fish proteins (HFP) are amorphous powders and they will cake rapidly at high temperature and relative humidity. An increase in relative humidity from 0 to 64% reduced the T_g of the material from 79.1 to −42.8°C. For other proteinated powders like dried soy sauce, Hamano and Sugimoto (1978) determined the caking onset at between 20 and 33% RH for soy sauce powder.

12.3.5. Caking Prevention

In addition to lump formation and flowability reduction, caking may also lead to poor rehydration and dispersibility of products, increase in lipid oxidation, loss of flavor and crispness, deterioration of organoleptic quality and shelf life, and reduction of process efficiency and yield. There is common agreement that a lumped product is considered to have poor quality and safety, it will not appeal to customers (Chung et al., 2000; Chuy and Labuza, 1994; Griffith, 1991). For example, moist and lumped products that accumulate in the dead corners of storage bins or window machines are focal points for sanitary hazard development where bacteria or fungi can grow.

A number of precautionary measures can be used to minimize the caking tendency in powders, including the use of drying processes, granulation, crystallization-habit modifiers to change crystal shape, and coating surfactants (Chen and Chou, 1993). The strict control of moisture content at low temperatures, when possible, is a key factor in minimizing the effects of caking. However, in many instances, anti-caking agents are added to hygroscopic food powders to improve the flowability and/or inhibit caking. The addition of anticaking agents will be explained in detail, as they have widespread importance in the food industry.

Anticaking agents are defined as substances added to finely powdered or crystalline food powders to prevent caking, lumping, or aggregation by improving their flowability (Hollenbach et al., 1983; Peleg and Hollenbach, 1984, Peleg, 1978). They are also known as flow conditioners, lubricants, glidants, or free-flowing agents because caking is equivalent to the loss of free-flowing properties during storage. Anti-caking agents are very fine powders (particle size in the order of microns) of silicon dioxide, silicates, insoluble phosphate and bi- or trivalent salts of stearic acid, talcum starches and modified carbohydrates (Hollenbach et al., 1982; Peleg and Hollenbach, 1984). Some anticaking agents are listed as nutrients or considered as food ingredients. Silicon and silicates are available in a range of different particle sizes (Peleg and Hollenbach, 1984). Commercial food grade anticaking agents and their applications in food uses are listed in Table 12.2.

Stringent requirements and limits are applied to anticaking agents used in food powders. First, anticaking agents must be inert, safe in specified amounts, and classified as "Generally Recognized as Safe" (GRAS). Second, they must be effective at low concentration, as their legally permitted concentration is limited to a restricted level (FDA, 1980), which in practice is usually within 1% or less (Hollenbach et al., 1982). In addition, an optimum concentration of anticaking agent to improve the host powder's flowability has been observed for most systems (York, 1975). Peleg and Mannheim (1973) established excess concentration ranges for optimal performance of conditioners aluminum silicate and calcium stearate of 1–3%. The conditioning effect was studied using bulk density and compressibility evaluation. Caking started to be observed at agent critical concentrations in the range of 0.5–1.0%. York (1975) used three glidants to improve the flowability of fine lactose powder and also found that a saturation situation existed beyond the reported level (about 1–2%), the flowability of the model material could not be further improved.

Anti-caking agents function in four different ways, and in order to compete with powders for moisture, they act as physical barriers between particles, as moisture protective barriers, or they increase T_g. Those effects are explained as follows:

- *Competing with host powder for moisture*: Usually anticaking agents have strong water adsorptive ability so that they can compete with the host particles for the available water, thus reducing the latter's hygroscopicity and consequently its tendency to cake. This mechanism works only when the moisture exposure time is short or the available moisture is limited. A most effective action is observed if the anticaking agent particles completely cover the surface of the host powder (Peleg and Hollenbach, 1984).

Table 12.2. Selected commercial food-grade anticaking agents and their applications in food uses (modified from Peleg and Hollenbach, 1984).

Major component	Trade name	Bulk density (g/cm^3)	Particle size (μm)	Recommended food uses
Silicon dioxide (SiO_2)	Syloid®	0.08–0.46	2–9	Flavors, egg, salts, dried soup and bases, cocoa, milk, spices
Sodium silicoaluminate	Zeolex®	0.256–0.288	18–45	Egg, dairy products, sauces, cake and dessert mixes
Calcium stearate [$Ca(C_{18}H_{35}O_2)_2 \cdot H_2O$]	—	0.088 (loose), 0.095 (tapped)	45×10^3	Meat dry-curing mixes, onion and garlic salts
Magnesium stearate [$Mg(C_{18}H_{35}O_2)_2 \cdot H_2O$]	—	0.117 (loose), 0.128 (tapped)	75×10^3	—
TCP [$3Ca_3(PO_4)_2 \cdot Ca(OH)_2$]	—	0.320	$1\% > 106 \times 10^3$; $99\% = (45–106) \times 10^3$	Sugar, salt, spices
TCP[$Ca_5(PO_4)_3OH$]	—	—	—	—
Depolymerized starch	Maltrin®	0.513–0.737	—	Spice blends, artificial sweeteners
Microcrystalline depolymerized cellulose	Avicel® (PH)		$(20–90) \times 10^3$	Imitation mozzarella cheese

- *Physical barriers between particles*: The anticaking agent can also interfere with liquid bridging (e.g., silicon dioxide) or decrease inter-particle friction completely by smoothing surfaces (e.g., stearates), reducing or neutralizing superficial molecular attractive forces, mainly electrostatic, and inhibiting crystal growth and altering its lattice pattern (Peleg and Hollenbach, 1984).
- *Increasing T_g*: The stability of various dairy-based products increases by increasing the concentration in the addition of high molecular weight carbohydrates, since the T_g of the sample is raised (Aguilera et al., 1995).
- *Acting as moisture-protective barriers*: Granules containing oily materials tend to cake because of the viscoelastic strength of the solidified oil. The addition of oily additives with a coagulation point lower than that of the oily material in the powder has an anticaking effect, since there is a reduction in the hardness of oily materials containing oily additives. The lower the coagulation point and molecular weight of the additive, the better the anticaking action (Wakiyama et al., 1994). Such agents are used on the surface, and do not enter the amorphous phase of the powder.

12.3.6. Laboratory Techniques and Test Procedures

Several thermal and thermo-mechanical based techniques have been used to study the glass transition process in food polymers (Peleg, 1993; Chuy and Labuza; 1994). The T_g is seen as a mean change in specific heat resulting in a step in the baseline (determined by using the exothermic peak and the baseline). The onset temperature T_g is the temperature at which the heat flux starts to rise from the baseline.

Differential scanning calorimetry (DSC) (Aguilera et al., 1993), differential thermal analysis (DTA) and thermal mechanical analysis (TMA) (Bhandary and Howes, 1999) can be used to determine glass transition T_g of amorphous matters. The former two methods determine changes in specific heat (C_p), while the third detects changes in elastic modulus. For products containing starch or flour, DSC does not seem to be sensitive enough to detect glass transition (Champion et al., 2000), but the primary relation of glass–liquid transition can be measured by mechanical spectroscopy (or dynamic mechanical thermal analysis).

Recently, nuclear magnetic resonance (NMR) spectroscopy has been recognized as another powerful method for the estimation of T_g of amorphous food materials. After glass transition temperature T_g is reached, changes in the free volume can be explained as increments of the average molecular distance given by molecular mobility and vibrations. These changes are directly measurable with NMR proton spin probes relaxation times as a function of T. Electronic paramagnetic resonance (EPR), which measures the mobility of nuclei (^{1}H and ^{13}C), has also been recognized as a potential tool for the study of glass–liquid transition (Bhandari and Howes, 1999; Champion et al., 2000). Impedance spectroscopy (or dielectrical thermal analysis, DETA) studies the variations of the dielectric constant as a function of T and/or frequency. This technique was demonstrated to be effective for the study of secondary relaxations in meals with low water contents (Champion et al., 2000).

The effectiveness of anti-caking agents can be investigated by methods used for flowability evaluation, because the main purpose of using anticaking agents is to improve flowability and/or inhibit caking. Those methods include measuring flow rate, angle of response, shear and tensile strength, inter-particle cohesion, size distribution particle morphology, unconfined yield stress, plots of whole function, and angle of internal friction (Peleg and Hollenbach, 1984; Aguilera et al., 1995). Hamano and Sugimoto (1978) measured caking as the level of hardness by measuring the pressure applied to the caked product at different moisture contents. Lloyd et al. (1996) measured the onset of viscous flow of amorphous lactose with plug density using a compaction apparatus. Caked product hardness can be assessed by scanning electron microscopy techniques in combination with uniaxial compression testing.

12.4. DETONATION AND DUST EXPLOSION

As defined by the Mine Safety and Health Administration, dust refers to finely divided solids that may become airborne from the original state without any chemical or physical changes other than fracture; dust consists of tiny solid particles carried by air currents (Mody and Jakhete, 1988). Among all the problems associated with dust emission, such as health hazards, abrasion damage to equipment, impaired visibility, unpleasant odors, material loss, and problems in community relations, the risks of dust explosion and fire are one of the greatest concerns in many industrial fields because of their destructive impact on both human beings and factory installations.

Dust explosion is a rapid chemical oxidation reaction of dust particles dispersed in air, which leads to a rapid energy release (Cross and Farrer, 1982). The system's temperature is so rapidly increased that a sharp pressure increase follows, causing this combustion reaction. Dust clouds often burn with explosive violence, even though the parent bulk material has no remarkable flammable hazard. Provided the particles are small enough, the majority of solids will be able to create dust explosion (Zeeuwen, 2001). The wide range of industrial materials that support an explosion includes chemicals, pharmaceuticals, coal, dyes, plastics, wood, many foodstuffs, agricultural products, and metal powders. In these areas, dusts may be produced either as end products or by-products. Food powders that are flammable, and therefore known to produce dust explosions, are fine powders of sugar, powdered milk, cocoa, starches, and flours, among other milling products.

A number of conditions must be satisfied simultaneously in order for the dust explosion to occur (Grossel, 1997): (a) the dust must be combustible; (b) there must be sufficient oxygen in the atmosphere to support combustion; (c) the dust must be a suspension with a particle size distribution that will propagate a flame; (d) a special dust concentration range must be reached; and (e) the suspension has to be in contact with an ignition source of sufficient energy. These conditions must be satisfied for the dust contained at specific vessel or chamber volumes and with a certain degree of turbulence.

In grain storage, the most troublesome operating hazard is grain dust that is produced by impact or abrasion of the grain. Grain dust is composed primarily of free starch granules and other small fragments of grain, such as bran flakes from the skin and finely broken hairs from the brush (Bailey, 1992; Hoseney and Faubion, 1992). Its ability to form an explosive mixture with air is the most serious problem generated by grain dust. During the 62-year period from 1860 to 1922 in the United States, there were 119 explosions in the grain and grain processing industry, which caused 215 deaths and 271 injuries.

A series of U.S. grain elevator explosions that happened within 8 days in 1979 killed 59 persons and injured 48, destroying 2.5% of the U.S. export elevators. This led to the initiation of extensive research programs to find prevention measures for grain elevator and mill dust explosions (Kauffman, 1987; Kent, 1983). In 1998, a grain dust explosion destroyed 14 concrete bins full of grain and claimed 14 lives; it is considered the largest explosion of this kind in the world (Wenzel, 1998).

The first recorded flour dust explosion happened in a Turin flour mill in 1785. Since then, the continued explosions have drawn scientific, economic, and even political attention. Other well-known industry explosions were related to such food powders as starch and cocoa (Cadle, 1965). Process steps where fine powders are heated have very strong association with dust explosion; examples include dilute pneumatic conveying and spray drying, which involves heat and a dilute suspension (Rhodes, 1998).

The number of dust explosions occurring in powder handling and processing is relatively low, especially considering that many installations are designed and operated with the appropriate measures in place. However, explosion accidents still occur with some frequency (Zeeuwen, 2001). There are always improvements that are needed to comply with preventive standardized requirements and to maintain a low level risk, as long as it is reasonably practicable. This section will cover the fundamentals of dust explosion mechanisms, measurement indices, and preventive methods for dust explosion hazard control.

12.4.1. Explosion Mechanisms

There are three different modes in which explosive dust can react with air or other gaseous oxidizers, i.e., smoldering, deflagration, and detonation. Each of these modes can be very dangerous and can cause significant human and property losses. During *smoldering mode*, the combustible material is slowly oxidized at speeds of millimeters or centimeters per hour. The *deflagration mode* requires speeds ranging from centimeters per second to hundreds of meters per second; the *detonation mode* explodes with a speed of kilometers per second. Deflagrative combustion of dust is the most common cause of explosion in many industries, so it is of the greatest interest in research (Kauffman et al., 1992).

During the development of an explosion, there is a rapid pressure rise in the atmosphere. The pressure variation with the time during the development of an explosion is shown in Fig. 12.16. The maximum pressure, as well as the maximum rate of pressure rise, must exceed critical limits to develop an explosion. A slow pressure rise allows the explosion to be released safely through adequate ventilation, while a rapid pressure rise gives the pressure wave an intensity high enough

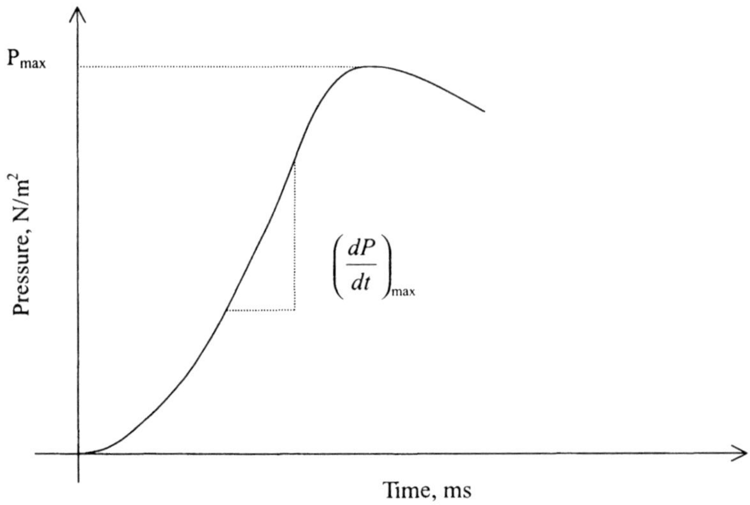

Figure 12.16. Pressure–time relationship during the development of an explosion (adapted from Shamlou, 1991).

to prevent the explosion from escaping through the normal ventilation system, causing devastating effects to nearby workers, equipment, and buildings (Shamlou, 1988).

A whole explosion process can be divided into two steps, the primary explosion and the secondary explosion. The primary explosion of the dust suspension is initiated by an ignition source. It is not very dangerous itself, except that it could result in a series of secondary explosions which generate and ignite further dust clouds by disturbing settled dusts on floors, roofs, ducts, and surfaces of machines all over the plant (Shamlou, 1988). The quantity of dust needed for a secondary dust explosion is very small. For most dusts, a concentration of 100 g/m^3 is well within the flammable range (Zeeuwen, 2001). For example, 500 g of dust on every m^2 of the floor would be sufficient to generate a flammable cloud 5 m high, if all dust was suspended uniformly. This height would be enough to fill most plant areas. Dust available for secondary dust explosions is often found on top of equipment, and on ridges, ledges, and pipelines. Grains and flour dust can be highly explosive under the following conditions: particle size smaller than 200 μm; particle concentration about 50 g/m^3; ignition source energy equal to 1 mJ; and temperature over 400°C (Shamlou, 1988).

At the onset of a dust explosion, a degree of turbulence already exists, especially in pneumatic conveying systems, and it will be increased as the flame front moves through the dust. It is extremely difficult to quantify turbulence in a dust explosion because it is likely to be non-uniform, thereby distorting the normal flow of a given process. Previous turbulence increases the access of oxygen to the active surface, resulting in an increased explosion hazard.

Generally speaking, an explosion occurs only where dust is dispersed in the air and a source of ignition is present. Dust clouds can be created during processing, such as grinding or fluidized drying. They can also be created during product handling, i.e., elevator emptying, pneumatic transportation, and filter vibration (Pineau, 1994). During pneumatic transfer, the fines may well remain suspended long after the granules have fallen, so that the dust concentration in the head space of the receiving vessel can be much higher than expected on the basis of the average dust content of the product. Another example is during the unusual case of arching in a silo, where conditions of high powder concentration and increased pressure in a closed system can be created. Unless oxygen is removed from the equipment (e.g., by inert gas flashing) a dust cloud might be formed.

A situation that must be considered separately is an explosion propagating in a long duct. In this case, the expansion of combustion products causes a flow in the dust, which increases the level of turbulence. Since turbulence enhances combustion rate, the explosion will accelerate continuously until a different type of explosion propagation is reached, i.e., a detonation. A detonation typically runs at 2 km/s (much more rapid than the speed of sound in the unburned mixture), has a higher pressure peak, and is more devastating and difficult to control than a normal dust explosion.

12.4.2. Factors Affecting Dust Explosions

Three main factors are summarized in the scheme called "triangle of fire" as shown in Fig. 12.17. These conditions must be satisfied for a certain dust explosion to occur, and they are described as follows: (a) *Dust air dispersion*: the combustible dust must be dispersed and mixed with air while contained within a volume; (b) *Minimum dust concentration boundary*: the concentration of dispersed dust must be above the minimum explosive concentration; and (c) *Ignition source and potential energy*: an ignition source with sufficient density and total energy to initiate the combustion wave must be present (Hertzberg and Cashdollar, 1987).

Particle size is important for dust explosion generation because it influences the combustion rate of the solids in air. Smaller particles will provide more surface area per unit mass of air available to react. In fact, the rate of heat generation by the combustion reaction is inversely proportional to dust particle size (Rhodes, 1998). Thus, the likelihood of flame propagation and explosion will increase with increasing particle size. Apart from the effect of surface area, smaller particles are more readily able to disperse in air, and consequently generate more heat per unit mass of air, increasing the heat-up rate. Furthermore, particle size distribution, surface properties, and moisture content all influence the potential fire hazard of the powder.

Food powders that produce dust clouds are generally very dry, fine, brittle, and fragile. Attrition usually occurs when they are subjected to impact or friction forces during processing and handling,

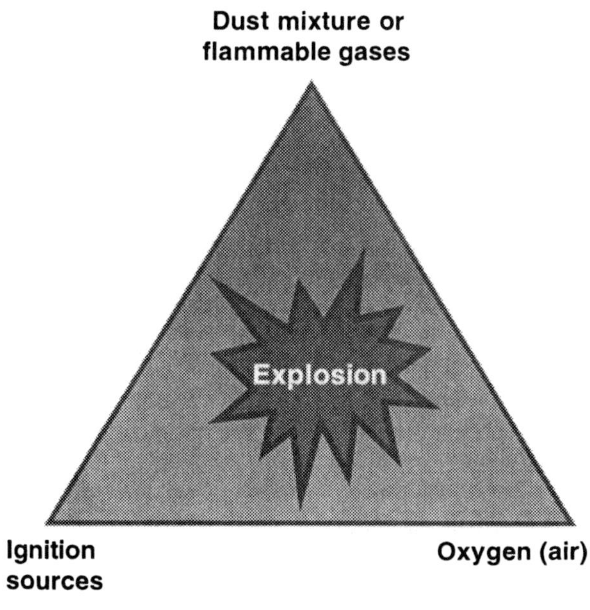

Figure 12.17. Explosion prevention measures based on the familiar hazard triangle (adapted from Jaeger and Siwek, 1999).

which in turn results in dust formation that may also develop into a dust explosion hazard (Peleg, 1983). There are several other particulate material properties related to dust dispersion into the air: the individual dust particle density, particle shape, cohesive properties with respect to each other, and the particle's adhesive properties with respect to supporting surfaces (Hertzberg and Cashdollar, 1987). Dust dispersibility is difficult to characterize, but in principle it depends on humidity and particle shape.

Various ignition sources can be characterized according to the types of energy they introduce to the system. The most common types of ignition sources for dust explosion are electrical, chemical, or purely thermal (Hertzberg and Cashdollar, 1987). In particular, naked flames (fire), welding and cutting, electrical equipment, mechanical friction and mechanical sparks, static electricity, hot surfaces, and self-heating or self-ignition can be listed (Zeeuwen, 2001). Sparks or flames may be caused in random cases by a shovel edge striking a concrete surface, matches, or a broken light bulb, among many particular situations. Other sources are hot surfaces, cutting tools, bearings, friction sparks, electric appliances, spontaneous heating, magnets, lightning, bins, and bucket elevators. Welding and hand lamps are the most frequently responsible ignition sources in the flour milling industry. Static electricity occurs whenever materials are rubbed together and then separated. This means that in most solids handling, a lot of static electricity is generated. When all material is conducting and earthed, however, the static charge will not be apparent. When the charge can accumulate and then discharge, a static ignition hazard arises. Some paints that contain a high percentage of metal powder (such as aluminum paint) can become a source of ignition when they are painted on a surface and struck by an object (Bailey, 1992; Kent, 1983; Land, 1991; Zeeuwen, 2001).

12.4.3. Explosion Hazard Evaluation

The evaluation of explosion hazards is an important task in industry because it can provide useful information about the potential danger of a dust and a guideline for explosion prevention. It depends on many factors, such as the explosive properties of the dust, kinds of oxidizers, explosive space characters, possible ignition sources, and other factors mentioned before. Among all these factors, the explosive properties of a dust are the most important since the dust's sensitivity to explosion and severity of explosion are directly related to those properties (Wolanski, 1992). The following parameters are included in this section to assess the potential explosibility of a dust cloud: minimum explosive concentration (MEC), minimum hazardous mass (MHM), minimum ignition energy (MIE), minimum ignition temperature (MIT), and maximum oxygen combustion (MOC). For some typical food powders, the values of those parameters are listed in Table 12.3.

12.4.3.1. Minimum Explosive Concentration (MEC)

MEC is usually defined as the minimum dust concentration to support self-sustaining flame propagation, and is closely related to the natural properties of a dust. Actually, there are two extreme concentration limits within which a dust explosion can occur, i.e., the minimum and maximum dust concentrations. To form an explosive dust cloud, the dust concentration should be higher than the MEC and lower than the maximum explosive concentration. Beyond those limits, the progress of deflagration is hindered because of the phlegmatization by the air (below the MEC) or by the dust (higher concentration) (Land, 1991). In real cases, only the MEC is considered since the maximum explosive concentration is usually very high.

The MEC value can be calculated based on explosion experiments carried out in a 20-liter spherical apparatus. The following conditions should be met in order to evaluate the MEC value

Table 12.3. Parameters for assessing explosibility of some food powders (modified from Jones and King, 1991).

Name	MEC (g/m^3)	MIE (J)	Ignition temperature (°C)		Limiting oxygen concentration (%)*
			Cloud	Layer	
Cellulose	55	0.080	480	270	C13
Cellulose (alpha)	45	0.040	410	300	—
Cocoa (natural)	75	0.10	510	240	—
Coffee (roasted)	85	0.16	720	270	C17
Corn	55	0.04	400	250	—
Corn starch	45	0.04	400	—	—
Grain dust	55	0.03	430	230	—
Milk (skimmed)	50	0.05	490	200	N15
Rice	85	0.10	510	450	—
Soy flour	60	0.10	550	340	C15
Sugar (powdered)	45	0.03	370	400	—
Wheat flour	50	0.06	440	440	—
Wheat starch (edible)	45	0.025	430	—	C12

*The prefix letter represents the dilute gas (C for carbon dioxide and N for nitrogen) and the numerical number for the oxygen content. For example, "C15" means diluting to an oxygen content of 15% with carbon dioxide as the dilution gas.

properly: (a) dust should be dispersed uniformly; (b) ignition energy must be sufficient to ignite the dust mixture (usually 10,000 J is supplied by a chemical ignitor), but not too high to enhance the combustion of lean mixtures; and (c) there should be a predetermined distance between the flame propagation and ignition source so that the influence of the ignition source can be neglected. Under these assumptions, the MEC is calculated by the following equation:

$$\text{MEC} = 15500 \times H_C \tag{12.48}$$

where H_C is the heat of combustion (kJ/m^3) and MEC has the unit of kg/m^3 (Wolanski, 1992).

The MEC for dust particle sizes lower than 100 μm is generally between 20 and 100 g/cm^3. Explosion experiments generally present important error sources such as the preparation of homogenous dust suspensions, variations in ignition energy, and constant closure volume settings. Therefore, figures concerning the minimal explosive concentration should be taken with skepticism (Pineau, 1994). Sometimes, when the average dust concentration in a confined space is lower than the MEC, the explosion hazard will still remain. This is because the concentration throughout the space is heterogeneous, and some localized dust might be much larger than the MEC (Wolanski, 1992).

12.4.3.2. Minimum Hazardous Mass (MHM)

MHM is defined as the amount of dust that can create a catastrophic pressure rise by explosive combustion in a given confined volume. It is strictly connected to the strength of the confinement and the natural properties of the dust. The importance of this parameter can be demonstrated in two cases, as shown in Fig. 12.18, where the same amount of dust is put into two cells of the same confined volume. Dust is dispersed uniformly in the first cell, while in the second cell dust occupies one-tenth of the cell's volume. The local dust concentration in the second cell is 10 times higher than the average dust concentration, which is higher than the MEC. When an ignition source is used, the dust cloud in the first cell will not combust and no rapid pressure rise will be caused, while the opposite will be expected to occur in the second one. The MHM can be calculated from the following

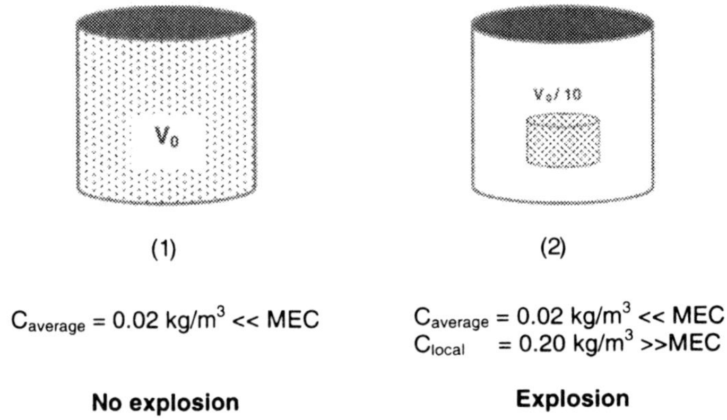

Figure 12.18. Different explosion estimations based on the average dust concentration and real local concentration because of uniformly and non-uniformly dispersed dusts (adapted from Wolanski, 1992).

equation (Wolanski, 1992):

$$\mathrm{MHM} = \frac{\Delta P C_V T_0 \theta}{H_C P_0} \tag{12.49}$$

where ΔP is the critical pressure rise to damage the confinement of a certain volume (Pa), θ is the average dust mixture density (kg/m^3), T_0 is the initial absolute temperature (K), P_0 is the initial absolute pressure, C_V is the specific heat capacity at a constant volume, and H_C is the heat of combustion of the dust (kJ/kg).

12.4.3.3. Minimum Ignition Energy (MIE)

MIE is the lowest value of electronic energy (stored in a capacitor and released upon discharge) that just suffices to ignite the most readily ignitable fuel/air/mixture at atmospheric pressure and temperature. Its value varies between the lower and higher limits, with the lower one representing the highest energy at which no ignition is found in at least 10 tests, while the higher one is the lowest energy at which the dust mixture is just ignited. Thus, energy at no ignition is greater than the MIE and lower than the energy at ignition (Jaeger and Siwek, 1999). Methods for evaluating the MIE can be found in a standard of the International Electrotechnical Commission (IEC, 1994).

The MIE value is of the greatest interest in dealing with electrostatic discharge (ESD) hazards in the processing, bulking, and shipping of plastic powders and many other materials in their manufacture because it provides a useful measure of the relative ESD-ignition risk of the material. For some food powders, their MIE values range from 25 to 2,000 mJ, as listed in Table 12.3. In principle, an MIE value of less than 100 mJ means that it would be possible to ignite the dust by an electrostatic discharge from a person. The British Standards Institute document mentions that most ignitions occur with dusts of an MIE value less than 25 mJ (Land, 1991; Jones and King, 1991).

12.4.3.4. Minimum Ignition Temperature (MIT) and Maximum Oxygen Combustion (MOC)

MIT is defined as the minimum temperature at which a dust suspension will autoignite. By passing a dust cloud through a laboratory furnace at various temperatures, the MIT value is expressed as the temperature at which the ignition takes place (indicated by visual observation of flame). MIT is affected by many factors, including the dust concentration, chemical nature of the dust, particle size, moisture content, surface area of the particles, and so on (Jaeger and Siwek, 1999). Such tests can be carried out for both dust layers and dust suspensions because they will show different behaviors.

MOC is used to determine the maximum permissible oxygen concentration when inerting is selected as the means of controlling dust explosions. Typical MOC values are around 10% by volume and they can be determined by the modified Hartmann apparatus (Jaeger and Siwek, 1999; Land, 1991).

Devices for determination of dust explosion generally include a vessel, which may be open or closed, an ignition source, which may be an electrical spark or electrically heated wire coil, and a supply of air for dispersion of dust. Three different apparatuses can be mainly used to measure dust explosion characteristics: a vertical tube apparatus, a 20-liter sphere, and the Goldbert–Greenwald furnace apparatus. The vertical apparatus is used for the classification test and for determination of MEC, MIE, and in a modified form for the MPOC (Rhodes, 1998). The 20-liter sphere method is used for the determination of the MEC and the maximum rate of pressure during explosion. The third basic test is used for the determination of MIT and the explosion characteristics at elevated temperatures. The apparatus includes a vertical electrically heated furnace tube that can reach temperatures up to 1,000°C.

12.4.4. Explosibility Classification

Dust explosibility can be classified in several ways, depending on different testing methods and standards used. An explosibility index E_{index} is defined as the product of the explosion severity and ignition sensitivity. Explosion severity is in turn defined as the ratio of the product of maximum explosion pressure and maximum rate of pressure rise for the sample dust to that of a standard material, i.e., Pittsburgh coal dust. Ignition sensitivity is defined as the ratio of the product of minimum ignition temperature, minimum ignition energy and minimum explosion concentration of the standard material to that of the sample dust (Shamlou, 1988). Explosion severity, ignition sensitivity, and explosibility of some food powders are listed in Table 12.4. The relative explosion evaluation based on this explosion index is listed in Table 12.5.

An alternative explosibility classification method is based upon the so-called "cubic law," which is the result of an idealized approach to the explosion dynamic in a closed vessel (van der Wel et al., 1991). It measures the explosion parameters of a combustible dust suspension in a closed sphere of

Table 12.4. Ignition sensitivity, explosion severity, and explosibility index for some food powders (adapted from Jones and King, 1991).

Name	Ignition sensitivity	Explosion severity	Explosibility index
Cellulose	1.0	2.8	2.8
Cellulose (alpha)	2.7	4.0	>10
Cocoa (natural)	0.5	1.1	0.6
Coffee (roasted)	0.2	0.1	<0.1
Corn	2.3	3.0	6.9
Corn starch	2.8	3.4	9.5
Grain dust	2.8	3.3	9.2
Milk (skimmed)	1.6	0.9	1.4
Rice	0.5	0.5	0.3
Soy flour	0.6	1.1	0.7
Sugar (powdered)	4.0	2.4	9.6
Wheat flour	1.5	2.7	4.1
Wheat starch (edible)	5.2	3.4	17.7

Table 12.5. Explosibility index E_{index} and classification of a dust explosion (adapted from Shamlou, 1988).

Explosion classification	Explosibillity index, E_{index}	Explosion severity	Ignition sensitivity
Weak	0.1	0.5	0.2
Moderate	0.1–1.0	0.5–1.0	0.2–1.0
Strong	1.0–10˜	1.0–2.0	1.0–5.0
Severe	10	2.0	5.0

Table 12.6. Classification of dust explosibility by a constant K_{ST} based upon the cubic law (modified from Shamlou, 1988).

St classification	K_{ST} (kN m/s)	Relative explosion rating
3	>300,000	Severe
2	200,000 < K < 300,000	Strong
1	0 < K < 200,000	Weak
0	0	No explosion

20-liter (or 1 m^3) capacity with a central ignition source (Jaeger and Siwek, 1999). From the pressure–time recording chart, shown in Fig. 12.16, the pressure change as a function of time is obtained and so is the maximum rate of pressure rise $(\mathrm{d}P/\mathrm{d}t)_{max}$. Thus, a size normalized explosibility index K_{ST} can be calculated according to the cubic law:

$$K_{ST} = \left(\frac{\mathrm{d}P}{\mathrm{d}t}\right)_{max} \times V \tag{12.50}$$

where V is the volume of the explosion vessel. The classification of explosibility based on the K_{ST} value is a measure of the explosion severity, and different rating values (from 0 to 3) are explained in Table 12.6.

Another way that has been found by most prevention specialists to qualify and predict dust explosion risks is to use probability measure (Jones and King, 1991). In this method, the overall explosion probability $P_{explosion}$ may be expressed as:

$$P_{explosion} = P_D \times P_I \tag{12.51}$$

where P_D is the existence probability of combustible dust/air mixture and P_I is the probability of simultaneous presence of adequate ignition source. The obvious advantage of this method is that the ignition risk (measured by P_I) can be viewed in isolation from other factors.

12.4.5. Dust Explosion Prevention

Dust explosion can be prevented with certainty if one of the three requirements mentioned above is reliably eliminated (Jaeger and Siwek, 1999): (a) the development of explosible mixtures (suspended particles, flammable gases); (b) the presence of atmospheric oxygen; and (c) the existence of effective ignition sources. In general, the most desirable strategy would be to change the process in order to eliminate the dust. When looking into designing a plant, equipment should be set to withstand the pressure generated by any explosion. The simplest and best way to minimize the explosion chance is by good maintenance practices, for example, by insuring that equipment, bins,

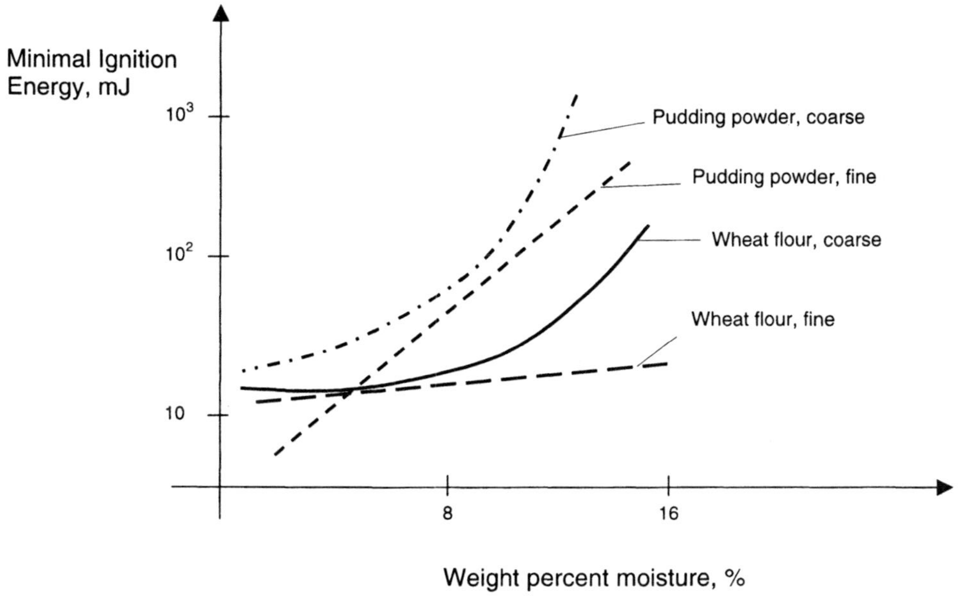

Figure 12.19. Effect of moisture content on the minimal ignition energy (MIE) of two food powders at different particle sizes (adapted from Grossel, 1997).

and transport systems are dust tight, or by cleaning and removing extraneous combustible material and solvent from work areas and equipment.

Furthermore, the use of vacuum systems at feed inlets, transfer points, or packing locations will assure dust–clean air in the production environment (Nagy and Verakis, 1983). In fact, vessels can be vented in order to relieve the pressure generated by the explosion (Rhodes, 1998). The principle of explosion venting is to discharge the vessel contents through an opening or vent to prevent the pressure from rising above the vessel design pressure. Venting is a relatively simple and inexpensive method of dust control. The mass and type of vent determine the pressure at which the vent opens and the delay before it is fully open. These factors, together with the size of the vent, determine the pressure rise and the maximum pressure reached after the vent opens.

Moreover, dust formation can be suppressed by light damping with water (about 1% by weight) on the dispersion. Therefore, moisture addition or other type of diluent to the dust can significantly decrease the risk of explosion. This preventive method is generally used in the flour milling process (Kent, 1983). Figure 12.19 illustrates the influence of moisture content on the minimum electric spark ignition energy (MIE) in wheat flour and pudding powders of different particle sizes. It can be observed that it is not actually possible to know in advance the exact moisture content needed in order to impede the explosion from occurring, as this varies with other factors as well, such as particle composition and particle size (Grossel, 1997). Water attaches to the dust particles, creating more sticky surfaces and thereby acting as an oxygen barrier impeding combustion. Furthermore, evaporation of water from the particles absorbs heat diminishing the explosibility in the system. Additionally, a dry dust is less cohesive and hence more likely to form a flammable cloud. The flour agglomeration process may also be used to decrease the dust explosion potential because its final agglomerated products are almost dust free (Kent, 1983).

When dust clouds are present, keeping the dust concentration outside the explosive range is rarely possible because of the sedimentation or whirling-up of fine particles. Thus, an explosive atmosphere can only be avoided by reducing the concentration of oxygen, by either using inert gases or working under vacuum conditions (Jaeger and Siwek, 1999). It has been shown that the use of inert gases (or inerting) is an effective way to prevent dust explosions; however, this practice is not applied very often because of its high cost and difficult implementation. In inerting, the oxygen is frequently replaced by nitrogen, carbon dioxide, and flue gas, and, therefore, it is reduced below the MOC. The explosibility of combustible dust can also be reduced by mixing inert agents into the dust cloud. Examples of explosion-inert dust powders are $CaCO_3$, $CaSO_4$, or $NaHCO_3$ (Tanaka, 1997).

Explosion problems are also prevented by suppressing ignition sources (Jaeger and Siwek, 1999). Ignition sources must be excluded as much as possible when explosive dusts are handled. Generally speaking, most dust ignition sources are attributed to poor maintenance and carelessness such as inadequate dust control, overheating of bearings, existence of open flames in the area, or arcing phenomena developed in improperly protected electrical instruments. However, there is still a substantial number of reported primary dust explosions that are caused by unknown sources (Jones and King, 1991; Shamlou, 1988). Explosive-proof equipment generally possesses materials that provide necessary isolation in order to avoid overheating or arcing development due to electrical tension. Eliminating ignition sources is sometimes not sufficient to prevent dust explosion; thereby, it should be combined with other supplementary techniques in order to assure prevention.

In some cases, when all measures have been taken in order to stop the formation and ignition of a dust cloud, a fire or primary explosion can still occur. In order to diminish the impact, an explosion suppression system can be used under certain conditions. This will prohibit the spreading of flames as well as the development of a secondary dust explosion. Figure 12.20 provides a diagram

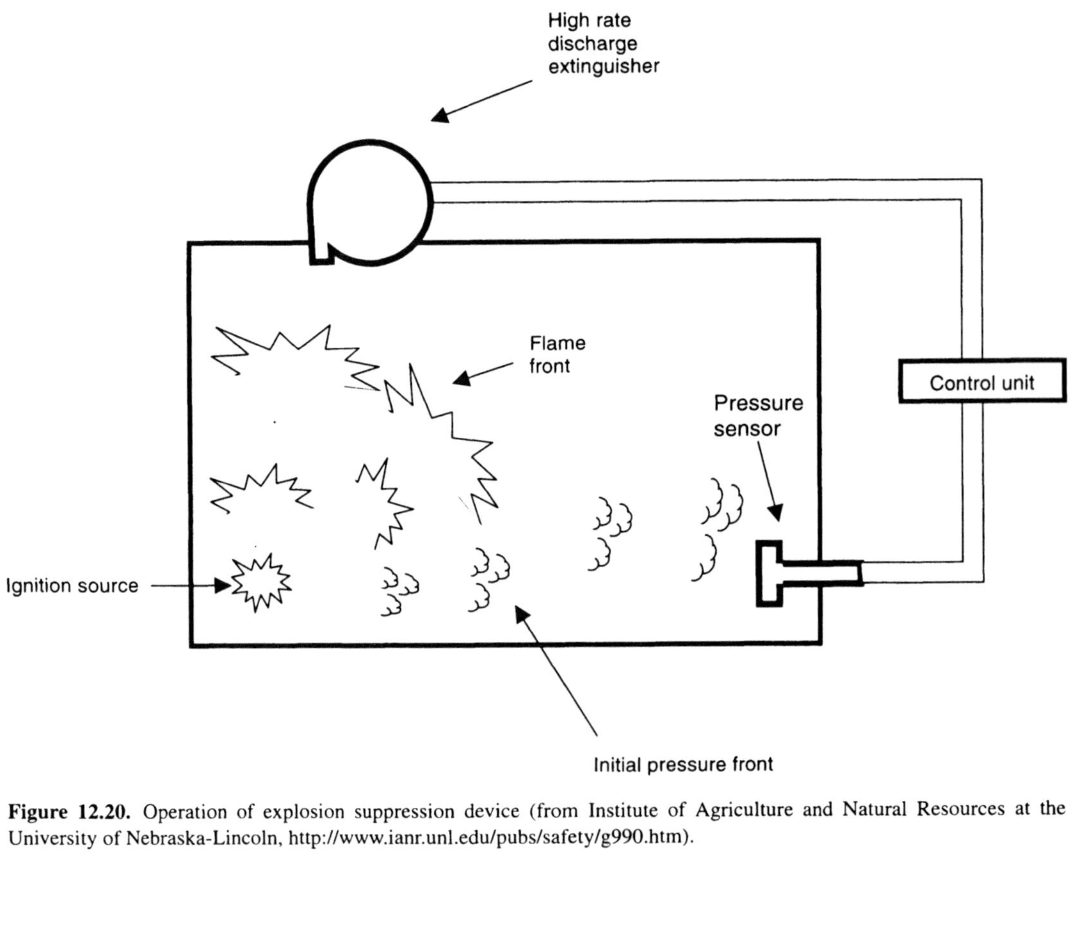

Figure 12.20. Operation of explosion suppression device (from Institute of Agriculture and Natural Resources at the University of Nebraska-Lincoln, http://www.ianr.unl.edu/pubs/safety/g990.htm).

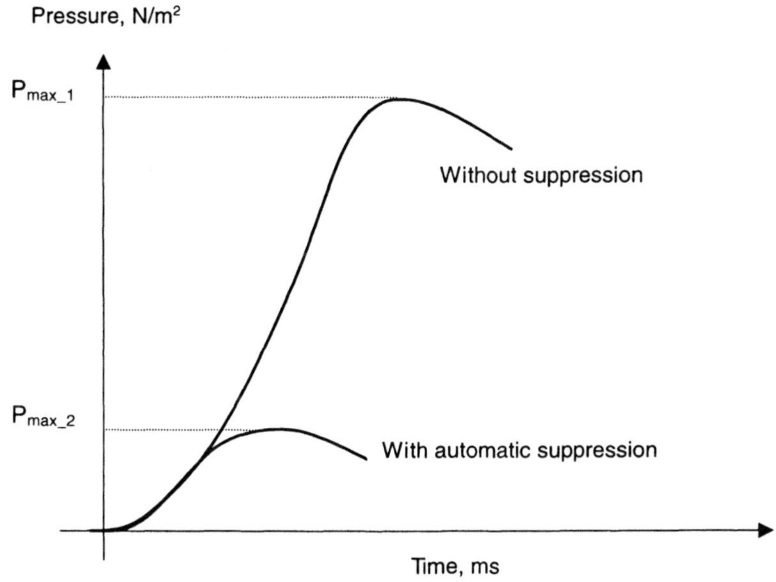

Figure 12.21. Comparison of the pressure–time relationships with and without the use of an automatic suppression system during the development of a dust explosion (adapted from Shamlou, 1991).

of a basic explosion suppression system. The system consists of an automatic detector that senses pressure changes within the plants. Once a pressure-wave (caused by the initial ignition) is detected, a suitable suppressant will be injected automatically at a speed high enough to suppress the explosion flames, and prevent the pressure from reaching its maximum explosion limits (Shamlou, 1988). The original development of the suppression system was started during the World War II, in an attempt to halt the consequent fuel-tank explosions caused by gun-fire; it found applications in the flash drying of starches and also the milling/drying of explosible materials (Land, 1991).

The explosion prevention effect in a suppression system is illustrated by the pressure–time relationship shown in Fig. 12.21. The suppressant is basically an inert substance in the form of gas, liquid, or powder. Powered monammonium phosphate and liquid halogenated hydrocarbons are good examples of well-known suppressants. Some passive stone dust and water barriers were used in the suppression studies of coal dust explosion in Poland (Land, 1991; Wolanski, 1992).

Due to the extreme complexity of many physical and chemical factors involved in dust explosion and the mechanisms of ignition, flame propagation, and solid combustion, there are still many unsolved problems about dust explosion, especially in accurate prediction of explosion characteristics and in reliable protection against dust explosions in various situations (Nomura, 1992). The processing and design engineers dealing with dusty powders should seek advice from the experts. More basic and applied research on dust explosion needs to be carried out to ensure better protection of human life and property against uncontrolled explosions in many industry fields, especially in the food and agricultural industries.

ATTRITION REFERENCES

Adams, M.J. and McKeown, R. (1996). Micromechanical analyses of the pressure–volume relationships for powders under confined uniaxial compression. *Powder Technol.* **88**: 155–163.

Adams, M.J., Mullier, M.A. and Seville, J.P.K. (1994). Agglomerate strength measurement using a uniaxial confined compression test. *Powder Technol.* **78**: 5–13.

Allen, M., Brown, G.J. and Miles, N.J. (1995). Measurement of boundary fractal dimensions: review of current techniques. *Powder Technol.* **84**: 1–14.

Austin, L.G., Bagga, P. and Cetic, M. (1981). Breakage properties of some materials in a laboratory. *Powder Technol.* **28**: 235–239.

Barbosa-Cánovas, G.V., Malave-López, J. and Peleg, M. (1985). Segregation in food powders. *Biotechnol. Prog.* **1**(2): 140–146.

Barletta, B.J. and Barbosa-Cánovas, G.V. (1993a). An attrition index to assess fines formation and particle size reduction in tapped agglomerated powders. *Powder Technol.* **77**: 89–93.

Barletta, B.J. and Barbosa-Cánovas, G.V. (1993b). Fractal analysis to characterize raggedness changes in tapped agglomerated food powders. *J. Food Sci.* **58**(5): 1030–1035, 1046.

Barletta, B.J., Knight, K.M. and Barbosa-Cánovas, G.V. (1993a). Review: attrition in agglomerated coffee. *Rev. Esp. Cienc. Tecnol. Aliment* **33**(1): 43–58.

Barletta, B.J., Knight, K.M. and Barbosa-Cánovas, G.V. (1993b). Compaction characteristics of agglomerated coffee during tapping. *J. Texture Studies* **24**: 253–268.

Bemrose, C.R. and Bridgwater, J. (1987). A review of attrition and attrition test methods. *Powder Technol.* **49**: 97–126.

Biscans, B., Chemini, R., Guiraud, P. and Laguerie, C. (1996). Design of an attrition experiment to simulate the effects of crystal-wall or crystal-stirrer impacts occurring in a crystallizer. *Powder Technol.* **86**: 155–161.

Chraibi, M.A. and Flamant, G. (1989). Kinetic, thermal and chemical attrition of manganese chloride particles in a fluidized bed. *Powder Technol.* **59**: 97–107.

Clark, N.N. (1986). Three techniques for implementing digital fractal analysis of particle shape. *Powder Technol.* **46**: 45–52.

Ghadiri, M. (1997). Particle impact and attrition. In *Powder Technology Handbook*, 2nd ed., Gotoh, K., Masuda, H. and Higashitani, K. (eds.). Marcel Dekker, Inc., New York, Chapter II.10, pp. 183–191.

Gerhards, C.H., Ulbricht, D.M. and Peleg, M. (1998). Mechanical characterization of individual instant coffee agglomerates. *J. Food Sci.* **63**(1): 140–142.

Gundepudi, M.K., Sankar, B.V., Mecholsky Jr., J.J. and Clupper, D.C. (1997). Stress analysis of brittle spheres under multiaxial loading. *Powder Technol.* **94**: 153–161.

Hayes, G.D. (1987). *Food Engineering Data Handbook.* John Wiley & Sons, Inc., New York, p. 83.

Hoornaert, F., Wauters, P.A.L., Meesters, G.M.H., Pratsinis, S.E. and Scarlett, B. (1996). Agglomeration behavior of powders in Lodige mixer granulor. *Powder Technol.* **96**: 116–128.

Kaye, B.H. (1989). A starting point for the random walk. In *A Random Walk Through Fractal Dimensions.* VCH Publishers, New York, Chapter 1, pp. 1–11.

Knight, P.C. and Bridgwater, J. (1985). Comparison of methods for assessing powder attrition. *Powder Technol.* **44**: 99–102.

Malave-López, J., Barbosa-Cánovas, G.V. and Peleg, M. (1985). Comparison of the compaction characteristics of selected food powders by vibration, tapping and mechanical compression. *J. Food Sci.* **50**: 1473–1476.

Malave-López, J. and Peleg, M. (1986). Mechanical attrition rate measurement in agglomerated instant coffee. *J. Food Sci.* **51**(3): 687–690, 697.

Malave-López, J., Barbosa-Cánovas, G.V. and Peleg, M. (1986). Kinetic models for the attrition of instant coffee. In *Particulate and Multiphase Process*, Ariman, T. and Veziroglu, T.N. (eds.). Hemisphere, New York.

Menacho, J.M. (1986). Some solutions for the kinetics of combined fracture and abrasion breakage. *Powder Technol.* **49**: 87–96.

Mort, P.R., Sabia, R., Niesz, D.E. and Riman, R.E. (1994). Automated generation and analysis of powder compaction diagram. *Powder Technol.* **78**: 111–119.

Mullier, M.A., Seville, J.P.K. and Adams, M.J. (1991). The effect of agglomerate strength on attrition during processing. *Powder Technol.* **65**: 321–333.

Neil, A.N. and Bridgwater, J. (1994). Attrition of particulate solids under shear. *Powder Technol.* **80**: 207–219.

Normand, M.D. and Peleg, M. (1986). Determination of the fractal dimension of a particle silhouette using image-processing techniques. *Powder Technol.* **45**: 271–275.

Nuebel, C. and Peleg, M. (1994). A research note: compressive stress–strain relationships of agglomerated instant coffee. *J. Food Proc. Eng.* **17**: 383–400.

Olivares-Francisco, C. and Barbosa-Cánovas, G.V. (1990). Characterization of the attrition process in agglomerated coffee by natural fractals. Presented at the IFT annual meeting, Anaheim, CA.

Park, B.S., Smith, D.M. and Thoma, S.G. (1993). Determination of agglomerate strength distributions, Part 4. Analysis of multimodal particle size distributions. *Powder Technol.* **76**: 125–133.

Peleg, M. (1983). Physical properties of foods. In *Physical Properties of Foods*, Peleg, M. and Bagley, E.B. (eds.). AVI Publishing Co., Westpoint, CT, Chapter 10, pp. 293–323.

Peleg, M. and Normand, M.D. (1985a). A research note: characterization of the ruggedness of instant coffee particle shape by natural fractals. *J. Food Sci.* **51**: 829–831.

Peleg, M. and Normand, M.D. (1985b). Mechanical stability as the limit to the fractal dimension of solid particle silhouettes. *Powder Technol.* **43**: 187–188.

Peleg, M. and Normand, M.D. (1987). Computer simulation of the attrition patterns of particulated and agglomerated foods. *J. Food Sci.* **52**(4): 943–947.

Popplewell, L.M., Campanella, O.H. and Peleg, M. (1988). Quantitative characterization of the particle size distribution of instant coffee during mechanical attrition. *J. Food Sci.* **53**(3): 877–881.

Popplewell, L.M., Campanella, O.H. and Peleg, M. (1989). Simulation of bimodal size distributions in aggregation and disintegration process. *Chem. Eng. Prog.* **85**(8): 56–62.

Popplewell, L.M. and Peleg, M. (1989). An 'Erosion index' to characterize fines production in size reduction processes. *Powder Technol.* **58**: 145–148.

Potapov, A.V. and Campbell, C.S. (1997). Computer simulation of shear-induced particle attrition. *Powder Technol.* **94**: 109–122.

Shipway, P.H. and Hutchings, L.M. (1993). Attrition of brittle spheres by fracture under compression and impact loading. *Powder Technol.* **76**: 23–30.

Simons, S.J.R. (1996). Modeling of agglomerating systems: from spheres to fractals. *Powder Technol.* **87**: 29–41.

Suzuki, M. (1991). Shearing force of powder bed. In *Powder Technology Handbook*, Linoya, K., Gotoh, K. and Higashitani, K. (eds.). Marcel Dekker, Inc., New York, pp. 89–98.

Swyngedau, S., Nussinovitch, A., Roy, I., Peleg, M. and Huang, V. (1991). Comparison of four models for the compressibility of breads and plastic foams. *J. Food Sci.* **56**(3): 756–759.

Yan, H. and Barbosa-Cánovas, G.V. (1997). Compression characteristics of agglomerated food powders: the effect of agglomerate size and water activity. *Food Sci. Technol. Int.* **3**: 351–359.

Yan, H. and Barbosa-Cánovas, G.V. (2001). Attrition evaluation for selected agglomerated food powders: the effect of agglomerate size and water activity. *J. Food Proc. Eng.* **24**(1): 37–49.

Yan, H. and Barbosa-Cánovas, G.V. (2002). The effect of padding foam on the compression characteristics of some agglomerated food powders. *Food Sci. Technol. Int.* **7**(5): 417–423.

SEGREGATION REFERENCES

Barbosa-Cánovas, G.V. (1985). *The Role of Surface Affinity and Interactions in the Segregation of Food Powders*. Ph.D. thesis, Amherst, MA.

Barbosa-Cánovas, G.V., Malavé-López, J. and Peleg, M. (1985). Segregation in food powders. *Biotechnol. Prog.* **1**(2): 140–146.

Barletta, B.J., Knight, K.M. and Barbosa-Cánovas, G.V. (1993). Review: attrition in agglomerated coffee. *Rev. Esp. Cienc. Tecnol. Aliment.* **33**(1): 43–85.

Brown, R.L. and Richards, J.C. (1970). Flow patterns and segregation. In *Principled of Powder Mechanics*. Pergamon Press, Oxford, Chapter 5, pp. 116–134.

Enstad, G.G. (2001). Segregation of powders—mechanisms, process and counteraction. In *Handbook of Conveying and Handling of Particulate Solids*, Levi, A. and Kalman, K. (eds.). Elsevier, Amsterdam, pp. 589–603.

Fan, L.T., Chen, S.J. and Watson, C.A. (1970). Solids mixing. *Indust. Eng. Chem.* **62**(7): 53–69.

Harris, J.F.G. and Hildon, A.M. (1970). Reducing segregation in binary powder mixtures with particular reference to oxygenated washing powders. *Ind. Eng. Chem. Process Des. Develop.* **9**(3): 363–367.

Lindley, L.T. (1991). Mixing processes for agricultural and food materials: 3. Powders and particulates. *J. Agric. Eng. Res.* **49**: 1–19.

McGlinchey, D. (1998). Assessment of segregation in industrial processes. *J. Powder/Bulk Solids Technol.* June: 54–56.

Olsen, J.L. and Rippie, E.G. (1964). Segregation kinetics of particulate solids systems. I: Influence of particle size and particle size distribution. *J Pharm. Sci.* **53**(2): 147–150.

Ott, R.L. (1992). *An Introduction to Statistical Methods and Data Analysis*, 4th ed. Duxbury Press, Belmont, CA.

Parsons, D.S. (1976). Particle segregation in fine powders by tapping as simulation of jostling during transportation. *Powder Technol.* **13**: 269–277.

Peleg, M. (1983). Physical properties of foods. In *Physical Properties of Foods*, Peleg, M. and Bagley, E.B. (eds.). AVI Publishing Co., Westpoint, CT, Chapter 10, pp. 293–323.

Popplewell, L.M., Campanella, O.H., Sapru, V. and Peleg, M. (1989). Theoretical comparison of two segregation indices for binary powder. *Powder Technol.* **58**: 55–61.

Rippie, E.G., Olsen, J.L. and Faiman, M.D. (1964). Segregation kinetics of particulate solids systems. II: Particle density, size interaction and wall effect. *J. Pharm. Sci.* **52**(11): 1360–1363.

Rollins, D.K., Faust, D.L. and Jabas, D.L. (1995). A superior approach to indices in determining mixture segregation. *Powder Technol.* **84**: 277–282.

Shinohara, K. (1997). Segregation of particles. In *Powder Technology Handbook*, 2nd ed., Gotoh, K., Masuda, H. and Higashitani, K. (eds.). Marcel Dekker, Inc., New York, Chapter III.12, pp. 383–393.

Shinohara, K., Kazunori, S. and Tanaka, T. (1972). Mechanisms of size segregation of particles in filling a hopper. *Ind. Eng. Chem. Proc. Des. Develop.* **11**(3): 369–373.

Sugimoto, M. and Yamamoto, K. (1976). Effect of particle permeation on segregation of solid mixture in rotating vessel. *J. Soc. Mater. Sci. Jpn.* **25**(274): 684–689.

Williams, J.C. (1976). The segregation of particulate materials: a review. *Powder Technol.* **15**: 245–251.

CAKING REFERENCES

Adhikari, B., Howes, T., Bhandari, B.R. and Truong, V. (2001). Stickiness in foods: a review of mechanisms and test methods. *Int. J. Food Properties* **4**(1): 1–33.

Aguilera, J.M., Levi, G. and Karel, M. (1993). Effect of water content on the glass transition and caking of fish protein hydrolyzates. *Biotechnol. Prog.* **9**(6): 651–654.

Aguilera, J.M., Valle, J.M. and Karel, M. (1995). Review: caking phenomena in food powders. *Trends Food Sci. Technol.* **6**: 149–154.

Barbosa-Cánovas, G.V. and Vega-Mercado, H. (1996). Physical, chemical and micorbiological characteristics of dehydrated foods. In *Dehydration of Foods*. Chapman & Hall, New York, Chapter 3, pp. 29–99.

Bhandari, B.R. and Howes, T. (1999). Implication of glass transition for the drying and stability of dried foods. *J. Food Eng.* **40**: 71–79.

Champion, D., Le Meste, M. and Simatos, D. (2000). Towards improved understanding of glass transition and relaxations in foods: molecular mobility in the glass transition range. *Trends Food Sci. Technol.* **11**: 41–55.

Chen, Y.L. and Chou, J.Y. (1993). Selection of anticaking agents through crystallization. *Powder Technol.* **77**: 1–6.

Chung, M.S., Ruan, R.R., Chen, P., Chung, S.H., Ahn, T.H. and Lee, K.H. (2000). Study of caking in powdered foods using nuclear magnetic resonance spectroscopy. *J. Food Sci.* **65**(1): 134–138.

Chuy, L.E. and Labuza, T.P. (1994). Caking and stickiness of dairy-based food powders as related to glass transition. *J. Food Sci.* **59**(1): 43–46.

FDA (Food and Drug Administration) (1980). Code of Federal Regulations. Parts 170.3, 172E. and 182.

Genin, N. and Rene, F.(1995). Analyse du role de la transition vitreause dans les procedes de conservation agro-alimentaire. *J. Food Eng.* **26**: 391–407.

Griffith, E.J. (1991). *Cake Formation in Particulate Systems*. VCH Publishers, Inc., New York.

Hamano, M. and Sugimoto, H. (1978). Water sorption, reduction of caking and improvement of free-flowingness of powdered soy sauce and miso. *J. Food Process. Preserv.* **2**: 185–196.

Hollenbach, A.M., Peleg, M. and Rufner, R. (1982). Effects of four anticaking agents on the bulk characteristics of ground sugar. *J. Food Sci.* **47**: 583–544.

Hollenbach, A.M., Peleg, M. and Rufner, R. (1983). Interparticle surface affinity and the bulk properties of conditioned powders. *Powder Technol.* **35**: 51–62.

Joupilla, K. and Roos, Y.H. (1994). Glass transitions and crystallization in milk powders. *J. Diary Sci.* **77**: 2907–2915.

Lai, H.M. and Schmidt, S.J. (1990). Lactose crystallization in skim milk powder observed by hydrodynamic equilibria, scanning electron microscopy and ^{2}H nuclear magnetic resonance. *J. Food Sci.* **55**(4): 994–999.

Levi, G. and Karel, M. (1995). The effect of phase transitions on release of *n*-propanol entrapped in carbohydrate glasses. *J. Food Eng.* **24**(1): 1–13.

Lloyd, R., Chen, D.X. and Hargreaves, J.B. (1996). Glass transition and caking of spray dried lactose. *Int. J. Food Sci. Technol.* **31**: 305–335.

Moreyra, R. and Peleg, M. (1981). Effect of equilibrium water activity on the bulk properties of selected food powders. *J. Food Sci.* **46**: 1918–1922.

Peleg, M. (1978). Flowability of food powders and methods for its evaluation. *J. Food Proc. Eng.* **1**: 303–328.

Peleg, M. (1983). Physical characteristics of food powders. In *Physical Properties of Foods*, Peleg, M. and Bagley, E.B. (eds.). AVI Publishing Company, Inc., Westport, CT, Chapter 10, pp. 293–323.

Peleg, M. (1992). On the use of WLF model in polymers and foods. *Crit. Rev. Food Sci. Nutr.* **32**(1): 59–66.

Peleg, M. (1994). A model of mechanical changes in biomaterials at and around their glass transition. *Biotechnol. Prog.* **10**: 385–388.

Peleg, M. (1996). On modeling changes in food and biosolids at and around the glass transition temperature range. *Crit. Rev. Food Sci. Nutr.* **36**(1/2): 49–67.

Peleg, M. and. Hollenbach, A.M. (1984). Flow conditioners and anticaking agents. *Food Technol.* March: 93–102,

Peleg, M. and Mannheim, C.H. (1973). Effect of conditioners on the flow properties of powdered sucrose. *Powder Technol.* **7**: 45–50.

Peleg, M. and Mannheim, C.H. (1977). The mechanism of caking of powdered onion. *J. Food Proc. Preserv.* **1**: 3–11.

Roos, Y.H. (1995). Time-dependent phenomena. In *Phase Transitions in Foods*. Academic Press, New York, Chapter 7, pp. 193–245.

Roos, Y.H. and Karel, M. (1991a). Phase transitions of mixtures of amorphous polysaccharides and sugars. *Biotechnol. Prog.* **7**: 49–53.

Roos, Y.H. and Karel, M. (1991b). Plasticizing effect of water on thermal behavior and crytallization of amorphous food models. *J. Food Sci.* **56**(1): 38–43.

Roos, Y.H. and Karel, M. (1991c). Water and molecular weight effects on glass transitions in amorphous carbohydrates and carbohydrate solutions. *J. Food Sci.* **56**(1): 1676–1681.

Roos, Y.H. and Karel, M. (1991d). Amorphous state delayed ice formation in sucrose solutions. *Int. J. Food Sci. Technol.* **26**: 553–566.

Schubert, H. (1981). Principles of agglomeration. *Int. Chem. Eng.* **21**(3): 363–377.

Soesanto, T. and Williams, M.C. (1981). Volumetric interpretation of viscosity concentrated and dilute sugar solutions. *J. Phys. Chem.* **85**: 3338–3341.

Troy, H.C. and Sharp, P.F. (1930). α and β-lactose in some milk products. *J. Diary Sci.* **13**: 140–157.

Tsourouflis, S., Flink, J.M. and Karel, M. (1976). Loss of structure in freeze-dried carbohydrates solutions: effect of temperature, moisture content. and composition. *J. Sci. Food Agric.* **27**: 509–519.

Wakiyama, N., Kusai, A. and Nishimura, K. (1994). Anti-caking action of oily additives on caking of granules containing oily material. *Yakuzaigaku Arch. Pract. Pharm.* **54**(3): 191.

Wallack, D.A. and King, J. (1988). Sticking and agglomeration of hygroscopic, amorphous carbohydrate and food powders. *Biotechnol. Prog.* **4**(1): 31–35.

York, P. (1975). The use of glidants to improve the flowability of fine lactose powder. *Powder Technol.* **11**: 197–198.

DUST EXPLOSION REFERENCES

Bailey, J.E. (1992). Whole grain storage. In *Storage of Cereal Grains and Their Products*, Sauer, D.B. (ed.). American Association of Cereal Chemists, St. Paul, MN, pp. 157–182.

Cadle, R.D. (1965). *Particle Size*. Reinhold Publishing Co., New York, pp. 51–158.

Cross, J. and Farrer, D. (1982). *Dust Explosion*. Plenum Press, New York, pp. 115–164.

Grossel, S. (1997). Fire and explosion hazards in powder handling and processing. In *Handook of Powder Science and Technology*, 2nd ed., Fayed, M.E. and Otten, L. (eds.). Chapman and Hall, New York, pp. 845–866.

Hertzberg, M. and Cashdollar, K.L. (1987). Introduction to dust explosions. In *Industrial Dust Explosions*, Cashdollar, K.L. and Hertzberg, M. (eds.). ASTM, Pittsburgh, pp. 5–32.

Hoseney, R.C. and Faubion, J.M. (1992). Physical properties of cereal grains. In *Storage of Cereal Grains and Their Products*, Sauer, D.B. (ed.). American Association of Cereal Chemists, Inc., St. Paul, MN, pp. 1–38.

IEC (International Electrotechnical Commission) (1994). *Electrical Apparatus for Use in the Presence of Ignitable Dust: Part 2: Test Methods, Section 3, Method for Determining the Minimum Ignition Energy of Dust/Air Mixtures*. IEC, Geneva.

Jaeger, N. and Siwek, R. (1999). Prevent explosions of combustible dusts. *Chem. Eng. Prog.* **95**(6): 25–37.

Jones, T.B. and King, J.L. (1991). *Powder Handling and Electrostatics*. Lewis Publishers Inc., Michigan, pp. 1–24.

Kauffman, C.W. (1987). Recent dust explosion experiences in the U.S. grain industry. In *Industrial Dust Explosions*, Cashdollar, K.L. and Hertzberg, M. (eds.). ASTM, Pittsburgh, pp. 243–264.

Kauffman, C.W., Sichel, M. and Wolanski, P. (1992). Research on dust explosions at the University of Michigan. *Powder Technol.* **71**: 119–134.

Kent, N.L. (1983). *Technology of Cereals*, 3rd ed. Pergamon Press, New York, pp. 1–196.

Land, C.M.V. (1991). Safeguarding drying. In *Industrial Drying Equipment*. Marcel Dekker, Inc., New York, Chapter 12, pp. 269–290.

Mody, V. and Jakhete, R. (1988). In *Dust Control Handbook*. Noyes Data Co., New Jersey, Chapter 1, pp. 1–6.

Nagy, J. and Verakis, H.C. (1983). *Development and Control of Dust Explosions*. Marcel Dekker, Inc., New York.

Nomura, S. (1992). Theoretical analysis of dust explosions. *Powder Technol.* **71**: 189–196.

Peleg, M. (1983). Physical characterization of food powders. In *Physical Properties of Foods*, Peleg, M., Bagley, E.B. (eds.). AVI Publishing Company, New York, pp. 293–323.

Pineau, J.P. (1994). The risk of dust explosion. In *Primary Cereal Processing*, Godon, B. and Willm, C. (eds.). VCH Publishers, Inc., New York, pp. 377–400.

Rhodes, M. (1998). Mixing and segregation. In *Introduction to Particle Technology*. John Wiley & Sons, West Sussex, England, pp. 283–300.

Shamlou, P.A. (1988). Safety in bulk solids handling. In *Handling of Bulk Solids*. Butterworths, London, Chapter 8.3, pp. 169–188.

Tanaka, T. (1997). Spontaneous ignition and dust explosion. In *Powder Technology Handbook*, Gotoh, K., Masuda, H. and Higashitani, K. (eds.). Marcel Dekker, Inc., New York, pp. 899–934.

van der Wel, P.G.J., van Veen, J.P.W., Lemkowitz, S.M., Scarlett, B. and van Wingerden, C.J.M. (1991). An interpretation of dust explosion phenomena on the basis of time scale. *Powder Technol.* **71**: 207–215.

Wenzel, B.J. (1998). Kansas grain dust explosion. *J. Fire Eng.* **151**(11): 65–75.

Wolanski, P. (1992). Dust explosion research in Poland. *Powder Technol.* **71**: 197–206.

Zeeuwen, P. (2001). Explosion hazards in powder handling and processing: the changes added. In *Powders and Solids. Development in Handling and Processing Technologies*, Hoyle, W. (ed.). The Royal Society of Chemistry, Cambridge, UK, pp. 11–28.

INDEX

Breinigsville, PA USA
18 March 2010
234431BV00012B/41/A